# Advances in Experimental Medicine and Biology

Volume 949

More information about this series at http://www.springer.com/series/5584

Rommy von Bernhardi
Editor

# Glial Cells in Health and Disease of the CNS

Springer

*Editor*
Rommy von Bernhardi
Faculty of Medicine
Pontificia Universidad Católica de Chile
Santiago
Chile

ISSN 0065-2598 ISSN 2214-8019 (electronic)
Advances in Experimental Medicine and Biology
ISBN 978-3-319-40762-3 ISBN 978-3-319-40764-7 (eBook)
DOI 10.1007/978-3-319-40764-7

Library of Congress Control Number: 2016948593



Printed on acid-free paper

This Springer imprint is published by Springer Nature
The registered company is Springer International Publishing AG
The registered company address is: Gewerbestrasse 11, 6330 Cham, Switzerland

# Contents

## Part II Glial Cell Dysfunction and Pathological Implications

# Contributors

**Beatriz Almolda** Unit of Histology, Department of Cell Biology, Physiology and Immunology, Institute of Neurosciences, Universitat Autònoma de Barcelona, Bellaterra, Spain

**Jaime Alvarez** Center for Integrative Biology, Universidad Mayor, Santiago, Geroscience Center for Brain Health and Metabolism, Santiago, Chile

**Rackele Amaral** Instituto de Ciências Biomédicas, Centro de Ciências da Saúde, Universidade Federal do Rio de Janeiro, Rio de Janeiro, RJ, Brazil

**Gonzalo Arboleda** Grupo de Neurociencias y Muerte Celular, Facultad de Medicina e Instituto de Genética, Universidad Nacional de Colombia, Bogotá, Colombia; Departamento de Patología, Facultad de Medicina, Universidad Nacional de Colombia, Bogotá, Colombia

**Luis Barbeito** Neurodegeneration Laboratory, Institut Pasteur, Montevideo, Uruguay

**Penha Cristina Barradas** Depto. Farmacologia e Psicobiologia, Instituto de Biologia—UERJ, Rio de Janeiro, RJ, Brazil

**Sebastián Beltrán-Castillo** Departamento de Biología, Facultad de Química y Biología, Universidad de Santiago de Chile (USACH), Santiago, Chile

**Mar Bosch-Queralt** Unit of Histology, Department of Cell Biology, Physiology and Immunology, Institute of Neurosciences, Universitat Autònoma de Barcelona, Bellaterra, Spain

**Gabriela Casanova** UMET, School of Sciences, UDELAR, Montevideo, Uruguay

**Bernardo Castellano** Unit of Histology, Department of Cell Biology, Physiology and Immunology, Institute of Neurosciences, Universitat Autònoma de Barcelona, Bellaterra, Spain

**María José Contreras** Grupo de Neurociencias y Muerte Celular, Facultad de Medicina e Instituto de Genética, Universidad Nacional de Colombia, Bogotá, Colombia

**Francisca Cornejo** Departamento de Neurología, Escuela de Medicina, Pontificia Universidad Católica de Chile, Santiago, Chile

**Felipe A. Court** Center for Integrative Biology, Universidad Mayor, Santiago; Geroscience Center for Brain Health and Metabolism, Santiago; Millenium Nucleus for Regenerative Biology, Pontificia Universidad Católica de Chile, Santiago, Chile

**Marta C. Cunha-Rodrigues** Depto. Farmacologia e Psicobiologia, Instituto de Biologia—UERJ, Rio de Janeiro, RJ, Brazil

**Alan P. da Costa** Depto. Farmacologia e Psicobiologia, Instituto de Biologia—UERJ, Rio de Janeiro, RJ, Brazil

**Anna Carolina Carvalho da Fonseca** Instituto de Ciências Biomédicas, Centro de Ciências da Saúde, Universidade Federal do Rio de Janeiro, Rio de Janeiro, RJ, Brazil

**Leda Dimou** Physiological Genomics, Biomedical Center, Ludwig-Maximilians-University, Munich, Germany

**Jaime Eugenín-von Bernhardi** Graduate School of Systemic Neuroscience, Physiological Genomics, Biomedical Center, Ludwig-Maximilians-University, Planegg-Martinsried, Munich, Germany

**Jaime Eugenín León** Departamento de Biología, Facultad de Química y Biología, Universidad de Santiago de Chile (USACH), Santiago, Chile

**Anabel Fernández** Neuroscience Division, Instituto de Investigaciones Biológicas Clemente Estable, (IIBCE), Montevideo; Comparative Neuroanatomy-Associated Unit, School of Sciences, UDELAR, Montevideo, Uruguay

**Carina Ferrari** Instituto de Ciencias Básicas y Medicina Experimental, Instituto Universitario del Hospital Italiano, Buenos Aires, Argentina

**Betsi Flores** Departamento de Neurología, Escuela de Medicina, Pontificia Universidad Católica de Chile, Santiago, Chile

**Celina Garcia** Instituto de Ciências Biomédicas, Centro de Ciências da Saúde, Universidade Federal do Rio de Janeiro, Rio de Janeiro, RJ, Brazil

**Luiz Henrique Geraldo** Instituto de Ciências Biomédicas, Centro de Ciências da Saúde, Universidade Federal do Rio de Janeiro, Rio de Janeiro, RJ, Brazil

**Berta González** Unit of Histology, Department of Cell Biology, Physiology and Immunology, Institute of Neurosciences, Universitat Autònoma de Barcelona, Bellaterra, Spain

**V. Gudi** Department of Neurology, Hannover Medical School, Hannover, Germany; Center for System Neurosciences, Hannover, Germany

**Florencia Heredia** Departamento de Neurología, Escuela de Medicina, Pontificia Universidad Católica de Chile, Santiago, Chile

**H.C. Hoyos** Department of Biological Chemistry, School of Pharmacy and Biochemistry, Institute of Chemistry and Biological Physical Chemistry (IQUIFIB), Buenos Aires, Argentina; School of Pharmacy and Biochemistry, University of Buenos Aires and National Research Council (CONICET), Buenos Aires, Argentina

**Eugenia Isasi** Cellular and Molecular Neurobiology, Instituto de Investigaciones Biológicas Clemente Estable (IIBCE), Montevideo, Uruguay

**Jenny Jaramillo** Grupo de Neurociencias y Muerte Celular, Facultad de Medicina e Instituto de Genética, Universidad Nacional de Colombia, Bogotá, Colombia

**Flavia Regina Souza Lima** Instituto de Ciências Biomédicas, Centro de Ciências da Saúde, Universidade Federal do Rio de Janeiro, Rio de Janeiro, Brazil

**Jonathan Julio Lorea-Hernández** Departamento de Neurobiología del Desarrollo y Neurofisiología, Instituto de Neurobiología, Universidad Nacional Autónoma de México, Querétaro, Mexico

**Alex C. Manhães** Depto. Ciências Fisiológicas; Instituto de Biologia Roberto Alcantara Gomes, Universidade do Estado do Rio de Janeiro, Rio de Janeiro, RJ, Brazil

**Mariel Marder** Department of Biological Chemistry, School of Pharmacy and Biochemistry, Institute of Chemistry and Biological Physical Chemistry (IQUIFIB), Buenos Aires, Argentina; School of Pharmacy and Biochemistry, University of Buenos Aires and National Research Council (CONICET), Buenos Aires, Argentina

**Diana Matias** Instituto de Ciências Biomédicas, Centro de Ciências da Saúde, Universidade Federal do Rio de Janeiro, Rio de Janeiro, RJ, Brazil

**Verónica Murta** Laboratorio de Neuropatología Molecular, Instituto de Biología Celular y Neurociencias, Universidad de Buenos Aires, Buenos Aires, Argentina

**Paola Muñoz** Departamento de Neurología, Escuela de Medicina, Pontificia Universidad Católica de Chile, Santiago, Chile

**Marco T. Núñez** Departamento de Biología, Facultad de Ciencias, Universidad de Chile, Santiago, Chile

**María José Olivares** Departamento de Biología, Facultad de Química y Biología, Universidad de Santiago de Chile (USACH), Santiago, Chile

**Silvia Olivera-Bravo** Cellular and Molecular Neurobiology, Instituto de Investigaciones Biológicas Clemente Estable (IIBCE), Montevideo, Uruguay

**Juan A. Orellana** Departamento de Neurología, Escuela de Medicina, Pontificia Universidad Católica de Chile, Santiago, Chile

**J.M. Pasquini** Department of Biological Chemistry, School of Pharmacy and Biochemistry, Institute of Chemistry and Biological Physical Chemistry (IQUIFIB), Buenos Aires, Argentina; School of Pharmacy and Biochemistry, University of Buenos Aires and National Research Council (CONICET), Buenos Aires, Argentina; Dpto. de Qca Biol, FFyB-UBA, Bs, Argentina

**L.A. Pasquini** Department of Biological Chemistry, School of Pharmacy and Biochemistry, Institute of Chemistry and Biological Physical Chemistry (IQUIFIB), Buenos Aires, Argentina; School of Pharmacy and Biochemistry, University of Buenos Aires and National Research Council (CONICET), Buenos Aires, Argentina; Dpto. de Qca Biol, FFyB-UBA, Bs, Argentina

**Fernando Peña-Ortega** Departamento de Neurobiología del Desarrollo y Neurofisiología, Instituto de Neurobiología, Universidad Nacional Autónoma de México, Querétaro, Mexico

**G.A. Rabinovich** Laboratory of Immunopathology, Institute of Biology and Experimental Medicine (IBYME; CONICET), Buenos Aires, Argentina; Laboratory of Functional Glycomics, Department of Biological Chemistry, Faculty of Exact and Natural Sciences, University of Buenos Aires, Buenos Aires, Argentina

**Ana Julia Rivera-Angulo** Departamento de Neurobiología del Desarrollo y Neurofisiología, Instituto de Neurobiología, Universidad Nacional Autónoma de México, Querétaro, Mexico

**Juan Carlos Rosillo** Neuroscience Division, Instituto de Investigaciones Biológicas Clemente Estable (IIBCE), Montevideo, Uruguay

**Alejandro D. Roth** Departamento de Biología, Facultad de Ciencias, Universidad de Chile, Santiago, Chile

**Nicole Salgado** Departamento de Neurología, Escuela de Medicina, Pontificia Universidad Católica de Chile, Santiago, Chile

**Adrián Sandoval-Hernández** Grupo de Neurociencias y Muerte Celular, Facultad de Medicina e Instituto de Genética, Universidad Nacional de Colombia, Bogotá, Colombia

**Tiago Savignon** Depto. Farmacologia e Psicobiologia, Instituto de Biologia—UERJ, Rio de Janeiro, RJ, Brazil; Depto. Farmacologia e Toxicologia, Instituto Nacional de Controle de Qualidade em Saúde, Rio de Janeiro, RJ, Brazil

**M. Stangel** Department of Neurology, Hannover Medical School, Hannover, Germany; Center for System Neurosciences, Hannover, Germany

**Frank Tenório** Depto. Farmacologia e Psicobiologia, Instituto de Biologia—UERJ, Rio de Janeiro, RJ, Brazil

**R. Ulrich** Department of Pathology, University of Veterinary Medicine Hannover, Hannover, Germany

**Juliana Vaillant** Depto. Farmacologia e Psicobiologia, Instituto de Biologia—UERJ, Rio de Janeiro, RJ, Brazil

**Nàdia Villacampa** Unit of Histology, Department of Cell Biology, Physiology and Immunology, Institute of Neurosciences, Universitat Autònoma de Barcelona, Bellaterra, Spain

**Rommy von Bernhardi** Departamento de Neurología, Escuela de Medicina, Pontificia Universidad Católica de Chile, Santiago, Chile

# Glial Cells and Integrity of the Nervous System

**Rommy von Bernhardi, Jaime Eugenín-von Bernhardi, Betsi Flores and Jaime Eugenín León**

**Abstract** Today, there is enormous progress in understanding the function of glial cells, including astroglia, oligodendroglia, Schwann cells, and microglia. Around 150 years ago, glia were viewed as a glue among neurons. During the course of the twentieth century, microglia were discovered and neuroscientists' views evolved toward considering glia only as auxiliary cells of neurons. However, over the last two to three decades, glial cells' importance has been reconsidered because of the evidence on their involvement in defining central nervous system architecture, brain metabolism, the survival of neurons, development and modulation of synaptic transmission, propagation of nerve impulses, and many other physiological functions. Furthermore, increasing evidence shows that glia are involved in the mechanisms of a broad spectrum of pathologies of the nervous system, including some psychiatric diseases, epilepsy, and neurodegenerative diseases to mention a few. It appears safe to say that no neurological disease can be understood without considering neuron–glia crosstalk. Thus, this book aims to show different roles played by glia in the healthy and diseased nervous system, highlighting some of their properties while considering that the various glial cell types are essential components not only for cell function and integration among neurons, but also for the emergence of important brain homeostasis.

R. von Bernhardi (✉) · B. Flores
Department of Neurology, School of Medicine, Pontificia Universidad Católica de Chile, Marcoleta 391, Santiago, Chile
e-mail: rvonb@med.puc.cl

J. Eugenín-von Bernhardi
Physiological Genomics, Biomedical Center, Ludwig-Maximilians-University, Pettenkoferstr.12, 80336 Munich, Germany

J. Eugenín-von Bernhardi
Graduate School of Systemic Neuroscience, Ludwig-Maximilians-University, 82152 Planegg-Martinsried, Munich, Germany

J.E. León
Department of Biology, Faculty of Chemistry and Biology, USACH, Santiago, Chile



R. von Bernhardi (ed.), *Glial Cells in Health and Disease of the CNS*,
Advances in Experimental Medicine and Biology 949,
DOI 10.1007/978-3-319-40764-7_1

**Keywords** Astrocytes · Microglia · Myelin · Development · Neuron–glia crosstalk · Neuroimmunity · NG-2 cells · Oligodendrocyte · Schwann cells · Tripartite synapses

## Abbreviations

| | |
|---|---|
| ADNF | Activity-dependent neurotrophic factor |
| ADAM10 | A disintegrin and metalloproteinase domain-containing protein 10 |
| AMPA | α-amino-3-hydroxy-5-methyl-4-isoxazolepropionic acid |
| ATP | Adenosine triphosphate |
| CNS | Central nervous system |
| EAAT | Excitatory amino acid transporters |
| EGF | Epidermal growth factor |
| ER | Endoplasmic reticulum |
| FcR | Receptor for the (*Fragment, crystallizable*) region of antibodies |
| GABA | γ-aminobutyric acid |
| GDNF | Glial cell-derived neurotrophic factor |
| GFAP | Glial fibrillary acidic protein |
| IGF-I | Insulin-like growth factor 1 |
| iNOS | Inducible nitric oxide synthase |
| $InsP_3$ | Inositol trisphosphate |
| MAG | Myelin associated glycoprotein |
| MBP | Myelin basic protein |
| MOG | Myelin oligodendrocyte glycoprotein |
| MS | Multiple sclerosis |
| NCAM | Neural cell adhesion molecule |
| NG-2 | Neuron–Glia antigen 2 |
| NMDA | N-methyl-D-aspartate |
| NO | Nitric oxide |
| OPCs | Oligodendrocyte precursor cells |
| PLP | Proteolipidprotein/DM20 |
| PNS | Peripheral nervous system |
| PMP22 | Peripheral myelin protein-22 |
| P0 | Protein zero |
| ROS | Reactive oxygen species |
| TNFα | Tumor necrosis factor α |
| VGLUT | Vesicular glutamate transporters |
| VNUT | Vesicular nucleotide transporters |

## The Concept of Neuron–Glia Crosstalk

The discovery of neurons as the basic functional unit of the nervous system towards the end of the nineteenth century was a milestone in neuroscience research in shaping our view of the functional organization of the nervous system. While neurons have been studied continuously since then, extensive glial research has been seemingly in steps, with the greatest understanding developed only over the last few decades. In the middle of the twentieth century, the first studies on glial physiology were done by Kuffler and Nicholls (1966). Recordings from the far smaller microglia came several decades later, and consistent with recent knowledge of their separate origin from the astrocytes and oligodendrocytes (Ginhoux et al. 2010; Rowitch and Kriegstein 2010), they are physiologically quite distinct. For nearly a century, the main view was that neurons were the only cells responsible for nearly all the complex functions of the nervous system, whereas glia were just relatively passive support cells (Halliday and Stevens 2011). Recent decades have seen a major shift in understanding of the roles particularly of astrocytes and microglia.

The development of animal models for many neurological and psychiatric diseases has enabled study of the cellular responses in various pathological stages. It appears that there are no neuropathological processes that occur without participation of glial cells, specifically microglia and astrocytes (Halliday and Stevens 2011; Quintanilla et al. 2012; Verkhratsky et al. 2014; Jha et al. 2015). Interest in glial cells has also been potentiated by the evidence showing glial responsiveness, leading to gliotransmission for the propagation of information among glial cells (Zorec et al. 2012) and eventually to neurons (Vesce et al. 1999; Araque 2008; Hamilton and Attwell 2010). This form of communication is much slower than the neuronal response; however, glial cells, and astrocytes in particular, appear to be able to modulate neuronal activity (Jirounek et al. 2002; Araque 2008) and thus brain function. Evidence gathered over the last couple of decades and data obtained earlier show that glia have important functions of their own in addition to their effects on neuronal function.

Based on their broad spectrum of characteristics and functions, glial cells are probably some of the most versatile cells in our body. Here we will discuss the participation of the NG2–glia progenitor cells, oligodendrocytes, Schwann cells, astrocytes, and microglia, both in the central (CNS) and peripheral nervous system (PNS). All are important players during development and for neural function in the adult animal. All cell types, including neurons and glia, show both passive and active support roles for each other (Fields and Stevens-Graham 2002; Ramirez et al. 2005; Vecino et al. 2015). For instance, considerable evidence now indicates that neurons provide trophic support and regulate glial activation (Ramírez et al. 2008; Biber et al. 2011).

## Astrocytes

Astrocytes, which interact with neurons, blood vessels, and many structures of the nervous system (Abbott et al. 2006; Cheslow and Alvarez 2016), are the most abundant glial cells in many parts of the CNS (Pakkenberg and Gundersen 1988). The ratio of glia to neurons varies from one brain region to another (Pakkenberg and Gundersen 1988), and it is still amply discussed (Hilgetag and Barbas 2009). Astrocytes are involved in synaptic transmission (Newman 2003; Croft et al. 2015; Gittis and Brasier 2015) and the regulation of neuronal processing. They remove extracellular potassium and neurotransmitters from the synaptic space, and participate in the energy metabolism (Araque et al. 1999; Perez-Alvarez and Araque 2013; Perea et al. 2009; Reichenbach and Bringmann 2013). They are key partners for synaptogenesis, synapse function, and synaptic plasticity (Chung et al. 2015; Gittis and Brasier 2015).

Although named for their star-like shape, astrocytes are difficult to characterize. They can have various shapes and expression of identity markers, including the characteristic intermediate filament glial fibrillary acidic protein (GFAP). Astrocytes display a remarkable heterogeneity in their morphology and function (Oberheim et al. 2012). Conceptually, astrocytes in different brain regions can have different physiological properties (Bayraktar et al. 2015; Schitine et al. 2015). For example, only brainstem astrocytes appear to have the ability to sense changes in $PCO_2$ and contribute as mediators of the respiratory response to hypercapnia in mammals (Gourine et al. 2010).

Two major morphologies are recognized:

(a) Protoplasmic astrocytes are in gray matter and have many fine processes, most of them elaborate and complex. Their processes contact blood capillaries, establishing perivascular endfeet, and contact neurons (Abbott et al. 2006; Hawkins and Davis 2005). The complex astrocyte–neurons–blood vessel is known as the neurovascular unit and form the blood–brain barrier (Hawkins and Davis 2005). Some protoplasmic astrocytes also form subpial endfeet at the pial surface.
(b) Fibrous astrocytes are in white matter (Hristova et al. 2010). Their processes are long (up to 300 μm), but much less elaborate than those of protoplasmic astroglia. The processes of fibrous astrocytes establish several perivascular or subpial endfeet. Fibrous astrocyte processes also send numerous extensions ('perinodal' processes) that contact oligodendroglia wrapped myelinated axons at nodes of Ranvier (Butt et al. 1994).

Other regions of the CNS contain distinctive populations of astroglial cells: velate astrocytes in the cerebellum, where they form a sheath surrounding granule neurons; interlaminar astrocytes in the cerebral cortex of higher primates; tanycytes in the periventricular organs, the hypophysis and the raphe nucleus; pituicytes in the neuro-hypophysis. Perivascular and marginal astrocytes are localized very close to the pia mater, where they form numerous endfeet, establishing the pial and perivascular glia limitans barrier, which assists in isolating the brain parenchyma from the vascular and subarachnoid compartments.

There are important groups of astroglia, including radial glia (Noctor et al. 2001; Hansen et al. 2010) in the developing brain, which are bipolar progenitor cells with one main process forming apical endfeet at the ventricular wall and the other, basal, at the pial surface (Noctor et al. 2001). Radial glia form a scaffold, which guides neuronal migration from the ventricular zone (Hartfuss et al. 2001) after they divide asymmetrically in a self-renewing fashion to generate neurons and then glia—first astrocytes and then oligodendroglia. In humans, they also give rise to progenitors that maintain only a basal process that accounts for the majority of neurons and glia in the cerebral cortex. It is believed that radial glial cells remain in the retina as Müller glia (Reichenbach and Bringmann 2013), and in the cerebellum as Bergmann glia. Some astrocytes in neurogenic niches of the adult brain, retain stem cell properties throughout life and are the source for the adult neuro- and gliogenesis (Riquelme et al. 2008; Shimada et al. 2012).

Astrocytes have many functions: they generate the brain environment, establish the microarchitecture of the parenchyma, maintain brain homeostasis, generate, store and distribute energy substrates, control the development of neurons, synaptogenesis and synaptic efficacy and maintenance, and participate in brain defense. Astrocytes define the microarchitecture of the parenchyma in the mammalian brain by dividing the grey matter, through the process called "tiling", into relatively independent structural units, within the limits of their processes. Furthermore, the individual astrocytes are integrated into functional syncytia through gap junctions between their processes (Wallraff et al. 2006).

## *Astrocytes and the Tripartite Synapsis*

The close relation between astrocytes and synapses as well as their functional expression of relevant receptors prompted the "tripartite synapse" hypothesis (Volterra and Meldolesi 2005; Santello et al. 2012). In the gray matter, astrocytes are closely associated with synaptic regions, enwrapping presynaptic terminals and postsynaptic structures (Vesce et al. 1999). Astrocyte membranes enwrap about 80 % of large synapses, being probably the most functionally active, but only about half of the small ones. In the cerebellum, nearly all of the synapses formed by parallel fibers on the Purkinje neuron are covered by the Bergmann glial cells.

According to the "tripartite synapse" hypothesis, synapses are comprised of three parts, the presynaptic terminal, postsynaptic membrane, and associated astrocyte. The astrocyte processes possess neurotransmitter receptors. The receptors expressed by astrocytes match the neurotransmitters released at the synapses they cover. In the cortex both pyramidal neurons and neighboring astrocytes express glutamate and purinergic receptors (Koizumi et al. 2005), whereas in the basal ganglia both neurons and astrocytes are sensitive to dopamine. Neurotransmitters released by the presynaptic terminal, which themselves have metabotropic receptors for the transmitters they release, activate receptors in both the postsynaptic neuron and the associated astrocyte. This results in the generation of a postsynaptic potential in the postsynaptic neuron and a $Ca^{2+}$ signal in the astrocyte (Vesce et al. 1999). $Ca^{2+}$ signals can propagate through the astrocytic syncytium, and could also trigger the release of neurotransmitters from neighbor astrocytes (Malarkey and Parpura 2008; Lalo et al. 2014), which in turn will signal onto both pre- and postsynaptic neuronal membranes.

Astrocytes regulate the extracellular concentration of potassium ions, neurotransmitters and metabolites, and water movements (Newman 2003). A key function of astrocytes is the control of extracellular $K^+$ (Kuffler and Nicholls 1966). Synaptic activity increases $K^+$ concentration from its resting level (3 mM) to 10–12 mM under physiological conditions, and to even higher concentrations under pathological conditions (Wallraff et al. 2006; Hansen and Nedergaard 1988). High extracellular $K^+$ increases neuronal reactivity.

Simultaneously, astrocytes remove most of the accumulated glutamate released at the synapses from the extracellular space through excitatory amino acid transporters (EAAT) (Anderson and Swanson 2000). Glutamate is the major excitatory neurotransmitter in the CNS of vertebrates. When released in excess, glutamate becomes excitotoxic and triggers neuronal cell death (Foran and Trotti 2009). Astrocyte uptake of glutamate is crucial for glutamatergic neurons. Glutamate is enzymatically converted into glutamine by the astrocyte-specific enzyme glutamine synthetase (Schousboe et al. 2014). Glutamine can be safely transported to presynaptic terminals; and after entering the neuron, it is transformed into glutamate by glutaminase. Astrocytes also possess the enzyme pyruvate carboxylase, and are a key source for de novo glutamate synthesis. Thus, astrocytes have the machinery to regulate the availability of glutamate, by both degrading it and generating new neurotransmitter.

## *Metabolic Support by Astrocytes*

Astrocytes provide some metabolic support for neurons. The glucose–lactate shuttle hypothesis (see Pellerin et al. 2007) proposes that astrocytes take up glucose, metabolize it to lactate, and release it as an energy substrate for neurons. In addition, astrocytes are the only brain cells that synthesize glycogen and thus govern the energy reservoir.

Moreover, astrocytes are a part of the neurovascular unit (Hawkins and Davis 2005) that coordinates neural activity with local blood flow according to the metabolic demands. Blood vessels are almost entirely covered by astrocyte endfeet, with one arm at the blood vessel, and the other at the neuron soma, synapse, or axon. Increased activity of neurons induces $Ca^{2+}$ signals in astrocytes, which could be integrated by the neurovascular unit, leading to the release of vasoactive agents that regulate local blood flow (Zonta et al. 2003; Takano et al. 2006). Astrocytes appear to link neuronal activity and blood perfusion, although reports are contradictory. Some groups report that astrocyte activity leads to local vasoconstriction, while others report vasodilatation.

## Astrocytes in Synaptogenesis and Synaptic Maintenance

Astrocytes regulate formation, maturation, maintenance, and stability of synapses. They secrete numerous factors, including thrombin, needed for synaptogenesis (Christopherson et al. 2005; Diniz et al. 2012). Synaptogenesis is also affected by glial signals regulating the expression of agrin, essential for synapse formation (Faissner et al. 2010). Later in life, astrocytes participate in the maturation of synapses. Several soluble molecules released by astrocytes are involved in the regulation of synapses, including tumor necrosis factor α (TNFα), which regulates the insertion of glutamate receptors and activity-dependent neurotrophic factor (ADNF), which increases the density of NMDA receptors in postsynaptic membranes (Slezak and Pfrieger 2003).

Astrocytes appear also to be capable of limiting the number of synapses. They ensheath the neuronal processes blocking the formation of synapses, as well as being involved in the elimination of synapses, which is the basis for the final tuning and plasticity of neuronal communication (Chung et al. 2013, 2015). Elimination of synapses is achieved by secretion of certain factors and proteolytic enzymes that degrade the extracellular matrix and reduce the stability of the synaptic contact. Subsequently, astrocyte processes invade the synaptic cleft and substitute for the missing neuronal synaptic element. This process is especially robust in neuropathological conditions.

## Signaling in Astrocyte Networks and Gliotransmission

Glutamate metabotropic receptors in astrocytes activate intracellular signaling cascades, providing a cell activation mechanism. Their excitability depends on the response of the $Ca^{2+}$ channels on the endoplasmic reticulum (ER): $InsP_3$ receptors and ryanodine receptors. Stimulation of astrocytes metabotropic receptors induces formation of $InsP_3$, which in turn triggers $Ca^{2+}$ release from the ER, increasing intracellular $Ca^{2+}$ levels (Hua et al. 2004). These $Ca^{2+}$ signals are generally associated with glial activation. Astrocytic $Ca^{2+}$ signals can propagate through astrocyte networks, generating intercellular $Ca^{2+}$ waves depending on the diffusion of $InsP_3$ through gap junctions, as well as the release of ATP from astrocytes (Beck et al. 2004; Scemes and Giaume 2006). Gap junctions may also be the signaling pathway in astrocyte networks, which involve various second messengers, metabolic substrates, and other molecules (Scemes and Giaume 2006; Wallraff et al. 2006).

As discussed in Chapters "Pharmacological Tools to Study the Role of Astrocytes in Neural Network Functions", "Physiological Functions of Glial Cell Hemichannels", and "Role of Astrocytes in Central Respiratory Chemoreception", astrocytes as well as other glia release a variety of gliotransmitters (Santello et al. 2012; Petrelli and Bezzi 2016), which include several molecules also utilized by neurons, such as glutamate, ATP, GABA, and D-serine, but also taurine and kynurenic acid, which appear to be exclusively released by glia (Rassoulpour et al. 2005). Different mechanisms of release of gliotransmitter have been described including: (i) diffusion through permeability channels, like volume-activated $Cl^-$ channels, unpaired connexin "hemichannels", pannexin channels, or $P2X_7$ pore-forming purinoceptors; (ii) through transporters, e.g., by reversal of EAAT or exchange via the cystine-glutamate antiporter or organic anion transporters; and (iii) through $Ca^{2+}$-dependent exocytosis.

Astrocytes have been proposed to undergo exocytosis because they express proteins involved in exocytosis such as synaptobrevin 2, syntaxin 1, and synaptosome-associated protein of 23 kDa (Ropert et al. 2015). Astrocytes also express transporters required for neutotransmitter accumulation in synaptic vesicles, including the vacuolar type of proton ATPase (V-ATPase), vesicular glutamate transporters (VGLUTs) 1, 2, and 3, and vesicular nucleotide transporters (VNUT) (Ni and Parpura 2009; Morel et al. 2014). Exocytotic release of transmitters by astrocytes is mediated by $Ca^{2+}$-dependent exocytotic glutamate and can affect several neuronal responses including increased neuronal $Ca^{2+}$ influx and generation in neurons of a slow inward current mediated by NMDA receptors (Montes de Oca Balderas and Aguilera 2015). Thus, glutamate released from astrocytes can affect neuronal excitability, modulate synaptic transmission, and synchronize synaptic events, therefore potentially modifying behavior (Oliveira et al. 2015).

## *Astrocytes in Neuropathology*

By the end of the nineteenth century, the neuropathological potential of glia was proposed by such prominent neuropathologists, as Carl Frommann, Franz Nissl, Alois Alzheimer, and Pio del Rio-Hortega. Nevertheless, the participation of glia remains poorly understood. As described, astrocytes are fundamental for brain homeostasis, and they also serve an important part of the nervous system defense system (Posada-Duque et al. 2014). Brain insults trigger a response of astrogliosis (Pekny and Pekna 2014), which is essential for limiting the area of damage by scar formation, for post-insult remodeling and for recovery of neural function. Astrocyte activation and dysfunction are apparent in all types of brain pathologies: acute lesions (trauma or stroke), developmental neurometabolic disorders (see Chapter "Astrocyte Dysfunction in Developmental Neurometabolic Diseases"), psychiatric diseases, and neurodegenerative processes, such as Alzheimer's disease, Parkinson's disease, and multiple sclerosis (MS).

## Oligodendrocytes and Schwann Cells

Oligodendrocytes in the CNS and Schwann cells in the PNS produce myelin that provides insulation for axons (Bercury and Macklin 2015), and are thus responsible for the high-speed information propagation in axons of vertebrates. Therefore, the cells are active partners for neurons for the propagation of information. Myelin sheaths are produced in response to neuronal activity (Grigoryan and Birchmeier 2015). In addition to providing axon insulation, the glia provide trophic support, affect the structure of axons, and modify their electrical properties by controlling their diameter, and the clustering of specific ion channels at the node and paranode region (Poliak and Peles 2003).

In the periphery, Schwann cells also serve key functions for the regeneration of axons, and at the neuromuscular junction (Love and Thompson 1998; Love et al. 2003; Lee et al. 2016), given the absence of astrocytes in the PNS, undertake similar functions at the synapses (Poliak and Peles 2003) (see Chapters "Oligodendrocytes: Functioning in a Delicate Balance Between High Metabolic Requirements and Oxidative Damage" and "Schwann Cell and Axon: An Interlaced Unit—From Action Potential to Phenotype Expression" for further reading on oligodendrocytes metabolism and the functional effect of the interaction axon–Schwann cell). There are also other specialized cells. Neuronal somata in sensory sympathetic and parasympathetic ganglia are surrounded by flattened satellite cells, and neuromuscular junctions are also covered by the terminal glia, a specialized Schwann cell (Connor and McMahan 1987; Kang et al. 2014).

Myelin insulation is needed for high-speed nerve conduction (up to 200 m/s) in vertebrates, although in an invertebrate, as the shrimp, could be even higher without the usual myelin (Xu and Terakawa 1999). Most oligodendrocytes are located in the white matter to form myelin. However, they are also found in gray matter and possibly regulate ionic homeostasis similar to astrocytes. Furthermore, axon–glia interaction is important for neuron regulation also in unmyelinated nerves (Jirounek et al. 2002).

Myelination is observed in all jawed vertebrates and can be traced back to 400 million years ago, and even a few invertebrates show "loose" myelin sheaths. In fact, the occurrence of myelin in evolution potentiated the neural development of vertebrates. Myelin plays a key role for allowing neurons to be interconnected in the complex fashion observed in the vertebrate nervous system (de Hoz and Simons 2015). In general and for biophysical reasons, only axons larger than 1 μ are myelinated (Rushton 1951; Waxman and Bennett 1972). Recent reports show that axons signal to the oligodendrocyte to determine the thickness of the myelin sheath (Bozzali and Wrabetz 2004). A key signaling mechanism provided by the axon is via the growth factor neuregulin-1, which binds to ErbB receptor tyrosine kinases, is expressed by oligodendrocytes (Lemke 2006). A similar signaling mechanism also exists in Schwann cells (Newbern and Birchmeier 2010; Salzer 2015).

Myelin forming oligodendrocytes have up to 40 processes, each forming a segment of myelin called an internode that is several hundred micrometers long around separate axons. Internodal segments are separated from each other by specialized regions known as the node of Ranvier, which spans for around 1 μm. At the node, the axon is not wrapped by myelin (Eshed-Eisenbach and Peles 2013). The end of intermodal segment contains more cytoplasm forming the paranodal loop creating special junctions with the axon that induce specific changes on the axon in the node region. In addition, astrocyte processes contact the axonal membrane at the nodal region in the CNS, whereas loose sheaths of Schwann cell cover the node region in the PNS (Eshed-Eisenbach and Peles 2013).

In contrast to oligodendrocytes that each can form a myelin sheath around several axons, each myelinating Schwann cell generates one segment of myelin sheath for only one axon. Due to compaction, myelin dry mass is about 70 % lipids and 30 % proteins. While the myelin structure formed by oligodendrocytes and Schwann cells has a similar ultrastructure, it is not composed of an identical set of proteins (Simons and Trotter 2007; Llorens et al. 2011). There are several highly specific proteins that are needed for the formation of myelin. The major proteins of the CNS myelin are myelin-associated glycoprotein (MAG), myelin basic protein (MBP), myelin oligodendrocyte glycoprotein (MOG), proteolipidprotein (PLP)/DM20, and peripheral myelin protein-22 (PMP22), whereas CNS and PNS myelin share the MBP, the PNS lacks MAG and PLP, but expresses the protein zero (P0) and PMP22 (Simons and Trotter 2007; Llorens et al. 2011).

## *Myelination and Saltatory Nerve Conduction*

Like astrocytes, oligodendrocytes are also interconnected by gap junctions formed by connexins, although connexin proteins for oligodendrocytes and astrocytes are distinct. Connexins are key components for the structure and function of myelin. In fact, mutations in certain connexin proteins lead to hypomyelination and are involved in several human pathologies (Kleopa et al. 2010).

The axon's node of Ranvier contains a high density of sodium channels, which allows for a fast inward $Na^+$ current capable of generating an action potential only at the node, in what is known as saltatory conduction (Schafer and Rasband 2006). As discussed in Chapter "Oligodendrocytes: Functioning in a Delicate Balance Between High Metabolic Requirements And Oxidative Damage", once the action potential is triggered at the node, it spreads passively and rapidly to the next node, where the next action potential is generated. This is not only faster, but consumes much less energy, since $Na^+$ gradients must be restored only at the node, reducing the amount of $Na^+$ to be transported out by the $Na^+/K^+$-ATPase. At the time of ensheathment, $Na^+$ channels start to cluster at the site of the future node of Ranvier, promoted by protein interactions between myelin and the axonal membrane involving cell adhesion molecules like gliomedin, neurofascin, and neural cell adhesion molecule (NCAM). $K^+$ channels, involved in repolarization, are concentrated very close, in the juxta paranodal region (Scherer 1999; Poliak and Peles 2003). In contrast, in non-myelinated axons, voltage sensitive sodium and potassium channels are distributed along the whole surface of the axon.

## *Myelination Impairment and Disease*

As discussed in Chapters "Peripheral Inflammation and Demyelinating Diseases", "Regulation of Oligodendrocyte Differentiation and Myelination by Nuclear Receptors: Role in Neurodegenerative Disorders", and "The Role of Galectin-3: From Oligodendroglial Differentiation and Myelination to Demyelination and Remyelination Processes in a Cuprizone-Induced", the most frequent disease affecting oligodendrocytes is MS. It is caused by the loss of myelin in areas of the brain and spinal cord resulting in the impairment of axonal action potential propagation (Olsen and Akirav 2015). Remyelination is observed, but often relapses occur leading to neurodegeneration as the disease progresses. The primary cause for the impairment of oligodendrocytes is still unknown. Demyelinated regions contain inflammatory cells, including infiltrating lymphocytes and macrophages, and activated microglia, which appear to potentiate or even initiate the damage cascade (Tanaka and Yoshida 2014). Most of the genetically determined pathologies are associated with mutations in myelin proteins or connexins, the molecular entities forming gap junctions. Similar to the CNS, mutations in Schwann cell myelin or gap junction proteins lead to neuropathies such as Charcot–Marie–Tooth disease (Kleopa 2011).

# NG2–Glia

As further discussed in Chapter "NG2–Glia, More Than Progenitor Cells", NG2–glia are progenitor cells in the immature and adult nervous system that give rise to oligodendrocytes. Therefore, they are also called oligodendrocyte precursor cells (OPCs) (Lopez Juarez et al. 2015). Although they appear to have the capacity to differentiate into astrocytes and oligodendrocytes, the main route seems to be confined to the oligodendrocyte lineage. However, they are found in the brain in numbers much greater than would be needed for that role, and in brain areas where oligodrendrogenesis does not often occur.

NG2–glia are a potential source for remyelination in demyelinating diseases such as MS (Grade et al. 2013). There are distinct markers that allow one to identify them, such as the transcription factor Olig-2 or the proteoglycan NG-2 (neuron/glia antigen 2), a membrane-spanning signaling protein that is exclusively found in NG2–glia in the CNS (Trotter 2005). NG2–glia appear to interact with neurons. They express glutamate receptors and are capable of sensing neuronal activity. This appears to be a potential mechanism for neurons to regulate the differentiation of NG2–glia into oligodendrocytes.

NG2–glia in culture may be induced by certain growth factors to become multipotent neural stem cells. It is important to note that most glial cells in the CNS are not neural stem cells, but NG2–glia are a distinct subset. Other investigators revealed the role of NG-2 in the communication between neurons and NG2–glia (Sakry et al. 2014). They reported that a NG-2 fragment that is shed into the extracellular matrix, is produced by activity-dependent cleavage of NG-2 by the secretase ADAM10 and other secretases. This process generates a large ectodomain, containing two neurexin-like domains, which is released to the extracellular medium, and two smaller pieces, which remain associated with the cell.

The physiological functions of NG-2 in the brain have been determined from research in mice lacking NG-2. In the absence of NG-2, neuronal AMPA receptors show altered subunit composition compared with normal mice, which is probably responsible for the impaired LTP observed in the KO animals (Sakry et al. 2014). When brain slices from NG-2 KO mice in culture are treated with a recombinant protein containing the extracellular domain of NG-2 that is shed by cleavage, the properties of neuronal AMPA receptors in the slices return to normal. Therefore, the neurexin-like domains appear to be crucial for the modulation of neuronal function by NG-2 cleavage.

NG-2 KO mice show behavioral deficits that depend on functioning somatosensory cortex (Yang et al. 2013), which shows the relevance of the regulatory crosstalk between neurons and NG2–glia. Although many questions remain to be answered, recent evidence indicates that in addition to receiving input from neurons, NG2–glia might modulate neuronal properties and, therefore, activity. These findings add weight to the notion that glia are far from being only support cells. At this point, it is becoming clear that neurons and glia talk constantly to each other.

## Microglia

Microglia are the immune cells of the CNS, corresponding to the endogenous brain defense system, and are responsible for CNS protection against diverse pathogenic factors (Kettenmann et al. 2011). They derive from progenitors of mesodermal origin that migrate into the nervous system halfway through development (Ginhoux et al. 2010). After invading the CNS, microglial precursors disseminate throughout the neural parenchyma and acquire a distinctive phenotype that clearly distinguishes them from the blood-derived monocytes. They are highly reactive and become activated in response to any changes in the nervous system (Hanisch and Kettenmann 2007). They migrate to the site of damage, proliferate, and become phagocytes. Microglia can also interact with the peripheral immune system by antigen presentation and sense the presence of brain tumors (see Chapter “Microglia in Cancer: For Good or for Bad?” for further reading on microglia–tumor interaction).

Under physiological conditions, microglia are the fastest moving cells in the nervous system and send multiple thin processes that extend in all directions. Similar to astrocytes, each microglial cell has its own territory, of around 50 μm diameter, showing very little overlap between neighboring microglia. The processes of resting microglial cells are constantly moving through its territory, with a speed of about 1.5 μm/min. In addition, processes also extend and retract small protrusions. Through these mechanisms, microglia scan through their territory. Considering the speed of this movement, the brain parenchyma can be completely scanned by microglia several times each day (Hanisch and Kettenmann 2007). Although microglia processes motility is not affected by neuronal activity per se, it is sensitive to activators (ATP) and inhibitors of purinoceptors (Koizumi et al. 2013). In contrast, focal neuronal damage induces a rapid and concerted movement of microglia toward the site of lesion (von Bernhardi and Muller 1995; Cornejo and von Bernhardi 2013; Duan et al. 2009). This injury-induced motility is regulated by activation of purinoceptors (Honda et al. 2001), and it is sensitive to the inhibition of gap junctions. Inhibition of gap junctions also affects motility of astrocytes processes, cells that also extend processes defining specific territories. Thus, both microglia and astrocytes show sophisticated scanning system that allows them to survey the environment. As further discussed in Chapter "Microglia Function in the Normal Brain", a feature that exemplifies that microglia are capable of cross talk with neurons is the microglial cell expression of fractalkine receptors. Fractalkine (CX3CL1), a 373 amino acids cytokine protein member of the CX3C chemokine family, is found particularly in neurons. CX3CL1 is upregulated in the hippocampus by spatial learning and is likely to regulate glutamate-mediated neurotransmission.

## *Activation of Microglia*

When an insult to the nervous system is detected by microglia, they launch a specific program that results in the rapid transformation from surveillance typical of ramified microglia into an ameboid shaped cell that may move toward a nearby site of injury, a process referred to as 'microglial activation', similarly observed in vertebrates (Raivich et al. 1999; Gerhard et al. 2006; Kettenmann et al. 2011; von Bernhardi et al. 2015b) and invertebrates, like the leech (Samuels et al. 2013; Dahl and Muller 2014). Activation involves multiple steps. In terms of morphological changes, microglia retract their processes, which become fewer and thicker, and increase the size of their cell bodies. Their motility is increased as they are recruited to move toward the injury. Microglia reportedly can also proliferate, increasing their numbers further at the injury site. In terms of function, microglia change the expression pattern of many enzymes and receptors, and the production of immune response molecules is induced (Colton 2009). If the damage is strong and persistent enough, microglia become phagocytes (see Chapter "Microglia Function in the Normal Brain"). Thus, microglial activation appears to be a complex and highly coordinated set of changes. Furthermore, in contrast to the traditional view of the polarized M1-M2 activation of microglia (Aguzzi et al. 2013; Crain et al. 2013), the process of activation most likely corresponds to a wide spectrum of many specific substates that will combine specific physiological features (Boche et al. 2003; von Bernhardi 2007; Kettenmann et al. 2011; Aguzzi et al. 2013; von Bernhardi et al. 2015b). In fact, activated microglia display heterogeneous properties depending on the pathological conditions and the region of the nervous system that is affected (Goings et al. 2006), similar to what has been described for astrocytes.

The precise nature of signals triggering microglial activation is not well understood (Kettenmann et al. 2011, 2013; Dahl and Muller 2014). It probably depends both on the lack of trophic/regulatory molecules released during normal CNS activity and the appearance of abnormal molecules (Biber et al. 2007). Neurotransmitters are good examples of signals regulating microglial activation (Hanisch and Kettenmann 2007; Pocock and Kettenmann 2007). Microglia express several neurotransmitter receptors, such as receptors for GABA, glutamate, dopamine, and noradrenaline. In general, activation of the receptors inhibits the inflammatory activation of microglia. It could be speculated that decreased neuronal activity could induce an "alerted" state in microglia. The other type of signals are the appearance of abnormal molecules or abnormal concentrations of physiological molecules, that indicate tissue deterioration (Inoue 2002; Minghetti et al. 2005; Biber et al. 2007). Damaged neurons can release high amounts of ATP, cytokines, neuropeptides, and growth factors. These factors can be sensed by microglia and trigger activation (Inoue 2002; Duan et al. 2009). Both types of signaling provide microglia information regarding the status of brain parenchyma (Samuels et al. 2013; Dahl and Muller 2014). Astrocytes (Orellana et al. 2013) and neurons (Bernhardi and Nicholls 1999; Neumann 2001; Ramírez et al. 2008) provide robust regulation of microglial activation.

Microglial activation may start with defense-oriented functions to fight off pathogens or to limit progression of tissue damage after an injury. However, after activation microglia may continue to show long-lasting changes. Epigenetic mechanisms organizing long-lasting adjustments may cause previously activated cells to behave differently after a second challenge (Orellana et al. 2014). Such microglia might facilitate the development of dysfunction or could result in what is observed as age-related impairment (Conde and Streit 2006; Block et al. 2007; von Bernhardi 2007; von Bernhardi et al. 2011). As further discussed in Chapter "Age-Dependent Changes in the Activation and Regulation of Microglia", age-related changes in the regulation of inflammatory activation could facilitate the development of neurodegenerative diseases (Minghetti et al. 2005; Hart et al. 2012; von Bernhardi et al. 2015a, b).

The functional heterogeneity of microglia must be considered. For example, a myelinated environment is a different setting for microglia than that of gray matter. White and gray matter microglia show different immunoregulation (Anderson et al. 2007). These differences could affect development and normal function, as well as the response to inflammation (Hristova et al. 2010; Hart et al. 2012). Regional heterogeneity of microglia is observed regarding their morphology, proliferative behavior (Ladeby et al. 2005b; Marshall et al. 2008), transcription, and translation of various constitutive and inducible molecules, including neurotrophins such as IGF-I (Elkabes et al. 1996), cytokines, and other inflammatory mediators (TNFα, IL6, nitric oxide (NO)), and membrane receptors (integrins, CD4, CD11c, CD34, CD40, CD45, CD86, MHC class II, FcR) (Ford et al. 1995; Ren et al. 1999; Ladeby et al. 2005a; Wirenfeldt et al. 2005; Sriram et al. 2006; Carson et al. 2007;

Davoust et al. 2008; Kawahara et al. 2009). Exposure to neurotransmitters, proximity to blood vessels, and properties of the blood–brain carrier controlling the microenvironment (Abbott et al. 2010) might be associated with differences observed in microglial morphology and function (Davoust et al. 2008).

## *NO Production and Oxidative Stress*

As previously mentioned, activated microglia can produce several inflammatory mediators, including NO. NO is a lipid soluble radical gas that freely crosses cell membranes and acts as a signaling molecule, participating in several biological processes (Govers and Oess 2004; Pacher et al. 2007). NO is synthesized from L-arginine and molecular oxygen by NO synthases (NOS). There are three isotypes of NOS: endothelial NOS (eNOS), neuronal NOS (nNOS), and inducible NOS (iNOS). The first two are constitutively expressed in endothelial cells and neurons, respectively, while the latter is induced in macrophages, microglia, astrocytes, and other cell types in response to inflammatory mediators including bacterial lipopolysaccharides (Forstermann and Sessa 2012; Govers and Oess 2004; Pacher et al. 2007).

NO has an extremely short half-life of only a few seconds due to its fast reaction with other molecules. Many of its oxidation products have been reported to have cytotoxic effects, and indeed its induction in the presence of bacteria reflects its bacteriocidal effect. Simultaneous production of NO and superoxide by microglia can form peroxynitrite, a powerful oxidant able to irreversibly inhibit mitochondrial respiration, cause DNA fragmentation and lipid oxidation, and induce neuronal death (Mander and Brown 2005; Moncada and Bolanos 2006; Brown 2010; Brown and Neher 2010). Importantly, oxidative stress produced by the accumulation of reactive oxygen species (ROS) increases as organisms age, favoring the onset of neurodegenerative disorders, as further discussed in Chapter "Age-Dependent Changes in the Activation and Regulation of Microglia".

## *Antigen Presentation*

Microglial cells are the principal antigen presenting cells in the CNS. Under basal conditions the expression of the molecular complex for presenting antigen, the major histocompatibility complex II (MHCII) and co-stimulatory molecules such as CD80, CD86 and CD40 are not detected. Upon stimulation the molecules are highly upregulated (Perry 1998). This upregulation has been observed in several pathologies (O'Keefe et al. 2002) including MS. By releasing cytokines such as CCL2, microglial cells are important for recruiting cells into the CNS. As discussed in Chapter "Microglia in Cancer: For Good or for Bad?", microglia can interact with infiltrating T lymphocytes (Ford et al. 1995) and, thus, mediate the immune response in the brain.

## *Damage Versus Protection*

Although research has mainly focused on the detrimental effects of microglia-mediated neuroinflammation and oxidative stress, it is now accepted that acute self-limited activation of microglia is essential for the structural and functional integrity of the CNS. Microglia participate actively in the remodeling of synapses and tissue repair (Trapp et al. 2007; Wake et al. 2009). They remove cell debris and myelin fragments and secrete neurotrophins and cytokines for the survival of injured neurons (Batchelor et al. 1999; Schwartz 2003). However, they can also produce cytotoxic factors and are involved in the pathogenesis of several neurodegenerative diseases (Brown 2010; Kim and de Vellis 2005). Therefore, their activity needs to be strictly regulated (see Chapters "Microglia Function in the Normal Brain" and "Purine Signaling and Microglial Wrapping" for further reading).

## Recapitulation

For a long time, glial cells were considered to be a cell population subordinated to neurons. However, this point of view has dramatically changed in recent decades and growing evidence indicates that each type of glial cell-astrocytes, oligodendrocytes, Schwann cells, NG-2 cells and microglia plays key roles in the normal functioning of the nervous system. It has also been demonstrated that they are involved in the pathogenesis of neurological diseases. Thus, we can state that every process carried out in our brain can only be understood if the interaction of neurons and glial cells are taken into account. This challenges the accepted paradigm that nervous system function results exclusively from neuronal network activity and suggests that glial cell activity influences the neuronal activity outcome. Although there are still several unresolved questions, current research is delivering interesting and promising results.

**Acknowledgments** Supported by Grants Fondo Nacional de Desarrollo Científico y Tecnológico (FONDECYT) 1130874 (JE) and 1131025 (RvB).

## References

Abbott NJ, Patabendige AA, Dolman DE, Yusof SR, Begley DJ (2010) Structure and function of the blood-brain barrier. Neurobiol Dis 37(1):13–25. doi:10.1016/j.nbd.2009.07.030

Abbott NJ, Ronnback L, Hansson E (2006) Astrocyte-endothelial interactions at the blood-brain barrier. Nat Rev Neurosci 7(1):41–53. doi:10.1038/nrn1824

Aguzzi A, Barres BA, Bennett ML (2013) Microglia: scapegoat, saboteur, or something else? Science 339(6116):156–161. doi:10.1126/science.1227901 [pii]

Anderson AC, Anderson DE, Bregoli L, Hastings WD, Kassam N, Lei C, Chandwaskar R, Karman J, Su EW, Hirashima M, Bruce JN, Kane LP, Kuchroo VK, Hafler DA (2007)

Promotion of tissue inflammation by the immune receptor Tim-3 expressed on innate immune cells. Science 318(5853):1141–1143. doi:10.1126/science.1148536
Anderson CM, Swanson RA (2000) Astrocyte glutamate transport: review of properties, regulation, and physiological functions. Glia 32(1):1–14
Araque A (2008) Astrocytes process synaptic information. Neuron Glia Biol 4(1):3–10. doi:10.1017/S1740925X09000064
Araque A, Parpura V, Sanzgiri RP, Haydon PG (1999) Tripartite synapses: glia, the unacknowledged partner. Trends Neurosci 22(5):208–215
Batchelor PE, Liberatore GT, Wong JY, Porritt MJ, Frerichs F, Donnan GA, Howells DW (1999) Activated macrophages and microglia induce dopaminergic sprouting in the injured striatum and express brain-derived neurotrophic factor and glial cell line-derived neurotrophic factor. J Neurosci 19(5):1708–1716
Bayraktar OA, Fuentealba LC, Alvarez-Buylla A, Rowitch DH (2015) Astrocyte development and heterogeneity. Cold Spring Harb Perspect Biol 7(1):a020362. doi:10.1101/cshperspect.a020362
Beck A, Nieden RZ, Schneider HP, Deitmer JW (2004) Calcium release from intracellular stores in rodent astrocytes and neurons in situ. Cell Calcium 35(1):47–58
Bercury KK, Macklin WB (2015) Dynamics and mechanisms of CNS myelination. Dev Cell 32(4):447–458. doi:10.1016/j.devcel.2015.01.016
Bernhardi RV, Nicholls JG (1999) Transformation of leech microglial cell morphology and properties following co-culture with injured central nervous system tissue. J Exp Biol 202(Pt 6):723–728
Biber K, Neumann H, Inoue K, Boddeke HW (2007) Neuronal 'On' and 'Off' signals control microglia. Trends Neurosci 30(11):596–602. doi:10.1016/j.tins.2007.08.007
Biber K, Tsuda M, Tozaki-Saitoh H, Tsukamoto K, Toyomitsu E, Masuda T, Boddeke H, Inoue K (2011) Neuronal CCL21 up-regulates microglia P2X4 expression and initiates neuropathic pain development. EMBO J 30(9):1864–1873. doi:10.1038/emboj.2011.89
Block ML, Zecca L, Hong JS (2007) Microglia-mediated neurotoxicity: uncovering the molecular mechanisms. Nat Rev Neurosci 8(1):57–69. doi:10.1038/nrn2038
Boche D, Cunningham C, Gauldie J, Perry VH (2003) Transforming growth factor-beta 1-mediated neuroprotection against excitotoxic injury in vivo. J Cereb Blood Flow Metab Off J Int Soc Cereb Blood Flow Metab 23(10):1174–1182. doi:10.1097/01.WCB.0000090080.64176.44
Bozzali M, Wrabetz L (2004) Axonal signals and oligodendrocyte differentiation. Neurochem Res 29(5):979–988
Brown GC (2010) Nitric oxide and neuronal death. Nitric Oxide Biol Chem/Off J Nitric Oxide Soc 23(3):153–165. doi:10.1016/j.niox.2010.06.001
Brown GC, Neher JJ (2010) Inflammatory neurodegeneration and mechanisms of microglial killing of neurons. Mol Neurobiol 41(2–3):242–247. doi:10.1007/s12035-010-8105-9
Butt AM, Duncan A, Berry M (1994) Astrocyte associations with nodes of Ranvier: ultrastructural analysis of HRP-filled astrocytes in the mouse optic nerve. J Neurocytol 23(8):486–499
Carson MJ, Bilousova TV, Puntambekar SS, Melchior B, Doose JM, Ethell IM (2007) A rose by any other name? The potential consequences of microglial heterogeneity during CNS health and disease. Neurother J Am Soc Exp Neurother 4(4):571–579. doi:10.1016/j.nurt.2007.07.002
Cheslow L, Alvarez JI (2016) Glial-endothelial crosstalk regulates blood-brain barrier function. Curr Opin Pharmacol 26:39–46. doi:10.1016/j.coph.2015.09.010
Christopherson KS, Ullian EM, Stokes CC, Mullowney CE, Hell JW, Agah A, Lawler J, Mosher DF, Bornstein P, Barres BA (2005) Thrombospondins are astrocyte-secreted proteins that promote CNS synaptogenesis. Cell 120(3):421–433. doi:10.1016/j.cell.2004.12.020
Chung WS, Allen NJ, Eroglu C (2015) Astrocytes control synapse formation, function, and elimination. Cold Spring Harb Perspect Biol 7(9):a020370. doi:10.1101/cshperspect.a020370
Chung WS, Clarke LE, Wang GX, Stafford BK, Sher A, Chakraborty C, Joung J, Foo LC, Thompson A, Chen C, Smith SJ, Barres BA (2013) Astrocytes mediate synapse elimination

through MEGF10 and MERTK pathways. Nature 504(7480):394–400. doi:10.1038/nature12776

Colton CA (2009) Heterogeneity of microglial activation in the innate immune response in the brain. J Neuroimmune Pharmacol Off J Soc Neuroimmune Pharmacol 4(4):399–418. doi:10.1007/s11481-009-9164-4

Conde JR, Streit WJ (2006) Effect of aging on the microglial response to peripheral nerve injury. Neurobiol Aging 27(10):1451–1461. doi:10.1016/j.neurobiolaging.2005.07.012

Connor EA, McMahan UJ (1987) Cell accumulation in the junctional region of denervated muscle. J Cell Biol 104(1):109–120

Cornejo F, von Bernhardi R (2013) Role of scavenger receptors in glia-mediated neuroinflammatory response associated with Alzheimer's disease. Mediators Inflamm 2013:895651. doi:10.1155/2013/895651

Crain JM, Nikodemova M, Watters JJ (2013) Microglia express distinct M1 and M2 phenotypic markers in the postnatal and adult central nervous system in male and female mice. J Neurosci Res 91(9):1143–1151. doi:10.1002/jnr.23242

Croft W, Dobson KL, Bellamy TC (2015) Plasticity of neuron-glial transmission: equipping glia for long-term integration of network activity. Neural Plast 2015:765792. doi:10.1155/2015/765792

Dahl G, Muller KJ (2014) Innexin and pannexin channels and their signaling. FEBS Lett 588 (8):1396–1402. doi:10.1016/j.febslet.2014.03.007

Davoust N, Vuaillat C, Androdias G, Nataf S (2008) From bone marrow to microglia: barriers and avenues. Trends Immunol 29(5):227–234. doi:10.1016/j.it.2008.01.010

de Hoz L, Simons M (2015) The emerging functions of oligodendrocytes in regulating neuronal network behaviour. Bioessays News Rev Mol Cell Dev Biol 37(1):60–69. doi:10.1002/bies.201400127

Diniz LP, Almeida JC, Tortelli V, Vargas Lopes C, Setti-Perdigao P, Stipursky J, Kahn SA, Romao LF, de Miranda J, Alves-Leon SV, de Souza JM, Castro NG, Panizzutti R, Gomes FC (2012) Astrocyte-induced synaptogenesis is mediated by transforming growth factor beta signaling through modulation of D-serine levels in cerebral cortex neurons. J Biol Chem 287 (49):41432–41445. doi:10.1074/jbc.M112.380824

Duan Y, Sahley CL, Muller KJ (2009) ATP and NO dually control migration of microglia to nerve lesions. Dev Neurobiol 69(1):60–72. doi:10.1002/dneu.20689

Elkabes S, DiCicco-Bloom EM, Black IB (1996) Brain microglia/macrophages express neurotrophins that selectively regulate microglial proliferation and function. J Neurosci 16 (8):2508–2521

Eshed-Eisenbach Y, Peles E (2013) The making of a node: a co-production of neurons and glia. Curr Opin Neurobiol 23(6):1049–1056. doi:10.1016/j.conb.2013.06.003

Faissner A, Pyka M, Geissler M, Sobik T, Frischknecht R, Gundelfinger ED, Seidenbecher C (2010) Contributions of astrocytes to synapse formation and maturation—potential functions of the perisynaptic extracellular matrix. Brain Res Rev 63(1–2):26–38. doi:10.1016/j.brainresrev.2010.01.001

Fields RD, Stevens-Graham B (2002) New insights into neuron-glia communication. Science 298 (5593):556–562. doi:10.1126/science.298.5593.556

Foran E, Trotti D (2009) Glutamate transporters and the excitotoxic path to motor neuron degeneration in amyotrophic lateral sclerosis. Antioxid Redox Signal 11(7):1587–1602. doi:10.1089/ars.2009.2444

Ford AL, Goodsall AL, Hickey WF, Sedgwick JD (1995) Normal adult ramified microglia separated from other central nervous system macrophages by flow cytometric sorting. Phenotypic differences defined and direct ex vivo antigen presentation to myelin basic protein-reactive CD4+ T cells compared. J Immunol 154(9):4309–4321

Forstermann U, Sessa WC (2012) Nitric oxide synthases: regulation and function. Eur Heart J 33 (7):829–837, 837a–837d. doi:10.1093/eurheartj/ehr304

Gerhard A, Pavese N, Hotton G, Turkheimer F, Es M, Hammers A, Eggert K, Oertel W, Banati RB, Brooks DJ (2006) In vivo imaging of microglial activation with [11C](R)-PK11195

PET in idiopathic Parkinson's disease. Neurobiol Dis 21(2):404–412. doi:10.1016/j.nbd.2005.08.002
Ginhoux F, Greter M, Leboeuf M, Nandi S, See P, Gokhan S, Mehler MF, Conway SJ, Ng LG, Stanley ER, Samokhvalov IM, Merad M (2010) Fate mapping analysis reveals that adult microglia derive from primitive macrophages. Science 330(6005):841–845. doi:10.1126/science.1194637
Gittis AH, Brasier DJ (2015) NEUROSCIENCE. Astrocytes tell neurons when to listen up. Science 349(6249):690–691. doi:10.1126/science.aad0678
Goings GE, Kozlowski DA, Szele FG (2006) Differential activation of microglia in neurogenic versus non-neurogenic regions of the forebrain. Glia 54(4):329–342. doi:10.1002/glia.20381
Gourine AV, Kasymov V, Marina N, Tang F, Figueiredo MF, Lane S, Teschemacher AG, Spyer KM, Deisseroth K, Kasparov S (2010) Astrocytes control breathing through pH-dependent release of ATP. Science 329(5991):571–575. doi:10.1126/science.1190721
Govers R, Oess S (2004) To NO or not to NO: 'where?' is the question. Histol Histopathol 19 (2):585–605
Grade S, Bernardino L, Malva JO (2013) Oligodendrogenesis from neural stem cells: perspectives for remyelinating strategies. Int J Dev Neurosci Off J Int Soc Dev Neurosci 31(7):692–700. doi:10.1016/j.ijdevneu.2013.01.004
Grigoryan T, Birchmeier W (2015) Molecular signaling mechanisms of axon-glia communication in the peripheral nervous system. Bioessays News Rev Mol Cellular Dev Biol 37(5):502–513. doi:10.1002/bies.201400172
Halliday GM, Stevens CH (2011) Glia: initiators and progressors of pathology in Parkinson's disease. Mov Disord Off J Mov Disord Soc 26(1):6–17. doi:10.1002/mds.23455
Hamilton NB, Attwell D (2010) Do astrocytes really exocytose neurotransmitters? Nat Rev Neurosci 11(4):227–238. doi:10.1038/nrn2803
Hanisch UK, Kettenmann H (2007) Microglia: active sensor and versatile effector cells in the normal and pathologic brain. Nat Neurosci 10(11):1387–1394. doi:10.1038/nn1997
Hansen AJ, Nedergaard M (1988) Brain ion homeostasis in cerebral ischemia. Neurochem Pathol 9:195–209
Hansen DV, Lui JH, Parker PR, Kriegstein AR (2010) Neurogenic radial glia in the outer subventricular zone of human neocortex. Nature 464(7288):554–561. doi:10.1038/nature08845
Hart AD, Wyttenbach A, Perry VH, Teeling JL (2012) Age related changes in microglial phenotype vary between CNS regions: grey versus white matter differences. Brain Behav Immun 26(5):754–765. doi:10.1016/j.bbi.2011.11.006 (pii: S0889-1591(11)00597-6)
Hartfuss E, Galli R, Heins N, Gotz M (2001) Characterization of CNS precursor subtypes and radial glia. Dev Biol 229(1):15–30. doi:10.1006/dbio.2000.9962
Hawkins BT, Davis TP (2005) The blood-brain barrier/neurovascular unit in health and disease. Pharmacol Rev 57(2):173–185. doi:10.1124/pr.57.2.4
Hilgetag CC, Barbas H (2009) Are there ten times more glia than neurons in the brain? Brain Struct Function 213(4–5):365–366. doi:10.1007/s00429-009-0202-z
Honda S, Sasaki Y, Ohsawa K, Imai Y, Nakamura Y, Inoue K, Kohsaka S (2001) Extracellular ATP or ADP induce chemotaxis of cultured microglia through Gi/o-coupled P2Y receptors. J Neurosci 21(6):1975–1982
Hristova M, Cuthill D, Zbarsky V, Acosta-Saltos A, Wallace A, Blight K, Buckley SM, Peebles D, Heuer H, Waddington SN, Raivich G (2010) Activation and deactivation of periventricular white matter phagocytes during postnatal mouse development. Glia 58(1):11–28. doi:10.1002/glia.20896
Hua X, Malarkey EB, Sunjara V, Rosenwald SE, Li WH, Parpura V (2004) $Ca^{2+}$-dependent glutamate release involves two classes of endoplasmic reticulum $Ca^{2+}$ stores in astrocytes. J Neurosci Res 76(1):86–97. doi:10.1002/jnr.20061
Inoue K (2002) Microglial activation by purines and pyrimidines. Glia 40(2):156–163. doi:10.1002/glia.10150

Jha MK, Lee WH, Suk K (2015) Functional polarization of neuroglia: implications in neuroinflammation and neurological disorders. Biochem Pharmacol. doi:10.1016/j.bcp.2015.11.003

Jirounek P, Bretton C, Dalle C (2002) Axon-glia interactions modulate axonal excitability in mammalian unmyelinated nerves. J Physiol Paris 96(3–4):237–241

Kang H, Tian L, Mikesh M, Lichtman JW, Thompson WJ (2014) Terminal Schwann cells participate in neuromuscular synapse remodeling during reinnervation following nerve injury. J Neurosci 34(18):6323–6333. doi:10.1523/JNEUROSCI.4673-13.2014

Kawahara K, Yoshida A, Koga K, Yokoo S, Kuniyasu A, Gotoh T, Sawada M, Nakayama H (2009) Marked induction of inducible nitric oxide synthase and tumor necrosis factor-alpha in rat CD40+ microglia by comparison to CD40− microglia. J Neuroimmunol 208(1–2):70–79. doi:10.1016/j.jneuroim.2009.01.007

Kettenmann H, Hanisch UK, Noda M, Verkhratsky A (2011) Physiology of microglia. Physiol Rev 91(2):461–553. doi:10.1152/physrev.00011.2010

Kettenmann H, Kirchhoff F, Verkhratsky A (2013) Microglia: new roles for the synaptic stripper. Neuron 77(1):10–18. doi:10.1016/j.neuron.2012.12.023 (pii: S0896-6273(12)01162-2)

Kim SU, de Vellis J (2005) Microglia in health and disease. J Neurosci Res 81(3):302–313

Kleopa KA (2011) The role of gap junctions in Charcot-Marie-Tooth disease. J Neurosci 31 (49):17753–17760. doi:10.1523/JNEUROSCI.4824-11.2011

Kleopa KA, Orthmann-Murphy J, Sargiannidou I (2010) Gap junction disorders of myelinating cells. Rev Neurosci 21(5):397–419

Koizumi S, Fujishita K, Inoue K (2005) Regulation of cell-to-cell communication mediated by astrocytic ATP in the CNS. Purinergic Signal 1(3):211–217. doi:10.1007/s11302-005-6321-y

Koizumi S, Ohsawa K, Inoue K, Kohsaka S (2013) Purinergic receptors in microglia: functional modal shifts of microglia mediated by P2 and P1 receptors. Glia 61(1):47–54. doi:10.1002/glia.22358

Kuffler SW, Nicholls JG (1966) The physiology of neuroglial cells. Ergebnisse der Physiologie, biologischen Chemie und experimentellen Pharmakologie 57:1–90

Ladeby R, Wirenfeldt M, Dalmau I, Gregersen R, Garcia-Ovejero D, Babcock A, Owens T, Finsen B (2005a) Proliferating resident microglia express the stem cell antigen CD34 in response to acute neural injury. Glia 50(2):121–131. doi:10.1002/glia.20159

Ladeby R, Wirenfeldt M, Garcia-Ovejero D, Fenger C, Dissing-Olesen L, Dalmau I, Finsen B (2005b) Microglial cell population dynamics in the injured adult central nervous system. Brain Res Brain Res Rev 48(2):196–206. doi:10.1016/j.brainresrev.2004.12.009

Lalo U, Palygin O, Rasooli-Nejad S, Andrew J, Haydon PG, Pankratov Y (2014) Exocytosis of ATP from astrocytes modulates phasic and tonic inhibition in the neocortex. PLoS Biol 12(1): e1001747. doi:10.1371/journal.pbio.1001747

Lee YI, Li Y, Mikesh M, Smith I, Nave KA, Schwab MH, Thompson WJ (2016) Neuregulin1 displayed on motor axons regulates terminal Schwann cell-mediated synapse elimination at developing neuromuscular junctions. Proc Natl Acad Sci USA 113(4):E479–487. doi:10.1073/pnas.1519156113

Lemke G (2006) Neuregulin-1 and myelination. Science's STKE Sig Transduct Knowl Environ 2006(325):pe11. doi:10.1126/stke.3252006pe11

Llorens F, Gil V, del Rio JA (2011) Emerging functions of myelin-associated proteins during development, neuronal plasticity, and neurodegeneration. FASEB J 25(2):463–475. doi:10.1096/fj.10-162792

Lopez Juarez A, He D, Richard LuQ (2015) Oligodendrocyte progenitor programming and reprogramming: Toward myelin regeneration. Brain Res. doi:10.1016/j.brainres.2015.10.051

Love FM, Son YJ, Thompson WJ (2003) Activity alters muscle reinnervation and terminal sprouting by reducing the number of Schwann cell pathways that grow to link synaptic sites. J Neurobiol 54(4):566–576. doi:10.1002/neu.10191

Love FM, Thompson WJ (1998) Schwann cells proliferate at rat neuromuscular junctions during development and regeneration. J Neurosci 18(22):9376–9385

Malarkey EB, Parpura V (2008) Mechanisms of glutamate release from astrocytes. Neurochem Int 52(1–2):142–154. doi:10.1016/j.neuint.2007.06.005
Mander P, Brown GC (2005) Activation of microglial NADPH oxidase is synergistic with glial iNOS expression in inducing neuronal death: a dual-key mechanism of inflammatory neurodegeneration. J Neuroinflammation 2:20. doi:10.1186/1742-2094-2-20
Marshall GP 2nd, Demir M, Steindler DA, Laywell ED (2008) Subventricular zone microglia possess a unique capacity for massive in vitro expansion. Glia 56(16):1799–1808. doi:10.1002/glia.20730
Minghetti L, Ajmone-Cat MA, De Berardinis MA, De Simone R (2005) Microglial activation in chronic neurodegenerative diseases: roles of apoptotic neurons and chronic stimulation. Brain Res Brain Res Rev 48(2):251–256. doi:10.1016/j.brainresrev.2004.12.015
Moncada S, Bolanos JP (2006) Nitric oxide, cell bioenergetics and neurodegeneration. J Neurochem 97(6):1676–1689. doi:10.1111/j.1471-4159.2006.03988.x
Montes de Oca Balderas P, Aguilera P (2015) A metabotropic-like flux-independent NMDA receptor regulates $Ca^{2+}$ exit from endoplasmic reticulum and mitochondrial membrane potential in cultured astrocytes. PLoS ONE 10(5):e012631410.1371/journal.pone.0126314
Morel L, Higashimori H, Tolman M, Yang Y (2014) VGluT1+ neuronal glutamatergic signaling regulates postnatal developmental maturation of cortical protoplasmic astroglia. J Neurosci 34 (33):10950–10962. doi:10.1523/JNEUROSCI.1167-14.2014
Neumann H (2001) Control of glial immune function by neurons. Glia 36(2):191–199
Newbern J, Birchmeier C (2010) Nrg1/ErbB signaling networks in Schwann cell development and myelination. Semin Cell Dev Biol 21(9):922–928. doi:10.1016/j.semcdb.2010.08.008
Newman EA (2003) New roles for astrocytes: regulation of synaptic transmission. Trends Neurosci 26(10):536–542. doi:10.1016/S0166-2236(03)00237-6
Ni Y, Parpura V (2009) Dual regulation of $Ca^{2+}$-dependent glutamate release from astrocytes: vesicular glutamate transporters and cytosolic glutamate levels. Glia 57(12):1296–1305. doi:10.1002/glia.20849
Noctor SC, Flint AC, Weissman TA, Dammerman RS, Kriegstein AR (2001) Neurons derived from radial glial cells establish radial units in neocortex. Nature 409(6821):714–720. doi:10.1038/35055553
O'Keefe GM, Nguyen VT, Benveniste EN (2002) Regulation and function of class II major histocompatibility complex, CD40, and B7 expression in macrophages and microglia: Implications in neurological diseases. J Neurovirol 8(6):496–512. doi:10.1080/13550280290100941
Oberheim NA, Goldman SA, Nedergaard M (2012) Heterogeneity of astrocytic form and function. Methods Mol Biol 814:23–45. doi:10.1007/978-1-61779-452-0_3
Oliveira JF, Sardinha VM, Guerra-Gomes S, Araque A, Sousa N (2015) Do stars govern our actions? Astrocyte involvement in rodent behavior. Trends Neurosci 38(9):535–549. doi:10.1016/j.tins.2015.07.006
Olsen JA, Akirav EM (2015) Remyelination in multiple sclerosis: cellular mechanisms and novel therapeutic approaches. J Neurosci Res 93(5):687–696. doi:10.1002/jnr.23493
Orellana JA, Busso D, Ramirez G, Campos M, Rigotti A, Eugenin J, von Bernhardi R (2014) Prenatal nicotine exposure enhances Cx43 and Panx1 unopposed channel activity in brain cells of adult offspring mice fed a high-fat/cholesterol diet. Front Cell Neurosci 8:403. doi:10.3389/fncel.2014.00403
Orellana JA, Montero TD, von Bernhardi R (2013) Astrocytes inhibit nitric oxide-dependent $Ca^{2+}$ dynamics in activated microglia: involvement of ATP released via pannexin 1 channels. Glia 61(12):2023–2037. doi:10.1002/glia.22573
Pacher P, Beckman JS, Liaudet L (2007) Nitric oxide and peroxynitrite in health and disease. Physiol Rev 87(1):315–424. doi:10.1152/physrev.00029.2006
Pakkenberg B, Gundersen HJ (1988) Total number of neurons and glial cells in human brain nuclei estimated by the disector and the fractionator. J Microsc 150(Pt 1):1–20
Pekny M, Pekna M (2014) Astrocyte reactivity and reactive astrogliosis: costs and benefits. Physiol Rev 94(4):1077–1098. doi:10.1152/physrev.00041.2013

Pellerin L, Bouzier-Sore AK, Aubert A, Serres S, Merle M, Costalat R, Magistretti PJ (2007) Activity-dependent regulation of energy metabolism by astrocytes: an update. Glia 55 (12):1251–1262. doi:10.1002/glia.20528

Perea G, Navarrete M, Araque A (2009) Tripartite synapses: astrocytes process and control synaptic information. Trends Neurosci 32(8):421–431. doi:10.1016/j.tins.2009.05.001

Perez-Alvarez A, Araque A (2013) Astrocyte-neuron interaction at tripartite synapses. Curr Drug Targets 14(11):1220–1224

Perry VH (1998) A revised view of the central nervous system microenvironment and major histocompatibility complex class II antigen presentation. J Neuroimmunol 90(2):113–121

Petrelli F, Bezzi P (2016) Novel insights into gliotransmitters. Curr Opin Pharmacol 26:138–145. doi:10.1016/j.coph.2015.11.010

Pocock JM, Kettenmann H (2007) Neurotransmitter receptors on microglia. Trends Neurosci 30 (10):527–535. doi:10.1016/j.tins.2007.07.007

Poliak S, Peles E (2003) The local differentiation of myelinated axons at nodes of Ranvier. Nat Rev Neurosci 4(12):968–980. doi:10.1038/nrn1253

Posada-Duque RA, Barreto GE, Cardona-Gomez GP (2014) Protection after stroke: cellular effectors of neurovascular unit integrity. Front Cell Neurosci 8:231. doi:10.3389/fncel.2014.00231

Quintanilla RA, Orellana JA, von Bernhardi R (2012) Understanding risk factors for Alzheimer's disease: interplay of neuroinflammation, connexin-based communication and oxidative stress. Arch Med Res 43(8):632–644. doi:10.1016/j.arcmed.2012.10.016

Raivich G, Bohatschek M, Kloss CU, Werner A, Jones LL, Kreutzberg GW (1999) Neuroglial activation repertoire in the injured brain: graded response, molecular mechanisms and cues to physiological function. Brain Res Brain Res Rev 30(1):77–105

Ramírez G, Rey S, von Bernhardi R (2008) Proinflammatory stimuli are needed for induction of microglial cell-mediated $A\beta PP_{244\text{-}C}$ and Aβ-neurotoxicity in hippocampal cultures. J Alzheimers Dis 15(1):45–59

Ramirez G, Toro R, Dobeli H, von Bernhardi R (2005) Protection of rat primary hippocampal cultures from A beta cytotoxicity by pro-inflammatory molecules is mediated by astrocytes. Neurobiol Dis 19(1–2):243–254. doi:10.1016/j.nbd.2005.01.007

Rassoulpour A, Wu HQ, Ferre S, Schwarcz R (2005) Nanomolar concentrations of kynurenic acid reduce extracellular dopamine levels in the striatum. J Neurochem 93(3):762–765. doi:10.1111/j.1471-4159.2005.03134.x

Reichenbach A, Bringmann A (2013) New functions of Muller cells. Glia 61(5):651–678. doi:10.1002/glia.22477

Ren L, Lubrich B, Biber K, Gebicke-Haerter PJ (1999) Differential expression of inflammatory mediators in rat microglia cultured from different brain regions. Brain Res Mol Brain Res 65 (2):198–205

Riquelme PA, Drapeau E, Doetsch F (2008) Brain micro-ecologies: neural stem cell niches in the adult mammalian brain. Philos Trans R Soc Lond B Biol Sci 363(1489):123–137. doi:10.1098/rstb.2006.2016

Ropert N, Jalil A, Li D (2015) Expression and cellular function of vSNARE proteins in brain astrocytes. Neuroscience. doi:10.1016/j.neuroscience.2015.10.036

Rowitch DH, Kriegstein AR (2010) Developmental genetics of vertebrate glial-cell specification. Nature 468(7321):214–222. doi:10.1038/nature09611

Rushton WA (1951) A theory of the effects of fibre size in medullated nerve. J Physiol 115 (1):101–122

Sakry D, Neitz A, Singh J, Frischknecht R, Marongiu D, Biname F, Perera SS, Endres K, Lutz B, Radyushkin K, Trotter J, Mittmann T (2014) Oligodendrocyte precursor cells modulate the neuronal network by activity-dependent ectodomain cleavage of glial NG2. PLoS Biol 12(11): e1001993. doi:10.1371/journal.pbio.1001993

Salzer JL (2015) Schwann cell myelination. Cold Spring Harb Perspect Biol 7(8):a020529. doi:10.1101/cshperspect.a020529

Samuels SE, Lipitz JB, Wang J, Dahl G, Muller KJ (2013) Arachidonic acid closes innexin/pannexin channels and thereby inhibits microglia cell movement to a nerve injury. Dev Neurobiol 73(8):621–631. doi:10.1002/dneu.22088
Santello M, Cali C, Bezzi P (2012) Gliotransmission and the tripartite synapse. Adv Exp Med Biol 970:307–331. doi:10.1007/978-3-7091-0932-8_14
Scemes E, Giaume C (2006) Astrocyte calcium waves: what they are and what they do. Glia 54 (7):716–725. doi:10.1002/glia.20374
Schafer DP, Rasband MN (2006) Glial regulation of the axonal membrane at nodes of Ranvier. Curr Opin Neurobiol 16(5):508–514. doi:10.1016/j.conb.2006.08.003
Scherer SS (1999) Nodes, paranodes, and incisures: from form to function. Ann N Y Acad Sci 883:131–142
Schitine C, Nogaroli L, Costa MR, Hedin-Pereira C (2015) Astrocyte heterogeneity in the brain: from development to disease. Front Cell Neurosci 9:76. doi:10.3389/fncel.2015.00076
Schousboe A, Scafidi S, Bak LK, Waagepetersen HS, McKenna MC (2014) Glutamate metabolism in the brain focusing on astrocytes. Adv Neurobiol 11:13–30. doi:10.1007/978-3-319-08894-5_2
Schwartz M (2003) Macrophages and microglia in central nervous system injury: are they helpful or harmful? J Cereb Blood Flow Metab Off J Int Soc Cereb Blood Flow Metab 23(4):385–394
Shimada IS, LeComte MD, Granger JC, Quinlan NJ, Spees JL (2012) Self-renewal and differentiation of reactive astrocyte-derived neural stem/progenitor cells isolated from the cortical peri-infarct area after stroke. J Neurosci 32(23):7926–7940. doi:10.1523/JNEUROSCI.4303-11.2012
Simons M, Trotter J (2007) Wrapping it up: the cell biology of myelination. Curr Opin Neurobiol 17(5):533–540. doi:10.1016/j.conb.2007.08.003
Slezak M, Pfrieger FW (2003) New roles for astrocytes: regulation of CNS synaptogenesis. Trends Neurosci 26(10):531–535. doi:10.1016/j.tins.2003.08.005
Sriram K, Matheson JM, Benkovic SA, Miller DB, Luster MI, O'Callaghan JP (2006) Deficiency of TNF receptors suppresses microglial activation and alters the susceptibility of brain regions to MPTP-induced neurotoxicity: role of TNF-α. FASEB J 20(6):670–682. doi:10.1096/fj.05-5106com
Takano T, Tian GF, Peng W, Lou N, Libionka W, Han X, Nedergaard M (2006) Astrocyte-mediated control of cerebral blood flow. Nat Neurosci 9(2):260–267. doi:10.1038/nn1623
Tanaka T, Yoshida S (2014) Mechanisms of remyelination: recent insight from experimental models. Biomol Concepts 5(4):289–298. doi:10.1515/bmc-2014-0015
Trapp BD, Wujek JR, Criste GA, Jalabi W, Yin X, Kidd GJ, Stohlman S, Ransohoff R (2007) Evidence for synaptic stripping by cortical microglia. Glia 55(4):360–368. doi:10.1002/glia.20462
Trotter J (2005) NG2-positive cells in CNS function and the pathological role of antibodies against NG2 in demyelinating diseases. J Neurol Sci 233(1–2):37–42. doi:10.1016/j.jns.2005.03.024
Vecino E, Rodriguez FD, Ruzafa N, Pereiro X, Sharma SC (2015) Glia-neuron interactions in the mammalian retina. Prog Retin Eye Res. doi:10.1016/j.preteyeres.2015.06.003
Verkhratsky A, Parpura V, Pekna M, Pekny M, Sofroniew M (2014) Glia in the pathogenesis of neurodegenerative diseases. Biochem Soc Trans 42(5):1291–1301. doi:10.1042/BST20140107
Vesce S, Bezzi P, Volterra A (1999) The active role of astrocytes in synaptic transmission. Cell Mol Life Sci 56(11–12):991–1000
Volterra A, Meldolesi J (2005) Astrocytes, from brain glue to communication elements: the revolution continues. Nat Rev Neurosci 6(8):626–640. doi:10.1038/nrn1722 [pii]
von Bernhardi R (2007) Glial cell dysregulation: a new perspective on Alzheimer disease. Neurotox Res 12(4):215–232
von Bernhardi R, Cornejo F, Parada GE, Eugenin J (2015a) Role of TGFβ signaling in the pathogenesis of Alzheimer's disease. Front Cell Neurosci 9:426. doi:10.3389/fncel.2015.00426
von Bernhardi R, Eugenin-von Bernhardi L, Eugenin J (2015b) Microglial cell dysregulation in brain aging and neurodegeneration. Front Aging Neurosci 7:124. doi:10.3389/fnagi.2015.00124

von Bernhardi R, Muller KJ (1995) Repair of the central nervous system: lessons from lesions in leeches. J Neurobiol 27(3):353–366. doi:10.1002/neu.480270308

von Bernhardi R, Tichauer J, Eugenin-von Bernhardi L (2011) Proliferating culture of aged microglia for the study of neurodegenerative diseases. J Neurosci Methods 202(1):65–69. doi:10.1016/j.jneumeth.2011.08.027

Wake H, Moorhouse AJ, Jinno S, Kohsaka S, Nabekura J (2009) Resting microglia directly monitor the functional state of synapses in vivo and determine the fate of ischemic terminals. J Neurosci 29(13):3974–3980. doi:10.1523/JNEUROSCI.4363-08.2009

Wallraff A, Kohling R, Heinemann U, Theis M, Willecke K, Steinhauser C (2006) The impact of astrocytic gap junctional coupling on potassium buffering in the hippocampus. J Neurosci 26 (20):5438–5447. doi:10.1523/JNEUROSCI.0037-06.2006

Waxman SG, Bennett MV (1972) Relative conduction velocities of small myelinated and non-myelinated fibres in the central nervous system. Nat New Biol 238(85):217–219

Wirenfeldt M, Babcock AA, Ladeby R, Lambertsen KL, Dagnaes-Hansen F, Leslie RG, Owens T, Finsen B (2005) Reactive microgliosis engages distinct responses by microglial subpopulations after minor central nervous system injury. J Neurosci Res 82(4):507–514. doi:10.1002/jnr.20659

Xu K, Terakawa S (1999) Fenestration nodes and the wide submyelinic space form the basis for the unusually fast impulse conduction of shrimp myelinated axons. J Exp Biol 202(Pt 15):1979–1989

Yang QK, Xiong JX, Yao ZX (2013) Neuron-NG2 cell synapses: novel functions for regulating NG2 cell proliferation and differentiation. BioMed Res Int 2013:402843. doi:10.1155/2013/402843

Zonta M, Angulo MC, Gobbo S, Rosengarten B, Hossmann KA, Pozzan T, Carmignoto G (2003) Neuron-to-astrocyte signaling is central to the dynamic control of brain microcirculation. Nat Neurosci 6(1):43–50. doi:10.1038/nn980

Zorec R, Araque A, Carmignoto G, Haydon PG, Verkhratsky A, Parpura V (2012) Astroglial excitability and gliotransmission: an appraisal of $Ca^{2+}$ as a signalling route. ASN Neuro 4(2). doi:10.1042/AN20110061

# Part I
# Glial Cell Function in the Healthy Nervous System

# NG2-glia, More Than Progenitor Cells

**Jaime Eugenín-von Bernhardi and Leda Dimou**

**Abstract** NG2-glia are a mysterious and ubiquitous glial population with a highly branched morphology. Initial studies suggested that their unique function is the generation and maintenance of oligodendrocytes in the central nervous system (CNS), important for proper myelination and therefore for axonal support and fast conduction velocity. Over the last years this simplistic notion has been dramatically changed: the wide and homogeneous distribution of NG2-glia within all areas of the developing CNS that is maintained during the whole lifespan, their potential to also differentiate into other cell types in a spatiotemporal manner, their active capability of maintaining their population and their dynamic behavior in altered conditions have raised the question: are NG2-glia simple progenitor cells or do they play further major roles in the normal function of the CNS? In this chapter, we will discuss some important features of NG2-glia like their homeostatic distribution in the CNS and their potential to differentiate into diverse cell types. Additionally, we will give some further insights into the properties that these cells have, like the ability to form synapses with neurons and their plastic behavior triggered by neuronal activity, suggesting that they may play a role specifically in myelin and more generally in brain plasticity. Finally, we will briefly review their behavior in disease models suggesting that their function is extended to repair the brain after insult.

**Keywords** NG2-glia · Myelination · Oligodendrocytes · Neuronal activity · NG2-glia neuronal synapse · Disease · Injury · Proliferation · Differentiation · Plasticity

J. Eugenín-von Bernhardi (✉) · L. Dimou (✉)
Physiological Genomics, Biomedical Center, Ludwig-Maximilians-University, Großhaderner Str. 9, 82152 Planegg-Martinsried, Germany
e-mail: Jaime.Eugenin@med.uni-muenchen.de

L. Dimou
e-mail: Leda.Dimou@lrz.uni-muenchen.de

J. Eugenín-von Bernhardi
Graduate School of Systemic Neuroscience, Ludwig-Maximilians-University, 82152 Planegg-Martinsried, Germany

 
R. von Bernhardi (ed.), *Glial Cells in Health and Disease of the CNS*,
Advances in Experimental Medicine and Biology 949,
DOI 10.1007/978-3-319-40764-7_2

## Abbreviations

| | |
|---|---|
| Aβ | Amyloid protein β |
| AD | Alzheimer's disease |
| AMPA | *α-amino-3-hydroxy-5-methyl-4-isoxazolepropionic acid* |
| AMPAR | AMPA receptor |
| Ascl1 | Achaete-scute homolog 1 |
| αScTX | A-scorpion toxin |
| BrdU | *5-bromo-2'-deoxyuridine* |
| $Ca_{V}s$ | Voltage-gated calcium channels |
| CC1 | Adenomatous polyposis coli |
| CNS | Central nervous system |
| DNQX | *6,7-dinitroquinoxaline-2,3-dione* |
| EAE | Experimental autoimmune encephalomyelitis |
| EdU | *5-ethynyl-2'-deoxyuridine* |
| EPSC | Excitatory postsynaptic current |
| $GABA_{A}R$ | *γ-aminobutyric acid* receptor |
| GPR17 | G-protein coupled receptor 17 |
| $K_{V}s$ | Voltage-gated potassium channels |
| LPC | *α-lysophosphatidylcholine* |
| Mash1 | Mammalian achaete-scute homolog 1 |
| MBP | Myelin basic protein |
| MCAO | Middle cerebral artery occlusion |
| mEPSC | Miniature EPSC |
| MS | Multiple sclerosis |
| $Na_{V}s$ | Voltage-gated sodium channels |
| NBQX | *2,3-dihydroxy-6-nitro-7-sulphamoyl-benzo(F)quinoxaline* |
| NG2 | Neuron/glia antigen 2 |
| NMDAR | *N-methyl-D-aspartate* receptor |
| OPCs | Oligodendrocyte progenitor cells |
| PDGF | Platelet-derived growth factor |
| PDGFR | PDGF receptor α |
| PFC | Prefrontral cortex |
| PLP | Proteolipid protein |
| PNS | Peripheral nervous system |
| PSD-95 | Postsynaptic density protein 95 |
| TeNT | Tetanus neurotoxin |
| TTX | Tetrodotoxin |

## NG2-glia in the Central Nervous System

In the mammalian central nervous system (CNS), oligodendrocytes that build the myelin, develop from a progenitor cell population during late gestational and early postnatal life (Miller 1996) (see Section "A Brief Introduction to Oligodendrocyte Structure and Function" in Chapter "Oligodendrocytes: Functioning in a Delicate Balance Between High Metabolic Requirements and Oxidative Damage," for a description on oliyodendro genesis). These progenitors are known as oligodendrocyte progenitor cells (OPCs) during development and as NG2-glia at later stages, becoming the fourth major group of glial cells in the CNS. The name NG2-glia derives from the expression of the chondroitin sulfate proteoglycan neuron/glia antigen 2 (NG2) (Fig. 1a–c) on their cell surface. To distinguish these over the pericytes of the CNS that also express NG2 (Ozerdem et al. 2001), we use the term NG2-glia instead of simply NG2-positive cells. NG2-glia can also be found in the literature as polydendrocytes, because of their branched morphology revealed by the immunolabeling for NG2 and the platelet-derived growth factor receptor α (PDGFRα) (Fig. 1a, b and d) and as OPCs due to their association to the generation and maintenance of the oligodendrocyte population under physiological and pathological conditions. However, restricting NG2-glia to be simply

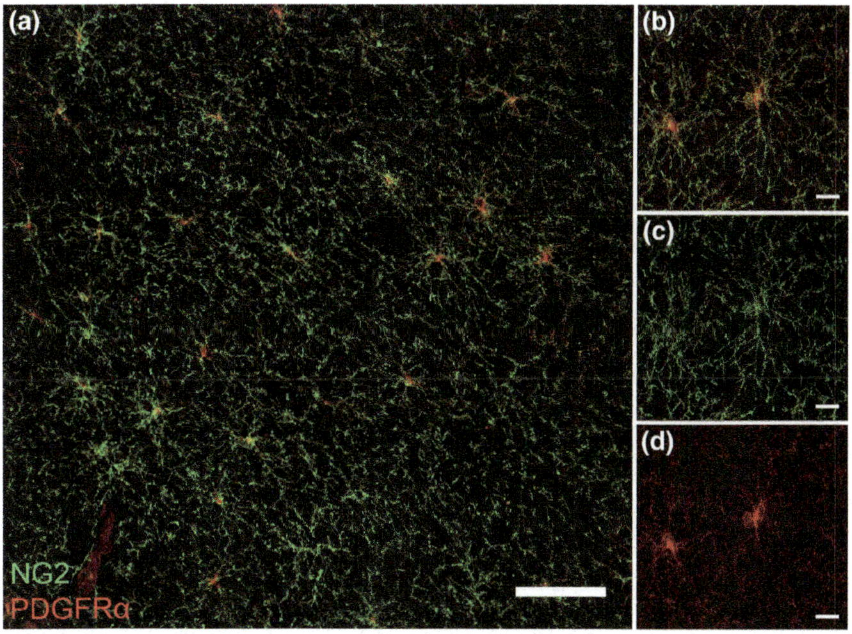


**Fig. 1** **a** Confocal microscopy image showing the multiprocess morphology of NG2-glia co-expressing NG2 and PDGFRα. Scale bar represents 50 μm. **b–d** Magnification of two NG2-glia; **b** co-expressing NG2 and PDGFRα, **c** expressing NG2 and **d** expressing PDGFRα. Scale bar represents 10 μm

oligodendrocyte progenitors does not give them enough credit. In fact, it has not only been shown that NG2-glia represent the major proliferative cell population in the healthy adult brain (Psachoulia et al. 2009; Simon et al. 2011; Dimou and Götz 2014) outside the neurogenic niches but also that they can self-renew. In vitro and in vivo fate mapping experiments suggest that they can also give rise to a sub-population of astrocytes in the ventrolateral forebrain during development (Raff et al. 1983; Zhu et al. 2008) and, strongly controversial, to neurons in the adult brain within restricted areas, like the piriform cortex and the hypothalamus (Guo et al. 2010; Robins et al. 2013) (Fig. 2). Nevertheless, the neuronal progeny of NG2-glia has been highly questioned in the field, as these results failed to replicate in several other mouse models genetically fate mapping NG2-glia (Dimou et al. 2008; Kang et al. 2010) leaving the question open if they are really capable to differentiate into neurons at least under physiological conditions.

NG2-glia represent around 5–10 % of the total cell population in the developing and adult brain and they are evenly distributed within the cerebral and cerebellar gray and white matter (Dawson et al. 2003) (Fig. 1a). Furthermore, NG2-glia can proliferate and differentiate into mature oligodendrocytes throughout life, although both their proliferation and differentiation rates depend on the area of the nervous system and decrease with age (Psachoulia et al. 2009; Kang et al. 2010; Zhu et al. 2011). Previous studies have supported the idea that NG2-glia can divide independently of extracellular signals to maintain their population and that they are

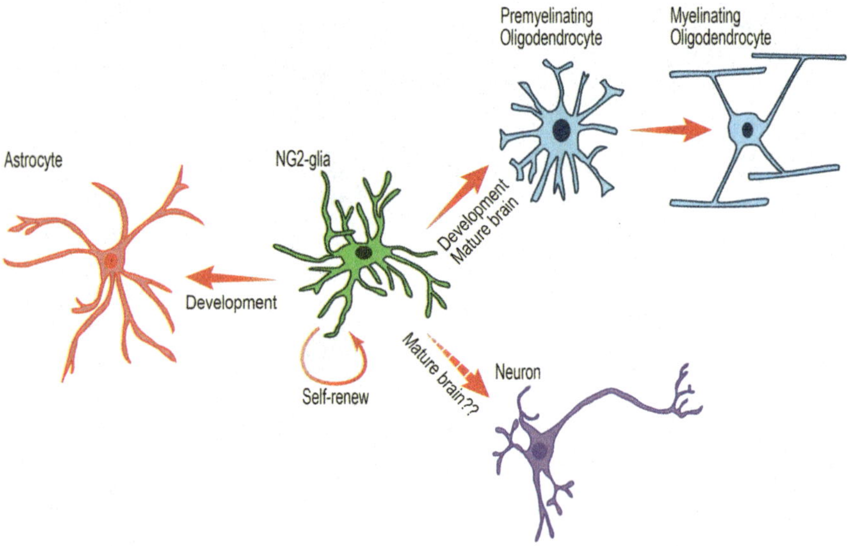


**Fig. 2** Scheme representing the different cell fates of NG2-glia. Canonically, NG2-glia have the capability to proliferate and differentiate into oligodendrocytes in the immature and mature brain. However, NG2-glia can also differentiate into astrocytes in the ventrolateral forebrain during development. Additionally, some studies have suggested that, in the mature brain, NG2-glia could also differentiate into neurons, nevertheless, these claim is still under strong criticism

capable to proliferate and migrate short distances within the intact brain parenchyma, processes that are highly regulated by self-repulsion among NG2-glia (Hughes et al. 2013). Moreover, when a NG2-glia differentiates into an oligodendrocyte or after focal laser ablation of individual NG2-glia, neighboring NG2-glia proliferate and/or migrate in order to fill the "NG2-glia-free gap" in the mammalian and zebrafish CNS (Kirby et al. 2006; Hughes et al. 2013; Birey and Aguirre 2015). Furthermore, NG2-glia can divide asymmetrically giving origin to one NG2-glia and one oligodendrocyte (Hill et al. 2014), a common mechanism that progenitors have to differentiate without altering their total number. This feature, that is homeostatically controlled, opens several questions, e.g., why is a constant population of NG2-glia needed in the adult brain, even in areas where only few oligodendrocytes exist? The assumption that NG2-glia solely act as progenitors for oligodendrocytes has therefore been strongly questioned over the last years and the hypothesis for a further active and functional role of NG2-glia in the healthy and diseased brain has been supported.

Unexpectedly, other important features have been assigned to NG2-glia that until some years ago were thought to be exclusive for neurons. In vitro and in vivo studies could show that NG2-glia express voltage-gated sodium channels ($Na_V$s) (Karadottir et al. 2008; De Biase et al. 2010) sensitive to tetrodotoxin (TTX), voltage-gated potassium channels ($K_V$s), as well as low and high voltage-gated calcium channels ($Ca_V$s) on their processes that get downregulated during their maturation into oligodendrocytes (for review see Verkhratsky and Steinhauser 2000), revealing these channels to be required solely during their progenitor stage.

This wide expression of voltage-gated channels in NG2-glia suggests that these cells may have dynamic electrical properties, and theoretically they have all the components needed to generate and propagate action potentials. As a matter of fact, electrophysiological characterization of white matter NG2-glia in the rat has brought a thorough discussion regarding their possible ability to generate action potentials. The study of Karadottir et al. (2008) could show the existence of two subpopulations of NG2-glia within the white matter of the rat brain, proposing a classification of spiking and non-spiking populations and suggesting the existence of NG2-glia capable of generating action potentials (Karadottir et al. 2008). However, spiking NG2-glia have not been found in other species besides the rat white matter (Bergles et al. 2000; Karadottir et al. 2008; De Biase et al. 2010; Clarke et al. 2012), suggesting that this feature could be specific for NG2-glia in certain species or during certain developmental stages or both (Clarke et al. 2012). Therefore, generation and propagation of action potentials by NG2-glia appear to have no plausible role for the moment. Nevertheless, NG2-glia with different electrical properties rise an interesting point in the field, the NG2-glia heterogeneity.

## NG2-glia Heterogeneity

Despite the even distribution of NG2-glia in the brain and their similar morphology, it is accepted in the last years that NG2-glia represent a highly heterogeneous population with diverse intrinsic properties and probably also distinct roles and functions. The heterogeneous nature of NG2-glia could be shown regarding different aspects of these cells, e.g., their morphology as well as their differentiation and proliferation properties. For instance, NG2-glia in the adult cerebral white matter can differentiate into oligodendrocytes faster and more efficiently than those located in the gray matter (Dimou et al. 2008). Subsequent homo- and heterotopic transplantation experiments revealed that these diverse differentiation properties between white and gray matter NG2-glia are the result of mainly intrinsic heterogeneity between these cells (Vigano et al. 2013). Interestingly, also their morphology and process distribution show differences between NG2-glia in these two areas (Vigano et al. 2013) (Fig. 3). NG2-glia in the white and gray matter have also been described to be different in regard to their cell cycle length. Although all NG2-glia can divide, gray matter cells have a longer cell cycle length than their white matter counterparts, further supporting the idea of the heterogeneity of NG2-glia (Psachoulia et al. 2009). Moreover, their response to extracellular signals is also variable. For example, NG2-glia in the white matter show a stronger proliferative response to platelet-derived growth factor (PDGF) than in the gray matter, despite that both populations express comparable levels of the receptor PDGFR$\alpha$ (Hill et al. 2013). Interestingly, in the last years NG2-glia heterogeneity has been suggested not only between different regions but also within the very same brain area. Indeed, some studies described the distinct expression of proteins only in subsets of NG2-glia located in the same region. For example, only 50 % of

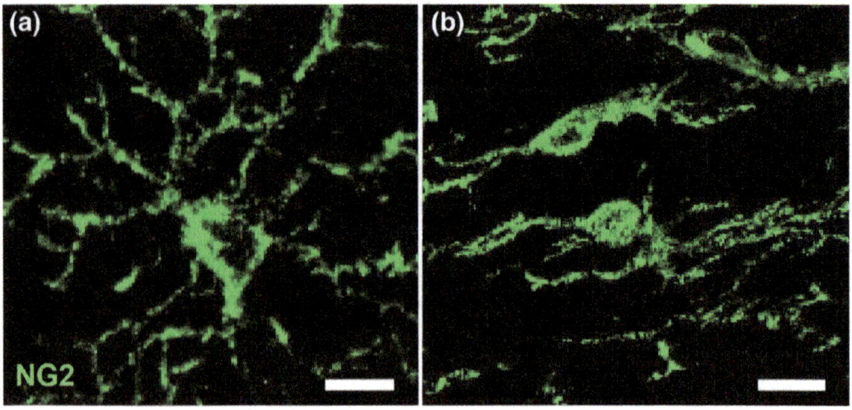


**Fig. 3** NG2-glia are heterogeneous in regard to their morphology. Magnification of confocal images showing NG2-glia from **a** the gray matter and **b** white matter of the brain, highlighting clear morphological differences between both populations. Scale bar represents 10 μm

NG2-glia in the cortical gray matter express the transcription factor achaete-scute homolog 1 or mammalian achaete-scute homolog 1 (Ascl1 or Mash1), an important factor for neuronal fate determination (Parras et al. 2007). However, by now no distinct roles could be assigned to these two populations. In the same line, also the G-protein coupled receptor 17 (GPR17), that is involved in the differentiation of NG2-glia (Boda et al. 2011; Chen et al. 2009), is only expressed in a subset of NG2-glia in different regions and at different ages. Interestingly, it could be shown that GPR17-positive cells represent a population of NG2-glia that differentiates very slowly in the intact brain. However, after a cerebral damage, they rapidly react and undergo maturation, suggesting a role as a "reserve pool" of adult progenitors that are maintained for repair processes (Vigano et al. 2016). Whether all these described differences between NG2-glia in distinct or even in the same area are the result solely of an intrinsic program or micro-environmental influences restricting or providing the cells special properties, is not clear; and the existence of evidences in favor and against both ideas keeps the question still unanswered (for review see Vigano and Dimou 2016).

## Synapses Between NG2-glia and Neurons

In addition to the controversial possibility of generating action potentials, NG2-glia have shown other features, which are also shared with neurons. The formation of synapses between NG2-glia and neurons is a captivating observation. These synapses were first observed in the mouse, where stimulation of neurons located in the CA3 region of the hippocampus triggered an evoked excitatory postsynaptic current (EPSC) of a sodium current nature in NG2-glia located in the CA1 region (Bergles et al. 2000). Furthermore, the spontaneous fusion of transmitter-filled vesicles that occurs at individual excitatory synapses between NG2-glia and neurons resulting in miniature EPSCs (mEPSCs) has also been shown, a feature not exclusively present in the hippocampus, but also in several other brain areas (Bergles et al. 2000; De Biase et al. 2010). In the same line, the brief application of the neurotoxin pardaxin, which enhances the frequency of vesicular release from nerve terminals, caused the appearance of high frequency bursts of mEPSCs in NG2-glia (Bergles et al. 2000), revealing a correlation between vesicular release and the electric properties of NG2-glia. In both studies, these currents were abolished by the application of the noncompetitive *α-amino-3-hydroxy-5-methyl-4-isoxazolepropionic acid* (AMPA)/Kainate receptor antagonist *2,3-dihydroxy-6-nitro-7-sulphamoyl-benzo(F)quino-xaline* (NBQX) (Bergles et al. 2000; De Biase et al. 2010), demonstrating the glutamatergic nature of these signals. Finally, the existence of synapses between NG2-glia and neurons has also been confirmed by transmission electron microscopy that revealed presynaptic release sites in the axon membrane close to NG2-glia membrane (Bergles et al. 2000; Ziskin et al. 2007).

Notably, the complexity of these NG2-glia-neuron synapses is not limited to the expression of the AMPA receptor (AMPAR), but extends to a wide variety of typical postsynaptic receptors such as *N-methyl-D-aspartate* receptor (NMDAR) (for review see Dzamba et al. 2013), acetylcholine receptor (De Angelis et al. 2012), *γ-aminobutyric acid* receptor ($GABA_AR$) (Von Blankenfeld et al. 1991; Williamson et al. 1998) and others. Moreover, not only receptors can be found in NG2-glia, but as transcriptome data also suggest, they may also express proteins important for the establishment of a postsynapse like the postsynaptic density protein 95 (PSD-95) (Sakry et al. 2011), one of the major components of the postsynaptic density and important for synapse formation, maturation, and remodeling (El-Husseini et al. 2000; Marrs et al. 2001). Nowadays, many questions still remain unanswered regarding these synaptic glia-neuron structures. First, it is neither clear if these synapses are only sending information from neurons to NG2-glia in an unidirectional fashion or if they could also be part of a bidirectional communication mechanism, nor if these synapses are present in all or in just a subpopulation of NG2-glia. For example, it could also be that NG2-glia express different channels and form synapses with neurons only at a specific timepoint of their life when they need neuronal signals to perform specific functions like, e.g., differentiation or neuronal support. Moreover, neither the molecular composition of these synapses nor the main functions that these glia-neuron structures may have, have been elucidated (for review see Dimou and Gallo 2016).

## Myelin Plasticity

Myelination of the nervous system enables fast electric conduction along the myelinated axon and reduces the metabolic cost of neuronal activity; features making myelination important for the proper function of the nervous system in gnathostomates vertebrates and some invertebrates (Davis et al. 1999). Myelin formation changes the electrical properties of axons by reducing the capacitance and increasing the peripheral resistance of the surrounded axon. Myelin sheaths do not cover the complete axon, but they are absent in short segments, structures known as nodes of Ranvier, that express high levels of $Na_v$s, important for the generation of action potentials in this area. This particular arrangement of myelin along axons provides neurons with the structural basis for the saltatory action potential propagation (Nave and Werner 2014).

In vertebrates, there are two cell types that have the unique ability to synthesize large amounts of membrane that wrap and compact around axons to form myelin: Schwann cells in the peripheral nervous system (PNS) and the oligodendrocytes in the CNS. Although both cell types have the capability to myelinate axons, they differ among other features enormously in regard to their developmental origin, their protein composition, the number of myelinated axonal segments each cell can establish, and the myelin periodicity. Therefore, this apparent structural and functional similarity of PNS and CNS myelin is a superb example of convergent cellular evolution of the nervous system (see Chapters “Oligodendrocytes: Functioning in a Delicate Balance Between High Metabolic Requirements and Oxidative Damage” and "Schwann Cell and Axon: An Interlaced Unit—From Action Potential to Phenotype Expression" for further reading on oligodendrocytes and Schwann cells).

In the past, the two-dimensional nature of electron microscopy images had promoted the notion that myelin is a static structure in our nervous system. Nevertheless, lately it has be shown that myelin in the rodent brain is constantly remodeling during lifespan, giving us a new perspective of highly plastic processes. Different signals trigger changes in the compaction of myelin sheaths, in the internode length (the areas of the myelinated axons flanked by the nodes of Ranvier), in the number of sheaths surrounding one axon, and in the number of axons that one oligodendrocyte can myelinate (Nave and Werner 2014).

Neuronal activity has shown to be an important factor in the remodeling of the myelin along the axon. The inhibition of action potentials with TTX, which blocks $Na_V$s currents, leads for example to a decrease in myelination in vitro and in the mouse optic nerve in vivo (Demerens et al. 1996). Conversely, the stimulation of neuronal activity with α-scorpion toxin (αScTX), which delays the inactivation of $Na_V$s, or direct electric stimulation of neurons with an electrode, triggers the increase of myelination in mixed oligodendrocyte-neuronal cultures (Demerens et al. 1996; Gary et al. 2012). Furthermore, social deprivated adult mice showed impaired myelination in the prefrontal cortex (PFC), an area which has been associated with complex emotional and cognitive behavior (Liu et al. 2012). Notably, neuronal activity dependent changes could also be observed in human white matter, a structure that primarily consists of myelinated axons, cells of the oligodendrocytic lineage, other glial cells, and no neuronal cell bodies. Diffusion tensor imaging (DTI) studies have shown that various complex visual-motor tasks, such as juggling or extensive piano playing, trigger changes in the white matter architecture, by significantly increasing its size (Bengtsson et al. 2005; Scholz et al. 2009). Amazingly, it appears that neuronal activity is not just limited in increasing the myelination of vertebrates' axons. Recent studies in the zebrafish larvae spinal cord have indeed shown that neuronal activity additionally provides a signal bias for which axons must be myelinated. By in vivo time lapse imaging of the zebrafish spinal cord, it was shown that while the initial axonal myelin wrapping is axon activity independent, the stabilization and extension of the prospective myelin sheaths depends on the activity-dependent secretion of axonal factors, like neurotransmitters and neurotrophic factors (Hines et al. 2015). Furthermore, blocking the synaptic vesicle release with tetanus neurotoxin (TeNT) decreases the number of myelinated axons and the total number of myelin sheaths per oligodendrocyte in the zebrafish spinal cord (Mensch et al. 2015). Interestingly, similar results could be obtained in rodents, where an activity independent myelination could be observed first, followed by an activity dependent myelination triggered by neuregulin (Lundgaard et al. 2013). Together, these studies show that neuronal activity has a direct effect on the oligodendrocytes population, defining which and how many axons must be myelinated.

Thus, one obvious potential role for NG2-glia-neuron synapses could be to regulate the proliferation and differentiation of NG2-glia. The outcome of these changes probably aims at improving or even modulating myelination of certain

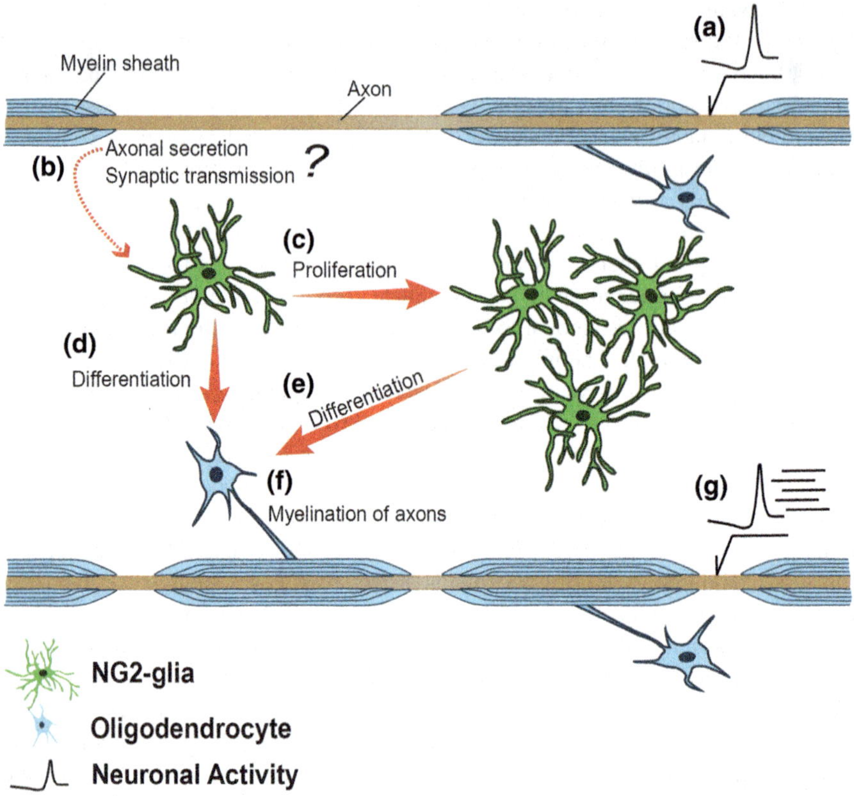


**Fig. 4** Scheme of the effect of neuronal activity on NG2-glia behavior. In response to **a** neuronal activity, **b** neurons could release signals secreted from the axon or the soma; or secreted into the synaptic cleft formed between neurons and NG2-glia. These signals could manipulate NG2-glia **c** proliferation or **d** differentiation into oligodendrocytes. It is also possible that **e** after proliferation the newborn NG2-glia differentiate. If new oligodendrocytes are generated, they could potentially **f** myelinate surrounding axons and **g** change the electrical properties of neurons and therefore modify the properties of the neuronal network

circuits, therefore enhancing the computational properties of networks required to improve performance in certain specific tasks (Fig. 4).

## NG2-glia Behavior and Dynamics Modulated by Neuronal Activity

Although many studies have been performed regarding the influence of neuronal activity on the cells of the oligodendrocyte lineage, it is still unclear whether this happens in a direct or indirect manner. Early work showed that intraocular injection of TTX diminishes the proliferation of NG2-glia close to axons of the retinal ganglion cells by preventing the neuronal activity-dependent release of mitogens (Barres and Raff 1993). NG2-glia can express a wide variety of purinergic receptors and it has additionally been shown that neuronal activity induces adenosine release by neurons and decreases NG2-glia proliferation (Stevens et al. 2002). It is therefore possible that other extra-synaptic molecules, whose release is activity dependent, like growth factors, also play a significant role in the modulation of the proliferation and differentiation of NG2-glia.

Apparently NG2-glia that are proliferating and differentiating could be also modulated by signals provided directly by the neuro-glia synapses. It has been shown that neurotransmitter release may have an important role in the development of the oligodendrocyte lineage. Experiments in organotypic cultures of cerebellum slices obtained from P6 mice have shown that the exposition to glutamate receptor agonists like kainate and AMPA decrease the proportion of NG2-glia positive to the proliferation marker *5-bromo-2'-deoxyuridine* (BrdU), a thymidine analogue incorporated into the cell during the S-phase of the cell cycle. In contrast, administration of the kainate and AMPA receptor antagonist *6,7-dinitroquinoxaline-2,3-dione* (DNQX) lead to an increase in cell proliferation (Yuan et al. 1998).

Interestingly, this evidence indicates that neuronal activity promotes changes in the behavior of NG2-glia in both, a direct and indirect way. Unfortunately, the investigation of this idea in in vivo experimental models can be very challenging. It could be shown that high-frequency electrical stimulation applied in the medullary pyramids of rats increases the proliferation and differentiation of NG2-glia in the contralateral dorsal corticospinal tract (Li et al. 2010). However, the experimental approach in this last study required the implantation of electrodes to promote neuronal activity in vivo that inevitably resulted in injury and subsequently in inflammation. This brings serious complications in the interpretation of these results as NG2-glia show morphological and functional changes in response to damage, by, e.g., increasing their proliferation rate, becoming hypertroph, polarizing toward the injury, and migrating into the injury (Dimou et al. 2008; Simon et al. 2011; Hughes et al. 2013; for review see Dimou and Götz 2014). Therefore, most experiments in this field have been performed by promoting neuronal activity through indirect methods.

Physical stimuli, provided by keeping mice under enriched environment conditions, e.g., by adding running wheels or other toys into cages for two weeks, trigger an increase in the differentiation of NG2-glia into oligodendrocytes, in the motor and somatosensory cortex and in the amygdala (Simon et al. 2011; Ehninger et al. 2011). Another strategy implemented in this field has been stimuli deprivation, which also leads to changes in the behavior of NG2-glia. NG2-glia from the barrel cortex of layer IV are functionally innervated by thalamocortical fibers coming from the ventral basal thalamus, determining the distribution of these cells in the walls of the barrel cortex and not in its core (Mangin et al. 2012). After cauterizing mouse whiskers at birth, NG2-glia show an aberrant homogenous distribution in the whole barrel cortex. Additionally, an increased fraction of NG2-glia in the core of the barrel cortex is positive for Ki67, an active proliferation marker. These results suggest the hypothesis that the thalamocortical fibers are inhibiting the proliferation of NG2-glia in the core of the barrel cortex (Mangin et al. 2012), resulting in the particular distribution of NG2-glia in the layer IV. Furthermore, in another experimental approach, clipping the whiskers of adult mice unilaterally induced a decrease in the total number of oligodendrocytes and an increased expression of activated caspase-3, a marker for apoptosis, in oligodendrocytes derived from proliferating NG2-glia in the somatosensory cortex ipsilateral to the clipped side (Hill et al. 2014). These results suggest that stimuli deprivation not only compromises the differentiation of NG2-glia into oligodendrocytes but also their survival (Hill et al. 2014).

New strategies to test how neuronal activity affects the dynamics of NG2-glia in vivo are emerging, promising to provide data while avoiding the collateral effects caused by CNS damage. The development of optogenetics, a technique in which light-sensitive ion channels can be selectively expressed in a specific subpopulation of neurons, which then can be activated through their exposure to a specific light wavelength, could result in great advances in this area. A recent study using a transgenic mouse line expressing the channel rhodopsin in neurons with an active Thy1 promoter, allowed the stimulation of neurons located in the cortical layer V and the analysis of the effects of this artificial stimulation in the dynamics of the oligodendrocyte lineage (Gibson et al. 2014). Indeed, stimulation of neurons lead to an increase in the number of cells positive for the thymidine analogue and proliferation marker, *5-ethynyl-2'-deoxyuridine* (EdU), in the neighborhood of the stimulated neurons. A fraction of these EdU-positive cells were positive for Olig2, a transcription factor used as a marker for the oligodendrocyte lineage (Gibson et al. 2014), for PDGFR$\alpha$, a marker for NG2-glia (Dimou and Gotz 2014), and for the adenomatous polyposis coli (CC1), a marker for mature oligodendrocytes (Gibson et al. 2014). These results highlight the relation between the experimental increase of firing rate and the increase of the proliferation and differentiation dynamics of the oligodendrocyte lineage. Moreover, the study went even further and showed that increased stimulation of neurons lead to an increase in myelination, specifically of the axons from the stimulated neurons, leading to the improvement of behavioral-motor function of these animals (Gibson et al. 2014).

The above described data, and the additional growing evidence consistently show that neuronal activity, triggered by different physiological environmental or artificial stimuli, efficiently promotes proliferation and differentiation of NG2-glia in in vitro as well as in in vivo models. It has also been shown by correlation that the modulation of this behavior can result in an improvement in the myelination of axons belonging to neurons with increased activity and, therefore, improvement of the fine tuning of neuronal networks which subsequently ameliorates tasks performance depending on this specific network. However, although signals provided by neurons to induce proliferation and differentiation and the consequences related to the change in the NG2-glia behavior have been identified, the cellular and molecular mechanisms being triggered by neuronal activity in the NG2-glia still remain unclear.

## NG2-glia reaction towards Central Nervous System Pathology

As we briefly mentioned before, NG2-glia change their dynamics and behavior when the CNS is damaged. Moreover, changes in NG2-glia in different models of injury and demyelination have suggested that NG2-glia may play an active role in repairing the brain under pathological conditions. Hypomyelinating mutant mouse lines like the *jimpy*, in which oligodendrocytes die as the result of a point mutation in the most abundant CNS myelin protein, the proteolipid protein (PLP), or the *shiverer*, in which oligodendrocytes are unable to form compact myelin sheaths due to the deletion of the myelin basic protein (MBP), have shown an increase in the proliferation and differentiation of NG2-glia in the spinal cord white matter in comparison to the control wild type groups (Wu et al. 2000; Bu et al. 2004). This shows an association between myelination defects and change of physiological NG2-glia behavior. Unfortunately, due to the mutation in important myelin proteins, both models are incapable to demonstrate if there is remyelination after NG2-glia reaction toward aberrant myelination. However, transplantation of healthy human NG2-glia into the *shiverer* mouse model resulted in improved remyelination of the brain, hinting into the capability of NG2-glia to repair demyelinated areas (Windrem et al. 2008) and therefore a probable strategy for the therapy of demyelinating diseases in the CNS (Franklin and Ffrench-Constant 2008) (see Chapter "Peripheral Inflammation and Demyelinating Diseases" for further reading on demyelinating diseases).

Other strategies have also been performed in this field to specifically injure brain white matter tracts and study the behavior and the remyelination properties of NG2-glia. Cultured cortical slices exposed to *α-lysophosphatidylcholine* (LPC), show damage in the corpus callosum, and subsequent demyelination of the affected area. After this acute lesion, an increase in proliferation of NG2-glia (Garay et al. 2011) and in their differentiation into myelinating oligodendrocytes (Gensert and Goldman 1997) could be observed. Two-photon in vivo imaging of postnatal cortical slices further confirmed these results and showed that this acute injury model changes the differentiation behavior of NG2-glia by increasing the asymmetric division of these cells, giving rise to one NG2-glia and one oligodendrocyte (Hill et al. 2014). Additionally, in the experimental autoimmune encephalomyelitis (EAE) mouse, a classic model for multiple sclerosis (MS; a disease that promotes demyelination and axonal loss in the CNS), or in antibody-induced demyelination, the number of NG2-glia also increased in those areas where extensive demyelination occurs, while the population of NG2-glia did not change in the tissue surrounding the lesions (Keirstead et al. 1998; Di Bello et al. 1999).

The role of NG2-glia during demyelination and the subsequent remyelination appears to be clear. The decrease in myelin triggers the proliferation and differentiation of NG2-glia in order to reestablish the loss of myelin in the CNS, making NG2-glia an excellent target for regenerative therapy. However, it is still unclear what the trigger for the reaction of NG2-glia after demyelination could be; it is e.g.

possible that NG2-glia possess the special capability to sense the lack of oligodendrocytes/myelin or that NG2-glia react to the aberrant neuronal activity promoted by the demyelination. It is known that the CNS of MS patients fails to remyelinate with disease progression. Interestingly, postmortem NG2 and PDGFRα immunolabeling of brains of these patients have revealed that NG2-glia are more frequently present in areas of the white matter undergoing active inflammatory demyelination, where commonly remyelination occurs, than in chronic lesions (Wilson et al. 2006), suggesting that NG2-glia are subjected to certain limitations for effective remyelination in long-term pathology, probably provided by an antagonistic environment or by an exhaustion of the NG2-glia limiting them to further proliferate and differentiate.

In contrast to de-/remyelination, the role of these cells in other brain insults like acute injury and neurodegeneration is not clear yet. After a stab wound injury in the cerebral cortex an increase in the proliferation and density of NG2-glia surrounding the lesion could be observed (Dimou et al. 2008; Simon et al. 2011). Moreover, in the APPPS1 mice, a mouse model for Alzheimer's disease (AD), where also demyelination occurs, NG2-glia react with increased proliferation and differentiation (Behrendt et al. 2013) and become hypertrophe in the gray matter cerebral cortex (Sirko et al. 2013). Li et al. (2013) have also reported that NG2-glia cluster around the amyloid plaques and as already suggested for other glial cells that they are capable of engulfing the amyloid protein β (Abeta) and degrade it by an autophagy-lysosomal pathway playing a role in clearing it from the affected brain (Li et al. 2013). Although the specific function of NG2-glia remains unknown, it could be possible that they may serve as orchestra directors recruiting and modulating the function of other glial and immune cells in the CNS in order to reestablish the homeostasis of the brain.

Another important question addressed in the field is whether the whole population or only a subset of NG2-glia are actively participating in repairing the brain damage. As discussed above, an interesting subset of NG2-glia that expresses GPR17 seems to resemble NG2-glia specialized for repair. GPR17 is a deorphanized receptor for both uracil nucleotides and cystein leukotrienes, cysLTs (e.g., UDP-glucose and $LTD_4$) (Lecca et al. 2008). Both ligands for GPR17 are secreted after brain injury and their extracellular concentration is increased suggesting a role of this receptor in "sensing" brain damage (Lecca et al. 2008). It has been shown that GPR17-positive NG2-glia in the adult brain increase their density after acute cortical stab wound injury in the gray matter surrounding the lesion and in the white matter underneath the lesion (Boda et al. 2011). Moreover, fate mapping studies of GPR17-positive cells have shown that although they have a limited differentiation capacity under physiological conditions, after stab wound injury or middle cerebral artery occlusion (MCAO), a model of ischemic stroke, these cells strongly differentiate into oligodendrocytes probably in order to repair the injured tissue. Interestingly, in the above described APP/PS1 mouse line, GPR17-positive NG2-glia were responsive specifically in the gray matter, but not in the white matter (Boda et al. 2011). This result highlights once again the potential heterogeneity of these cells and their functional differences in the brain.

## Concluding Remarks

We have shown in this chapter that NG2-glia are a population of cells that are widely represented within the CNS during the late stage of development and in the whole adult lifespan. Their potential to possibly originate different cell types, their long cell cycle length and maintenance of their own population size resembles more the characteristics of neuronal stem cells rather than simple oligodendrocyte progenitors. Moreover, special features like the expression of channels and receptors, their response to neurotransmitters and growth factors and the synapse formation with neurons make them a unique group of glial cells. Additionally, the capability of NG2-glia to modulate their behavior and dynamics in response to neuronal activity and disease, suggest an important role not only for myelin maintenance and remodeling under physiological but also for repair under pathological conditions. There are still many mysteries around the specific NG2-glia functions in the healthy and diseased brain. The understanding of these cells may lead to significant improvement not only of our global knowledge of the complexity of the brain, but also of the treatment of traumatic and neurodegenerative diseases and the improvement of the quality of life during normal aging, which also has been related to an impairment of myelin (Sturrock 1976; Peters 2002; Lasiene et al. 2009; Lu et al. 2011; Lu et al. 2013).

## References

Barres BA, Raff MC (1993) Proliferation of oligodendrocyte precursor cells depends on electrical activity in axons. Nature 361(6409):258–260. doi:10.1038/361258a0

Behrendt G, Baer K, Buffo A, Curtis MA, Faull RL, Rees MI, Gotz M, Dimou L (2013) Dynamic changes in myelin aberrations and oligodendrocyte generation in chronic amyloidosis in mice and men. Glia 61(2):273–286. doi:10.1002/glia.22432

Bengtsson SL, Nagy Z, Skare S, Forsman L, Forssberg H, Ullen F (2005) Extensive piano practicing has regionally specific effects on white matter development. Nat Neurosci 8 (9):1148–1150. doi:10.1038/nn1516

Bergles DE, Roberts JD, Somogyi P, Jahr CE (2000) Glutamatergic synapses on oligodendrocyte precursor cells in the hippocampus. Nature 405(6783):187–191. doi:10.1038/35012083

Birey F, Aguirre A (2015) Age-dependent Netrin-1 signaling regulates NG2+ glial cell spatial homeostasis in normal adult gray matter. J Neurosci 35(17):6946–6951. doi:10.1523/JNEUROSCI.0356-15.2015

Boda E, Vigano F, Rosa P, Fumagalli M, Labat-Gest V, Tempia F, Abbracchio MP, Dimou L, Buffo A (2011) The GPR17 receptor in NG2 expressing cells: focus on in vivo cell maturation and participation in acute trauma and chronic damage. Glia 59(12):1958–1973. doi:10.1002/glia.21237

Bu J, Banki A, Wu Q, Nishiyama A (2004) Increased NG2(+) glial cell proliferation and oligodendrocyte generation in the hypomyelinating mutant shiverer. Glia 48(1):51–63. doi:10.1002/glia.20055

Chen Y, Wu H, Wang S, Koito H, Li J, Ye F, Hoang J, Escobar SS, Gow A, Arnett HA, Trapp BD, Karandikar NJ, Hsieh J, Lu QR (2009) The oligodendrocyte-specific G

protein-coupled receptor GPR17 is a cell-intrinsic timer of myelination. Nat Neurosci 12 (11):1398–1406. doi:10.1038/nn.2410
Clarke LE, Young KM, Hamilton NB, Li H, Richardson WD, Attwell D (2012) Properties and fate of oligodendrocyte progenitor cells in the corpus callosum, motor cortex, and piriform cortex of the mouse. J Neurosci 32(24):8173–8185. doi:10.1523/JNEUROSCI.0928-12.2012
Davis AD, Weatherby TM, Hartline DK, Lenz PH (1999) Myelin-like sheaths in copepod axons. Nature 398(6728):571. doi:10.1038/19212
Dawson MR, Polito A, Levine JM, Reynolds R (2003) NG2-expressing glial progenitor cells: an abundant and widespread population of cycling cells in the adult rat CNS. Mol Cell Neurosci 24(2):476–488
De Angelis F, Bernardo A, Magnaghi V, Minghetti L, Tata AM (2012) Muscarinic receptor subtypes as potential targets to modulate oligodendrocyte progenitor survival, proliferation, and differentiation. Dev Neurobiol 72(5):713–728. doi:10.1002/dneu.20976
De Biase LM, Nishiyama A, Bergles DE (2010) Excitability and synaptic communication within the oligodendrocyte lineage. J Neurosci 30(10):3600–3611. doi:10.1523/JNEUROSCI.6000-09.2010
Demerens C, Stankoff B, Logak M, Anglade P, Allinquant B, Couraud F, Zalc B, Lubetzki C (1996) Induction of myelination in the central nervous system by electrical activity. Proc Natl Acad Sci USA 93(18):9887–9892
Di Bello IC, Dawson MR, Levine JM, Reynolds R (1999) Generation of oligodendroglial progenitors in acute inflammatory demyelinating lesions of the rat brain stem is associated with demyelination rather than inflammation. J Neurocytol 28(4–5):365–381
Dimou L, Gallo V (2015) NG2-glia and their functions in the central nervous system. Glia 63 (8):1429–1451. doi:10.1002/glia.22859
Dimou L, Götz M (2014) Glial cells as progenitors and stem cells: new roles in the healthy and diseased brain. Physiol Rev 94(3):709–737. doi:10.1152/physrev.00036.2013
Dimou L, Simon C, Kirchhoff F, Takebayashi H, Gotz M (2008) Progeny of Olig2-expressing progenitors in the gray and white matter of the adult mouse cerebral cortex. J Neurosci 28 (41):10434–10442. doi:10.1523/JNEUROSCI.2831-08.2008
Dzamba D, Honsa P, Anderova M (2013) NMDA receptors in glial cells: pending questions. Curr Neuropharmacol 11(3):250–262. doi:10.2174/1570159X11311030002
Ehninger D, Wang LP, Klempin F, Romer B, Kettenmann H, Kempermann G (2011) Enriched environment and physical activity reduce microglia and influence the fate of NG2 cells in the amygdala of adult mice. Cell Tissue Res 345(1):69–86. doi:10.1007/s00441-011-1200-z
El-Husseini AE, Schnell E, Chetkovich DM, Nicoll RA, Bredt DS (2000) PSD-95 involvement in maturation of excitatory synapses. Science 290(5495):1364–1368
Franklin RJ, Ffrench-Constant C (2008) Remyelination in the CNS: from biology to therapy. Nat Rev Neurosci 9(11):839–855. doi:10.1038/nrn2480
Garay L, Tungler V, Deniselle MC, Lima A, Roig P, De Nicola AF (2011) Progesterone attenuates demyelination and microglial reaction in the lysolecithin-injured spinal cord. Neuroscience 192:588–597. doi:10.1016/j.neuroscience.2011.06.065
Gary DS, Malone M, Capestany P, Houdayer T, McDonald JW (2012) Electrical stimulation promotes the survival of oligodendrocytes in mixed cortical cultures. J Neurosci Res 90(1):72–83. doi:10.1002/jnr.22717
Gensert JM, Goldman JE (1997) Endogenous progenitors remyelinate demyelinated axons in the adult CNS. Neuron 19(1):197–203
Gibson EM, Purger D, Mount CW, Goldstein AK, Lin GL, Wood LS, Inema I, Miller SE, Bieri G, Zuchero JB, Barres BA, Woo PJ, Vogel H, Monje M (2014) Neuronal activity promotes oligodendrogenesis and adaptive myelination in the mammalian brain. Science 344 (6183):1252304. doi:10.1126/science.1252304
Guo F, Maeda Y, Ma J, Xu J, Horiuchi M, Miers L, Vaccarino F, Pleasure D (2010) Pyramidal neurons are generated from oligodendroglial progenitor cells in adult piriform cortex. J Neurosci 30(36):12036–12049. doi:10.1523/JNEUROSCI.1360-10.2010

Hill RA, Patel KD, Medved J, Reiss AM, Nishiyama A (2013) NG2 cells in white matter but not gray matter proliferate in response to PDGF. J Neurosci 33(36):14558–14566. doi:10.1523/JNEUROSCI.2001-12.2013

Hill RA, Patel KD, Goncalves CM, Grutzendler J, Nishiyama A (2014) Modulation of oligodendrocyte generation during a critical temporal window after NG2 cell division. Nat Neurosci 17(11):1518–1527. doi:10.1038/nn.3815

Hines JH, Ravanelli AM, Schwindt R, Scott EK, Appel B (2015) Neuronal activity biases axon selection for myelination in vivo. Nat Neurosci 18(5):683–689. doi:10.1038/nn.3992

Hughes EG, Kang SH, Fukaya M, Bergles DE (2013) Oligodendrocyte progenitors balance growth with self-repulsion to achieve homeostasis in the adult brain. Nat Neurosci 16(6):668–676. doi:10.1038/nn.3390

Kang SH, Fukaya M, Yang JK, Rothstein JD, Bergles DE (2010) NG2+ CNS glial progenitors remain committed to the oligodendrocyte lineage in postnatal life and following neurodegeneration. Neuron 68(4):668–681. doi:10.1016/j.neuron.2010.09.009

Karadottir R, Hamilton NB, Bakiri Y, Attwell D (2008) Spiking and nonspiking classes of oligodendrocyte precursor glia in CNS white matter. Nat Neurosci 11(4):450–456. doi:10.1038/nn2060

Keirstead HS, Levine JM, Blakemore WF (1998) Response of the oligodendrocyte progenitor cell population (defined by NG2 labelling) to demyelination of the adult spinal cord. Glia 22 (2):161–170

Kirby BB, Takada N, Latimer AJ, Shin J, Carney TJ, Kelsh RN, Appel B (2006) In vivo time-lapse imaging shows dynamic oligodendrocyte progenitor behavior during zebrafish development. Nat Neurosci 9(12):1506–1511. doi:10.1038/nn1803

Lasiene J, Matsui A, Sawa Y, Wong F, Horner PJ (2009) Age-related myelin dynamics revealed by increased oligodendrogenesis and short internodes. Aging Cell 8(2):201–213. doi:10.1111/j.1474-9726.2009.00462.x

Lecca D, Trincavelli ML, Gelosa P, Sironi L, Ciana P, Fumagalli M, Villa G, Verderio C, Grumelli C, Guerrini U, Tremoli E, Rosa P, Cuboni S, Martini C, Buffo A, Cimino M, Abbracchio MP (2008) The recently identified P2Y-like receptor GPR17 is a sensor of brain damage and a new target for brain repair. PLoS ONE 3(10):e3579. doi:10.1371/journal.pone.0003579

Li Q, Brus-Ramer M, Martin JH, McDonald JW (2010) Electrical stimulation of the medullary pyramid promotes proliferation and differentiation of oligodendrocyte progenitor cells in the corticospinal tract of the adult rat. Neurosci Lett 479(2):128–133. doi:10.1016/j.neulet.2010.05.043

Li W, Tang Y, Fan Z, Meng Y, Yang G, Luo J, Ke ZJ (2013) Autophagy is involved in oligodendroglial precursor-mediated clearance of amyloid peptide. Mol Neurodegener 8:27. doi:10.1186/1750-1326-8-27

Liu J, Dietz K, DeLoyht JM, Pedre X, Kelkar D, Kaur J, Vialou V, Lobo MK, Dietz DM, Nestler EJ, Dupree J, Casaccia P (2012) Impaired adult myelination in the prefrontal cortex of socially isolated mice. Nat Neurosci 15(12):1621–1623. doi:10.1038/nn.3263

Lu PH, Lee GJ, Raven EP, Tingus K, Khoo T, Thompson PM, Bartzokis G (2011) Age-related slowing in cognitive processing speed is associated with myelin integrity in a very healthy elderly sample. J Clin Exp Neuropsychol 33(10):1059–1068. doi:10.1080/13803395.2011.595397

Lu PH, Lee GJ, Tishler TA, Meghpara M, Thompson PM, Bartzokis G (2013) Myelin breakdown mediates age-related slowing in cognitive processing speed in healthy elderly men. Brain Cogn 81(1):131–138. doi:10.1016/j.bandc.2012.09.006

Lundgaard I, Luzhynskaya A, Stockley JH, Wang Z, Evans KA, Swire M, Volbracht K, Gautier HO, Franklin RJ, Charles F-C, Attwell D, Karadottir RT (2013) Neuregulin and BDNF induce a switch to NMDA receptor-dependent myelination by oligodendrocytes. PLoS Biol 11 (12):e1001743. doi:10.1371/journal.pbio.1001743

Mangin JM, Li P, Scafidi J, Gallo V (2012) Experience-dependent regulation of NG2 progenitors in the developing barrel cortex. Nat Neurosci 15(9):1192–1194. doi:10.1038/nn.3190

Marrs GS, Green SH, Dailey ME (2001) Rapid formation and remodeling of postsynaptic densities in developing dendrites. Nat Neurosci 4(10):1006–1013. doi:10.1038/nn717

Mensch S, Baraban M, Almeida R, Czopka T, Ausborn J, El Manira A, Lyons DA (2015) Synaptic vesicle release regulates myelin sheath number of individual oligodendrocytes in vivo. Nat Neurosci 18(5):628–630. doi:10.1038/nn.3991

Miller RH (1996) Oligodendrocyte origins. Trends Neurosci 19(3):92–96

Nave KA, Werner HB (2014) Myelination of the nervous system: mechanisms and functions. Annu Rev Cell Dev Biol 30:503–533. doi:10.1146/annurev-cellbio-100913-013101

Ozerdem U, Grako KA, Dahlin-Huppe K, Monosov E, Stallcup WB (2001) NG2 proteoglycan is expressed exclusively by mural cells during vascular morphogenesis. Dev Dyn 222(2):218–227. doi:10.1002/dvdy.1200

Parras CM, Hunt C, Sugimori M, Nakafuku M, Rowitch D, Guillemot F (2007) The proneural gene Mash1 specifies an early population of telencephalic oligodendrocytes. J Neurosci 27 (16):4233–4242. doi:10.1523/JNEUROSCI.0126-07.2007

Peters A (2002) The effects of normal aging on myelin and nerve fibers: a review. J Neurocytol 31 (8–9):581–593

Psachoulia K, Jamen F, Young KM, Richardson WD (2009) Cell cycle dynamics of NG2 cells in the postnatal and ageing brain. Neuron Glia Biol 5(3–4):57–67. doi:10.1017/S1740925X09990354

Raff MC, Miller RH, Noble M (1983) A glial progenitor cell that develops in vitro into an astrocyte or an oligodendrocyte depending on culture medium. Nature 303(5916):390–396

Robins SC, Trudel E, Rotondi O, Liu X, Djogo T, Kryzskaya D, Bourque CW, Kokoeva MV (2013) Evidence for NG2-glia derived, adult-born functional neurons in the hypothalamus. PLoS ONE 8(10):e78236. doi:10.1371/journal.pone.0078236

Sakry D, Karram K, Trotter J (2011) Synapses between NG2 glia and neurons. J Anat 219(1):2–7. doi:10.1111/j.1469-7580.2011.01359.x

Scholz J, Klein MC, Behrens TE, Johansen-Berg H (2009) Training induces changes in white-matter architecture. Nat Neurosci 12(11):1370–1371. doi:10.1038/nn.2412

Simon C, Gotz M, Dimou L (2011) Progenitors in the adult cerebral cortex: cell cycle properties and regulation by physiological stimuli and injury. Glia 59(6):869–881. doi:10.1002/glia.21156

Sirko S, Behrendt G, Johansson PA, Tripathi P, Costa M, Bek S, Heinrich C, Tiedt S, Colak D, Dichgans M, Fischer IR, Plesnila N, Staufenbiel M, Haass C, Snapyan M, Saghatelyan A, Tsai LH, Fischer A, Grobe K, Dimou L, Gotz M (2013) Reactive glia in the injured brain acquire stem cell properties in response to sonic hedgehog [corrected]. Cell Stem Cell 12 (4):426–439. doi:10.1016/j.stem.2013.01.019

Stevens B, Porta S, Haak LL, Gallo V, Fields RD (2002) Adenosine: a neuron-glial transmitter promoting myelination in the CNS in response to action potentials. Neuron 36(5):855–868

Sturrock RR (1976) Changes in neurologia and myelination in the white matter of aging mice. J Gerontol 31(5):513–522

Verkhratsky A, Steinhauser C (2000) Ion channels in glial cells. Brain Res Brain Res Rev 32(2–3):380–412

Vigano F, Dimou L (2016) The heterogeneous nature of NG2-glia. Brain Res. doi:10.1016/j.brainres.2015.09.012

Vigano F, Mobius W, Gotz M, Dimou L (2013) Transplantation reveals regional differences in oligodendrocyte differentiation in the adult brain. Nat Neurosci 16(10):1370–1372. doi:10.1038/nn.3503

Vigano F, Schneider S, Cimino M, Bonfanti E, Gelosa P, Sironi L, Abbracchio MP, Dimou L (2016) GPR17 expressing NG2-Glia: Oligodendrocyte progenitors serving as a reserve pool after injury. Glia 64(2):287–299. doi:10.1002/glia.22929

Von Blankenfeld G, Trotter J, Kettenmann H (1991) Expression and developmental regulation of a GABAA receptor in cultured murine cells of the oligodendrocyte lineage. Eur J Neurosci 3 (4):310–316

Williamson AV, Mellor JR, Grant AL, Randall AD (1998) Properties of GABA(A) receptors in cultured rat oligodendrocyte progenitor cells. Neuropharmacology 37(7):859–873
Wilson HC, Scolding NJ, Raine CS (2006) Co-expression of PDGF alpha receptor and NG2 by oligodendrocyte precursors in human CNS and multiple sclerosis lesions. J Neuroimmunol 176 (1–2):162–173. doi:10.1016/j.jneuroim.2006.04.014
Windrem MS, Schanz SJ, Guo M, Tian GF, Washco V, Stanwood N, Rasband M, Roy NS, Nedergaard M, Havton LA, Wang S, Goldman SA (2008) Neonatal chimerization with human glial progenitor cells can both remyelinate and rescue the otherwise lethally hypomyelinated shiverer mouse. Cell Stem Cell 2(6):553–565. doi:10.1016/j.stem.2008.03.020
Wu Q, Miller RH, Ransohoff RM, Robinson S, Bu J, Nishiyama A (2000) Elevated levels of the chemokine GRO-1 correlate with elevated oligodendrocyte progenitor proliferation in the jimpy mutant. J Neurosci 20(7):2609–2617
Yuan X, Eisen AM, McBain CJ, Gallo V (1998) A role for glutamate and its receptors in the regulation of oligodendrocyte development in cerebellar tissue slices. Development 125 (15):2901–2914
Zhu X, Bergles DE, Nishiyama A (2008) NG2 cells generate both oligodendrocytes and gray matter astrocytes. Development 135(1):145–157. doi:10.1242/dev.004895
Zhu X, Hill RA, Dietrich D, Komitova M, Suzuki R, Nishiyama A (2011) Age-dependent fate and lineage restriction of single NG2 cells. Development 138(4):745–753. doi:10.1242/dev.047951
Ziskin JL, Nishiyama A, Rubio M, Fukaya M, Bergles DE (2007) Vesicular release of glutamate from unmyelinated axons in white matter. Nat Neurosci 10(3):321–330. doi:10.1038/nn1854

# Pharmacological Tools to Study the Role of Astrocytes in Neural Network Functions

**Fernando Peña-Ortega, Ana Julia Rivera-Angulo and Jonathan Julio Lorea-Hernández**

**Abstract** Despite that astrocytes and microglia do not communicate by electrical impulses, they can efficiently communicate among them, with each other and with neurons, to participate in complex neural functions requiring broad cell-communication and long-lasting regulation of brain function. Glial cells express many receptors in common with neurons; secrete gliotransmitters as well as neurotrophic and neuroinflammatory factors, which allow them to modulate synaptic transmission and neural excitability. All these properties allow glial cells to influence the activity of neuronal networks. Thus, the incorporation of glial cell function into the understanding of nervous system dynamics will provide a more accurate view of brain function. Our current knowledge of glial cell biology is providing us with experimental tools to explore their participation in neural network modulation. In this chapter, we review some of the classical, as well as some recent, pharmacological tools developed for the study of astrocyte's influence in neural function. We also provide some examples of the use of these pharmacological agents to understand the role of astrocytes in neural network function and dysfunction.

**Keywords** Astrocyte · Microglia · Aconitase · Fluoroacetate · Fluorocitrate · Glutamine synthetase

## Abbreviations and Acronyms

| | |
|---|---|
| ADP | Adenosine-diphosphate |
| ATP | Adenosine-triphosphate |
| BAPTA-AM | 1,2-Bis(2-aminophenoxy)ethane-N,N,N′,N′-tetraacetic acid tetrakis (acetoxymethyl ester) |
| BBB | Blood–Brain Barrier |
| cAMP | Cyclic adenosine monophosphate |

F. Peña-Ortega (✉) · A.J. Rivera-Angulo · J.J. Lorea-Hernández
Departamento de Neurobiología del Desarrollo y Neurofisiología,
Instituto de Neurobiología, Universidad Nacional Autónoma de México,
Boulevard Juriquilla 3001, Querétaro 76230, Mexico
e-mail: jfpena@unam.mx



R. von Bernhardi (ed.), *Glial Cells in Health and Disease of the CNS*,
Advances in Experimental Medicine and Biology 949,
DOI 10.1007/978-3-319-40764-7_3

| | |
|---|---|
| CGa | Cystine–glutamate antiporter |
| CNS | Central Nervous System |
| Cox-2 | Cyclooxygenese-2 |
| Cx30 | Connexin 30 |
| Cx43 | Connexin 43 |
| FA | Fluoroacetate |
| FC | Fluorocitrate |
| GABA | Gamma-aminobutyric acid |
| GFAP | Glial Fibrillary Acidic Protein |
| GLAST | Glutamate Aspartate Transporter |
| GLT-1 | Glial Glutamate Transporter 1 |
| GS | Glutamine Synthetase |
| iNOS | Inducible Nitric Oxide Synthase |
| L5 | Lumbar segment 5 |
| L-AAA | L-alpha-aminoadipic acid |
| LTP | Long-Term Potentiation |
| MCT1 | Monocarboxylate Transporter 1 |
| MCT4 | Monocarboxylate Transporter 4 |
| MPTP | 1-methyl-4-phenyl-1,2,3,6-tetrahydropyridine |
| MSO | L-methionine-S-sulfoximine |
| NMDA | N-methyl-D-aspartate |
| ONO-2506 | (2R)-2-Propyloctanoic acid |
| PAR1 | Protease-activated receptor 1 |
| pH | -log $[H^+]$ |
| S100B | S100 $Ca^{2+}$-binding protein B |
| TCAC | Tricaboxylic acid cycle |
| TeNT | Tetanus Neurotoxin |
| TgAPP(sw) | Transgenic mice carrying the amyloid precursor protein with the Swedish mutation |
| TFLLR | L-threonyl-L-phenylalanyl-L-leucyl-L-leucyl-L-argininamide |

## Introduction

Neurons are specialized to produce various firing patterns and synaptic responses that allow them to interact with each other and to produce complex neural network dynamics (Ramirez et al. 2004; Peña-Ortega 2012). However, glial cells (mainly astrocytes and microglia), which do not communicate using electrical impulses, are well suited to participate in complex neural functions requiring broad spatial integration and long-term temporal regulation (Peña-Ortega 2012; Fields et al. 2014). Glial cells have biological properties that influence all types of neuronal networks, including the respiratory rhythm generator that produces stereotyped network outputs (Hülsmann et al. 2000; Lorea-Hernández et al. 2016) and cortical networks involved in complex neuronal processing (Perea et al. 2014). Astrocytes, for instance, can modulate synaptic transmission and may couple multiple neurons and synapses into functional assemblies (Sasaki et al. 2014), whereas microglia can remove synapses in an activity-dependent manner, thereby altering neural network activity in real time (Tremblay and Majewska 2011). Thus, the incorporation of glial cell function into the understanding of nervous system function may provide a more accurate view of brain function.

The ratio of glial cells to neurons has increased during evolution (Reichenbach 1989; Nedergaard et al. 2003). For instance, nematodes have 302 neurons and 50 glial cells (a ratio of 6–1) (Sulston et al. 1983; Nedergaard et al. 2003). In rodents there are only 2–3 neurons per glia (Bass et al. 1971; Ren et al. 1992; Nedergaard et al. 2003; Bandeira et al. 2009). In contrast, in primates there might be equal numbers or even slightly more glial cells than neurons (Dombrowski et al. 2001; Nedergaard et al. 2003; Sherwood et al. 2006; Azevedo et al. 2009). Glia have many of the same receptors as neurons, secrete neurotransmitters and neurotrophic and neuroinflammatory factors, control clearance of neurotransmitters from synaptic clefts, and are intimately involved in synaptic plasticity (Van Wagoner et al. 1999). The prevailing view of the synapse as a structure involving pre- and postsynaptic connections between two or more neurons has limited our thinking about synaptic function and plasticity, and their role in cognition. The emerging concepts of the "tri-partite" synapse (Araque et al. 1999) (see Chapter "Glial Cells and Integrity of the Nervous System") and even the "quad-partite" synapse (Tremblay and Majewska 2011) may better describe recent evidence indicating that multiple cell types work together at synapses and contribute to network function (Araque et al. 1999; Tremblay and Majewska 2011; Sasaki et al. 2014). Our understanding of glial cell biology is providing us with the experimental tools to explore glial participation in neural network function. In this chapter, we review some of the classical, as well as some recent, pharmacological tools developed for this purpose (Hülsmann et al. 2000; Henneberger et al. 2010; Bélanger et al. 2011; Sasaki et al. 2014; Carlsen and Perrier 2014), without ignoring the fact that the use of animals with specific genetic modifications is now also contributing to this effort (Theis et al. 2003; Suzuki et al. 2011; Pannasch et al. 2011, 2014; Perea et al. 2014).

## Astrocytes and the Pharmacological Tools to Study Them

Virchow described the so-called "neuroglia," which originally were thought to provide physical support for neurons (Somjen 1988). Later, it was established that astrocytes constitute an integral part of the blood–brain barrier (BBB), helping to limit the influx of potentially toxic factors into the CNS (Janzer and Raff 1987). Astrocytes also absorb excess potassium ions and excitotoxic neurotransmitters such as glutamate (Ridet et al. 1997; Nedergaard et al. 2003;

Ransom et al. 2003) and help control the homeostasis of surrounding synapses, with a fundamental role in energy metabolite supply (Hassel et al. 1995; Westergaard et al. 1995; Allaman et al. 2011). Moreover, astrocytes enveloping synapses also control the volume of extracellular space, and hence, the extracellular levels and diffusion of neuroactive substances (Nagelhus and Ottersen 2013) (see Chapter "Glial Cells and Integrity of the Nervous System" for further reading on astrocyte functions).

Today, we know that one astrocyte can contact many thousands of synapses via its processes; it has been estimated that a single astrocyte can influence up to 140,000 synapses in the adult hippocampus of rodents (Bushong et al. 2002) and from 270,000 to 2 million synapses in the human cortex (Oberheim et al. 2009). Moreover, we know that astrocytes express receptors for many different neurotransmitters, and in response to the activation of few of them, astrocytes use $Ca^{2+}$ signaling to propagate their excitation (Duffy and MacVicar 1995; Bezzi et al. 1998; Shelton and Mccarthy 2000; Araque et al. 2002; Agulhon et al. 2008). This provides a mechanism by which astrocytes can monitor and respond to ongoing synaptic transmission by releasing gliotransmitters, such as glutamate, ATP, D-serine, growth factors, and cytokines (Santello et al. 2012; Ji et al. 2013) (Fig. 1).

Astrocytes are interconnected with each other through connexin 30 (Cx30) and connexin 43 (Cx43) (Pannasch et al. 2011, 2014), which help them to communicate via $Ca^{2+}$ waves (Barres et al. 1990; McCarthy and Salm 1991, Perea and Araque 2005) that subsequently control the coordinated release of the gliotransmitters (Moraga-Amaro et al. 2014) (see Chapters "Physiological Functions of Glial Cell Hemichannels" and "Role of Astrocytes in Central Respiratory Chemoreception" for further information) (Fig. 1). Interestingly, inhibiting gliotransmission by sleep deprivation attenuates neural network activity and impairs cognition (Yoo et al. 2007), as does genetic removal of Cx30 and Cx43 (Theis et al. 2003; Suzuki et al. 2011; Pannasch et al. 2011, 2014). Stimulation of astrocytes has also been achieved mechanically (Liu et al. 2011) or through the intra-astroglial release of caged $Ca^{2+}$ (Agulhon et al. 2010; Belanger et al. 2011; Sasaki et al. 2014). In contrast, $Ca^{2+}$ buffering can be used to inhibit astrocyte activity (Henneberger et al. 2010; Belanger et al. 2011; Sasaki et al. 2014; Carlsen and Perrier 2014), but the physiological relevance of such treatments, and even of their influence on gliotransmitter release, is still debated (Hamilton and Attwell 2010).

There have been several attempts to assess behavioral effects of genetically manipulating the major astroglial proteins (Roder et al. 1996; Shibuki et al. 1996; Nishiyama et al. 2002; Suzuki et al. 2011). The effects of over-expression of S100 $Ca^{2+}$-binding protein B (S100B) (Fig. 1), previously known as SB100β, on exploratory behaviors were studied in transgenic mice. A significant difference in the spatial and temporal exploratory pattern was observed between control and S100B mutants (Roder et al. 1996). Mutant mice with a deletion of S100B developed normally, with the cytoarchitecture of the brain preserved, but they exhibited enhanced long-term potentiation (LTP) in the hippocampal CA1 region, better performance in the Morris water maze, and greater contextual fear

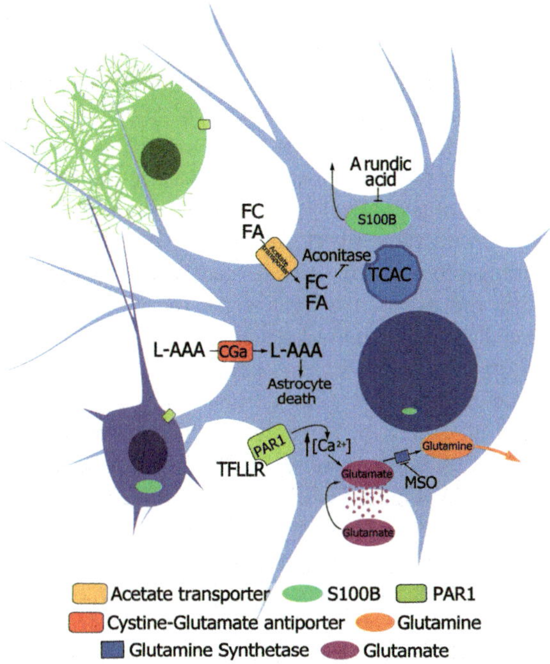


**Fig. 1** Cellular and molecular targets of the drugs used for astrocyte modulation. Astrocytes (represented by the *blue cell*) can be activated by the application of L-threonyl-L-phenylalanyl-L-leucyl-L-leucyl-L-argininamide (TFLLR), which is an agonist of the protease-activated receptor 1 (PAR1). Note that both microglia (represented by the *green cell*) and neurons (represented by the *purple cell*) can also express PAR1. Astrocytes can be killed by the toxin L-alpha-aminoadipic acid (L-AAA); it is selectively incorporated into astrocytes by the cystine–glutamate antiporter (CGa), which is not present either in microglia or in neurons. Astroglial function can also be evaluated by using L-methionine-S-sulfoximine (MSO) to inhibit glutamine synthetase (GS), which is mainly expressed in astrocytes. Astroglial metabolism can also be inhibited by fluoroacetate (FA) and fluorocitrate (FC), which affect the tricaboxylic acid cycle (TCAC) by inhibiting the enzyme aconitase. FA and FC are selectively taken into astrocytes by the acetate transporter, which is absent in both microglia and neurons. Arundic acid, also known as ONO-256, inhibits the production and release of the S100 $Ca^{2+}$-binding protein B (S100B), which is mainly produced by astrocytes but can also be produced by some populations of neurons

conditioning (Nishiyama et al. 2002). Affecting astroglial function through the disruption of the astroglial lactate transporters, the monocarboxylate transporter 4 (MCT4) or 1 (MCT1) produced amnesia and LTP impairment, which was rescued by L-lactate but not by equicaloric glucose (Suzuki et al. 2011). Consistent with this, inhibition of astroglial glycogenolysis with 1,4-dideoxy-1,4-imino-d-arabinitol impaired memory, and the impairment was rescued by lactate (Newman et al. 2011). Also, the selective photostimulation of astrocytes in genetically modified mice with channelrhodopsin-2 enhanced both excitatory and inhibitory synaptic transmission in the primary visual cortex, which either increased or decreased baseline visual responses together with complementary changes in orientation selectivity (Perea et al. 2014). The selective expression of tetanus neurotoxin (TeNT) in astrocytes significantly reduced the duration of carbachol-induced gamma oscillations in hippocampal slices without affecting synaptic transmission (Lee et al. 2014). Moving from the great advantages of using genetically modified mice to understand astrocyte physiology and its role in brain function, this chapter shall now focus on pharmacological approaches to study the role of glial cells, in this case astrocytes, in neural network function.

One of the pharmacological strategies to study astrocytes involvement in neural network function is the use of the agonist for the protease-activated receptor 1 (PAR1), L-threonyl-L-phenylalanyl-L-leucyl-L-leucyl-L-argininamide (TFLLR; Lalo et al. 2014a, b) (Fig. 1). TFLLR is considered a useful astroglial tool since PAR1 localizes preferentially in glial fibrillary acidic protein (GFAP) positive astrocytes (Wang et al. 2002a, b; Boven et al. 2003; Sorensen et al. 2003; Junge et al. 2004; Lee et al. 2007; Shigetomi et al. 2008; Hermann et al. 2009; Han et al. 2011; Shavit et al. 2011) (Fig. 1). PAR1 activation triggers astrocyte proliferation (Grabham and Cunningham 1995; Wang et al. 2002a, b) and stellation (Scarisbrick et al. 2012). Interestingly, PAR1 activation considerably increases astroglial $Ca^{2+}$ concentration (Wang et al. 2002a, b; Lee et al. 2007; Shigetomi et al. 2008; Vandell et al. 2008; Hermann et al. 2009; Han et al. 2011; Oh et al. 2012; Wang et al. 2013) (Fig. 1), which can later increase the $[Ca^{2+}]$ in neurons by releasing glutamate (Lee et al. 2007; Hermann et al. 2009; Oh et al. 2012; Han et al. 2013) (Fig. 1). In fact, astroglial release of glutamate induced by TFLLR occurs via the $Ca^{2+}$-activated anion channel Bestrophin 1 (Oh et al. 2012; Woo et al. 2012; Han et al. 2013), and the neuronal $Ca^{2+}$ influx takes place through the NMDA receptor (Lee et al. 2007; Woo et al. 2012; Han et al. 2013). It is also known that PAR1 receptor-mediated astrocyte excitation causes gliotransmitter release by exocytosis (Bowser and Khakh 2007). Activating PAR1 receptors with TFLLR reduces synaptic transmission by releasing astroglial ATP, which is extracellularly transformed into adenosine that activates presynaptic A1 receptors (Carlsen and Perrier 2014).

Despite some evidence that PAR1 is expressed neither in neurons (Hermann et al. 2009) nor in microglia (Ishida et al. 2006), there are other reports that PAR1 is expressed in subpopulations of neurons (Yang et al. 1997; Gingrich et al. 2000; Striggow et al. 2001; Ishida et al. 2006; Vellani et al. 2010; Han et al. 2011) (Fig. 1) and that it can be detected in microglia after chronic insults (Henrich-Noack et al. 2006; Laskowski et al. 2007; Pompili et al. 2011) (Fig. 1). In fact, there is evidence that PAR1 participates in microglia activation (Suo et al. 2002) and that PAR1 is functional in microglial cultures (Möller et al. 2000; Suo et al. 2002; Balcaitis et al. 2003; Fabrizi et al. 2009) (Fig. 1). In spite of the evidence that TFLLR can increase the intracellular $Ca^{2+}$ concentration in dentate neurons as well as in non-neuronal cells (Han et al. 2011), it was demonstrated that this increase can be prevented by adding the gliotoxin fluoroacetate or by blocking astroglial $Ca^{2+}$ waves with BAPTA-AM (Shigetomi et al. 2008), both of which are known to specifically affect astroglial function (Fonnum et al. 1997; Liu et al. 2004).

Arundic acid ((2R)-2-propyloctanoic acid) is becoming a promising pharmacological tool to study astrocyte function (Asano et al. 2005). Also known as ONO-2506, arundic acid was identified as an astroglial modulator through screening tests searching for an agent that could inhibit astroglial S100B (Asano et al. 2005) (Fig. 1). S100B is an acidic $Ca^{2+}$-binding protein found in the cytoplasm and nucleus of astrocytes (Fig. 1). Released S100B produces dual effects (Rothermundt et al. 2003; Yasuda et al. 2004). At nanomolar concentrations, it stimulates the outgrowth of neural processes and enhances the survival of neurons after tissue damage or ischemia (Rothermundt et al. 2003; Yasuda et al. 2004), and at micromolar levels, it stimulates expression of proinflammatory cytokines and induces apoptosis (Petrova et al. 2000; Lam et al. 2001; Rothermundt et al. 2003). The S100B protein is commonly found in the CNS in astrocytes (Ludwin et al. 1976; Friend et al. 1992; Reeves et al. 1994; Muramatsu et al. 2003; Romero-Alemán Mdel et al. 2003; Kortvely et al. 2003; Vives et al. 2003; Shapiro et al. 2008) (Fig. 1), but never in microglia (Lillo et al. 2002; Muramatsu et al. 2003; Shapiro et al. 2008; Trias et al. 2013). Despite the evidence that neurons lack S100B (Reeves et al. 1994; Romero-Alemán Mdel et al. 2003; Kortvely et al. 2003), there are some reports that S100B can be found in certain neurons (Friend et al. 1992; Vives et al. 2003) (Fig. 1) and in oligodendrocytes (Gonçalves et al. 2008). As mentioned, arundic acid can inhibit the production and release of S100B protein from astrocytes (Hu and Van Eldik 1996; Asano et al. 2005; Mori et al. 2005; Wajima et al. 2013; Wang et al. 2013; Hanada et al. 2014) (Fig. 1). Arundic acid inhibits spontaneous epileptic discharges in Cacna1atm2Nobs/tm2Nobs mice, without affecting maximal electroshock seizures or pentylenetetrazole-induced seizures, by increasing basal glial release of kynurenic acid but not of L-glutamate, D-serine or GABA (Yamamura et al. 2013). Furthermore, arundic acid inhibits excitation-induced release of L-glutamate, D-serine, GABA, and kynurenic in a fluorocitrate-sensitive (thus, astrocyte-dependent) manner (Yamamura et al. 2013). The previous findings suggest that arundic acid can be a good anticonvulsant since it enhances glial inhibitory transmitter release without affecting excitatory transmitter release at rest and it also inhibits glial transmitter release induced by hyperactivation (Yamamura et al. 2013). In addition, arundic acid inhibits the expression of cyclooxygenese-2 (Cox-2), nerve growth factor-beta, and inducible nitric oxide synthase (iNOS) induced by lipopolysaccharide in glial cultures (Shimoda et al. 1998; Tateishi et al. 2002; Shinagawa et al. 1999). Arundic acid does not act on neuronal death in pure neuron cultures, but in cultured astrocytes it suppresses injury-induced changes, such as the increase in S100B content, the secretion of nerve growth factor, the reduction of glutamate transporter (GLT-1 and GLAST) expression, and the disappearance of $GABA_A$ receptors, in a dose-dependent manner and without affecting GFAP expression (Katsumata et al. 1999; Shinagawa et al. 1998, 1999). On the other hand, arundic acid administration significantly inhibits both S100B and GFAP overproduction in the brain of spontaneous hypertensive rats (Higashino et al. 2009).

By suppressing astroglial activation and the production of S100B, the administration of arundic acid reduces the expression of GFAP in activated astrocytes and reduces infarct volume in injured brain (Tateishi et al. 2002; Mori et al. 2005; Asano et al. 2005; Higashino et al. 2009). Administration of arundic acid improves motor function and reduces S100B production while inhibiting the expansion of secondary injury in rats with subdural hematomas (Hanada et al. 2014). Arundic acid decreases the expression of S100B protein produced by activated astrocytes around ischemic lesions (Asano et al. 2005; Mori et al. 2005; Wajima et al. 2013), and it prevents the depletion of dopamine in the striatum and the loss of dopaminergic neurons in the substantia nigra in a mouse model of Parkinson Disease caused by the administration of the toxin 1-methyl-4-phenyl-1,2,3,6-tetrahydropyridine (MPTP) (Kato et al. 2003). Furthermore, arundic acid can ameliorate neurological deficits caused by the MPTP treatment (Kato et al. 2004). Arundic acid can reduce the beta-amyloid deposits and the increase of soluble amyloid-beta peptide, S100B, and beta-amyloid plaque-associated reactive gliosis (astrocytosis and microgliosis) in the transgenic mouse TgAPP(sw), which exhibits some features of Alzheimer's Disease (Mori et al. 2005).

Aside from astrocyte "modulators," a broad variety of "gliotoxins" have been used to probe the role astrocytes in brain function by either killing them (Olney et al. 1971) or by just inhibiting their tricarboxylic acid cycle (Clarke et al. 1970; Hassel et al. 1995) or their glutamine synthesis (Tanigami et al. 2005a, b). L-alpha-aminoadipic acid (L-AAA) is a six-carbon homologue of L-glutamate that has gliotoxic effects both in vivo and in vitro (Olney et al. 1971; Huck et al. 1984; Xu et al. 2004; Leffler et al. 2006; Banasr and Duman 2008) (Fig. 1). The initial observation that L-AAA can kill astrocytes was made by Olney et al. (1971), and later it was reported that astrocytes, but not neurons, can take up L-AAA through the cystine-glutamate antiporter (Huck et al. 1984; Pow 2001) (Fig. 1). Once in the astroglial cytoplasm, L-AAA induces cell death by an unknown mechanism (Brown and Kretzschmar 1998) (Fig. 1). As expected, astrocyte death is reflected as a reduction in cerebral glutamate and glutamine (Lee et al. 2013) and as a loss of astroglial integrity at the ultrastructural level (Takada and Hattori 1986; Rodríguez et al. 2004). Immunohistochemical staining also shows that L-AAA decreases the number of astrocytes but not of neurons (Sun et al. 2013). Astroglial death after intracerebral injection of L-AAA induces anhedonia, anxiety, and helplessness (Banasr and Duman 2008; Domin et al. 2014). L-AAA also affects attention, working memory, and reversal learning (Lima et al. 2014). At the cellular level, L-AAA enhances tonic NMDA responses and neuronal activity in vitro (Fleming et al. 2011) and can alter the somatosensory-evoked potentials and the multiunit sensory-evoked spike rates in the barrel cortex (Sun et al. 2013).

Fluoroacetate (FA) and its toxic metabolite fluorocitrate (FC), both aconitase inhibitors (Clarke et al. 1970) (Fig. 1), are preferentially taken up by glial cells via their acetate transporters (Clarke et al. 1970) (Fig. 1), and they specifically inhibit the glial tricarboxylic acid cycle (Clarke et al. 1970; Paulsen et al. 1987; Clarke 1991; Martín et al. 2007) (Fig. 1). Both substances have been extensively used to study the function of glial cells in the nervous system (Clarke et al. 1970; Paulsen et al. 1987; Clarke 1991; Hülsmann et al. 2000; Martín et al. 2007; Huxtable et al. 2010; Lorea-Hernández et al. 2016). Although FA does not affect neuronal ultrastructure, it causes reversible ultrastructural alterations in astrocytes (Paulsen et al. 1987). These structural changes are accompanied by a temporary decrease in glutamine, glutamate, and aspartate levels and an increase in alanine concentration in the brain (Paulsen et al. 1987). Accordingly, with its astrocyte-specific effect, FC does not affect the activity of other cellular enzyme markers, such as choline acetyltransferase, GABA-2-oxoglutarate amino-transferase, glutamic acid decarboxylase, glutamine synthetase, acetylcholinesterase, and cholinesterase (Paulsen et al. 1987). Even at high concentrations (up to 20 mM), FA only starts to affect pyramidal cells after 1 h of incubation (Canals et al. 2008). This specificity for glial cells arises from the uptake of FA and FC by the acetate transporter, which is selectively expressed by glial cells and not by neurons (Clarke et al. 1970). FA causes a marked increase in extracellular adenosine concentration, exceeding 1 μM, which is sufficient to almost completely inhibit synaptic transmission by activating A1 receptors (Canals et al. 2008; Wall and Dale 2013). It has been reported that this increase in adenosine occurs in the absence of neurons, that it can be observed in pure glial cultures, and that it results from the blockade of the tricarboxylic acid cycle within glial cells (Canals et al. 2008) (Fig. 1). Interestingly, injection of FA in chicks affects long-term memory (Gibbs and Bowser 2009; Gibbs et al. 2011) and abolishes the increase in memory induced by a PAR1 agonist (Gibbs et al. 2011). Inhibition of astroglial metabolism with FA also causes spike-and-wave discharges and absence seizures in various species, including mice, cats, dogs, and rabbits (Ward 1947; Chenoweth and St. John 1947; Goldberg et al. 1966; Hornfeldt and Larson 1990). In addition, FA can inhibit the respiratory rhythm generation in brainstem slices (Hülsmann et al. 2000; Huxtable et al. 2010). This effect can be reversed by glutamine (Hülsmann et al. 2000; Huxtable et al. 2010).

FC is a metabolite of FA that also inhibits aconitase (Clarke et al. 1970) (Fig. 1) and suppresses the tricarboxylic acid cycle in glia (Clarke et al. 1970) (Fig. 1), causing citrate accumulation, reducing glutamine production, and depriving astrocytes of energy (Hassel et al. 1995; Fonnum et al. 1997). FC also interferes with the glutamate–glutamine cycle (Paulsen et al. 1987; Swanson and Graham 1994; Largo et al. 1996). As demonstrated for FA, FC affects just astrocyte metabolism and not that of neurons (Keyser and Pellmar 1994; Willoughby et al. 2003) (Fig. 1). Accordingly, FC reduces extracellular kynurenic acid (Yamamura et al. 2013) and glutamate levels (Tanahashi et al. 2012) without affecting levels of GABA (Yamamura et al. 2013). Moreover, FC reduces the $K^+$-evoked release of kynurenic acid (Yamamura et al. 2013) and D-serine (Tanahashi et al. 2012) but not GABA release (Yamamura et al. 2013). Experiments conducted in astrocyte-brain endothelial cell co-cultures and brain tissue cultures show that the primary effect of FC is on astrocytes (Gesuete et al. 2011; Sá Santos et al. 2011) (Fig. 1). Thus, FC prevents the gliosis induced by oxaliplatin but not the alterations in neuronal morphology or the microgliosis produced by this compound (Di Cesare et al. 2014). Interestingly, FC treatment decreases the levels of several memory-related proteins, such as AMPA receptor GluR1/2, postsynaptic density protein 93/95, Arc, and phosphorylated cAMP response element binding proteins (Shang et al. 2015), while it increases synaptophysin and synapsin I levels in the hippocampus (Shang et al.

2015). FC treatment also increases the levels of phosphorylated Tau at multiple, Alzheimer-related phosphorylation sites (Shang et al. 2015). These effects correlate with the activation of glycogen synthase kinase-3beta and the inactivation of protein phosphatase-2A (Shang et al. 2015). Similar effects are also observed in the primary hippocampal neurons cultured with the conditioned media from FC-treated primary astrocytes (Shang et al. 2015). Accordingly, intracerebroventricular injection of FC impairs memory (Wang et al. 2009b). FC abolishes the uptake of sulforhodamine 101 and reduces GABAergic transmission (Christian and Huguenard 2013) and some other types of synaptic transmission (Bonansco et al. 2011; Christian and Huguenard 2013). FC also alters tissue pH and respiratory output (Erlichman et al. 1998; Holleran et al. 2001). Recent evidence suggests that, in addition to inhibiting astrocyte function, FC may also reduce satellite glial cell activation and function (Liu et al. 2012). Infusion of FC into the dorsal root ganglion (L5) inhibits satellite glial cell activation and reduces spinal nerve injury-induced pain behavior, suggesting the participation of these cells in the genesis of neuropathic pain (Liu et al. 2012).

L-methionine-S-sulfoximine (MSO) is a glutamate analogue that inhibits glutamine synthetase (GS) (Tanigami et al. 2005a, b), which is present only in astrocytes (Norenberg and Martinez-Hernandez 1979; Tanigami et al. 2005a, b). Thus, the effects of MSO can be counteracted by exogenous application of glutamine (Bacci et al. 2002; Blin et al. 2002; Gibbs and Hertz 2005; Tanigami et al. 2005a, b; Liang et al. 2006; Okada-Ogawa et al. 2009). GS is the only enzyme in mammals known to effectively synthesize glutamine (Petroff et al. 2002), which is critical for producing the most abundant excitatory and inhibitory neurotransmitters: glutamate and GABA (Petroff et al. 2002). GS is also important for the clearance of glutamate released during excitatory synaptic transmission and for the metabolism of brain ammonia (Albrecht and Jones 1999; Martinez-Hernandez et al. 1977). A large proportion of extracellular glutamate is taken up by high-affinity excitatory amino acid transporters present on the plasma membrane of astrocytes (Danbolt 2001). Once in the astrocyte, glutamate binds to GS along with ammonia and ATP for enzymatic conversion to glutamine and ADP (Otis and Jahr 1998). The stoichiometry of glutamate transport across the astroglial plasma membrane suggests that rapid metabolism of intracellular glutamate via GS is a prerequisite for efficient glutamate clearance from the extracellular space (Otis and Jahr 1998). A loss of GS in astrocytes is therefore likely to perturb the homeostasis of glutamine, glutamate, GABA, and ammonia in the brain.

Blockade of GS not only prevents the synthesis of glutamine from glutamate, but also results in the loading of astrocytes with glycogen (D'Amelio et al. 1987; Gutierrez and Norenberg 1977; Hevor et al. 1985). After systemic administration of MSO, brain astrocytes show cytoplasmic swelling, a doubling in the number of mitochondria, and glycogen deposition, whereas neurons and oligodendroglia show no ultrastructural abnormalities (Gutierrez and Norenberg 1977). MSO does not affect neuronal levels of choline acetyltransferase or glutamate decarboxylase (Somers and Beckstead 1990). Interestingly, hyperammonemia increases the number of swollen astrocytes in the cortex, and MSO reduces this increase to control values (Tanigami et al. 2005a, b). Furthermore, the number of GFAP-immunopositive cells in the cortex is greater in hyperammonemic rats, and the increase in superficial cortical layers is attenuated by MSO (Tanigami et al. 2005a, b).

In contrast to the finding that MSO reduces epileptiform discharge in hippocampal slices (Bacci et al. 2002), there are reports that MSO reduces GABAergic transmission (Liang et al. 2006) and produces convulsions, hypothermia, and ataxia (Stransky 1969; Ginefri-Gayet and Gayet 1988). Furthermore, chronic infusion of MSO in the hippocampus produces an epileptic phenotype (Wang et al. 2009a; Dhaher et al. 2014). MSO also induces depressive behavior in animals, which is reversed by glutamine (Lee et al. 2014), and it induces inhibition of memory consolidation in a glutamine-sensitive manner (Gibbs and Hertz 2005). Similar to the other gliotoxins, MSO also inhibits respiratory rhythm generation in vivo (Young et al. 2005) and in vitro (Hülsmann et al. 2000; Huxtable et al. 2010), which is reversed by glutamine (Hülsmann et al. 2000; Huxtable et al. 2010).

In summary, we have presented several pharmacological tools that either increase (activate) or inhibit astroglial function and found that despite some possible non-specific effects, these drugs have been very useful for the study of astrocytes' contribution to neuronal network activity. In our opinion, the controlled use of these pharmacological tools in combination with the transgenic animals that allow astroglial function to be modified in a more specific manner will be particularly powerful for the understanding of astroglial modulation of brain function. This approach will support the idea that astrocytes are important regulators of neural network physiology and pathology. Furthermore, the use of these pharmacological tools to modulate astroglial function is likely to reveal cellular and molecular targets to treat diseases caused by astrocytes or that are sustained by them.

**Acknowledgments** We thank Dr. Dorothy Pless for editorial comments. Ana Rivera-Angulo and Jonathan-Julio Lorea-Hernández are graduate students at UNAM and received fellowships from CONACyT. This study was supported by CONACyT Grants 235789, 24688, 117 and 181323; and by DGAPA-UNAM Grant IN200715.

## References

Agulhon C, Petravicz J, McMullen AB, Sweger EJ, Minton SK, Taves SR, Casper KB, Fiacco TA, McCarthy KD (2008) What is the role of astrocyte calcium in neurophysiology? Neuron 59 (6):932–946

Albrecht J, Jones EA (1999) Hepatic encephalopathy: molecular mechanisms underlying the clinical syndrome. J Neurol Sci 170(2):138–146

Allaman I, Bélanger M, Magistretti PJ (2011) Astrocyte-neuron metabolic relationships: for better and for worse. Trends Neurosci 34(2):76–87

Araque A, Parpura V, Sanzgiri RP, Haydon PG (1999) Tripartite synapses: glia, the unacknowledged partner. Trends Neurosci 22(5):208–215

Araque A, Martín ED, Perea G, Arellano JI, Buño W (2002) Synaptically released acetylcholine evokes $Ca^{2+}$ elevations in astrocytes in hippocampal slices. J Neurosci 22(7):2443–2450

Asano T, Mori T, Shimoda T, Shinagawa R, Satoh S, Yada N, Katsumata S, Matsuda S, Kagamiishi Y, Tateishi N (2005) Arundic acid (ONO-2506) ameliorates delayed ischemic brain damage by preventing astrocytic overproduction of S100B. Curr Drug Targets CNS Neurol Disord 4(2):127–142
Azevedo FA, Carvalho LR, Grinberg LT, Farfel JM, Ferretti RE, Leite RE, Jacob Filho W, Lent R, Herculano-Houzel S (2009) Equal numbers of neuronal and nonneuronal cells make the human brain an isometrically scaled-up primate brain. J Comp Neurol 513(5):532–541
Bacci A, Sancini G, Verderio C, Armano S, Pravettoni E, Fesce R, Franceschetti S, Matteoli M (2002) Block of glutamate-glutamine cycle between astrocytes and neurons inhibits epileptiform activity in hippocampus. J Neurophysiol 88(5):2302–2310
Balcaitis S, Xie Y, Weinstein JR, Andersen H, Hanisch UK, Ransom BR, Möller T (2003) Expression of proteinase-activated receptors in mouse microglial cells. NeuroReport 14 (18):2373–2377
Banasr M, Duman RS (2008) Glial loss in the prefrontal cortex is sufficient to induce depressive-like behaviors. Biol Psychiatry 64(10):863–870
Bandeira F, Lent R, Herculano-Houzel S (2009) Changing numbers of neuronal and non-neuronal cells underlie postnatal brain growth in the rat. Proc Natl Acad Sci USA 106(33):14108–14113
Barres BA, Koroshetz WJ, Chun LL, Corey DP (1990) Ion channel expression by white matter glia: the type-1 astrocyte. Neuron 5(4):527–544
Bass NH, Hess HH, Pope A, Thalheimer C (1971) Quantitative cytoarchitectonic distribution of neurons, glia, and DNa in rat cerebral cortex. J Comp Neurol 143(4):481–490
Bélanger M, Allaman I, Magistretti PJ (2011) Brain energy metabolism: focus on astrocyte-neuron metabolic cooperation. Cell Metab 14(6):724–738
Bezzi P, Carmignoto G, Pasti L, Vesce S, Rossi D, Rizzini BL, Pozzan T, Volterra A (1998) Prostaglandins stimulate calcium-dependent glutamate release in astrocytes. Nature 391 (6664):281–285
Blin M, Crusio WE, Hévor T, Cloix JF (2002) Chronic inhibition of glutamine synthetase is not associated with impairment of learning and memory in mice. Brain Res Bull 57(1):11–15
Bonansco C, Couve A, Perea G, Ferradas CÁ, Roncagliolo M, Fuenzalida M (2011) Glutamate released spontaneously from astrocytes sets the threshold for synaptic plasticity. Eur J Neurosci 33(8):1483–1492
Boven LA, Vergnolle N, Henry SD, Silva C, Imai Y, Holden J, Warren K, Hollenberg MD, Power C (2003) Up-regulation of proteinase-activated receptor 1 expression in astrocytes during HIV encephalitis. J Immunol 170(5):2638–2646
Bowser DN, Khakh BS (2007) Two forms of single-vesicle astrocyte exocytosis imaged with total internal reflection fluorescence microscopy. Proc Natl Acad Sci USA 104(10):4212–4217
Brown DR, Kretzschmar HA (1998) The glio-toxic mechanism of alpha-aminoadipic acid on cultured astrocytes. J Neurocytol 27(2):109–118
Bushong EA, Martone ME, Jones YZ, Ellisman MH (2002) Protoplasmic astrocytes in CA1 stratum radiatum occupy separate anatomical domains. J Neurosci 22(1):183–192
Canals S, Larrosa B, Pintor J, Mena MA, Herreras O (2008) Metabolic challenge to glia activates an adenosine-mediated safety mechanism that promotes neuronal survival by delaying the onset of spreading depression waves. J Cereb Blood Flow Metab 28(11):1835–1844
Carlsen EM, Perrier JF (2014) Purines released from astrocytes inhibit excitatory synaptic transmission in the ventral horn of the spinal cord. Front Neural Circuits 8:60
Chenoweth MB, St. John EF (1947) Studies on the pharmacology of fluoroacetate; effects on the central nervous systems of dogs and rabbits. J Pharmacol Exp Ther 90(1):76–82
Christian CA, Huguenard JR (2013) Astrocytes potentiate GABAergic transmission in the thalamic reticular nucleus via endozepine signaling. Proc Natl Acad Sci USA 110(50): 20278–20283
Clarke DD (1991) Fluoroacetate and fluorocitrate: mechanism of action. Neurochem Res 16 (9):1055–1058

Clarke DD, Nicklas WJ, Berl S (1970) Tricarboxylic acid-cycle metabolism in brain. Effect of fluoroacetate and fluorocitrate on the labelling of glutamate, aspartate, glutamine and gamma-aminobutyrate. Biochem J 120(2):345–351

D'Amelio FE, Mehler WR, Gibbs MA, Eng LF, Wu JY (1987) Immunocytochemical localization of glutamic acid decarboxylase (GAD) and glutamine synthetase (GS) in the area postrema of the cat. Light and electron microscopy. Brain Res 410(2):232–244

Danbolt NC (2001) Glutamate uptake. Prog Neurobiol 65(1):1–105

Dhaher R, Damisah EC, Wang H, Gruenbaum SE, Ong C, Zaveri HP, Gruenbaum BF, Eid T (2014) 5-aminovaleric acid suppresses the development of severe seizures in the methionine sulfoximine model of mesial temporal lobe epilepsy. Neurobiol Dis 67:18–23

Di Cesare Mannelli L, Pacini A, Micheli L, Tani A, Zanardelli M, Ghelardini C (2014) Glial role in oxaliplatin-induced neuropathic pain. Exp Neurol 261:22–33

Dombrowski SM, Hilgetag CC, Barbas H (2001) Quantitative architecture distinguishes prefrontal cortical systems in the rhesus monkey. Cereb Cortex 11(10):975–988

Domin H, Szewczyk B, Woźniak M, Wawrzak-Wleciał A, Śmiałowska M (2014) Antidepressant-like effect of the mGluR5 antagonist MTEP in an astroglial degeneration model of depression. Behav Brain Res 273:23–33

Duffy S, MacVicar BA (1995) Adrenergic calcium signaling in astrocyte networks within the hippocampal slice. J Neurosci 15(8):5535–5550

Erlichman JS, Li A, Nattie EE (1998) Ventilatory effects of glial dysfunction in a rat brain stem chemoreceptor region. J Appl Physiol (1985) 85(5):1599-1604

Fabrizi C, Pompili E, Panetta B, Nori SL, Fumagalli L (2009) Protease-activated receptor-1 regulates cytokine production and induces the suppressor of cytokine signaling-3 in microglia. Int J Mol Med 24(3):367–371

Fields RD, Araque A, Johansen-Berg H, Lim SS, Lynch G, Nave KA, Nedergaard M, Perez R, Sejnowski T, Wake H (2014) Glial biology in learning and cognition. Neuroscientist 20 (5):426–431

Fonnum F, Johnsen A, Hassel B (1997) Use of fluorocitrate and fluoroacetate in the study of brain metabolism. Glia 21(1):106–113

Friend WC, Clapoff S, Landry C, Becker LE, O'Hanlon D, Allore RJ, Brown IR, Marks A, Roder J, Dunn RJ (1992) Cell-specific expression of high levels of human S100 beta in transgenic mouse brain is dependent on gene dosage. J Neurosci 12(11):4337–4346

Gesuete R, Orsini F, Zanier ER, Albani D, Deli MA, Bazzoni G, De Simoni MG (2011) Glial cells drive preconditioning-induced blood-brain barrier protection. Stroke 42(5):1445–1453

Gibbs ME, Bowser DN (2009) Astrocytes and interneurons in memory processing in the chick hippocampus: roles for G-coupled protein receptors, GABA(B) and mGluR1. Neurochem Res 34(10):1712–1720

Gibbs ME, Hertz L (2005) Importance of glutamate-generating metabolic pathways for memory consolidation in chicks. J Neurosci Res 81(2):293–300

Gibbs ME, Shleper M, Mustafa T, Burnstock G, Bowser DN (2011) ATP derived from astrocytes modulates memory in the chick. Neuron Glia Biol 7(2–4):177–186

Ginefri-Gayet M, Gayet J (1988) Study of the hypothermia induced by methionine sulfoximine in the rat. Pharmacol Biochem Behav 31(4):797–802

Gingrich MB, Junge CE, Lyuboslavsky P, Traynelis SF (2000) Potentiation of NMDA receptor function by the serine protease thrombin. J Neurosci 20(12):4582–4595

Goldberg ND, Passonneau JV, Lowry OH (1966) Effects of changes in brain metabolism on the levels of citric acid cycle intermediates. J Biol Chem 241(17):3997–4003

Gonçalves CA, Leite MC, Nardin P (2008) Biological and methodological features of the measurement of S100B, a putative marker of brain injury. Clin Biochem 41(10–11):755–763

Grabham P, Cunningham DD (1995) Thrombin receptor activation stimulates astrocyte proliferation and reversal of stellation by distinct pathways: involvement of tyrosine phosphorylation. J Neurochem 64(2):583–591

Gutierrez JA, Norenberg MD (1977) Ultrastructural study of methionine sulfoximine-induced Alzheimer type II astrocytosis. Am J Pathol 86(2):285–300

Hamilton NB, Attwell D (2010) Do astrocytes really exocytose neurotransmitters? Nat Rev Neurosci 11(4):227–238
Han KS, Mannaioni G, Hamill CE, Lee J, Junge CE, Lee CJ, Traynelis SF (2011) Activation of protease activated receptor 1 increases the excitability of the dentate granule neurons of hippocampus. Mol Brain 4:32
Han KS, Woo J, Park H, Yoon BJ, Choi S, Lee CJ (2013) Channel-mediated astrocytic glutamate release via Bestrophin-1 targets synaptic NMDARs. Mol Brain 6:4
Hanada M, Shinjo R, Miyagi M, Yasuda T, Tsutsumi K, Sugiura Y, Imagama S, Ishiguro N, Matsuyama Y (2014) Arundic acid (ONO-2506) inhibits secondary injury and improves motor function in rats with spinal cord injury. J Neurol Sci 337(1–2):186–192
Hassel B, Westergaard N, Schousboe A, Fonnum F (1995) Metabolic differences between primary cultures of astrocytes and neurons from cerebellum and cerebral cortex. Effects of fluorocitrate. Neurochem Res 20(4):413–420
Henneberger C, Papouin T, Oliet SH, Rusakov DA (2010) Long-term potentiation depends on release of D-serine from astrocytes. Nature 463(7278):232–236
Henrich-Noack P, Riek-Burchardt M, Baldauf K, Reiser G, Reymann KG (2006) Focal ischemia induces expression of protease-activated receptorl (PAR1) and AR3 on microglia and enhances PAR4 labeling in the penumbra. Brain Res 1070(1):232–241
Hermann GE, Van Meter MJ, Rood JC, Rogers RC (2009) Proteinase-activated receptors in the nucleus of the solitary tract: evidence for glial-neural interactions in autonomic control of the stomach. J Neurosci 29(29):9292–9300
Hevor TK, Delorme P, Gayet J (1985) Glycogen content and fructose-1,6-biphosphatase activity in methionine sulfoximine epileptogenic mouse brain and liver after protein synthesis inhibition. Neuropathol Appl Neurobiol 11(2):129–139
Higashino H, Niwa A, Satou T, Ohta Y, Hashimoto S, Tabuchi M, Ooshima K (2009) Immunohistochemical analysis of brain lesions using S100B and glial fibrillary acidic protein antibodies in arundic acid- (ONO-2506) treated stroke-prone spontaneously hypertensive rats. J Neural Transm 116(10):1209–1219
Holleran J, Babbie M, Erlichman JS (2001) Ventilatory effects of impaired glial function in a brain stem chemoreceptor region in the conscious rat. J Appl Physiol (1985) 90(4):1539–1547
Hornfeldt CS, Larson AA (1990) Seizures induced by fluoroacetic acid and fluorocitric acid may involve chelation of divalent cations in the spinal cord. Eur J Pharmacol 179(3):307–313
Hu J, Van Eldik LJ (1996) S100 beta induces apoptotic cell death in cultured astrocytes via a nitric oxide-dependent pathway. Biochim Biophys Acta 1313(3):239–245
Huck S, Grass F, Hörtnagl H (1984) The glutamate analogue alpha-aminoadipic acid is taken up by astrocytes before exerting its gliotoxic effect in vitro. J Neurosci 4(10):2650–2657
Hülsmann S, Oku Y, Zhang W, Richter DW (2000) Metabolic coupling between glia and neurons is necessary for maintaining respiratory activity in transverse medullary slices of neonatal mouse. Eur J Neurosci 12(3):856–862
Huxtable AG, Zwicker JD, Alvares TS, Ruangkittisakul A, Fang X, Hahn LB, Posse de Chaves E, Baker GB, Ballanyi K, Funk GD (2010) Glia contribute to the purinergic modulation of inspiratory rhythm-generating networks. J Neurosci 30(11):3947–3958
Ishida Y, Nagai A, Kobayashi S, Kim SU (2006) Upregulation of protease-activated receptor-1 in astrocytes in Parkinson disease: astrocyte-mediated neuroprotection through increased levels of glutathione peroxidase. J Neuropathol Exp Neurol 65(1):66–77
Janzer RC, Raff MC (1987) Astrocytes induce blood-brain barrier properties in endothelial cells. Nature 325(6101):253–257
Ji K, Miyauchi J, Tsirka SE (2013) Microglia: an active player in the regulation of synaptic activity. Neural Plast 2013:627325
Junge CE, Lee CJ, Hubbard KB, Zhang Z, Olson JJ, Hepler JR, Brat DJ, Traynelis SF (2004) Protease-activated receptor-1 in human brain: localization and functional expression in astrocytes. Exp Neurol 188(1):94–103
Kato H, Kurosaki R, Oki C, Araki T (2004) Arundic acid, an astrocyte-modulating agent, protects dopaminergic neurons against MPTP neurotoxicity in mice. Brain Res 1030(1):66–73

Kato H, Araki T, Imai Y, Takahashi A, Itoyama Y (2003) Protection of dopaminergic neurons with a novel astrocyte modulating agent (R)-(-)-2-propyloctanoic acid (ONO-2506) in an MPTP-mouse model of Parkinson's disease. J Neurol Sci 208(1–2):9–15

Katsumata S, Tateishi N, Kagamiishi Y, Shintaku K, Hayakawa T, Shimoda T, Shinagawa R, Akiyama T, Katsube N (1999) Inhibitory effect of ONO-2506 on GABAA receptor disappearance in cultured astrocytes and ischemic brain. Soc Neurosci Abs 25(843):11

Keyser DO, Pellmar TC (1994) Synaptic transmission in the hippocampus: critical role for glial cells. Glia 10(4):237–243

Kortvely E, Varszegi S, Palfi A, Gulya K (2003) Intracellular targeting of calmodulin mRNAs in primary hippocampal cells. J Histochem Cytochem 51(4):541–544

Lalo U, Rasooli-Nejad S, Pankratov Y (2014a) Exocytosis of gliotransmitters from cortical astrocytes: implications for synaptic plasticity and aging. Biochem Soc Trans 42(5):1275–1281

Lalo U, Palygin O, Rasooli-Nejad S, Andrew J, Haydon PG, Pankratov Y (2014b) Exocytosis of ATP from astrocytes modulates phasic and tonic inhibition in the neocortex. PLoS Biol 12(1): e1001747

Lam AG, Koppal T, Akama KT, Guo L, Craft JM, Samy B, Schavocky JP, Watterson DM, Van Eldik LJ (2001) Mechanism of glial activation by S100B: involvement of the transcription factor NFkappaB. Neurobiol Aging 22(5):765–772

Largo C, Cuevas P, Somjen GG, Martín del Río R, Herreras O (1996) The effect of depressing glial function in rat brain in situ on ion homeostasis, synaptic transmission, and neuron survival. J Neurosci 16(3):1219–1229

Laskowski A, Reiser G, Reymann KG (2007) Protease-activated receptor-1 induces generation of new microglia in the dentate gyrus of traumatised hippocampal slice cultures. Neurosci Lett 415(1):17–21

Lee CJ, Mannaioni G, Yuan H, Woo DH, Gingrich MB, Traynelis SF (2007) Astrocytic control of synaptic NMDA receptors. J Physiol 581(Pt 3):1057–1081

Lee Y, Son H, Kim G, Kim S, Lee DH, Roh GS, Kang SS, Cho GJ, Choi WS, Kim HJ (2013) Glutamine deficiency in the prefrontal cortex increases depressive-like behaviours in male mice. J Psychiatry Neurosci 38(3):183–191

Lee HS, Ghetti A, Pinto-Duarte A, Wang X, Dziewczapolski G, Galimi F, Huitron-Resendiz S, Piña-Crespo JC, Roberts AJ, Verma IM, Sejnowski TJ, Heinemann SF (2014) Astrocytes contribute to gamma oscillations and recognition memory. Proc Natl Acad Sci USA 111(32): E3343–E3352

Leffler CW, Parfenova H, Fedinec AL, Basuroy S, Tcheranova D (2006) Contributions of astrocytes and CO to pial arteriolar dilation to glutamate in newborn pigs. Am J Physiol Heart Circ Physiol 291(6):H2897–H2904

Liang SL, Carlson GC, Coulter DA (2006) Dynamic regulation of synaptic GABA release by the glutamate-glutamine cycle in hippocampal area CA1. J Neurosci 26(33):8537–8548

Lillo C, Velasco A, Jimeno D, Cid E, Lara JM, Aijón J (2002) The glial design of a teleost optic nerve head supporting continuous growth. J Histochem Cytochem 50(10):1289–1302

Lima A, Sardinha VM, Oliveira AF, Reis M, Mota C, Silva MA, Marques F, Cerqueira JJ, Pinto L, Sousa N, Oliveira JF (2014) Astrocyte pathology in the prefrontal cortex impairs the cognitive function of rats. Mol Psychiatry 19(7):834–841

Liu QS, Xu Q, Arcuino G, Kang J, Nedergaard M (2004) Astrocyte-mediated activation of neuronal kainate receptors. Proc Natl Acad Sci USA 101(9):3172–3177

Liu T, Sun L, Xiong Y, Shang S, Guo N, Teng S, Wang Y, Liu B, Wang C, Wang L, Zheng L, Zhang CX, Han W, Zhou Z (2011) Calcium triggers exocytosis from two types of organelles in a single astrocyte. J Neurosci 31(29):10593–10601

Liu FY, Sun YN, Wang FT, Li Q, Su L, Zhao ZF, Meng XL, Zhao H, Wu X, Sun Q, Xing GG, Wan Y (2012) Activation of satellite glial cells in lumbar dorsal root ganglia contributes to neuropathic pain after spinal nerve ligation. Brain Res 1427:65–77

Lorea-Hernández JJ, Morales T, Rivera-Angulo AJ, Alcantara-Gonzalez D, Peña-Ortega F (2016) Microglia modulate respiratory rhythm generation and autoresuscitation. Glia 64(4):603-619

Ludwin SK, Kosek JC, Eng LF (1976) The topographical distribution of S-100 and GFA proteins in the adult rat brain: an immunohistochemical study using horseradish peroxidase-labelled antibodies. J Comp Neurol 165(2):197–207

Martín ED, Fernández M, Perea G, Pascual O, Haydon PG, Araque A, Ceña V (2007) Adenosine released by astrocytes contributes to hypoxia-induced modulation of synaptic transmission. Glia 55(1):36–45

Martinez-Hernandez A, Bell KP, Norenberg MD (1977) Glutamine synthetase: glial localization in brain. Science 195(4284):1356–1358

McCarthy KD, Salm AK (1991) Pharmacologically-distinct subsets of astroglia can be identified by their calcium response to neuroligands. Neuroscience 41(2–3):325–333

Möller T, Hanisch UK, Ransom BR (2000) Thrombin-induced activation of cultured rodent microglia. J Neurochem 75(4):1539–1547

Moraga-Amaro R, Jerez-Baraona JM, Simon F, Stehberg J (2014) Role of astrocytes in memory and psychiatric disorders. J Physiol Paris 108(4–6):240–251

Mori T, Town T, Tan J, Tateishi N, Asano T (2005) Modulation of astrocytic activation by arundic acid (ONO-2506) mitigates detrimental effects of the apolipoprotein E4 isoform after permanent focal ischemia in apolipoprotein E knock-in mice. J Cereb Blood Flow Metab 25 (6):748–762

Muramatsu Y, Kurosaki R, Watanabe H, Michimata M, Matsubara M, Imai Y, Araki T (2003) Expression of S-100 protein is related to neuronal damage in MPTP-treated mice. Glia 42 (3):307–313

Nagelhus EA, Ottersen OP (2013) Physiological roles of aquaporin-4 in brain. Physiol Rev 93 (4):1543–1562

Nedergaard M, Ransom B, Goldman SA (2003) New roles for astrocytes: redefining the functional architecture of the brain. Trends Neurosci 26(10):523–530

Newman LA, Korol DL, Gold PE (2011) Lactate produced by glycogenolysis in astrocytes regulates memory processing. PLoS ONE 6(12):e28427

Nishiyama H, Knopfel T, Endo S, Itohara S (2002) Glial protein S100B modulates long-term neuronal synaptic plasticity. Proc Natl Acad Sci USA 99(6):4037–4042

Norenberg MD, Martinez-Hernandez A (1979) Fine structural localization of glutamine synthetase in astrocytes of rat brain. Brain Res 161(2):303–310

Oberheim NA, Takano T, Han X, He W, Lin JH, Wang F, Xu Q, Wyatt JD, Pilcher W, Ojemann JG, Ransom BR, Goldman SA, Nedergaard M (2009) Uniquely hominid features of adult human astrocytes. J Neurosci 29(10):3276–3287

Oh SJ, Han KS, Park H, Woo DH, Kim HY, Traynelis SF, Lee CJ (2012) Protease activated receptor 1-induced glutamate release in cultured astrocytes is mediated by Bestrophin-1 channel but not by vesicular exocytosis. Mol Brain 5:38

Okada-Ogawa A, Suzuki I, Sessle BJ, Chiang CY, Salter MW, Dostrovsky JO, Tsuboi Y, Kondo M, Kitagawa J, Kobayashi A, Noma N, Imamura Y, Iwata K (2009) Astroglia in medullary dorsal horn (trigeminal spinal subnucleus caudalis) are involved in trigeminal neuropathic pain mechanisms. J Neurosci 29(36):11161–11171

Olney JW, Ho OL, Rhee V (1971) Cytotoxic effects of acidic and sulphur containing amino acids on the infant mouse central nervous system. Exp Brain Res 14(1):61–76

Otis TS, Jahr CE (1998) Anion currents and predicted glutamate flux through a neuronal glutamate transporter. J Neurosci 18(18):7099–7110

Pannasch U, Vargová L, Reingruber J, Ezan P, Holcman D, Giaume C, Syková E, Rouach N (2011) Astroglial networks scale synaptic activity and plasticity. Proc Natl Acad Sci USA 108 (20):8467–8472

Pannasch U, Freche D, Dallérac G, Ghézali G, Escartin C, Ezan P, Cohen-Salmon M, Benchenane K, Abudara V, Dufour A, Lübke JH, Déglon N, Knott G, Holcman D, Rouach N (2014) Connexin 30 sets synaptic strength by controlling astroglial synapse invasion. Nat Neurosci 17(4):549–558

Paulsen RE, Contestabile A, Villani L, Fonnum F (1987) An in vivo model for studying function of brain tissue temporarily devoid of glial cell metabolism: the use of fluorocitrate. J Neurochem 48(5):1377–1385
Peña-Ortega F (2012) Tonic neuromodulation of the inspiratory rhythm generator. Front Physiol 3:253
Perea G, Araque A (2005) Glial calcium signaling and neuron-glia communication. Cell Calcium 38(3–4):375–382
Perea G, Yang A, Boyden ES, Sur M (2014) Optogenetic astrocyte activation modulates response selectivity of visual cortex neurons in vivo. Nat Commun 5:3262
Petroff OA, Errante LD, Rothman DL, Kim JH, Spencer DD (2002) Glutamate-glutamine cycling in the epileptic human hippocampus. Epilepsia 43(7):703–710
Petrova TV, Hu J, Van Eldik LJ (2000) Modulation of glial activation by astrocyte-derived protein S100B: differential responses of astrocyte and microglial cultures. Brain Res 853(1):74–80
Pompili E, Fabrizi C, Nori SL, Panetta B, Geloso MC, Corvino V, Michetti F, Fumagalli L (2011) Protease-activated receptor-1 expression in rat microglia after trimethyltin treatment. J Histochem Cytochem 59(3):302–311
Pow DV (2001) Visualising the activity of the cystine-glutamate antiporter in glial cells using antibodies to aminoadipic acid, a selectively transported substrate. Glia 34(1):27–38
Ramirez JM, Tryba AK, Peña F (2004) Pacemaker neurons and neuronal networks: anintegrative view. Curr Opin Neurobiol 14(6):665–674
Ransom B, Behar T, Nedergaard M (2003) New roles for astrocytes (stars at last). Trends Neurosci 26(10):520–522
Reeves RH, Yao J, Crowley MR, Buck S, Zhang X, Yarowsky P, Gearhart JD, Hilt DC (1994) Astrocytosis and axonal proliferation in the hippocampus of S100b transgenic mice. Proc Natl Acad Sci USA 91(12):5359–5363
Reichenbach A (1989) Glia:neuron index: review and hypothesis to account for different values in various mammals. Glia 2(2):71–77
Ren JQ, Aika Y, Heizmann CW, Kosaka T (1992) Quantitative analysis of neurons and glial cells in the rat somatosensory cortex, with special reference to GABAergic neurons and parvalbumin-containing neurons. Exp Brain Res 92(1):1–14
Ridet JL, Malhotra SK, Privat A, Gage FH (1997) Reactive astrocytes: cellular and molecular cues to biological function. Trends Neurosci 20(12):570–577
Roder JK, Roder JC, Gerlai R (1996) Memory and the effect of cold shock in the water maze in S100 beta transgenic mice. Physiol Behav 60(2):611–615
Rodríguez MJ, Martínez-Sánchez M, Bernal F, Mahy N (2004) Heterogeneity between hippocampal and septal astroglia as a contributing factor to differential in vivo AMPA excitotoxicity. J Neurosci Res 77(3):344–353
Romero-Alemán Mdel M, Monzón-Mayor M, Yanes C, Arbelo-Galván JF, Lang D, Renau-Piqueras J, Negrín-Martínez C (2003) S100 immunoreactive glial cells in the forebrain and midbrain of the lizard Gallotia galloti during ontogeny. J Neurobiol 57(1):54–66
Rothermundt M, Peters M, Prehn JH, Arolt V (2003) S100B in brain damage and neurodegeneration. Microsc Res Tech 60(6):614–632
Sá Santos S, Sonnewald U, Carrondo MJ, Alves PM (2011) The role of glia in neuronal recovery following anoxia: In vitro evidence of neuronal adaptation. Neurochem Int 58(6):665–675
Santello M, Calì C, Bezzi P (2012) Gliotransmission and the tripartite synapse. Adv Exp Med Biol 970:307–331
Sasaki T, Ishikawa T, Abe R, Nakayama R, Asada A, Matsuki N, Ikegaya Y (2014) Astrocyte calcium signalling orchestrates neuronal synchronization in organotypic hippocampal slices. J Physiol 592(Pt 13):2771–2783
Scarisbrick IA, Radulovic M, Burda JE, Larson N, Blaber SI, Giannini C, Blaber M, Vandell AG (2012) Kallikrein 6 is a novel molecular trigger of reactive astrogliosis. Biol Chem 393(5): 355–367

Shang XL, Wang QB, Liu XP, Yao XQ, Cao FY, Wang Q, Zhang JY, Wang JZ, Liu GP (2015) Fluorocitrate induced the alterations of memory-related proteins and tau hyperphosphorylation in SD rats. Neurosci Lett 584:230–235

Shapiro LA, Wang L, Ribak CE (2008) Rapid astrocyte and microglial activation following pilocarpine-induced seizures in rats. Epilepsia 49(Suppl 2):33–41

Shavit E, Michaelson DM, Chapman J (2011) Anatomical localization of protease-activated receptor-1 and protease-mediated neuroglial crosstalk on peri-synaptic astrocytic endfeet. J Neurochem 119(3):460–473

Shelton MK, McCarthy KD (2000) Hippocampal astrocytes exhibit Ca2 + -elevating muscarinic cholinergic and histaminergic receptors in situ. J Neurochem 74(2):555–563

Sherwood CC, Stimpson CD, Raghanti MA, Wildman DE, Uddin M, Grossman LI, Goodman M, Redmond JC, Bonar CJ, Erwin JM, Hof PR (2006) Evolution of increased glia-neuron ratios in the human frontal cortex. Proc Natl Acad Sci USA 103(37):13606–13611

Shibuki K, Gomi H, Chen L, Bao S, Kim JJ, Wakatsuki H, Fujisaki T, Fujimoto K, Katoh A, Ikeda T, Chen C, Thompson RF, Itohara S (1996) Deficient cerebellar long-term depression, impaired eyeblink conditioning, and normal motor coordination in GFAP mutant mice. Neuron 16(3):587–599

Shigetomi E, Bowser DN, Sofroniew MV, Khakh BS (2008) Two forms of astrocyte calcium excitability have distinct effects on NMDA receptor-mediated slow inward currents in pyramidal neurons. J Neurosci 28(26):6659–6663

Shimoda T, Tateishi K, Shintaku K, Shintaku K, Yada N, Katagi J, Akiyama T, Maekawa H, Shinagawa R, Kondo K (1998) ONO-2506, a novel astrocyte modulating agent, suppresses the increase of COX-2 and iNOSmRNA expression in cultured astrocytes and ischemic brain. Soc Neurosci 24(384):13

Shinagawa R, Tateishi N, Shimoda T, Shintaku K, Yada N, Honjo K, Kagamiishi Y, Kondo K (1998) ONO-2506 ameliorates neurodegeneration through inhibition of reduction of GLT-1 expression. Soc Neurosci Abs 24(384):14

Shinagawa R, Tateishi N, Shimoda T, Maekawa H, Yada N, Akiyama T, Matsuda S, Katsube N (1999) Modulating effects of ONO-2506 on astrocytic activation in cultured astrocytes from rat cerebrum. Soc Neurosci Abs 25(843):10

Somers DL, Beckstead RM (1990) Chronic methionine sulfoximine administration reduces synaptosomal aspartate and glutamate in rat striatum. Neurosci Lett 115(2–3):335–340

Somjen GG (1988) Nervenkitt: notes on the history of the concept of neuroglia. Glia 1(1):2–9

Sorensen SD, Nicole O, Peavy RD, Montoya LM, Lee CJ, Murphy TJ, Traynelis SF, Hepler JR (2003) Common signaling pathways link activation of murine PAR1, LPA, and S1P receptors to proliferation of astrocytes. Mol Pharmacol 64(5):1199–1209

Stransky Z (1969) Time course of rat brain GABA levels following methionine sulphoximine treatment. Nature 224(5219):612–613

Striggow F, Riek-Burchardt M, Kiesel A, Schmidt W, Henrich-Noack P, Breder J, Krug M, Reymann KG, Reiser G (2001) Four different types of protease-activated receptors are widely expressed in the brain and are up-regulated in hippocampus by severe ischemia. Eur J Neurosci 14(4):595–608

Sulston JE, Schierenberg E, White JG, Thomson JN (1983) The embryonic cell lineage of the nematode Caenorhabditis elegans. Dev Biol 100(1):64–119

Sun ML, Yu HX, Han Y, Tian J, Yu YQ (2013) Glial modulation of vibrissal sensory processing in rats. Chin J Physiol 56(6):309–317

Suo Z, Wu M, Ameenuddin S, Anderson HE, Zoloty JE, Citron BA, Andrade-Gordon P, Festoff BW (2002) Participation of protease-activated receptor-1 in thrombin-induced microglial activation. J Neurochem 80(4):655–666

Suzuki A, Stern SA, Bozdagi O, Huntley GW, Walker RH, Magistretti PJ, Alberini CM (2011) Astrocyte-neuron lactate transport is required for long-term memory formation. Cell 144 (5):810–823

Swanson RA, Graham SH (1994) Fluorocitrate and fluoroacetate effects on astrocyte metabolism in vitro. Brain Res 664(1–2):94–100

Takada M, Hattori T (1986) Fine structural changes in the rat brain after local injections of gliotoxin, alpha-aminoadipic acid. Histol Histopathol 1(3):271–275

Tanahashi S, Yamamura S, Nakagawa M, Motomura E, Okada M (2012) Clozapine, but not haloperidol, enhances glial D-serine and L-glutamate release in rat frontal cortex and primary cultured astrocytes. Br J Pharmacol 165(5):1543–1555

Tanigami H, Rebel A, Martin LJ, Chen TY, Brusilow SW, Traystman RJ, Koehler RC (2005a) Effect of glutamine synthetase inhibition on astrocyte swelling and altered astroglial protein expression during hyperammonemia in rats. Neuroscience 131(2):437–449

Tanigami H, Rebel A, Martin LJ, Chen TY, Brusilow SW, Traystman RJ, Koehler RC (2005b) Effect of glutamine synthetase inhibition on astrocyte swelling and altered astroglial protein expression during hyperammonemia in rats. Neuroscience 131(2):437–449

Tateishi N, Mori T, Kagamiishi Y, Satoh S, Katsube N, Morikawa E, Morimoto T, Matsui T, Asano T (2002) Astrocytic activation and delayed infarct expansion after permanent focal ischemia in rats. Part II: suppression of astrocytic activation by a novel agent (R)-(-)-2-propyloctanoic acid (ONO-2506) leads to mitigation of delayed infarct expansion and early improvement of neurologic deficits. J Cereb Blood Flow Metab 22(6):723–734

Theis M, Jauch R, Zhuo L, Speidel D, Wallraff A, Döring B, Frisch C, Söhl G, Teubner B, Euwens C, Huston J, Steinhäuser C, Messing A, Heinemann U, Willecke K (2003) Accelerated hippocampal spreading depression and enhanced locomotory activity in mice with astrocyte-directed inactivation of connexin43. J Neurosci 23(3):766–776

Tremblay MÈ, Majewska AK (2011) A role for microglia in synaptic plasticity? Commun Integr Biol 4(2):220–222

Trias E, Díaz-Amarilla P, Olivera-Bravo S, Isasi E, Drechsel DA, Lopez N, Bradford CS, Ireton KE, Beckman JS, Barbeito L (2013) Phenotypic transition of microglia into astrocyte-like cells associated with disease onset in a model of inherited ALS. Front Cell Neurosci 7:274

Van Wagoner NJ, Oh JW, Repovic P, Benveniste EN (1999) Interleukin-6 (IL-6) production by astrocytes: autocrine regulation by IL-6 and the soluble IL-6 receptor. J Neurosci 19(13):5236–5244

Vandell AG, Larson N, Laxmikanthan G, Panos M, Blaber SI, Blaber M, Scarisbrick IA (2008) Protease-activated receptor dependent and independent signaling by kallikreins 1 and 6 in CNS neuron and astroglial cell lines. J Neurochem 107(3):855–870

Vellani V, Kinsey AM, Prandini M, Hechtfischer SC, Reeh P, Magherini PC, Giacomoni C, McNaughton PA (2010) Protease activated receptors 1 and 4 sensitize TRPV1 in nociceptive neurones. Mol Pain 6:61

Vives V, Alonso G, Solal AC, Joubert D, Legraverend C (2003) Visualization of S100B-positive neurons and glia in the central nervous system of EGFP transgenic mice. J Comp Neurol 457 (4):404–419

Wajima D, Nakagawa I, Nakase H, Yonezawa T (2013) Neuroprotective effect of suppression of astrocytic activation by arundic acid on brain injuries in rats with acute subdural hematomas. Brain Res 1519:127–135

Wall MJ, Dale N (2013) Neuronal transporter and astrocytic ATP exocytosis underlie activity-dependent adenosine release in the hippocampus. J Physiol 591(Pt 16):3853–3871

Wang H, Ubl JJ, Stricker R, Reiser G (2002a) Thrombin (PAR1)-induced proliferation in astrocytes via MAPK involves multiple signaling pathways. Am J Physiol Cell Physiol 283(5): C1351–C1364

Wang H, Ubl JJ, Reiser G (2002b) Four subtypes of protease-activated receptors, co-expressed in rat astrocytes, evoke different physiological signaling. Glia 37(1):53–63

Wang Y, Zaveri HP, Lee TS, Eid T (2009a) The development of recurrent seizures after continuous intrahippocampal infusion of methionine sulfoximine in rats: a video-intracranial electroencephalographic study. Exp Neurol 220(2):293–302

Wang L, Li CC, Wang GW, Cai JX (2009b) The effects of centrally administered fluorocitrate via inhibiting glial cells on working memory in rats. Sci China C Life Sci 52(8):701–709

Wang H, Zhang L, Zhang IY, Chen X, Da Fonseca A, Wu S, Ren H, Badie S, Sadeghi S, Ouyang M, Warden CD, Badie B (2013) S100B promotes glioma growth through chemoattraction of myeloid-derived macrophages. Clin Cancer Res 19(14):3764–3775

Ward AA Jr (1947) Convulsive activity induced by fluoroacetate. J Neurophysiol 10(2):105–111

Westergaard N, Sonnewald U, Schousboe A (1995) Metabolic trafficking between neurons and astrocytes: the glutamate/glutamine cycle revisited. Dev Neurosci 17(4):203–211

Willoughby JO, Mackenzie L, Broberg M, Thoren AE, Medvedev A, Sims NR, Nilsson M (2003) Fluorocitrate-mediated astroglial dysfunction causes seizures. J Neurosci Res 74(1):160–166

Woo DH, Han KS, Shim JW, Yoon BE, Kim E, Bae JY, Oh SJ, Hwang EM, Marmorstein AD, Bae YC, Park JY, Lee CJ (2012) TREK-1 and Best1 channels mediate fast and slow glutamate release in astrocytes upon GPCR activation. Cell 151(1):25–40

Xu HL, Koenig HM, Ye S, Feinstein DL, Pelligrino DA (2004) Influence of the glia limitans on pial arteriolar relaxation in the rat. Am J Physiol Heart Circ Physiol 287(1):H331–H339

Yamamura S, Hoshikawa M, Dai K, Saito H, Suzuki N, Niwa O, Okada M (2013) ONO-2506 inhibits spike-wave discharges in a genetic animal model without affecting traditional convulsive tests via gliotransmission regulation. Br J Pharmacol 168(5):1088–1100

Yang Y, Akiyama H, Fenton JW 2nd, Brewer GJ (1997) Thrombin receptor on rat primary hippocampal neurons: coupled calcium and cAMP responses. Brain Res 761(1):11–18

Yasuda Y, Tateishi N, Shimoda T, Satoh S, Ogitani E, Fujita S (2004) Relationship between S100beta and GFAP expression in astrocytes during infarction and glial scar formation after mild transient ischemia. Brain Res 1021(1):20–31

Yoo SS, Hu PT, Gujar N, Jolesz FA, Walker MP (2007) A deficit in the ability to form new human memories without sleep. Nat Neurosci 10(3):385–392

Young JK, Dreshaj IA, Wilson CG, Martin RJ, Zaidi SI, Haxhiu MA (2005) An astrocyte toxin influences the pattern of breathing and the ventilatory response to hypercapnia in neonatal rats. Respir Physiol Neurobiol 147(1):19–30

# Microglia Function in the Normal Brain

**Rommy von Bernhardi, Florencia Heredia, Nicole Salgado and Paola Muñoz**

**Abstract** The activation of microglia has been recognized for over a century by their morphological changes. Long slender microglia acquire a short sturdy ramified shape when activated. During the past 20 years, microglia have been accepted as an essential cellular component for understanding the pathogenic mechanism of many brain diseases, including neurodegenerative diseases. More recently, functional studies and imaging in mouse models indicate that microglia are active in the healthy central nervous system. It has become evident that microglia release several signal molecules that play key roles in the crosstalk among brain cells, i.e., astrocytes and oligodendrocytes with neurons, as well as with regulatory immune cells. Recent studies also reveal the heterogeneous nature of microglia diverse functions depending on development, previous exposure to stimulation events, brain region of residence, or pathological state. Subjects to approach by future research are still the unresolved questions regarding the conditions and mechanisms that render microglia protective, capable of preventing or reducing damage, or deleterious, capable of inducing or facilitating the progression of neuropathological diseases. This novel knowledge will certainly change our view on microglia as therapeutic target, shifting our goal from their general silencing to the generation of treatments able to change their activation pattern.

**Keywords** Central nervous system · Cytokines · Development · Glia · Neuroinflammation

## Abbreviations

| | |
|---|---|
| 5-HT | Serotonin |
| Aβ | β-amyloid |
| AD | Alzheimer's disease |
| ALS | Amyotrophic lateral sclerosis |
| ATP | Adenosine triphosphate |

R. von Bernhardi (✉) · F. Heredia · N. Salgado · P. Muñoz
Escuela de Medicina. Departamento de Neurología,
Pontificia Universidad Católica de Chile, Marcoleta 391, Santiago, Chile
e-mail: rvonb@med.puc.cl



R. von Bernhardi (ed.), *Glial Cells in Health and Disease of the CNS*,
Advances in Experimental Medicine and Biology 949,
DOI 10.1007/978-3-319-40764-7_4

| | |
|---|---|
| BDNF | Brain derived neurotrophic factor |
| CNS | Central nervous system |
| CNTF | Ciliary neurotrophic factor |
| DAMPs | Damage- or Danger-associated molecular patterns |
| EP2 | Prostanoid receptor subtype 2 |
| GABA | Gamma aminobutyric acid |
| GDNF | Glia derived neurotrophic factor |
| GM-CSF | Granulocyte/macrophage colony stimulating factor |
| HIV-1 | Human immunodeficiency virus |
| IFNγ | Interferon gamma |
| IGF1 | Insulin-like growth factor 1 |
| IL1 | Interleukin 1 |
| iNOS | Inducible nitric oxide synthase |
| InsP3 | Inositol trisphosphate |
| LPS | Lipopolysaccharides |
| LTP | Long time Potentiation |
| M-CSF | Macrophage colony-stimulating factor |
| MHC | Class I molecules of histocompatibility major complex |
| NGF | Nerve growth factor |
| NMDA | *N*-methyl-d-aspartate |
| NO | Nitric Oxide |
| NT | Neurotrophin |
| PAMPs | Pathogen-associated molecular patterns |
| PGE2 | Prostaglandin E2 |
| PRRs | Pattern recognition receptors |
| RANTES | Regulated on activation, normal T cell expressed and secreted—chemokine CCL5 |
| RNS | Reactive nitrogen species |
| ROS | Reactive oxygen species |
| Rs | Receptors |
| SDF-1α | Stromal cell-derived factor |
| SIRPα | Signal regulatory protein α |
| SRs | Scavenger receptors |
| TGFβ | Transforming growth factor-β |
| TLRs | Toll-like receptors |
| TNFα | Tumor necrosis factor α |
| TSPs | Thrombospondins |

## Introduction

Microglia are the resident immune cells of the central nervous system (CNS), accounting for approximately 10 % of the total cell number in the healthy mammalian brain (Prinz and Priller 2014). They derive from myeloid progenitors, being related to peripheral monocyte-macrophages (Ginhoux et al. 2010). Microglial cell progenitors originating in the yolk sac migrate and colonize the CNS during embryonic development, before the blood–brain barrier is established, differentiating and becoming confined into the CNS. Throughout the life, microglia appear to be capable of local self-renewal. Adult healthy animals show very little exchange between blood and brain parenchyma (Mildner et al. 2007). Thus, maintenance of their population normally does not depend on recruitment of circulating progenitors. However, monocytes invading the brain have been observed under pathological conditions, such as blood–brain barrier damage by trauma or severe inflammation and ischemic vascular damage (Ajami et al. 2007; Casano and Peri 2015; Kierdorf et al. 2013), where they transform into microglia with a ramified phenotype (Mildner et al. 2007). The environment provided by the brain parenchyma appears to be key for microglia phenotype. Astrocytes-conditioned medium induces morphological and functional changes of microglia and blood monocytes in culture (Sievers et al. 1994; Ramirez et al. 2005; Tichauer et al. 2007; von Bernhardi and Ramírez 2001; Orellana et al. 2013), an effect that is at least partly mimicked by adenosine triphosphate (ATP) or adenosine. Other mediators capable of modifying microglia activation are cytokines released from astrocytes, including transforming growth factor β (TGFβ), macrophage colony-stimulating factor (M-CSF), and granulocyte/macrophage colony stimulating factor (GM-CSF) (Schilling et al. 2001; Alarcón et al. 2005; Flores and von Bernhardi 2012; Herrera-Molina et al. 2012; Tichauer et al. 2014; Tichauer and von Bernhardi 2012).

Microglia, as member of the monocyte-macrophage family, function as mononuclear phagocytes, recognizing and scavenging dead cells, pathogens and several endogenous, and exogenous compounds. As mentioned in Chapter "Glial Cells and Integrity of the Nervous System," under physiological conditions and in the absence of inflammatory stimuli, microglia are found in a "surveillance state," morphologically defined by having a small soma with long fine-ramified processes. Surveillance microglia are highly dynamic, retracting and extending their processes in response to environmental cues, interacting with blood vessels, neurons, ependymal cells, and other glial cells (Nimmerjahn et al. 2005; Ramirez et al. 2005; Chen and Trapp 2015; Heneka et al. 2015). They express constitutive markers like Iba-1, and several other markers, such as MHC-I, MHC-II, FcR, CD68, depending on the environmental cues they sense (Fig. 1), being involved in antigen presentation, cytotoxic activation, phagocytosis, antibody-associated phagocytosis, etc. An astrocyte labeled with antibody against glial fibrillary acidic protein (GFAP) is also included to compare their morphologies. Microglial cell surveillance is highly relevant for CNS development and function throughout life.

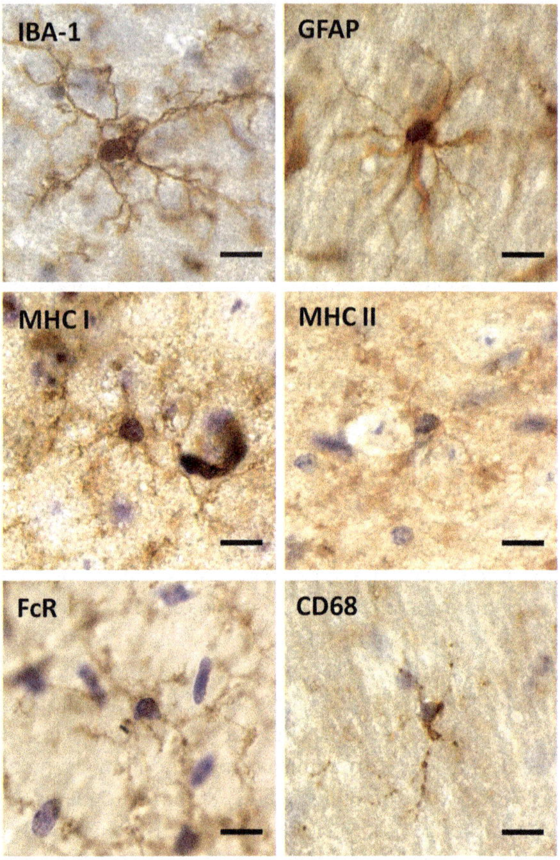


**Fig. 1** Labeling of glia by activation markers. Immunohistochemical detection of glia activation markers in hipocampal cryosections obtained from unstimulated adult mice counterstaining with Harris Haematoxylin. Iba-1 and GFAP antibodies identified microglia and activated astrocytes, respectively. MHC-I, MHC-II, FcR, and CD68 identify microglia populations that are functionally different, showing differences in their morphology as well as in the labeling pattern. Scale bar = 10 μm

## Microglia Motility and Migration

Microglia exhibit two types of motility: the active movement of their processes, sensing the environment, and their translocation in the brain parenchyma. Migration is frequently observed during development, when invading cells migrate into the CNS, and when recruited after an insult and migrate to the site of injury/stimulation. As discussed in Chapters "Glial Cells and Integrity of the Nervous System and Purine Signaling and Microglial Wrapping," many molecules appear to signal for microglia migration, including ATP, cannabinoids, chemokines, lysophosphatidic acid, bradykinin, ion channels, and transporters (Davalos et al. 2005; Walter et al. 2003; Schwab 2001; Rappert et al. 2002; Schilling et al. 2004; Ifuku et al. 2007).

Although under nonstimulated conditions they do not migrate, real-time imaging reveals that microglial processes are constantly moving (Davalos et al. 2005; Nimmerjahn et al. 2005). Processes move rapidly toward an injury. Time-lapse microscopy of brain slices from adult mice shows extensive migration of microglia 24 h after an injury (Carbonell et al. 2005).

Recruitment of microglia to a lesion involves several factors including chemokines released from both neurons and glial cells, among others. Most of the chemokines are released as soluble factors that form chemotactic gradients for cell migration, although CX3CL1 (fractalkine) occurs also as a surface-bound molecule. Microglial activation following neuropathological challenges affects the expression of chemokine receptors (Kremlev et al. 2004), acting as a source and a target of chemokines in an auto/paracrine fashion. Activation of CXCR3 receptor by chemokine CCL21 is linked to microglial migration (Rappert et al. 2002), and the CCL2/CCR2 system appears to be crucial for the recruitment of peripheral monocytes to the CNS, where they become microglia (Davoust et al. 2008; Mildner et al. 2007; Prinz and Priller 2010). Expression of CCR, receptor for CCL2 ligands, identifies functional subsets of microglia. The CX3CR1, the receptor for fractalkine, is also a key molecule for the CNS-relevant macrophage subclassification (Prinz and Priller 2010).

It is especially interesting that chemokines including CCL2, CCL21, or CX3CL1 also appear to serve as signals from endangered neurons to microglia (Biber et al. 2008). It has been suggested that CX3CL1 expressed by neurons could provide a constitutive calming influence on CX3CR1-expressing microglia, thus representing a neuron-to-microglia signaling system similar to those described for CD200/CD200R or CD47/SIRP-1α. Interruption of this regulatory mechanism could facilitate enhanced responses to activating signals. In fact, deficiency in fractalkine signaling results in enhanced severity of CNS damage in several disease models (Cardona et al. 2006; Prinz and Priller 2010). Similarly, activation of CCR5 by the chemokine CCL5 "regulated on activation, normal T cell expressed and secreted" (RANTES), supresses lipopolysaccharide (LPS)-induced expression of inflammatory cytokines, such as interleukin (IL)1β, IL6 and tumor necrosis factor (TNF)α, and inducible nitric oxide synthase (iNOS) in microglia. In contrast, motor neuron death after nerve injury is accelerated in CCR5 knock-out animals, suggesting that CCR5-mediated suppression of microglia toxicity protects neurons (Gamo et al. 2008).

## Microglia-Mediated Phagocytosis

Microglia are the professional phagocytes of the CNS. Phagocytosis is a key function during development as well as in the normal and pathological adult brain (Neumann et al. 2009). During development, microglia remove apoptotic cells, mediated by an "eat me" signal produced by apoptotic cells to microglia (Marin-Teva et al. 2004). They are also involved in synapse removal (Stevens et al. 2007) and in pruning synapses in the developing and postnatal brain (see Chapter "Purine Signaling and Microglial Wrapping" for a complete description on microglial wrapping).

Phagocytosis depends on different mechanisms (Table 1). Pathogens are recognized by Toll-like receptors (TLRs), and apoptotic neurons are recognized by various receptor systems, including asialoglycoprotein-like-, vitronectin-, and phosphatidylserine receptors (Witting et al. 2000). Multiple factors regulate phagocytosis, including ATP, through the metabotropic P2Y6 receptor (Inoue et al. 2009). The P2Y6 receptor is upregulated when neurons are damaged and could be a trigger for phagocytosis (Koizumi et al. 2007). In contrast, activation of P2X7 receptors suppresses phagocytosis, whereas inhibition of P2X7 expression by shRNA or oxATP/BBG restores phagocytosis (Fang et al. 2009). The ciliary neurotrophic factor (CNTF), glia derived neurotrophic factor (GDNF), and M-CSF potentiates phagocytic by microglia (Chang et al. 2006; Lee et al. 2009; Mitrasinovic and Murphy 2003). Substrate-bound complement component C1q enhance both FcR and CR1-mediated phagocytosis (Webster et al. 2000), whereas the prostanoid receptor subtype 2 (EP2), downregulates phagocytosis (Liang et al. 2005; Shie et al. 2005).

**Table 1** Receptors and regulatory molecules associated with microglial cell functions

| Function | | Microglial receptors | Regulatory molecules | References |
|---|---|---|---|---|
| Phagocytosis | Apoptotic cells | Asialoglycoprotein-like-,vutronectin- & phophatidyiserine Rs | | Witting et al. (2000) |
| | | Metabotropic P1 adenosine Rs, metabotropic P2Y & ionotropic P2X purinoRS | ATP | Inoue et al. (2009); Koizumi et al. (2007); Fang et al. (2009); Kirischuk et al. (1995); Lalo et al. (2008) |
| | | GDNF Rs | GDNF, NO | Chang et al. (2006) |
| | | CNTFRα | CNTF | Lee et al. (2009) |
| | Pathogens | TLRs | Inflammatory cytokines and chemokines | Olson and Miller (2004) |
| Development | Neurogenesis (genesis, differentiation & migration) | TLRs | IL-1β, IL-6, IFNγ | Shigemoto-Mogami et al. (2014); Aarum et al. (2003); Walton et al. (2006); Nakanishi et al. (2007); Cepko et al. (1996) |
| | Programmed cell death (phagocytosis) | TNFα Rs1 (TNFR1) | NGF, superoxide ions, TNFα | Frade and Barde (1998); Marin-Teva et al. (2004); Sedel et al. (2004) |
| | Synaptogenesis | IL-10 receptors | TSPs, anti-inflammatory cytokine IL-10 | Chamak et al. (1995); Moller et al. (1996); Lim et al. (2013) |
| | Synaptic maturation | KARAP\DAP12 | | Roumier et al. (2004) |
| | Synapse removal (synaptic pruning) | Fractalkine receptor (CX3CR1) | CX3CL1 | Paolicelli et al. (2011) |
| | | Complement Rs3 (CR3) | MHC1, complement components (C3, C1q) | Corriveau et al. (1998); Goddard et al. (2007); Schafer et al. (2012); Stevens et al. (2007) |

(continued)

**Table 1** (continued)

| Function | | Microglial receptors | Regulatory molecules | References |
|---|---|---|---|---|
| Adult life | Modulation of neuronal activity | Rs fir beurotransmitters, neuropeptides & neuromodulators | Cytokines & RNs (TGFβ-1, NO) | Li et al. (2012); Herrera-Molina and von Bernhardi (2005); Tichauer et al. (2007) |
| | Neuronal surveillance | Fractalkine Rs (CX3CR1), purinergic Rs P2Y12 | ATP & gap junction proteins | Davalos et al. (2005); Liang et al. (2009); Haynes et al. (2006) |
| | Synaptic plasticity (involved in learning & behavior) | Fractalkine receptor (CX3CR1) | CX3CL1 | Paolicelli et al. (2011); Rogers et al. (2011) |
| | | | NT, inflammatory cytokines (IL-1β, TNFα) | Schmid et al. (2009); Goshen et al. (2007); Beattie and Malenka (2002); Loscher et al. (2003); Avital et al. (2003); Labrousse et al. (2009) |
| | Neurogenesis in adult brain | Neurotransmitter Rs | NT & regulatory cytokines (IGF1, BDNF, IL4) | Butovsky et al. (2006); Parkhurst et al. (2013); Ribeiro Xavier et al. (2015) |
| | | | Inflammatory cytokines (IL1-β, IL-6 TNFα) (inhibition) | Ribeiro Xavier et al. (2015); Ben-Hur et al. (2003); Monje et al. (2003); Koo and Duman (2008) |
| | | TLRs (TLR2, TLR4) | | Rolls et al. (2007) |
| | Synaptic stripping | MHC class F receptors | NGF, NT-4/5, TGFβ1, GDNF, FGF, IL-3 | Nakajima et al. (2007); Trapp et al. (2007); Oliveira et al. (2004); Huh et al. (2000) |
| | | | TNFα, IL-6, NO | Nakajima et al. (2005) |
| Pathophysiological conditions | Neurodegeneration (phagocytosis, production factors with inflammatory and immunoregulatory effect) | SRs | Chemokines (CCL2, CCL21, CX3CL1, CXCL10, CXCL12) | Rappert et al. (2004); Koenigsknecht and Landret (2004); Alarcón et al. (2005); Murgas et al. (2012); Bezzi et al. (2001); Stewart et al. (2010); van Weering et al. (2011) |
| | | TLRs (TLR2, TLR4,TLR9) | Inflammatory cytokines (IL1-, IL-6, TNFα, IFNγ) | Murgas et al. (2012); Bezzi et al. (2001); Mount et al. (2007); Chakrabarty et al. (2010) |

(continued)

**Table 1** (continued)

| Function | | Microglial receptors | Regulatory molecules | References |
|---|---|---|---|---|
| | Hypoxia, cerebral ischemia, autoimmunity | Chemokine Rs (CXCR3, CCR3, CCR5, CXCR4) | ROS | Block et al. (2006) |
| | | Cytokine Rs | RNS (NO) | Murgas et al. (2012) |
| | | M-CSFR | | Mitrasinovic and Murphy (2002) |
| | | TLRs | | Stewart et al. (2010); Lotz et al. (2005); Tahara et al. (2006) |

As the resident immune cells of the CNS, microglia are the first line of defense against exogenous threats. The pattern recognition receptors (PRRs), abundantly expressed in microglia, detect infectious agents and assist in the control of the adaptive immunity and the cooperative activities of effector cells (Beurel et al. 2010; Hanisch et al. 2008; Padovan et al. 2007). In addition to pathogen detection by pathogen-associated molecular patterns (PAMPs), several PRRs, including TLRs, also bind endogenous molecules that are generated or modified upon tissue injury. These molecules are classified as damage- or danger-associated molecular patterns (DAMPs) (Bianchi 2007; Kono and Rock 2008; Matzinger 2007). The TLRs 1–9 and co-receptors, like CD14 are widely expressed in cells of the innate as well as adaptive immune system, but also in nonimmune cells (Schaffler et al. 2007). In the brain, TLRs are mainly expressed in glia, although some has been detected in neurons (Aravalli et al. 2007b; Carpentier et al. 2008; Hanisch et al. 2008; Okun et al. 2009; Konat et al. 2006).

The stimulation of TLRs triggers various programs of microglial activation and activates secretion of cytokines and chemokines (Aravalli et al. 2007b; Okun et al. 2009). Several reports indicate the importance of TLRs in various CNS diseases including infection, trauma, stroke, neurodegeneration, and autoimmunity (Babcock et al. 2006; Caso et al. 2007; Lehnardt et al. 2002; Nau and Bruck 2002; Nguyen et al. 2004).

## Participation of Microglia in Development

Microglia are intimately involved in the development of the nervous system (Table 1). They have roles both in neurogenesis and neuronal death. Microglia appears to have both detrimental and supportive effects on neurogenesis (Ekdahl et al. 2009), which could depend in the activation state of microglia (Schwartz et al. 2006).

Differentiation of neural precursors in culture requires the presence of soluble factors secreted by microglia (Nakanishi et al. 2007; Walton et al. 2006). Those factors are also involved in directing migration of newly generated neural cells (Aarum et al. 2003).

The role of microglia for neuronal loss by programmed cell death during development has been described in several brain regions, including the retina, where the pro-apoptotic action of microglia is mediated through nerve growth factor (NGF) (Frade and Barde 1998). Similarly, microglia induce apoptotic death of Purkinje neurons by releasing superoxide ions (Marin-Teva et al. 2004), and motoneurons apoptosis via secretion of TNFα in the embryo (Sedel et al. 2004).

In early postnatal development, elimination of excess synapses—known as synaptic pruning—appears also to be a microglia-mediated mechanism. Mice lacking fractalkine receptor (CX3CR1), have reduced numbers of brain microglia, and show impairment of synaptic pruning, resulting in an abnormally high number of synaptic spines (Paolicelli et al. 2011).

On the other hand, microglia are also involved in the formation of new synapses, especially in the early postnatal brain. Microglia stimulate synaptogenesis by secreting the extracellular matrix proteins thrombospondins (TSPs) (Moller et al. 1996), which are also produced by astrocytes (Christopherson et al. 2005). TSP1 interacts with the integrin-associated protein CD47, which is regulated by signal regulatory protein (SIRP)α, a transmembrane protein expressed by neurons and macrophages (Matozaki et al. 2009). The SIRPα-CD47 complex is involved in the regulation of migration and phagocytosis, immune homeostasis, and neuronal networks, playing homeostatic roles in the immune system, and participating in synaptic patterning (Umemori and Sanes 2008). Microglia also serve roles on the functional maturation of synapses (Paolicelli and Gross 2011). Behavioral abnormalities, including impairment of social interaction and autistic-like behavior (Tang et al. 2014; Zhan et al. 2014) have been reported on several models of microglial cell dysfunction.

## Participation of Microglia in Adult Life

As the name implies, surveillance microglia actively survey the parenchyma, to rapidly activate upon appearance of a threat to the CNS. Microglial activation in response to various stimuli correlates with conspicuous morphological changes. Microglia reduce the complexity and shortens their branched processes (Fig. 2). Several stages can be identified, including process withdrawal, and formation of new processes allowing mobility in the tissue (Lynch 2009; Stence et al. 2001; Streit et al. 2005).

Microglia is a nonhomogeneous population, their activation being a highly regulated process. Thus, activated microglia can acquire distinct functional states (Hanisch and Kettenmann 2007; Perry et al. 2007; Perry and Holmes 2014; Schwartz et al. 2006). Activation is not an all-or-none process, but varies depending

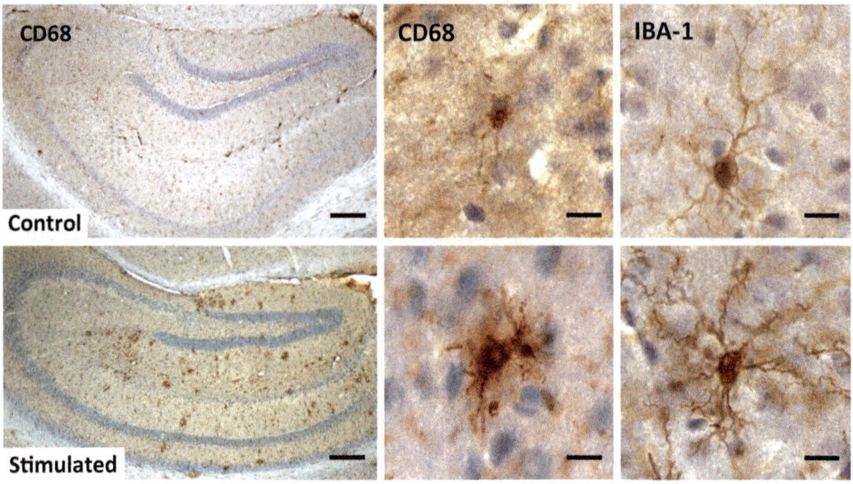


**Fig. 2** Inflammatory activation-dependent morphological changes of microglia. Immunohistochemical labeling of the constitutive identity marker Iba-1 and the phagocytic activation-specific marker CD68, and counterstaining with Harris Haematoxylin, of hipocampal cryosections obtained from inflammatory unstimulated and stimulated young mice. Low (4×) magnification microphotographs of hippocampal section labeled with CD68 show slender-shaped microglia evenly distributed. In contrast, the distribution of microglia in the hippocampus becomes more cluster-like. At high magnification, activated microglia shows shorter sturdier processes. CD68-labeled microglia show an amoeba-like shape with a big cell body and very short processes, whereas Iba-1 shows many cells with long, although studier processes than those observed in unstimulated animales. Scale bar = 100 μm in the right panel and 10 μm in the higher magnification microphotographs at the left

on the stimulation context (Hanisch and Kettenmann 2007; von Bernhardi et al. 2015b; Areschoug and Gordon 2009). Multiple signals converge to maintain or change their functional state and to regulate their specific functional repertoire (Table 1). Activation is triggered when microglia detect the appearance, abnormal concentration, or altered format of molecules that serve as signals (Block and Hong 2005; Block et al. 2007; Hanisch and Kettenmann 2007). The involvement of two signaling systems has been proposed, an "on" receptor-mediated signaling corresponding to a novel molecule that is recognized by microglia triggering activation; and an "off" receptor-mediated signaling that persistently signals to maintain microglia in a certain default activation state (Biber et al. 2007; Hanisch and Kettenmann 2007; Kettenmann et al. 2011).

"On" signals include structures associated with bacterial cell walls, viral envelopes, or their DNAs and RNAs, typically identified as signs of infection. Pathogen structures are sensed through PRRs, such as TLRs (Hanisch et al. 2008) and Scavenger Receptors (SRs) (Ozeki et al. 2006; Godoy et al. 2012; Murgas et al. 2014). Molecules released after tissue damage are also signals, and they induce especially robust microglial responses (Nimmerjahn et al. 2005; Lu et al. 2010; Napoli and Neumann 2009). Intracellular proteins or serum factors can activate microglia when they are induced upon stress, appear in new compartments or suffer biochemical modifications (Hanisch et al. 2008; Lehnardt et al. 2008; Rubartelli and Lotze 2007), as well as some neurotransmitters indicating impaired neuronal activity (Boucsein et al. 2003; Haynes et al. 2006). This, both pathogen- and damage-associated molecular patterns (PAMPs/DAMPs, respectively), activate microglia.

The "off" receptor-mediated signaling is due to the loss of constitutive control signaling in the normal CNS, as observed with ligand-receptor systems CD200-CD200R, CX3CL1- CX3CR1, and CD172a-CD47 (Barclay et al. 2002; Brooke et al. 2004; Cardona et al. 2006; Hoek et al. 2000). Thus, the "on signals" are identified as a sign of threat to the CNS homeostasis. Whereas in the "off signal," the loss of regulation is the signal.

The CNS show regional variations in glial and neuronal cell populations as well as in their environment. For example, the different vulnerability of CA1 versus CA3 neurons depends on the regional microglia response upon stimulation (Hanisch and Kettenmann 2007), with hippocampal neuronal cell death and glial activation depending on the chemokine/receptor system CXCL10/CXCR3 (van Weering et al. 2011).

As discussed in Chapter "Glial Cells and Integrity of the Nervous System," acute self-limited activation of microglia should be deemed as protective, given microglia primarily support and protect the structural and functional integrity of the CNS. Although research has mostly focused on the detrimental consequences of microglia-mediated neuroinflammation, and their potentiation of neuronal damage, it is now accepted that microglia activation is important for protection and repair of the diseased and injured brain. However, the final outcome will depend on the environmental context and timeframe of action (Hellwig et al. 2013; Kierdorf and Prinz 2013; von Bernhardi et al. 2015b). When encountering a mild injury or impairment, microglia could act immediately to repair and offer trophic support, and even reduce activating synaptic input by remodeling synapses (Trapp et al. 2007; Wake et al. 2009). However, the everyday activity of microglia is very difficult to assess (Hanisch and Kettenmann 2007). Thus, in general there is much more evidence on the failure and harmful contributions of microglia than on their physiological roles.

Microglia serve several functional roles, modulating neuronal activity and viability in culture and in the adult brain through direct contact with neurons (Li et al. 2012; Kohman et al. 2013) and through their release of soluble mediators, including cytokines and reactive species (Herrera-Molina and von Bernhardi 2005; Ramírez et al. 2008; Ramirez et al. 2005; Tichauer et al. 2007; von Bernhardi and Eugenin 2004; Glass et al. 2010; Di Filippo et al. 2010; von Bernhardi and Eugenin 2012).

Microglia play an active role in the functional integrity of the CNS and its normal physiological performance even affecting learning and behavior (Ziv et al. 2006; Ziv and Schwartz 2008), through their effect, together with T cells, at various levels. Both synaptic contacts and neuron trophism could depend on factors produced by activated microglia. Microglia express several neurotrophins (Elkabes et al. 1996; Kim and de Vellis 2005; Ferrini and De Koninck 2013), releasing many factors with powerful neurotrophic actions (Morgan et al. 2004). A number of cytokines appear to have roles in the maturating CNS. In the adult CNS, IL1 drives astrocytes proliferation in response to injury (Giulian et al. 1988).

Microglia also contribute to the plasticity of the CNS through support of neurogenesis in adult individuals (Butovsky et al. 2006; Ziv et al. 2006; McPherson et al. 2011; Ekdahl et al. 2003), which appears to depend on certain subpopulations of microglia (Ribeiro Xavier et al. 2015). Microglia have neurotransmitter receptors and are responsive to serotonin (5-HT) (Pocock and Kettenmann 2007) and cytokine levels, and can influence precursor cells, showing a positive regulation of neurogenesis by 5-HT and a negative regulation by stress and elevated glucocorticoids (Kempermann 2002; Kempermann and Kronenberg 2003). Thus, under certain conditions, microglia can adopt a pro-neurogenic phenotype, which involves the expression of neurotrophins and regulatory cytokines, such as insulin-like growth factor 1 (IGF1), BDNF, and IL4 (Parkhurst et al. 2013; Chen and Trapp 2015; Ribeiro Xavier et al. 2015). However, in inflammatory activation states, microglia consistently appears to inhibit neurogenesis (Monje et al. 2003; Nakanishi et al. 2007).

Similar to the synaptic pruning observed during development, microglia keep a structural role in circuit refinement throughout life. The role of microglia in removing synapses, is known as "synaptic stripping." It is also observed in response to focal inflammation (Trapp et al. 2007). The "stripping" predominantly removes excitatory glutamatergic synapses, thus limiting neuronal excitability and glutamate excitotoxicity (Linda et al. 2000). Microglia scan synapses, establishing contacts with them that last a few minutes. In ischemia, contacts become longer, lasting for around an hour (Wake et al. 2009). These long lasting interactions often result in the disappearance of that synaptic contact. Any abnormalities in synaptic performance could activate microglia. However, the signal for microglia to remove a synapsis is poorly understood. The specificity of this action is associated with major histocompatibility complex (MHC) class F receptors, which are present in both neurons and microglia (Cullheim and Thams 2007) (see Chapter "Purine Signaling and Microglial Wrapping" for further reading on synaptic stripping).

## Participation of Microglia in Pathophysiological Conditions

Both the absence of protective functions served by microglia, or their abnormal or excessive activation (von Bernhardi 2007; von Bernhardi et al. 2015b), could led to functional impairment and eventually to development of a disease of the CNS. The relevance of microglia activation and subsequent proliferation in aging, in which condition they adopt an "activated-like" morphology (Fig. 3; see Chapter "Age-dependent Changes in the Activation and Regulation of Microglia" for further reading on aging) (Conde and Streit 2006; Gavilan et al. 2007; von Bernhardi 2007; von Bernhardi et al. 2015b) as well as in many pathological contexts have been discussed over the past years (von Bernhardi et al. 2010; Heneka et al. 2014; Perry

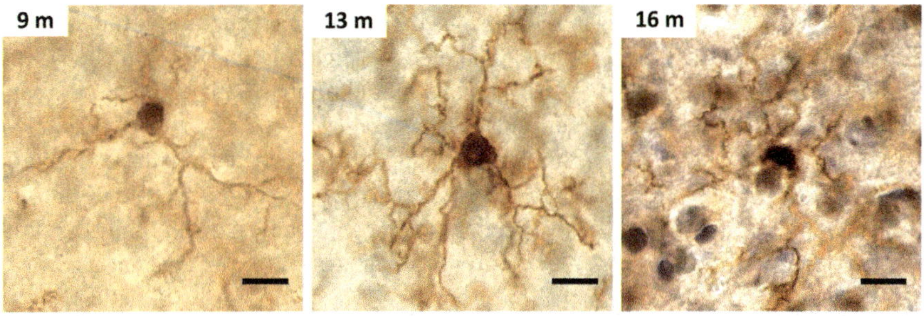


**Fig. 3** Activation of hippocampal microglia with aging. Hippocampal cross section obtained from 9-, 13-, and 16-month old mice were labeled for Iba-1 and counterstaining with Harris Haematoxylin, a monocyte-macrophage identity marker that labels constitutively microglia, to compare the morphological features as the animal ages. At 9-months-old animals, microglia have long and ramified processes, which persisted at 13 months of age. In contrast, microglia from 16-month-old mice begins to shorten their processes, which become sturdier and the cell body increases in size. Scale bar = 10 μm

and Holmes 2014; Heneka et al. 2015; Hu et al. 2015; von Bernhardi et al. 2015a; Yirmiya et al. 2015).

Any disturbance on brain homeostasis, as observed in infection, trauma, ischemia, altered neuronal activity, and both acute and chronic neurological injuries and diseases, induces profound changes in microglial cell shape, gene expression and functional behavior in which is defined as microglial cell activation (Hanisch and Kettenmann 2007; Block et al. 2007; Colton and Wilcock 2010; Colton 2009; Davoust et al. 2008; Graeber and Streit 2010; Streit et al. 2005; van Rossum and Hanisch 2004). Activated microglia show enlarged cell bodies and short and sturdy processes (Fig. 2). They can become motile and be actively recruited to the injury site following chemotactic gradients, and can also increase their proliferation. This phenotype is also correlated with functional changes occurring in complex and broad spectrum responses (Table 1). The range of microglial cell activities covers induction and release of multiple factors with inflammatory and immunoregulatory effects, phagocytotic activities to clear debris, damaged cells, or pathogens, production of neurotrophins and interaction with damaged neurons. Inflammatory response goes from responses centered around the production and release of inflammatory cytokines, such as TNFα, IL1β, and IL6 to release of factors with an anti-inflammatory effect (Casano and Peri 2015; Hu et al. 2015; Chen and Trapp 2015). Although some authors consider inflammatory microglia as detrimental, and anti-inflammatory regulatory microglia as neuroprotective, this rigid classification fails to recognize the complexity of microglial cell function and regulation (Fenn et al. 2014). Furthermore, regulatory microglia do not show always neuroprotective effects (Cherry et al. 2014).

The role of microglia-mediated phagocytosis in neurodegeneration has been established by several experimental approaches. Microglia are needed for removal of the dendritic trees of interneurons in the dentate gyrus after entorhinal cortex lesions (Rappert et al. 2004). In response to the lesion, microglia accumulate at the molecular layer in the dentate gyrus, mediated by signaling through the chemokine receptor CXCR3. Deletion of CXCR3 results in the failure of microglia recruitment, and the dendritic trees of interneurons are preserved.

Microglia also phagocytose molecules and debris such as myelin or amyloid deposits. Several studies report that Aβ is taken up by microglia in culture through mechanisms depending on Scavenger Receptors (Koenigsknecht and Landreth 2004; Alarcón et al. 2005; Cornejo and von Bernhardi 2013; Murgas et al. 2012), among others.

As discussed in Chapter "Age-dependent Changes in the Activation and Regulation of Microglia," there is increasing evidence for altered chemokine signaling in diverse CNS diseases such as Alzheimer's disease (AD) or multiple sclerosis (Gebicke-Haerter et al. 2001; Trebst et al. 2008) which may involve microglia activation (Stewart et al. 2010). Microglial cells from AD brains may have elevated levels of CCR3 and CCR5 receptors (Gebicke-Haerter et al. 2001). CXCL10 and its receptor CXCR3 have been linked to various CNS pathologies (van Weering et al. 2011). Studying the mechanisms by which this system mediates *N*-methyl-d-aspartate (NMDA)-induced neuronal toxicity in the hippocampus, the authors demonstrated that astrocytes and microglia cooperate to deliver the effect and that the deficiency in either the ligand or the receptor diminished or enhanced cell death depending on the tissue subregion and that microglia was the responsible cellular element by which this difference in neuronal vulnerability is organized.

A mechanism involving microglia, astrocytes, and chemokines has been proposed for glutamate toxicity (Bezzi et al. 2001). Binding of CXCL12 (stromal cell-derived factor, SDF-1α) to its receptor CXCR4 in astrocytes, results in Inositol trisphosphate (InsP3) production, $[Ca^{2+}]i$ increase, and release of TNFα. The binding of TNFα to its receptor triggers signaling, through autocrine a paracrine mechanism that causes prostaglandin E2 (PGE2) production. The PGE2, in turn, induces the release of glutamate, which can participate in glia-glia or glia-neuron communication, but can also initiate neurotoxicity. In the latter situation, SDF-1α would also act on microglia, thus driving enhanced TNFα release from both glial populations and ultimately causing massive glutamate release.

AD is associated with a significant elevation of TLR expression in the brain (Letiembre et al. 2009; Walter et al. 2007). Treatment with Aβ potentiated TLR2 and TLR4-mediated responses, while inhibiting TLR9 in mouse microglia cultures, (Lotz et al. 2005). At the same time, all three receptors (TLR2, TLR4, and TLR9) stimulated the uptake of Aβ by microglia (Tahara et al. 2006). The levels of TLRs in CNS are generally upregulated in many neurodegenerative diseases, including multiple sclerosis, Parkinson's disease, and amyotrophic lateral sclerosis (ALS) (Okun et al. 2009).

Microglia activation, in turn, upregulates the synthesis of TLRs (Kielian et al. 2005; McKimmie and Fazakerley 2005). Similarly, the levels of TLRs in microglia are increased following hypoxia (Ock et al. 2007) and cerebral ischemia (Ziegler et al. 2007) and by inflammatory processes; for example, TNFα stimulates expression of TLR2 in cultured mouse microglia (Syed et al. 2007).

TLRs also regulate microglial death following pathological activation. TLR4 triggers microglial apoptosis via autocrine production of interferon gamma (IFNγ), whereas TLR2 is coupled to caspase-8-dependent apoptotic pathways (Lehnardt et al. 2007). Similarly, TLR2 participate in microglial apoptosis following human immunodeficiency virus (HIV-1) infection (Aravalli et al. 2007a, 2008).

A link of neurodegenerative processes in AD to microglial TLR4 is suggested because Aβ fibers bind to CD14, the co-receptor for LPS signaling via TLR4 (Fassbender et al. 2004). CD14 and TLR-dependent mechanisms appear to promote Aβ clearance and participate in inflammatory responses of microglia (Fassbender et al. 2004; Landreth and Reed-Geaghan 2009; Reed-Geaghan et al. 2009; Tahara et al. 2006). Pronounced CD14 immunoreactivity is observed in microglia close to AD lesion sites in AD brains (Liu et al. 2005). Importantly, a microglial CD36-TLR4-TLR6 complex appears to promote inflammation in response to Aβ (Stewart et al. 2010).

However, TLR signaling can also be neuroprotective, by both driving clearance of infectious agents, and by organizing CNS-intrinsic as well as immune system-mediated support of neural cell survival, tissue preservation, and CNS functioning (Glezer et al. 2007; Hanisch et al. 2008). Thus, a critical issue is to understand the mechanisms by which TLRs could engage in detrimental or in beneficial programs.

## Concluding Remarks

Microglia affects the development, structure, and function of neuronal networks. They constantly monitor the status of synaptic contacts and receive information from neuronal activity. Multiple activation states of microglia may allow for the existence of microglia with different functions, which dynamically interact with neurons and potentiate their plastic capabilities. Furthermore, they appear to be also able to remodel neuronal connectivity and thus participate in physiological processes.

Commitment to distinct reactive phenotypes depending on their activation profile would then have a variable effect on neurons. It will be important to identify the nature of such instructing signals as they govern functional orientations of microglia. Little is also known about the heterogeneity of microglia, i.e., the differences in functional capacities of individual microglial populations within different CNS regions. Finally, in pathological situations with blood-derived monocytes/macrophages infiltrating the CNS, features and functions of resident microglia and the newly invading cells may complement each other, with both detrimental and beneficial consequences (Shechter et al. 2009; Simard et al. 2006). Understanding this various issues will be especially interesting to develop migroglia-based strategies for the management of several impairments of the nervous system.

**Acknowledgments** This work was supported by grant FONDECYT 1131025 to RvB.

# References

Aarum J, Sandberg K, Haeberlein SL, Persson MA (2003) Migration and differentiation of neural precursor cells can be directed by microglia. Proc Natl Acad Sci USA 100(26):15983–15988. doi:10.1073/pnas.2237050100

Ajami B, Bennett JL, Krieger C, Tetzlaff W, Rossi FM (2007) Local self-renewal can sustain CNS microglia maintenance and function throughout adult life. Nat Neurosci 10(12):1538–1543. doi:10.1038/nn2014

Alarcón R, Fuenzalida C, Santibanez M, von Bernhardi R (2005) Expression of scavenger receptors in glial cells. Comparing the adhesion of astrocytes and microglia from neonatal rats to surface-bound beta-amyloid. J Biol Chem 280(34):30406–30415. doi:10.1074/jbc.M414686200

Aravalli RN, Hu S, Lokensgard JR (2007a) Toll-like receptor 2 signaling is a mediator of apoptosis in herpes simplex virus-infected microglia. J Neuroinflammation 4:11. doi:10.1186/1742-2094-4-11

Aravalli RN, Hu S, Lokensgard JR (2008) Inhibition of toll-like receptor signaling in primary murine microglia. J Neuroimmu Pharmacol Off J Soc NeuroImmu Pharmacol 3(1):5–11. doi:10.1007/s11481-007-9097-8

Aravalli RN, Peterson PK, Lokensgard JR (2007b) Toll-like receptors in defense and damage of the central nervous system. J Neuroimmu Pharmacol Off J Soc NeuroImmu Pharmacol 2 (4):297–312. doi:10.1007/s11481-007-9071-5

Areschoug T, Gordon S (2009) Scavenger receptors: role in innate immunity and microbial pathogenesis. Cell Microbiol 11(8):1160–1169. doi:10.1111/j.1462-5822.2009.01326.x

Avital A, Goshen I, Kamsler A, Segal M, Iverfeldt K, Richter-Levin G, Yirmiya R (2003) Impaired interleukin-1 signaling is associated with deficits in hippocampal memory processes and neural plasticity. Hippocampus 13:826–834

Babcock AA, Wirenfeldt M, Holm T, Nielsen HH, Dissing-Olesen L, Toft-Hansen H, Millward JM, Landmann R, Rivest S, Finsen B, Owens T (2006) Toll-like receptor 2 signaling in response to brain injury: an innate bridge to neuroinflammation. J Neurosci 26 (49):12826–12837. doi:10.1523/JNEUROSCI.4937-05.2006

Barclay AN, Wright GJ, Brooke G, Brown MH (2002) CD200 and membrane protein interactions in the control of myeloid cells. Trends Immunol 23(6):285–290

Beattie MS, Malenka RC (2002) Control of synaptic strength by glial TNFalpha. Science 295:2282–2285

Ben-Hur T, Ben-Menachem O, Furer V, Einstein O, Mizrachi-Kol R, Grigoriadis N (2003) Effects of proinflammatory cytokines on the growth, fate, and motility of multipotential neural precursor cells. Mol Cell Neurosci 24:623–631

Beurel E, Michalek SM, Jope RS (2010) Innate and adaptive immune responses regulated by glycogen synthase kinase-3 (GSK3). Trends Immunol 31(1):24–31. doi:10.1016/j.it.2009.09.007

Bezzi P, Domercq M, Brambilla L, Galli R, Schols D, De Clercq E, Vescovi A, Bagetta G, Kollias G, Meldolesi J, Volterra A (2001) CXCR4-activated astrocyte glutamate release via TNFalpha: amplification by microglia triggers neurotoxicity. Nat Neurosci 4(7):702–710. doi:10.1038/89490

Bianchi ME (2007) DAMPs, PAMPs and alarmins: all we need to know about danger. J Leukoc Biol 81(1):1–5. doi:10.1189/jlb.0306164

Biber K, Neumann H, Inoue K, Boddeke HW (2007) Neuronal 'On' and 'Off' signals control microglia. Trends Neurosci 30(11):596–602. doi:10.1016/j.tins.2007.08.007

Biber K, Vinet J, Boddeke HW (2008) Neuron-microglia signaling: chemokines as versatile messengers. J Neuroimmunol 198(1–2):69–74. doi:10.1016/j.jneuroim.2008.04.012
Block ML, Hong JS (2005) Microglia and inflammation-mediated neurodegeneration: multiple triggers with a common mechanism. Prog Neurobiol 76(2):77–98. doi:10.1016/j.pneurobio.2005.06.004
Block ML, Li G, Qin L , Wu X, Pei Z, Wang T et al (2006) Potent regulation of microglia-derived oxidative stress and dopaminergic neuron survival: substance P vs. dynorphin. FASEB J 20:251–258
Block ML, Zecca L, Hong JS (2007) Microglia-mediated neurotoxicity: uncovering the molecular mechanisms. Nat Rev Neurosci 8(1):57–69. doi:10.1038/nrn2038
Boucsein C, Zacharias R, Farber K, Pavlovic S, Hanisch UK, Kettenmann H (2003) Purinergic receptors on microglial cells: functional expression in acute brain slices and modulation of microglial activation in vitro. Eur J Neurosci 17(11):2267–2276
Brooke G, Holbrook JD, Brown MH, Barclay AN (2004) Human lymphocytes interact directly with CD47 through a novel member of the signal regulatory protein (SIRP) family. J Immunol 173(4):2562–2570
Butovsky O, Ziv Y, Schwartz A, Landa G, Talpalar AE, Pluchino S, Martino G, Schwartz M (2006) Microglia activated by IL-4 or IFN-gamma differentially induce neurogenesis and oligodendrogenesis from adult stem/progenitor cells. Mol Cell Neurosci 31(1):149–160. doi:10.1016/j.mcn.2005.10.006
Carbonell WS, Murase S, Horwitz AF, Mandell JW (2005) Migration of perilesional microglia after focal brain injury and modulation by CC chemokine receptor 5: an in situ time-lapse confocal imaging study. J Neurosci 25(30):7040–7047. doi:10.1523/JNEUROSCI.5171-04.2005
Cardona AE, Pioro EP, Sasse ME, Kostenko V, Cardona SM, Dijkstra IM, Huang D, Kidd G, Dombrowski S, Dutta R, Lee JC, Cook DN, Jung S, Lira SA, Littman DR, Ransohoff RM (2006) Control of microglial neurotoxicity by the fractalkine receptor. Nat Neurosci 9(7):917–924. doi:10.1038/nn1715
Carpentier PA, Duncan DS, Miller SD (2008) Glial toll-like receptor signaling in central nervous system infection and autoimmunity. Brain Behav Immun 22(2):140–147. doi:10.1016/j.bbi.2007.08.011
Casano AM, Peri F (2015) Microglia: multitasking specialists of the brain. Dev Cell 32(4):469–477. doi:10.1016/j.devcel.2015.01.018
Caso JR, Pradillo JM, Hurtado O, Lorenzo P, Moro MA, Lizasoain I (2007) Toll-like receptor 4 is involved in brain damage and inflammation after experimental stroke. Circulation 115(12):1599–1608. doi:10.1161/CIRCULATIONAHA.106.603431
Cepko CL, Austin CP, Yang X, Alexiades M, Ezzeddine D (1996) Cell fate determination in the vertebrate retina. Proc Natl Acad Sci USA 93:589–595
Chakrabarty P, Jansen-West K, Beccard A, Ceballos-Diaz C, Levites Y, Verbeeck C (2010) Massive gliosis induced by interleukin-6 suppresses Abeta deposition in vivo: evidence against inflammation as a driving forcé for amyloid deposition. FASEB J 24:548–559
Chamak B, Dobbertin A, Mallat M (1995) Immunohistochemical detection of thrombospondin in microglia in the developing rat brain. Neuroscience 69:177–187
Chang YP, Fang KM, Lee TI, Tzeng SF (2006) Regulation of microglial activities by glial cell line derived neurotrophic factor. J Cell Biochem 97(3):501–511. doi:10.1002/jcb.20646
Chen Z, Trapp BD (2015) Microglia and neuroprotection. J Neurochem. doi:10.1111/jnc.13062
Cherry JD, Olschowka JA, O'Banion MK (2014) Neuroinflammation and M2 microglia: the good, the bad, and the inflamed. J Neuroinflammation 11:98. doi:10.1186/1742-2094-11-98
Christopherson KS, Ullian EM, Stokes CC, Mullowney CE, Hell JW, Agah A, Lawler J, Mosher DF, Bornstein P, Barres BA (2005) Thrombospondins are astrocyte-secreted proteins that promote CNS synaptogenesis. Cell 120(3):421–433. doi:10.1016/j.cell.2004.12.020
Colton C, Wilcock DM (2010) Assessing activation states in microglia. CNS Neurol Disord: Drug Targets 9(2):174–191

Colton CA (2009) Heterogeneity of microglial activation in the innate immune response in the brain. J Neuroimmu Pharmacol Off J Soc NeuroImmun Pharmacol 4(4):399–418. doi:10.1007/s11481-009-9164-4
Conde JR, Streit WJ (2006) Microglia in the aging brain. J Neuropathol Exp Neurol 65(3):199–203. doi:10.1097/01.jnen.0000202887.22082.63
Cornejo F, von Bernhardi R (2013) Role of scavenger receptors in glia-mediated neuroinflammatory response associated with Alzheimer's disease. Mediat Inflamm 2013:895651. doi:10.1155/2013/895651
Corriveau RA, Huh GS, Shatz CJ (1998) Regulation of class I MHC gene expression in the developing and mature CNS by neural activity. Neuron 21:505–520
Cullheim S, Thams S (2007) The microglial networks of the brain and their role in neuronal network plasticity after lesion. Brain Res Rev 55(1):89–96. doi:10.1016/j.brainresrev.2007.03.012
Davalos D, Grutzendler J, Yang G, Kim JV, Zuo Y, Jung S, Littman DR, Dustin ML, Gan WB (2005) ATP mediates rapid microglial response to local brain injury in vivo. Nat Neurosci 8 (6):752–758. doi:10.1038/nn1472
Davoust N, Vuaillat C, Androdias G, Nataf S (2008) From bone marrow to microglia: barriers and avenues. Trends Immunol 29(5):227–234. doi:10.1016/j.it.2008.01.010
Di Filippo M, Chiasserini D, Tozzi A, Picconi B, Calabresi P (2010) Mitochondria and the link between neuroinflammation and neurodegeneration. J Alzheimers Dis 20(Suppl 2):S369–S379. doi:10.3233/JAD-2010-100543
Ekdahl CT, Claasen JH, Bonde S, Kokaia Z, Lindvall O (2003) Inflammation is detrimental for neurogenesis in adult brain. Proc Natl Acad Sci USA 100(23):13632–13637. doi:10.1073/pnas.2234031100
Ekdahl CT, Kokaia Z, Lindvall O (2009) Brain inflammation and adult neurogenesis: the dual role of microglia. Neuroscience 158(3):1021–1029. doi:10.1016/j.neuroscience.2008.06.052
Elkabes S, DiCicco-Bloom EM, Black IB (1996) Brain microglia/macrophages express neurotrophins that selectively regulate microglial proliferation and function. J Neurosci 16 (8):2508–2521
Fang KM, Yang CS, Sun SH, Tzeng SF (2009) Microglial phagocytosis attenuated by short-term exposure to exogenous ATP through P2X receptor action. J Neurochem 111(5):1225–1237. doi:10.1111/j.1471-4159.2009.06409.x
Fassbender K, Walter S, Kuhl S, Landmann R, Ishii K, Bertsch T, Stalder AK, Muehlhauser F, Liu Y, Ulmer AJ, Rivest S, Lentschat A, Gulbins E, Jucker M, Staufenbiel M, Brechtel K, Walter J, Multhaup G, Penke B, Adachi Y, Hartmann T, Beyreuther K (2004) The LPS receptor (CD14) links innate immunity with Alzheimer's disease. FASEB J 18(1):203–205. doi:10.1096/fj.03-0364fje
Fenn AM, Hall JC, Gensel JC, Popovich PG, Godbout JP (2014) IL-4 signaling drives a unique arginase+/IL-1beta+microglia phenotype and recruits macrophages to the inflammatory CNS: consequences of age-related deficits in IL-4Ralpha after traumatic spinal cord injury. J Neurosci 34(26):8904–8917. doi:10.1523/JNEUROSCI.1146-14.2014
Ferrini F, De Koninck Y (2013) Microglia control neuronal network excitability via BDNF signalling. Neural Plast 2013:429815. doi:10.1155/2013/429815
Flores B, von Bernhardi R (2012) Transforming growth factor beta1 modulates amyloid beta-induced glial activation through the Smad3-dependent induction of MAPK phosphatase-1. J Alzheimer Dis 32(2):417–429. doi:10.3233/JAD-2012-120721 Y05258137G173L5 V[pii]
Frade JM, Barde YA (1998) Microglia-derived nerve growth factor causes cell death in the developing retina. Neuron 20(1):35–41
Gamo K, Kiryu-Seo S, Konishi H, Aoki S, Matsushima K, Wada K, Kiyama H (2008) G-protein-coupled receptor screen reveals a role for chemokine receptor CCR5 in suppressing microglial neurotoxicity. J Neurosci 28(46):11980–11988. doi:10.1523/JNEUROSCI.2920-08.2008
Gavilan MP, Revilla E, Pintado C, Castano A, Vizuete ML, Moreno-Gonzalez I, Baglietto-Vargas D, Sanchez-Varo R, Vitorica J, Gutierrez A, Ruano D (2007) Molecular and cellular

characterization of the age-related neuroinflammatory processes occurring in normal rat hippocampus: potential relation with the loss of somatostatin GABAergic neurons. J Neurochem 103(3):984–996. doi:10.1111/j.1471-4159.2007.04787.x
Gebicke-Haerter PJ, Spleiss O, Ren LQ, Li H, Dichmann S, Norgauer J, Boddeke HW (2001) Microglial chemokines and chemokine receptors. Prog Brain Res 132:525–532. doi:10.1016/S0079-6123(01)32100-3
Ginhoux F, Greter M, Leboeuf M, Nandi S, See P, Gokhan S, Mehler MF, Conway SJ, Ng LG, Stanley ER, Samokhvalov IM, Merad M (2010) Fate mapping analysis reveals that adult microglia derive from primitive macrophages. Science 330(6005):841–845. doi:10.1126/science.1194637
Giulian D, Young DG, Woodward J, Brown DC, Lachman LB (1988) Interleukin-1 is an astroglial growth factor in the developing brain. J Neurosci 8(2):709–714
Glass CK, Saijo K, Winner B, Marchetto MC, Gage FH (2010) Mechanisms underlying inflammation in neurodegeneration. Cell 140(6):918–934. doi:10.1016/j.cell.2010.02.016
Glezer I, Simard AR, Rivest S (2007) Neuroprotective role of the innate immune system by microglia. Neuroscience 147(4):867–883. doi:10.1016/j.neuroscience.2007.02.055
Goddard CA, Butts DA, Shatz CJ (2007) Regulation of CNS synapses by neuronal MHC class I. Proc Natl Acad Sci USA 104:6828–6833
Godoy B, Murgas P, Tichauer J, Von Bernhardi R (2012) Scavenger receptor class A ligands induce secretion of IL1beta and exert a modulatory effect on the inflammatory activation of astrocytes in culture. J Neuroimmunol 251(1–2):6–13. doi:10.1016/j.jneuroim.2012.06.004
Goshen I, Kreisel T, Ounallah-Saad H, Renbaum P, Zalzstein Y, Ben-Hur T, Levy-Lahad E, Yirmiya R (2007) A dual role for interleukin-1 in hippocampal dependent memory processes. Psychoneuroendocrinology 32:1106–1115
Graeber MB, Streit WJ (2010) Microglia: biology and pathology. Acta Neuropathol 119(1):89–105. doi:10.1007/s00401-009-0622-0
Hanisch UK, Johnson TV, Kipnis J (2008) Toll-like receptors: roles in neuroprotection? Trends Neurosci 31(4):176–182. doi:10.1016/j.tins.2008.01.005
Hanisch UK, Kettenmann H (2007) Microglia: active sensor and versatile effector cells in the normal and pathologic brain. Nat Neurosci 10(11):1387–1394. doi:10.1038/nn1997
Haynes SE, Hollopeter G, Yang G, Kurpius D, Dailey ME, Gan WB, Julius D (2006) The P2Y12 receptor regulates microglial activation by extracellular nucleotides. Nat Neurosci 9(12):1512–1519. doi:10.1038/nn1805
Hellwig S, Heinrich A, Biber K (2013) The brain's best friend: microglial neurotoxicity revisited. Front Cell Neurosci 7:71. doi:10.3389/fncel.2013.00071
Heneka MT, Golenbock DT, Latz E (2015) Innate immunity in Alzheimer's disease. Nat Immunol 16(3):229–236. doi:10.1038/ni.3102
Heneka MT, Kummer MP, Latz E (2014) Innate immune activation in neurodegenerative disease. Nat Rev 14(7):463–477. doi:10.1038/nri3705
Herrera-Molina R, Flores B, Orellana JA, von Bernhardi R (2012) Modulation of interferon-gamma-induced glial cell activation by transforming growth factor beta1: a role for STAT1 and MAPK pathways. J Neurochem 123(1):113–123. doi:10.1111/j.1471-4159.2012.07887.x
Herrera-Molina R, von Bernhardi R (2005) Transforming growth factor-beta 1 produced by hippocampal cells modulates microglial reactivity in culture. Neurobiol Dis 19(1–2):229–236. doi:10.1016/j.nbd.2005.01.003
Hoek RM, Ruuls SR, Murphy CA, Wright GJ, Goddard R, Zurawski SM, Blom B, Homola ME, Streit WJ, Brown MH, Barclay AN, Sedgwick JD (2000) Down-regulation of the macrophage lineage through interaction with OX2 (CD200). Science 290(5497):1768–1771
Hu X, Leak RK, Shi Y, Suenaga J, Gao Y, Zheng P, Chen J (2015) Microglial and macrophage polarization-new prospects for brain repair. Nat Rev Neurol 11(1):56–64. doi:10.1038/nrneurol.2014.207
Huh GS, Boulanger LM, Du H, Riquelme PA, Brotz TM, Shatz CJ (2000) Functional requirement for class I MHC in CNS development and plasticity. Science 290:2155–2159

Ifuku M, Farber K, Okuno Y, Yamakawa Y, Miyamoto T, Nolte C, Merrino VF, Kita S, Iwamoto T, Komuro I, Wang B, Cheung G, Ishikawa E, Ooboshi H, Bader M, Wada K, Kettenmann H, Noda M (2007) Bradykinin-induced microglial migration mediated by B1-bradykinin receptors depends on $Ca^{2+}$ influx via reverse-mode activity of the $Na^{+}/Ca^{2+}$ exchanger. J Neurosci 27(48):13065–13073. doi:10.1523/JNEUROSCI.3467-07.2007

Inoue K, Koizumi S, Kataoka A, Tozaki-Saitoh H, Tsuda M (2009) P2Y(6)-evoked microglial phagocytosis. Int Rev Neurobiol 85:159–163. doi:10.1016/S0074-7742(09)85012-5

Kempermann G (2002) Regulation of adult hippocampal neurogenesis—implications for novel theories of major depression. Bipolar Disord 4(1):17–33

Kempermann G, Kronenberg G (2003) Depressed new neurons–adult hippocampal neurogenesis and a cellular plasticity hypothesis of major depression. Biol Psychiatry 54(5):499–503

Kettenmann H, Hanisch UK, Noda M, Verkhratsky A (2011) Physiology of microglia. Physiol Rev 91(2):461–553. doi:10.1152/physrev.00011.2010

Kielian T, Esen N, Bearden ED (2005) Toll-like receptor 2 (TLR2) is pivotal for recognition of S. aureus peptidoglycan but not intact bacteria by microglia. Glia 49(4):567–576. doi:10.1002/glia.20144

Kierdorf K, Katzmarski N, Haas CA, Prinz M (2013) Bone marrow cell recruitment to the brain in the absence of irradiation or parabiosis bias. PLoS One 8(3):e58544. doi:10.1371/journal.pone.0058544

Kierdorf K, Prinz M (2013) Factors regulating microglia activation. Front Cell Neurosci 7:44. doi:10.3389/fncel.2013.00044

Kim SU, de Vellis J (2005) Microglia in health and disease. J Neurosci Res 81(3):302–313

Kirischuk S, Moller T, Voitenko N, Kettenmann H, Verkhratsky A (1995) ATP-induced cytoplasmic calcium mobilization in Bergmann glial cells. J Neurosci 15:7861–7871

Koenigsknecht J, Landreth G (2004) Microglial phagocytosis of fibrillar beta-amyloid through a beta1 integrin-dependent mechanism. J Neurosci 24(44):9838–9846. doi:10.1523/JNEUROSCI.2557-04.2004

Kohman RA, Bhattacharya TK, Kilby C, Bucko P, Rhodes JS (2013) Effects of minocycline on spatial learning, hippocampal neurogenesis and microglia in aged and adult mice. Behav Brain Res 242:17–24. doi:10.1016/j.bbr.2012.12.032

Koizumi S, Shigemoto-Mogami Y, Nasu-Tada K, Shinozaki Y, Ohsawa K, Tsuda M, Joshi BV, Jacobson KA, Kohsaka S, Inoue K (2007) UDP acting at P2Y6 receptors is a mediator of microglial phagocytosis. Nature 446(7139):1091–1095. doi:10.1038/nature05704

Konat GW, Kielian T, Marriott I (2006) The role of Toll-like receptors in CNS response to microbial challenge. J Neurochem 99(1):1–12. doi:10.1111/j.1471-4159.2006.04076.x

Kono H, Rock KL (2008) How dying cells alert the immune system to danger. Nat Rev 8(4):279–289. doi:10.1038/nri2215

Koo JW, Duman RS (2008) IL-1beta is an essential mediator of the antineurogenic and anhedonic effects of stress. Proc Natl Acad Sci USA 105:751–756

Kremlev SG, Roberts RL, Palmer C (2004) Differential expression of chemokines and chemokine receptors during microglial activation and inhibition. J Neuroimmunol 149(1–2):1–9. doi:10.1016/j.jneuroim.2003.11.012

Labrousse VF, Costes L, Aubert A, Darnaudery M, Ferreira G, Amedee T, Laye S (2009) Impaired interleukin-1beta and c-Fos expression in the hippocampus is associated with a spatial memory deficit in P2X(7) receptor-deficient mice. PLoS One 4:e6006

Lalo U, Pankratov Y, Wichert SP, Rossner MJ, North RA, Kirchhoff F, Verkhratsky A (2008) P2X1 and P2X5 subunits form the functional P2X receptor in mouse cortical astrocytes. J Neurosci 28:5473–5480

Landreth GE, Reed-Geaghan EG (2009) Toll-like receptors in Alzheimer's disease. Curr Top Microbiol Immunol 336:137–153. doi:10.1007/978-3-642-00549-7_8

Lee TI, Yang CS, Fang KM, Tzeng SF (2009) Role of ciliary neurotrophic factor in microglial phagocytosis. Neurochem Res 34(1):109–117. doi:10.1007/s11064-008-9682-0

Lehnardt S, Lachance C, Patrizi S, Lefebvre S, Follett PL, Jensen FE, Rosenberg PA, Volpe JJ, Vartanian T (2002) The toll-like receptor TLR4 is necessary for lipopolysaccharide-induced oligodendrocyte injury in the CNS. J Neurosci 22(7):2478–2486. doi:20026268

Lehnardt S, Schott E, Trimbuch T, Laubisch D, Krueger C, Wulczyn G, Nitsch R, Weber JR (2008) A vicious cycle involving release of heat shock protein 60 from injured cells and activation of toll-like receptor 4 mediates neurodegeneration in the CNS. J Neurosci 28 (10):2320–2331. doi:10.1523/JNEUROSCI.4760-07.2008

Lehnardt S, Wennekamp J, Freyer D, Liedtke C, Krueger C, Nitsch R, Bechmann I, Weber JR, Henneke P (2007) TLR2 and caspase-8 are essential for group B Streptococcus-induced apoptosis in microglia. J Immunol 179(9):6134–6143

Letiembre M, Liu Y, Walter S, Hao W, Pfander T, Wrede A, Schulz-Schaeffer W, Fassbender K (2009) Screening of innate immune receptors in neurodegenerative diseases: a similar pattern. Neurobiol Aging 30(5):759–768. doi:10.1016/j.neurobiolaging.2007.08.018

Li Y, Du XF, Liu CS, Wen ZL, Du JL (2012) Reciprocal regulation between resting microglial dynamics and neuronal activity in vivo. Dev Cell 23(6):1189–1202. doi:10.1016/j.devcel.2012.10.027

Liang X, Wang Q, Hand T, Wu L, Breyer RM, Montine TJ, Andreasson K (2005) Deletion of the prostaglandin E2 EP2 receptor reduces oxidative damage and amyloid burden in a model of Alzheimer's disease. J Neurosci 25(44):10180–10187. doi:10.1523/JNEUROSCI.3591-05.2005

Liang KJ, Lee JE, Wang YD, Ma W, Fontainhas AM, Fariss RN, Wong WT (2009) Regulation of dynamic behavior of retinal microglia by CX3CR1 signaling. Investig Ophthalmol Vis Sci 50:4444–4451

Lim SH, Park E, You B, Jung Y, Park AR, Park SG, Lee JR (2013) Neuronal synapse formation induced by microglia and interleukin 10. PLoS One 8:e81218 astrocytes. J Neurosci 28:5473–5480

Linda H, Shupliakov O, Ornung G, Ottersen OP, Storm-Mathisen J, Risling M, Cullheim S (2000) Ultrastructural evidence for a preferential elimination of glutamate-immunoreactive synaptic terminals from spinal motoneurons after intramedullary axotomy. J Comp Neurol 425(1):10–23

Liu Y, Walter S, Stagi M, Cherny D, Letiembre M, Schulz-Schaeffer W, Heine H, Penke B, Neumann H, Fassbender K (2005) LPS receptor (CD14): a receptor for phagocytosis of Alzheimer's amyloid peptide. Brain 128(Pt 8):1778–1789. doi:10.1093/brain/awh531

Loscher CE, Mills KH, Lynch MA (2003) Interleukin-1 receptor antagonist exerts agonist activity in the hippocampus independent of the interleukin-1 type I receptor. J Neuroimmunol 137:117–124

Lotz M, Ebert S, Esselmann H, Iliev AI, Prinz M, Wiazewicz N, Wiltfang J, Gerber J, Nau R (2005) Amyloid beta peptide 1-40 enhances the action of Toll-like receptor-2 and -4 agonists but antagonizes Toll-like receptor-9-induced inflammation in primary mouse microglial cell cultures. J Neurochem 94(2):289–298. doi:10.1111/j.1471-4159.2005.03188.x

Lu C, Hua F, Liu L, Ha T, Kalbfleisch J, Schweitzer J, Kelley J, Kao R, Williams D, Li C (2010) Scavenger receptor class-A has a central role in cerebral ischemia-reperfusion injury. J Cereb Blood Flow Metab Off J Int Soc Cereb Blood Flow Metab 30(12):1972–1981. doi:10.1038/jcbfm.2010.59

Lynch MA (2009) The multifaceted profile of activated microglia. Mol Neurobiol 40(2):139–156. doi:10.1007/s12035-009-8077-9

Marin-Teva JL, Dusart I, Colin C, Gervais A, van Rooijen N, Mallat M (2004) Microglia promote the death of developing Purkinje cells. Neuron 41(4):535–547

Matozaki T, Murata Y, Okazawa H, Ohnishi H (2009) Functions and molecular mechanisms of the CD47-SIRPalpha signalling pathway. Trends Cell Biol 19(2):72–80. doi:10.1016/j.tcb.2008.12.001

Matzinger P (2007) Friendly and dangerous signals: is the tissue in control? Nat Immunol 8(1):11–13. doi:10.1038/ni0107-11

McKimmie CS, Fazakerley JK (2005) In response to pathogens, glial cells dynamically and differentially regulate Toll-like receptor gene expression. J Neuroimmunol 169(1–2):116–125. doi:10.1016/j.jneuroim.2005.08.006

McPherson CA, Kraft AD, Harry GJ (2011) Injury-induced neurogenesis: consideration of resident microglia as supportive of neural progenitor cells. Neurotox Res 19(2):341–352. doi:10.1007/s12640-010-9199-6

Mildner A, Schmidt H, Nitsche M, Merkler D, Hanisch UK, Mack M, Heikenwalder M, Bruck W, Priller J, Prinz M (2007) Microglia in the adult brain arise from Ly-6ChiCCR2+ monocytes only under defined host conditions. Nat Neurosci 10(12):1544–1553. doi:10.1038/nn2015

Mitrasinovic OM, Murphy GM Jr (2002) Accelerated phagocytosis of amyloid-B by mouse and human microglia overexpressing the macrophage colony-stimulating factor receptor. J Biol Chem 277:29889–29896

Mitrasinovic OM, Murphy GM Jr (2003) Microglial overexpression of the M-CSF receptor augments phagocytosis of opsonized Abeta. Neurobiol Aging 24(6):807–815

Moller JC, Klein MA, Haas S, Jones LL, Kreutzberg GW, Raivich G (1996) Regulation of thrombospondin in the regenerating mouse facial motor nucleus. Glia 17(2):121–132. doi:10.1002/(SICI)1098-1136(199606)17:2<121:AID-GLIA4>3.0.CO;2-5

Monje ML, Toda H, Palmer TD (2003) Inflammatory blockade restores adult hippocampal neurogenesis. Science 302(5651):1760–1765. doi:10.1126/science.1088417

Morgan SC, Taylor DL, Pocock JM (2004) Microglia release activators of neuronal proliferation mediated by activation of mitogen-activated protein kinase, phosphatidylinositol-3-kinase/Akt and delta-Notch signalling cascades. J Neurochem 90(1):89–101. doi:10.1111/j.1471-4159.2004.02461.x

Mount MP, Lira A, Grimes D, Smith PD, Faucher S, Slack R, et al (2007) Involvement of interferon-gamma in microglial-mediated loss of dopaminergic neurons. J Neurosci 27:3328–3337

Murgas P, Cornejo FA, Merino G, von Bernhardi R (2014) SR-A regulates the inflammatory activation of astrocytes. Neurotox Res 25(1):68–80. doi:10.1007/s12640-013-9432-1

Murgas P, Godoy B, von Bernhardi R (2012) Abeta potentiates inflammatory activation of glial cells induced by scavenger receptor ligands and inflammatory mediators in culture. Neurotox Res 22(1):69–78. doi:10.1007/s12640-011-9306-3

Nakajima Y, Tohyama T, Kurihara S, Kohsaka S (2005) Axotomy-dependent urokinase induction in the rat facial nucleus: possible stimulation of microglia by neurons. Neurochem Int 46:107–116

Nakajima K, Tohyama Y, Maeda S, Kohsaka S, Kurihara T (2007) Neuronal regulation by which microglia enhance the production of neurotrophic factors for GABAergic, catecholaminergic, and cholinergic neurons. Neurochem Int 50:807–820

Nakanishi M, Niidome T, Matsuda S, Akaike A, Kihara T, Sugimoto H (2007) Microglia-derived interleukin-6 and leukaemia inhibitory factor promote astrocytic differentiation of neural stem/progenitor cells. Eur J Neurosci 25(3):649–658. doi:10.1111/j.1460-9568.2007.05309.x

Napoli I, Neumann H (2009) Microglial clearance function in health and disease. Neuroscience 158(3):1030–1038. doi:10.1016/j.neuroscience.2008.06.046

Nau R, Bruck W (2002) Neuronal injury in bacterial meningitis: mechanisms and implications for therapy. Trends Neurosci 25(1):38–45

Neumann H, Kotter MR, Franklin RJ (2009) Debris clearance by microglia: an essential link between degeneration and regeneration. Brain 132(Pt 2):288–295. doi:10.1093/brain/awn109

Nguyen MD, D'Aigle T, Gowing G, Julien JP, Rivest S (2004) Exacerbation of motor neuron disease by chronic stimulation of innate immunity in a mouse model of amyotrophic lateral sclerosis. J Neurosci 24(6):1340–1349. doi:10.1523/JNEUROSCI.4786-03.2004

Nimmerjahn A, Kirchhoff F, Helmchen F (2005) Resting microglial cells are highly dynamic surveillants of brain parenchyma in vivo. Science 308(5726):1314–1318. doi:10.1126/science.1110647

Ock J, Jeong J, Choi WS, Lee WH, Kim SH, Kim IK, Suk K (2007) Regulation of Toll-like receptor 4 expression and its signaling by hypoxia in cultured microglia. J Neurosci Res 85 (9):1989–1995. doi:10.1002/jnr.21322

Okun E, Griffioen KJ, Lathia JD, Tang SC, Mattson MP, Arumugam TV (2009) Toll-like receptors in neurodegeneration. Brain Res Rev 59(2):278–292. doi:10.1016/j.brainresrev.2008.09.001

Oliveira AL, Thams S, Lidman O, Piehl F, Hökfelt T, Kärre K, Lindå H, Cullheim S (2004) A role for MHC class I molecules in synaptic plasticity and regeneration of neurons after axotomy. Proc Natl Acad Sci USA 101(51):17843–17848

Olson JK, Miller SD (2004) Microglia initiate central nervous system innate and adaptive immune responses through multiple TLRs. J Immunol 173:3916–3924

Orellana JA, Montero TD, von Bernhardi R (2013) Astrocytes inhibit nitric oxide-dependent Ca(2 +) dynamics in activated microglia: involvement of ATP released via pannexin 1 channels. Glia 61(12):2023–2037. doi:10.1002/glia.22573

Ozeki Y, Tsutsui H, Kawada N, Suzuki H, Kataoka M, Kodama T, Yano I, Kaneda K, Kobayashi K (2006) Macrophage scavenger receptor down-regulates mycobacterial cord factor-induced proinflammatory cytokine production by alveolar and hepatic macrophages. Microb Pathog 40(4):171–176. doi:10.1016/j.micpath.2005.12.006

Padovan E, Landmann RM, De Libero G (2007) How pattern recognition receptor triggering influences T cell responses: a new look into the system. Trends Immunol 28(7):308–314. doi:10.1016/j.it.2007.05.002

Paolicelli RC, Bolasco G, Pagani F, Maggi L, Scianni M, Panzanelli P, Giustetto M, Ferreira TA, Guiducci E, Dumas L, Ragozzino D, Gross CT (2011) Synaptic pruning by microglia is necessary for normal brain development. Science 333(6048):1456–1458. doi:10.1126/science.1202529

Paolicelli RC, Gross CT (2011) Microglia in development: linking brain wiring to brain environment. Neuron Glia Biol 7(1):77–83. doi:10.1017/S1740925X12000105

Parkhurst CN, Yang G, Ninan I, Savas JN, Yates JR 3rd, Lafaille JJ, Hempstead BL, Littman DR, Gan WB (2013) Microglia promote learning-dependent synapse formation through brain-derived neurotrophic factor. Cell 155(7):1596–1609. doi:10.1016/j.cell.2013.11.030

Perry VH, Cunningham C, Holmes C (2007) Systemic infections and inflammation affect chronic neurodegeneration. Nat Rev 7(2):161–167. doi:10.1038/nri2015 nri2015[pii]

Perry VH, Holmes C (2014) Microglial priming in neurodegenerative disease. Nature reviews Neurology 10(4):217–224. doi:10.1038/nrneurol.2014.38

Pocock JM, Kettenmann H (2007) Neurotransmitter receptors on microglia. Trends Neurosci 30 (10):527–535. doi:10.1016/j.tins.2007.07.007

Prinz M, Priller J (2010) Tickets to the brain: role of CCR2 and CX3CR1 in myeloid cell entry in the CNS. J Neuroimmunol 224(1–2):80–84. doi:10.1016/j.jneuroim.2010.05.015

Prinz M, Priller J (2014) Microglia and brain macrophages in the molecular age: from origin to neuropsychiatric disease. Nat Rev Neurosci 15(5):300–312. doi:10.1038/nrn3722

Ramírez G, Rey S, von Bernhardi R (2008) Proinflammatory stimuli are needed for induction of microglial cell-mediated AbetaPP_{244-C} and Abeta-neurotoxicity in hippocampal cultures. J Alzheimer Dis 15(1):45–59

Ramirez G, Toro R, Dobeli H, von Bernhardi R (2005) Protection of rat primary hippocampal cultures from A beta cytotoxicity by pro-inflammatory molecules is mediated by astrocytes. Neurobiol Dis 19(1–2):243–254. doi:10.1016/j.nbd.2005.01.007

Rappert A, Bechmann I, Pivneva T, Mahlo J, Biber K, Nolte C, Kovac AD, Gerard C, Boddeke HW, Nitsch R, Kettenmann H (2004) CXCR3-dependent microglial recruitment is essential for dendrite loss after brain lesion. J Neurosci 24(39):8500–8509. doi:10.1523/JNEUROSCI.2451-04.2004

Rappert A, Biber K, Nolte C, Lipp M, Schubel A, Lu B, Gerard NP, Gerard C, Boddeke HW, Kettenmann H (2002) Secondary lymphoid tissue chemokine (CCL21) activates CXCR3 to trigger a Cl- current and chemotaxis in murine microglia. J Immunol 168(7):3221–3226

Reed-Geaghan EG, Savage JC, Hise AG, Landreth GE (2009) CD14 and toll-like receptors 2 and 4 are required for fibrillar A{beta}-stimulated microglial activation. J Neurosci 29(38):11982–11992. doi:10.1523/JNEUROSCI.3158-09.2009 29/38/11982[pii]

Ribeiro Xavier AL, Kress BT, Goldman SA, Lacerda de Menezes JR, Nedergaard M (2015) A distinct population of microglia supports adult neurogenesis in the subventricular zone. J Neurosci 35(34):11848–11861. doi:10.1523/JNEUROSCI.1217-15.2015

Rogers JT, Morganti JM, Bachstetter AD, Hudson CE, Peters MM, Grimmig BA, Weeber EJ, Bickford PC, Gemma C (2011) CX3CR1 deficiency leads to impairment of hippocampal cognitive function and synaptic plasticity. J Neurosci 31:16241–16250

Rolls A, Shechter R, London A, Ziv Y, Ronen A, Levy R, Schwartz M (2007) Toll-like receptors modulate adult hippocampal neurogenesis. Nat Cell Biol 9:1081–1088

Roumier A, Bechade C, Poncer JC, Smalla KH, Tomasello E, Vivier E, Gundelfinger ED, Triller A, Bessis A (2004) Impaired synaptic function in the microglial KARAP/DAP12-deficient mouse. J Neurosci 24:11421–11428

Rubartelli A, Lotze MT (2007) Inside, outside, upside down: damage-associated molecular-pattern molecules (DAMPs) and redox. Trends Immunol 28(10):429–436. doi:10.1016/j.it.2007.08.004

Schafer DP, Lehrman EK, Kautzman AG, Koyama R, Mardinly AR, Yamasaki R, Ransohoff RM, Greenberg ME, Barres BA, Stevens B (2012) Microglia sculpt postnatal neural circuits in an activity and complement-dependent manner. Neuron 74:691–705

Schaffler A, Scholmerich J, Salzberger B (2007) Adipose tissue as an immunological organ: Toll-like receptors, C1q/TNFs and CTRPs. Trends Immunol 28(9):393–399. doi:10.1016/j.it.2007.07.003

Schilling T, Nitsch R, Heinemann U, Haas D, Eder C (2001) Astrocyte-released cytokines induce ramification and outward K+ channel expression in microglia via distinct signalling pathways. Eur J Neurosci 14(3):463–473

Schilling T, Stock C, Schwab A, Eder C (2004) Functional importance of $Ca^{2+}$-activated $K^+$ channels for lysophosphatidic acid-induced microglial migration. Eur J Neurosci 19(6):1469–1474. doi:10.1111/j.1460-9568.2004.03265.x

Schmid AW, Lynch MA, Herron CE (2009) The effects of IL-1 receptor antagonist on beta amyloid mediated depression of LTP in the rat CA1 in vivo. Hippocampus 19:670–676

Schwab A (2001) Ion channels and transporters on the move. News Physiol Sci Int J Physiol Produc Join Int Union Physiol Sci Am Physiol Soc 16:29–33

Schwartz M, Butovsky O, Bruck W, Hanisch UK (2006) Microglial phenotype: is the commitment reversible? Trends Neurosci 29(2):68–74. doi:10.1016/j.tins.2005.12.005

Sedel F, Bechade C, Vyas S, Triller A (2004) Macrophage-derived tumor necrosis factor alpha, an early developmental signal for motoneuron death. J Neurosci 24(9):2236–2246. doi:10.1523/JNEUROSCI.4464-03.2004

Shechter R, London A, Varol C, Raposo C, Cusimano M, Yovel G, Rolls A, Mack M, Pluchino S, Martino G, Jung S, Schwartz M (2009) Infiltrating blood-derived macrophages are vital cells playing an anti-inflammatory role in recovery from spinal cord injury in mice. PLoS Med 6(7):e1000113. doi:10.1371/journal.pmed.1000113

Shie FS, Breyer RM, Montine TJ (2005) Microglia lacking E prostanoid receptor subtype 2 have enhanced Abeta phagocytosis yet lack Abeta-activated neurotoxicity. Am J Pathol 166(4):1163–1172

Shigemoto-Mogami Y, Hoshikawa K, Goldman JE, Sekino Y, Sato K (2014) Microglia enhance neurogenesis and oligodendrogenesis in the early postnatal subventricular zone. J Neurosci 34:2231–2243

Sievers J, Schmidtmayer J, Parwaresch R (1994) Blood monocytes and spleen macrophages differentiate into microglia-like cells when cultured on astrocytes. Ann Anat Anatom Anz Off Organ Anatom Ges 176(1):45–51

Simard AR, Soulet D, Gowing G, Julien JP, Rivest S (2006) Bone marrow-derived microglia play a critical role in restricting senile plaque formation in Alzheimer's disease. Neuron 49(4):489–502. doi:10.1016/j.neuron.2006.01.022

Stence N, Waite M, Dailey ME (2001) Dynamics of microglial activation: a confocal time-lapse analysis in hippocampal slices. Glia 33(3):256–266

Stevens B, Allen NJ, Vazquez LE, Howell GR, Christopherson KS, Nouri N, Micheva KD, Mehalow AK, Huberman AD, Stafford B, Sher A, Litke AM, Lambris JD, Smith SJ, John SW, Barres BA (2007) The classical complement cascade mediates CNS synapse elimination. Cell 131(6):1164–1178. doi:10.1016/j.cell.2007.10.036

Stewart CR, Stuart LM, Wilkinson K, van Gils JM, Deng J, Halle A, Rayner KJ, Boyer L, Zhong R, Frazier WA, Lacy-Hulbert A, El Khoury J, Golenbock DT, Moore KJ (2010) CD36 ligands promote sterile inflammation through assembly of a Toll-like receptor 4 and 6 heterodimer. Nat Immunol 11(2):155–161. doi:10.1038/ni.1836

Streit WJ, Conde JR, Fendrick SE, Flanary BE, Mariani CL (2005) Role of microglia in the central nervous system's immune response. Neurol Res 27(7):685–691. doi:10.1179/016164105X49463

Syed MM, Phulwani NK, Kielian T (2007) Tumor necrosis factor-alpha (TNF-alpha) regulates Toll-like receptor 2 (TLR2) expression in microglia. J Neurochem 103(4):1461–1471. doi:10.1111/j.1471-4159.2007.04838.x

Tahara K, Kim HD, Jin JJ, Maxwell JA, Li L, Fukuchi K (2006) Role of toll-like receptor signalling in Abeta uptake and clearance. Brain 129(Pt 11):3006–3019. doi:10.1093/brain/awl249

Tang G, Gudsnuk K, Kuo SH, Cotrina ML, Rosoklija G, Sosunov A, Sonders MS, Kanter E, Castagna C, Yamamoto A, Yue Z, Arancio O, Peterson BS, Champagne F, Dwork AJ, Goldman J, Sulzer D (2014) Loss of mTOR-dependent macroautophagy causes autistic-like synaptic pruning deficits. Neuron 83(5):1131–1143. doi:10.1016/j.neuron.2014.07.040

Tichauer J, Saud K, von Bernhardi R (2007) Modulation by astrocytes of microglial cell-mediated neuroinflammation: effect on the activation of microglial signaling pathways. NeuroImmunoModulation 14(3–4):168–174

Tichauer JE, Flores B, Soler B, Eugenin-von Bernhardi L, Ramirez G, von Bernhardi R (2014) Age-dependent changes on TGFbeta1 Smad3 pathway modify the pattern of microglial cell activation. Brain Behav Immun 37:187–196. doi:10.1016/j.bbi.2013.12.018

Tichauer JE, von Bernhardi R (2012) Transforming growth factor-beta stimulates beta amyloid uptake by microglia through Smad3-dependent mechanisms. J Neurosci Res 90(10):1970–1980. doi:10.1002/jnr.23082

Trapp BD, Wujek JR, Criste GA, Jalabi W, Yin X, Kidd GJ, Stohlman S, Ransohoff R (2007) Evidence for synaptic stripping by cortical microglia. Glia 55(4):360–368. doi:10.1002/glia.20462

Trebst C, Konig F, Ransohoff R, Bruck W, Stangel M (2008) CCR5 expression on macrophages/microglia is associated with early remyelination in multiple sclerosis lesions. Mult Scler 14(6):728–733. doi:10.1177/1352458508089359

Umemori H, Sanes JR (2008) Signal regulatory proteins (SIRPS) are secreted presynaptic organizing molecules. J Biol Chem 283(49):34053–34061. doi:10.1074/jbc.M805729200

van Rossum D, Hanisch UK (2004) Microglia. Metab Brain Dis 19(3–4):393–411

van Weering HR, Boddeke HW, Vinet J, Brouwer N, de Haas AH, van Rooijen N, Thomsen AR, Biber KP (2011) CXCL10/CXCR3 signaling in glia cells differentially affects NMDA-induced cell death in CA and DG neurons of the mouse hippocampus. Hippocampus 21(2):220–232. doi:10.1002/hipo.20742

von Bernhardi R (2007) Glial cell dysregulation: a new perspective on Alzheimer disease. Neurotox Res 12(4):215–232

von Bernhardi R, Cornejo F, Parada GE, Eugenin J (2015a) Role of TGFbeta signaling in the pathogenesis of Alzheimer's disease. Front Cell Neurosci 9:426. doi:10.3389/fncel.2015.00426

von Bernhardi R, Eugenin-von Bernhardi L, Eugenin J (2015b) Microglial cell dysregulation in brain aging and neurodegeneration. Front Aging Neurosci 7:124. doi:10.3389/fnagi.2015.00124 von Bernhardi R, Eugenin J (2004) Microglial reactivity to beta-amyloid is modulated by astrocytes and proinflammatory factors. Brain Res 1–2:186–193. doi:10.1016/j.brainres.2004.07.084 (1025)

von Bernhardi R, Eugenin J (2012) Alzheimer's disease: redox dysregulation as a common denominator for diverse pathogenic mechanisms. Antioxid Redox Signal 16(9):974–1031. doi:10.1089/ars.2011.4082

von Bernhardi R, Ramírez G (2001) Microglia-astrocyte interaction in Alzheimer's disease: friends or foes for the nervous system? Biol Res 34(2):123–128

von Bernhardi R, Tichauer JE, Eugenin J (2010) Aging-dependent changes of microglial cells and their relevance for neurodegenerative disorders. J Neurochem 112(5):1099–1114. doi:10.1111/j.1471-4159.2009.06537.x

Wake H, Moorhouse AJ, Jinno S, Kohsaka S, Nabekura J (2009) Resting microglia directly monitor the functional state of synapses in vivo and determine the fate of ischemic terminals. J Neurosci 29(13):3974–3980. doi:10.1523/JNEUROSCI.4363-08.2009

Walter L, Franklin A, Witting A, Wade C, Xie Y, Kunos G, Mackie K, Stella N (2003) Nonpsychotropic cannabinoid receptors regulate microglial cell migration. J Neurosci 23 (4):1398–1405

Walter S, Letiembre M, Liu Y, Heine H, Penke B, Hao W, Bode B, Manietta N, Walter J, Schulz-Schuffer W, Fassbender K (2007) Role of the toll-like receptor 4 in neuroinflammation in Alzheimer's disease. Cell Physiol Biochem Int J Exp Cell Physiol Biochem Pharmacol 20 (6):947–956. doi:10.1159/000110455

Walton NM, Sutter BM, Laywell ED, Levkoff LH, Kearns SM, Marshall GP 2nd, Scheffler B, Steindler DA (2006) Microglia instruct subventricular zone neurogenesis. Glia 54(8):815–825. doi:10.1002/glia.20419

Webster SD, Yang AJ, Margol L, Garzon-Rodriguez W, Glabe CG, Tenner AJ (2000) Complement component C1q modulates the phagocytosis of Abeta by microglia. Exp Neurol 161(1):127–138. doi:10.1006/exnr.1999.7260

Witting A, Muller P, Herrmann A, Kettenmann H, Nolte C (2000) Phagocytic clearance of apoptotic neurons by Microglia/Brain macrophages in vitro: involvement of lectin-, integrin-, and phosphatidylserine-mediated recognition. J Neurochem 75(3):1060–1070

Yirmiya R, Rimmerman N, Reshef R (2015) Depression as a microglial disease. Trends Neurosci 38(10):637–658. doi:10.1016/j.tins.2015.08.001

Zhan Y, Paolicelli RC, Sforazzini F, Weinhard L, Bolasco G, Pagani F, Vyssotski AL, Bifone A, Gozzi A, Ragozzino D, Gross CT (2014) Deficient neuron-microglia signaling results in impaired functional brain connectivity and social behavior. Nat Neurosci 17(3):400–406. doi:10.1038/nn.3641

Ziegler G, Harhausen D, Schepers C, Hoffmann O, Rohr C, Prinz V, Konig J, Lehrach H, Nietfeld W, Trendelenburg G (2007) TLR2 has a detrimental role in mouse transient focal cerebral ischemia. Biochem Biophys Res Commun 359(3):574–579. doi:10.1016/j.bbrc.2007.05.157

Ziv Y, Ron N, Butovsky O, Landa G, Sudai E, Greenberg N, Cohen H, Kipnis J, Schwartz M (2006) Immune cells contribute to the maintenance of neurogenesis and spatial learning abilities in adulthood. Nat Neurosci 9(2):268–275. doi:10.1038/nn1629

Ziv Y, Schwartz M (2008) Immune-based regulation of adult neurogenesis: implications for learning and memory. Brain Behav Immun 22(2):167–176. doi:10.1016/j.bbi.2007.08.006

# Physiological Functions of Glial Cell Hemichannels

**Juan A. Orellana**

**Abstract** The brain performs exceptionally complex and dynamic tasks that depend on the coordinated interaction of neurons, glial cells, endothelial cells, pericytes, smooth muscle cells, ependymal cells, and circulating blood cells. Among these cells, glial cells have emerged as crucial protagonists in the regulation of synaptic transmission and neural function. Indeed, these cells express a wide range of receptors that enable them to sense changes in neuronal activity and the microenvironment by responding locally via the release of bioactive molecules known as gliotransmitters. In the central nervous system (CNS), a novel mechanism that allows gliotransmission via the opening of hemichannels has been proposed. These channels are composed of six protein subunits consisting of connexins or pannexins, which are two highly conserved protein families that are encoded by 21 and 3 genes, respectively, in humans. Typically, glial cell hemichannels exhibit low levels of activity, but this activity is sufficient to ensure the release of a broad spectrum of gliotransmitters, including ATP, D-serine, glutamate, adenosine, and glutathione. Here, we briefly review the current findings regarding the effects of the hemichannel-dependent release of gliotransmitters on the physiology of the CNS.

**Keywords** Connexins · Pannexins · (ATP) adenosine triphosphate · Astroglial signaling · Gliotransmitters

### Abbreviations and acronyms

| | |
|---|---|
| Arc | Arcuate |
| ATP | Adenosine triphosphate |
| BLA | Basolateral amygdala |
| $[Ca^{2+}]_i$ | Intracellular free $Ca^{2+}$ concentration |
| CSF | Cerebrospinal fluid |
| CNS | Central nervous system |
| Cx26 | Connexin26 |

J.A. Orellana (✉)
Departamento de Neurología, Escuela de Medicina, Pontificia Universidad Católica de Chile, Marcoleta #391, Santiago, Chile
e-mail: jaorella@uc.cl



R. von Bernhardi (ed.), *Glial Cells in Health and Disease of the CNS*, Advances in Experimental Medicine and Biology 949,
DOI 10.1007/978-3-319-40764-7_5

| | |
|---|---|
| Cx36 | Connexin36 |
| Cx43 | Connexin43 |
| GFAP | Glial fibrillary acidic protein |
| GJCs | Gap junction channels |
| kDa | Kilodalton |
| KO | Knockout |
| $NAD^+$ | Nicotinamide adenine dinucleotide |
| NMDA | N-methyl-D-aspartate |
| MAP2 | Microtubule-associated protein 2 |
| MBH | Mediobasal hypothalamus |
| MI | Metabolic inhibition |
| Panx1 | Pannexin1 |
| Panx2 | Pannexin2 |
| $PGE_2$ | Prostaglandin E2 |
| RTN | Retrotrapezoid nucleus |
| siRNA | Small interfering ribonucleic acid |
| VMH | Ventromedial hypothalamic nuclei |
| VMS | Ventral medullary surface |
| TDCS | Transcranial direct current stimulation |

## Introduction

To achieve a coordinated response to an external stimulus, many phyla have developed complex and integrative neural structures that progressively enabled the sophisticated analysis of information. Indeed, as the central nervous system (CNS) evolved from a basic netlike structure to compacted ganglia and centralized brains, a new cell type with nonneuronal features emerged in several organisms, that is the glia (Barres 2008). Possibly driven by positive Darwinian selection, glial cells have persisted in high densities and have acquired greater diversity in high mammals. Indeed, they are the most abundant brain cell population—10 times more frequent than neurons—and constitute approximately 50 % of the total cell mass of the CNS (Verkhratsky and Toescu 2006). There are two major classifications of the glial family, macroglia (oligodendrocytes, astrocytes, NG2 cells, and ependymoglial cells) and microglia, which belong to neuroectodermal and mesenchymal origin, respectively.

Although often overlooked for much of the twentieth century when discussing brain function, in the last two decades, glial cells have emerged as crucial protagonists in the processing of highly complex information. This is particularly true for astrocytes, which constitute a far-reaching syncytial network that anatomically and functionally connects neuronal synapses with brain blood vessels (Volterra and Meldolesi 2005). Through its processes, each astrocyte may contact a numbers of neuronal chemical synapses that range from thousands to millions depending on the mammalian species (Oberheim et al. 2006). Thus, in a delicate physical and functional interaction, astroglial processes, together with pre- and postsynaptic neuronal structures, comprise the “tripartite synapse” (see Chap. 1). Embedded in this structure, astrocytes sense neuronal activity and respond locally through the release of bioactive molecules termed “gliotransmitters” (e.g., glutamate, adenosine triphosphate [ATP] and D-serine) (Perea et al. 2009). In addition to surrounding the synaptic cleft, astrocytes project specialized terminal processes, which are known as “endfeet”, toward capillaries, intracerebral asterioles, and venules and thus cover 99 % of the abluminal vascular surface (Simard et al. 2003). This resulting communication with neurons and vascular cells provides astrocytes with an incomparable architectural advantage that facilitates the local and long-distance release of gliotransmitters and vasoactive factors and thereby modulates different neuronal circuits and networks.

Microglia are not far behind astrocytes regarding the regulation of neuronal chemical synapses. In addition to their well-known roles on brain immunity and inflammatory responses, microglia are now recognized as essential players in the integration and consolidation of neuronal circuits. Various studies have shown that microglia constantly extend toward and retract from synapses, participating in broad range of previously undiscovered functions, including neuronal surveillance, synapse elimination, regulation of cell death, and others (Schafer et al. 2013; Tremblay et al. 2010; Wake et al. 2013). Indeed, some authors have proposed a shift of the current concept of tripartite synapse to a “quad-partite synapse” (Schafer et al. 2013). Interestingly, neurotransmitter release by neurons can modify various aspects of glial cell function, including cellular migration, phagocytosis, intercellular $Ca^{2+}$ wave generation, metabolic coupling, blood flow control, gliotransmitter release, and others (Fields and Burnstock 2006; Fields and Stevens-Graham 2002; Inoue et al. 2007). This reciprocal modulation closes a permanent feedback loop of interactions between neurons and glial cells denominated “neuron-glia crosstalk”.

As previously discussed in Chap. 1, gliotransmission is part of the basis of neuron–glia crosstalk. Multiple mechanisms of the mediation of gliotransmitter release have been described, including $Ca^{2+}$-dependent exocytosis (Bezzi et al. 2004; Imura et al. 2013; Zhang et al. 2004), carrier membrane transport (Rossi et al. 2000) and the opening of a wide range of channels, such as $P2X_7$ channels (Duan et al. 2003; Hamilton et al. 2008; Suadicani et al. 2006), volume-regulated anion channels (Kimelberg et al. 1990; Lee et al. 2010; Rudkouskaya et al. 2008; Takano et al. 2005) and connexin hemichannels (Fig. 1); (Iglesias et al. 2009; Stout et al. 2002; Ye et al. 2003). Each hemichannel is composed of six protein subunits termed connexins. The latter belong to a highly conserved protein family encoded by 21 genes in humans and 20 in mice, with orthologs in other vertebrate species (Fig. 2); (Abascal and Zardoya 2013). Connexins are abundantly expressed in cells of the CNS (Orellana et al. 2009), and they are named after their predicted molecular mass expressed in kilodaltons (kDa), for example, connexin43 (Cx43) has a molecular mass of $\sim$43 kDa.

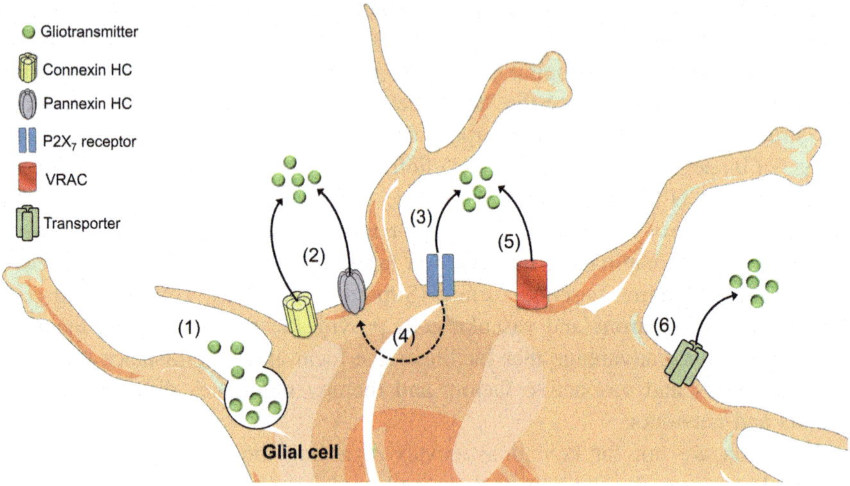


**Fig. 1** Mechanisms of gliotransmitter release. Glial cells release gliotransmitters (e.g., glutamate, D-serine and ATP) through $Ca^{2+}$- and SNARE-dependent exocytosis (*1*) in addition to the release that occurs through alternative non-exocytotic pathways (2–8). Depolarization, reductions in extracellular divalent cation concentrations, increases in intracellular $Ca^{2+}$ and posttranslational modifications might result in the opening of connexin and pannexin hemichannels (HCs) and thus allow the release of gliotransmitters (*2*). Long-lasting activation of $P2X_7$ by ATP might lead to the appearance of large currents and the rapid exchange of large molecules, including the release of gliotransmitters (*3*). One theory states that $P2X_7$ receptor conductance dilates over the time and thereby facilitates the passage of large molecules; however, another theory states that ATP activates a second non-selective permeabilization pathway (Baroja-Mazo et al. 2012). Recently, it was shown that Panx1 hemichannels might mediate this permeability for large molecules in astrocytes (*4*) (Iglesias et al. 2009). Additionally, gliotransmitter release may occur through volume-regulated anion channels (VRAC) (*5*) and different carriers and/or co-transporters acting normally or in reverse (*6*) (e.g., excitatory amino acid transporters, the cystine-glutamate antiporter and the D-serine/chloride co-transporter)

For many years, the main function attributed to connexin hemichannels was providing the building blocks of gap junction channels (GJCs), which allow direct but selective cytoplasmic continuity and molecular exchange between contacting cells (Saez et al. 2003); (Fig. 2). Nonetheless, in the last decade, the presence of functional connexin hemichannels in "nonjunctional" membranes has been demonstrated by several experimental approaches (Saez and Leybaert 2014). These channels act like aqueous pores that are permeable to ions and small molecules and thus provide a diffusional route of exchange between the intra- and extracellular milieu (Saez et al. 2010; Wang et al. 2013). Therefore, these channels allow the cellular release of relevant quantities of autocrine and paracrine signaling molecules (e.g., ATP, glutamate, nicotinamide adenine dinucleotide [$NAD^+$] and prostaglandin E2 [$PGE_2$]) to the extracellular milieu (Saez et al. 2010).

Recently, another gene family encoding a set of three membrane proteins termed pannexins was identified (Bruzzone et al. 2003); (Fig. 2). Currently, the majority of

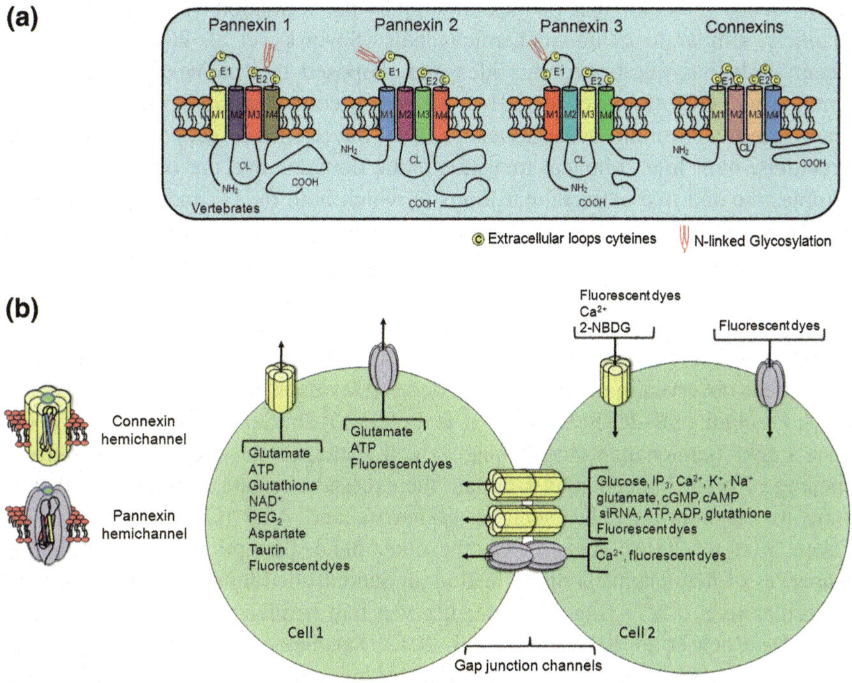


**Fig. 2** Basic structure and function of connexin- and pannexin-based channels. **a** Pannexins and connexins share a similar membrane topology that includes four α-helical transmembrane domains (M1-M4) connected by two extracellular loops (E1 and E2) and one cytoplasmic loop (CL) in which both the amino (NH2)- and carboxy (COOH)-termini are intracellular. The relative positions of the extracellular loop cysteines and glycosylated asparagines are also shown. **b** Vertebrate hemichannels are formed from six connexins or pannexins consisting of six subunits each. Recently, a band pattern that is more consistent with an octamer than with a hexamer was observed for Panx2 in cross-linking studies and native gels of purified homomeric full-length and C-terminal truncation mutants (Ambrosi et al. 2010). Each hemichannel is formed by connexins or pannexins that oligomerize laterally, resulting in a central pore in the activated state. Under resting conditions, the hemichannels preferentially remain closed, but they may be activated by diverse physiological and pathological conditions to provide a diffuse transmembrane route between the intra- and extracellular milieu. The middle center of the schematic shows connexin gap junction channels (GJCs) in close contact between two cells. These channels permit the intercellular exchange of metabolites, second messengers and ions and allow for the transmission of intercellular $Ca^{2+}$ waves, the spread of electrotonic potentials, local blood flow regulation and spatial buffering of ions and metabolites. Although most evidence indicates that pannexins do not form GJCs, one recent study suggested that Panx1 and Panx3 can form GJCs with pharmacological and permeability properties that are distinct from those of channels comprised of Cx43 (Sahu et al. 2014)

the evidence indicates that pannexins support the formation of single membrane channels, similar to connexin hemichannels (Sosinsky et al. 2011); however, a recent study has questioned this idea and proposed that pannexin channels also function as GJCs (Sahu et al. 2014). Although connexins and pannexins do not share significant amino acid sequences, they have similar secondary and tertiary structures, with four α-helical transmembrane domains that are connected by one cytoplasmic and two extracellular loops in which both the N- and C-termini are in the intracellular compartment. Unlike the other pannexins and connexins that oligomerize into hexameric hemichannels, pannexin2 (Panx2) seems to form octamers (Ambrosi et al. 2010).

It is broadly accepted that impairments of the permeability properties of hemichannels might be critical to the initiation and maintenance of the homeostatic imbalances observed in diverse brain diseases (Davidson et al. 2013; De Vuyst et al. 2011; Fasciani et al. 2013; Orellana et al. 2009, 2012b; Salameh et al. 2013, 2009). How might hemichannels contribute to cell damage? At one end, uncontrolled opening of hemichannels may facilitate the excessive release of molecules that are toxic to neighboring cells, such as glutamate and ATP (Orellana et al. 2011; Takeuchi et al. 2006); (Fig. 3). On the other hand, disruptions of the regulatory properties of hemichannels might lead to an uncontrolled influx of potentially toxic molecules (e.g., $Ca^{2+}$) (Fig. 3). Indeed, given that hemichannels are permeable to $Ca^{2+}$ (De Bock et al. 2012; Fiori et al. 2012; Sanchez et al. 2009; Schalper et al. 2010) and that their opening is controlled by intracellular free $Ca^{2+}$ concentration ($[Ca^{2+}]_i$) (De Vuyst et al. 2006, 2009; Ponsaerts et al. 2010), it is possible that hemichannel activity in concert with N-methyl-D-aspartate [NMDA] or P2 receptor activation might result in $Ca^{2+}$ overload and the subsequent generation of free radicals, lipid peroxidation, and plasma membrane damage (Fig. 3). Another important mechanism of cell damage that is triggered by hemichannel opening could rely on the osmotic and ionic imbalances induced by the uncontrolled influx of $Na^{2+}$ and $Cl^{-}$, which would result in further cell swelling and plasma membrane breakdown (Islam et al. 2012; Paul et al. 1991); (Fig. 3).

Currently, the majority of the evidence supports the notion that hemichannels play important roles in physiological brain functions, including ischemic tolerance (Lin et al. 2008; Schock et al. 2008), establishment of adhesive interactions (Cotrina et al. 2008), fear memory consolidation (Stehberg et al. 2012), synaptic transmission (Chever et al. 2014; Klaassen et al. 2011; Prochnow et al. 2012), spontaneous electrical activity (Moore et al. 2014), glucose sensing (Orellana et al. 2012a), chemoreception (Wenker et al. 2012), blood–brain barrier permeability (De Bock et al. 2011; Kaneko et al. 2015), neuronal migration (Liu et al. 2010, 2012), and metabolic autocrine regulation (Kawamura et al. 2010). Given that previous articles have extensively reviewed the role of hemichannels under pathological conditions (Bosch and Kielian 2014; Davidson et al. 2013; Shestopalov and Slepak 2014), we focus here on presenting a brief collection of the most recent findings regarding the physiological implications of hemichannels in the CNS.

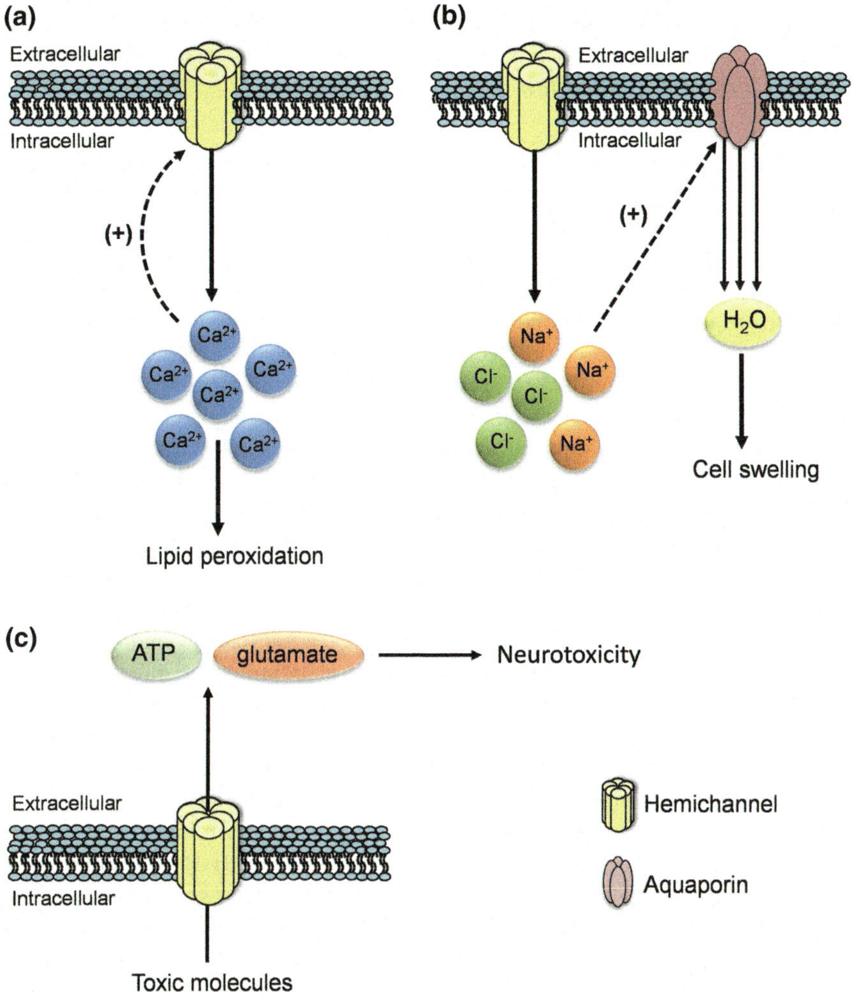


**Fig. 3** Hemichannel opening is associated with cell death. Under normal conditions, hemichannels (*yellow* channels) exhibit low levels of activity that are sufficient to ensure the release of several paracrine molecules. However, upon exposure to pathological stimuli, the hemichannels undergo dysregulation and an exacerbated opening that results in cellular damage via the following different mechanisms: **a** $Ca^{2+}$ entry via hemichannels may activate phospholipase $A_2$ and subsequently induce the generation of arachidonic acid and the activation of the cyclooxygenase/lipoxygenase pathways, which further leads to increased levels of free radicals, lipid peroxidation and plasma membrane damage; **b** $Na^{2+}$ and $Cl^{-}$ entry via hemichannels triggers cellular swelling due to an increased influx of $H_2O$ via aquaporins (*red* channels); and **c** the release of large amounts of transmitters via hemichannels (e.g., glutamate and ATP) might reduce the viability of healthy neighboring cells

## Hemichannel Function in the Physiology of the CNS

### *Brain Ischemic Tolerance*

Although molecular mechanisms of ischemic tolerance have been widely investigated (Gidday 2006), most of these studies have focused on neurons. Nevertheless, recent findings suggest that astroglial signaling via hemichannels may be relevant to this process. Typically, brief episodes of sublethal ischemia known as preconditioning induce resistance to subsequent more severe insults (Gidday 2006). This phenomenon is often called ischemic tolerance and represent a physiological process that provides robust neuroprotection and remains as a key medical challenge to the understanding of neuronal repair after brain injury. Pioneering studies by Lin and colleagues showed that preconditioning reduces the degradation of Cx43 in astrocytes, leading to a prominent increase in surface levels of Cx43 hemichannels (Lin et al. 2008). Underscoring the involvement of hemichannels in preconditioning responses, Cx43 null mice have been found to be insensitive to hypoxic preconditioning, whereas wild-type littermate mice exhibit prominent reductions in infarct volume after the induction of preconditioning via the occlusion of the middle cerebral artery (Lin et al. 2008). The implicated mechanism involves the release of ATP through astroglial Cx43 hemichannels and the subsequent hydrolyzation of this ATP to adenosine, which results in a potent neuroprotective effect.

In the same year of the above study, it was proposed that connexin36 (Cx36) hemichannels could participate in the ischemic tolerance phenomenon (Schock et al. 2008). This study showed that prior depolarization with KCl reduces the neuronal death triggered by metabolic inhibition (MI) in an in vitro model of ischemia that is characterized by the suppression of anaerobic glycolysis and mitochondrial oxidative phosphorylation (Schock et al. 2008). Prior depolarization with KCl triggers a Cx36 hemichannel opening in cerebellar granule neurons that underpins the release of ATP, and the action of ATP on P2Y receptors provides a potent neuroprotective effect against MI (Schock et al. 2008). Interestingly, pharmacological blockade or downregulation of Cx36 reduces the KCl-induced ischemic tolerance and ATP release and thus reveals a crucial role of Cx36 hemichannels in this response. Although Panx1 hemichannels have been linked to ischemic tolerance in the heart and to the release of cardioprotectants (Vessey et al. 2010, 2011), their involvement in brain preconditioning responses remains unknown.

### *Brain Chemoreception*

When the nervous system detects changes in the peripheral pH (or $CO_2$), it triggers a coordinated response that modulates breathing control to maintain arterial $CO_2$ levels. This phenomenon is typically known as central chemoreception and involves the activation of acid-sensitive neurons via several as-yet unidentified acid-sensitive ion channels in different hindbrain sites, including the retrotrapezoid nucleus (RTN), medullary raphe and locus coeruleus (Funk 2010) (see Chap. 6). A few years ago, a crucial role of glial cell hemichannels in the central control of breathing was uncovered. Findings from Huckstepp and colleagues demonstrated that Cx26 hemichannels participate in the $CO_2$-dependent release of ATP in the ventral medullary surface (VMS) (Huckstepp et al. 2010b). Using ex vivo dye uptake measurements, these authors observed that a glial fibrillary acidic protein (GFAP) positive population but not a microtubule-associated protein 2 (MAP2) positive population exhibited a $Pco_2$-dependent increase in hemichannel activity in the VMS and that this increase was correlated with high levels of connexin26 (Cx26) immunoreactivity. Indeed, the $CO_2$-dependent dye uptake occurred in the subpial astrocytes and the astrocytes that ensheath the penetrating blood vessels; all of these cells exhibited prominent Cx26 expression. Interestingly, hemichannel blockers diminished the $CO_2$-dependent increase in the breathing response, as measured with phrenic nerve discharge recordings (Huckstepp et al. 2010b). These data suggest that a $CO_2$-sensitive mechanism that is linked to Cx26 hemichannel-dependent release of ATP may play an important role in central respiratory chemosensitivity. In agreement with this idea, recent in vitro experiments in HeLa cells showed that Cx26 transfection is sufficient to confer the ability to release ATP and exhibit hemichannel current events upon $CO_2$ stimulation to these cells (Huckstepp et al. 2010a). Further studies found that a carbamate bridge between Lys125 and Arg104 might act as a $CO_2$ sensor in Cx26 (Meigh et al. 2013). Notwithstanding this evidence, further pharmacological (e.g., mimetic peptides) and molecular [e.g., tissue-specific inducible knockouts (KO)] studies are required to completely elucidate the contributions of glial Cx26 hemichannels to central chemoreception and breathing control.

## *Brain Glucose Sensing*

The hypothalamus controls energy homeostasis and feeding/satiety behaviors based on its ability to sense peripheral glucose levels. A plethora of recent studies have focused on the unexpected role of glial cells in these processes (Tonon et al. 2013). The majority of these studies have focused on astrocytes (McDougal et al. 2013), and the involvement of other glial cells in hypothalamic glucose sensing has received little attention. Tanycytes are among the glial cells that have recently been proposed to mediate the above-mentioned phenomena (Langlet 2014). These cells are ependymoglial cells that are localized in the III–V ventricle of the basal hypothalamus, and their cellular processes contact the cerebrospinal fluid (CSF) and neurons from regions involved in energy balance, such as arcuate (Arc) and ventromedial hypothalamic nuclei (VMH) (Rodriguez et al. 2005). Pioneering studies from Frayling and colleagues (Frayling et al. 2011) showed that non-metabolizable analogues of glucose increase $[Ca^{2+}]_i$ in hypothalamic tanycytes, specifically those located in the most dorsal regions of the hypothalamus. Moreover, this phenomenon is dependent on P2Y1R activation, indicating the involvement of ATP or its metabolites (Frayling et al. 2011). Supporting this idea, it was recently reported that glucose stimulation evokes the release of ATP in tanycytes, which triggers changes in $[Ca^{2+}]_i$ that depend on a glucose sensing mechanism similar to that observed in β-pancreatic cells (Orellana et al. 2012a). Ethidium uptake and whole-cell patch clamp recordings have been used to demonstrate that acute glucose stimulation triggers the opening of Cx43 hemichannels, which results in the further release of ATP and autocrine activation of $P2Y_1Rs$ (Orellana et al. 2012a). In agreement with the notion that Cx43 participates in glucose sensing and feeding behaviors in the brain, recent in vivo experiments have shown that its expression in the mediobasal hypothalamus (MBH) is decreased by fasting and is augmented by glucose injection-induced hyperglycemia (Allard et al. 2014). Interestingly, small interfering ribonucleic acid (siRNA) injection-induced downregulation of Cx43 in the MBH results in an impaired insulin response in rats that were subjected to an intracarotid glucose load (Allard et al. 2014). Further in vivo experiments are required to determine how Cx43 hemichannels regulate the insulin response to peripheral glucose and the mechanisms involved in tanycyte-hypothalamic neuron communication.

## *Cognition and Behavior*

The majority of cognitive and behavioral studies have centrally positioned neurons as the primary cells that are responsible for higher brain processing. Nonetheless, evidence strongly suggests that glial cells might crucially participate in cognitive tasks and behavior and even contribute to different psychiatric disorders (Banasr et al. 2010; Perea et al. 2009). Recently, Stehberg et al. (2012) determined that the release of gliotransmitters via Cx43 hemichannels is critical for fear memory consolidation in the basolateral amygdala (BLA), which is a brain region that is essential for anxiety and emotional memory processing. These authors microinjected the BLA with TAT-L2, a peptide that specifically inhibits Cx43 hemichannels without affecting gap junctional communication among neighboring astrocytes (Ponsaerts et al. 2010). After 24 h of an auditory fear conditioning protocol, the rats treated with TAT-L2 peptide exhibited impaired fear memory consolidation and complete amnesia toward the auditory fear conditioning training (Stehberg et al. 2012). These responses were transitory and did not influence short-term memory, locomotion, or shock reactivity. Importantly, the TAT-L2-induced inhibition of fear memory consolidation was blunted by co-injection of TAT-L2 with an exogenous cocktail of the following gliotransmitters: glutamate, glutamine, lactate, D-serine, glycine, and ATP (Stehberg et al. 2012).

A few studies have linked pannexins with a physiological role in cognition and behavior. For example, Prochnow and colleagues recently showed that pannexin1 $(Panx1)^{-/-}$ mice exhibit cognitive alterations that include exacerbated anxiety behaviors, deficient object recognition, and impaired spatial learning memory (Prochnow et al. 2012). Followup work in Panx1$^{-/-}$ mice employing cookie-finding tests, revealed that their reduced ability to detect odorant cues is possibly linked to impairments in learning capabilities, because the processing of olfactory information appeared to be unaffected in these animals (Kurtenbach et al. 2014). Despite these observations, Panx1$^{-/-}$/Panx2$^{-/-}$ mice do not exhibit changes in anxiety or exploratory behaviors compared with their wild-type littermates (Bargiotas et al. 2012). Heterologous expression of Panx2 in Xenopus oocytes does not produce functional hemichannels (Bruzzone et al. 2001), but neurons seem to express these channels even in the absence of Panx1 (Bargiotas et al. 2011). The latter finding has motivated speculation that, as a consequence of posttranslational modifications or dimerization with other partners, Panx2 can form hemichannels in mammalian cells in the absence of Panx1 (Bargiotas et al. 2011). Whether compensatory regulations between Panx1 and Panx2 account for the differences in the behaviors of Panx1$^{-/-}$ and Panx1$^{-/-}$/Panx2$^{-/-}$ mice is a subject for future investigations, and the use of cell tissue-specific KO of Panx1 and/or Panx2 (e.g., neurons versus astrocytes) will be required for such studies.

Pannexins have been implicated in alterations in cognition and behavior during brain recovery after injury. Indeed, Panx1 hemichannel opening worsens seizures and behavioral manifestations of status epilepticus in vivo (Santiago et al. 2011). Santiago and colleagues determined that Panx1 hemichannels underpin membrane permeabilization and ATP release during kainic acid-induced seizures. Relevantly, pharmacological (mefloquine) and genetic approaches (Panx1$^{-/-}$ mice) have been used to elucidate the contribution of Panx1 hemichannels to the amelioration of seizure outcomes (Santiago et al. 2011). These data agreed with the protective effects of pharmacological Panx1 hemichannel blockade in a pilocarpine-induced seizure model (Kim and Kang 2011). The beneficial effect observed in Panx1$^{-/-}$ mice may result from reduced ATP release, which might limit the activation of excitatory P2X receptors and thus the progression of status epilepticus. Others researchers have proposed that reduced Panx1 function might prevent the total loss of cellular ATP and thus impair recovery from prolonged seizures in animals (Santiago et al. 2011). Accordingly, the inhibition of Panx1 hemichannels with probenecid ameliorates the spatial learning deficits of aged rats (Mawhinney et al. 2011), whereas transcranial direct current stimulation (TDCS), which is a therapeutic approach for several neurological disorders (Williams et al. 2009), decreases Panx1 expression and augments spine density following brain ischemia (Jiang et al. 2011). Moreover, recent observations have shown that Panx1$^{-/-}$/Panx2$^{-/-}$ mice exhibit improved functional outcomes and smaller infarcts than wild-type mice when subjected to ischemic stroke (Bargiotas et al. 2011). Followup work revealed that Panx1$^{-/-}$/Panx2$^{-/-}$ mice exhibit reductions in impairments in parameters such as exploration, anxiety, sensorimotor function, and behavioral symmetry compared with their wild-type littermates (Bargiotas et al. 2012). Because some evidence indicates that Panx1 likely forms GJCs (Sahu et al. 2014), future efforts to uncover the roles of pannexins in behavior need to consider that the molecular ablation of Panx1 might affect the functions of both channels.

## Concluding Remarks

The initial characterization of non-junctional hemichannel currents involved large depolarizations and cell lysis; thus, hemichannel opening was considered to be incompatible with normal cell life (Paul et al. 1991). Despite these early beliefs, current evidence supports the notion that hemichannels seem to be active under physiological conditions. Apparently, these channels are less open in normal than in pathological states, but they are sufficiently open to ensure paracrine communication in different tissues, including the CNS. Do the permeability properties of hemichannels remain unaltered during inflammation? How do changes in the permeabilities of hemichannels to $Ca^{2+}$ and different gliotransmitters influence brain diseases? Which posttranslational modifications are responsible of these changes? These are some of the puzzling questions that future studies should attempt to address. Characterization of the primary elements that specifically regulate hemichannel function in physiological and pathophysiological conditions will enable the identification of future therapies for neurological disorders.

**Acknowledgments** This work was partially supported by the Fondo Nacional de Desarrollo Científico y Tecnológico (FONDECYT) Grant 11121133 (to JAO), 1160710 (to JAO), the Comisión Nacional de Investigación Científica y Tecnológica (CONICYT) and Programa de Investigación Asociativa (PIA) Grant Anillo de Ciencia y Tecnología ACT1411 (to JAO). We apologize to the authors and groups whose work we did not cite due to space limitations.

## References

Abascal F, Zardoya R (2013) Evolutionary analyses of gap junction protein families. Biochem Biophys Acta 1828:4–14

Allard C, Carneiro L, Grall S, Cline BH, Fioramonti X, Chrétien C, Baba-Aissa F, Giaume C, Pénicaud L, Leloup C (2014) Hypothalamic astroglial connexins are required for brain glucose sensing-induced insulin secretion. J Cereb Blood Flow Metab 34(2):339–346

Ambrosi C, Gassmann O, Pranskevich JN, Boassa D, Smock A, Wang J, Dahl G, Steinem C, Sosinsky GE (2010) Pannexin1 and Pannexin2 channels show quaternary similarities to connexons and different oligomerization numbers from each other. J Biol Chem 285: 24420–24431

Banasr M, Chowdhury GM, Terwilliger R, Newton SS, Duman RS, Behar KL, Sanacora G (2010) Glial pathology in an animal model of depression: reversal of stress-induced cellular, metabolic and behavioral deficits by the glutamate-modulating drug riluzole. Mol Psychiatry 15:501–511

Bargiotas P, Krenz A, Hormuzdi SG, Ridder DA, Herb A, Barakat W, Penuela S, von Engelhardt J, Monyer H, Schwaninger M (2011) Pannexins in ischemia-induced neurodegeneration. Proc Natl Acad Sci USA 108:20772–20777

Bargiotas P, Krenz A, Monyer H, Schwaninger M (2012) Functional outcome of pannexin-deficient mice after cerebral ischemia. Channels (Austin) 6:453–456

Barres BA (2008) The mystery and magic of glia: a perspective on their roles in health and disease. Neuron 60:430–440

Bezzi P, Gundersen V, Galbete JL, Seifert G, Steinhauser C, Pilati E, Volterra A (2004) Astrocytes contain a vesicular compartment that is competent for regulated exocytosis of glutamate. Nat Neurosci 7:613–620

Bosch M, Kielian T (2014) Hemichannels in neurodegenerative diseases: is there a link to pathology? Front Cell Neurosci 8:242
Bruzzone R, Hormuzdi SG, Barbe MT, Herb A, Monyer H (2003) Pannexins, a family of gap junction proteins expressed in brain. Proc Natl Acad Sci USA 100(23):13644–13649
Bruzzone S, Guida L, Zocchi E, Franco L, DE Flora A (2001) Connexin 43 hemi channels mediate Ca2+ regulated transmembrane NAD+ fluxes in intact cells. FASEB J 15(1):10–12
Cotrina ML, Lin JH, Nedergaard M (2008) Adhesive properties of connexin hemichannels. Glia 56:1791–1798
Chever O, Lee CY, Rouach N (2014) Astroglial connexin43 hemichannels tune basal excitatory synaptic transmission. J Neurosci 34:11228–11232
Davidson JO, Green CR, Bennet L, Nicholson LF, Danesh-Meyer H, O'Carroll SJ, Gunn AJ (2013) A key role for connexin hemichannels in spreading ischemic brain injury. Curr Drug Targets 14:36–46
De Bock M, Culot M, Wang N, Bol M, Decrock E, De Vuyst E, da Costa A, Dauwe I, Vinken M, Simon AM (2011) Connexin channels provide a target to manipulate brain endothelial calcium dynamics and blood-brain barrier permeability. J Cereb Blood Flow Metab 31:1942–1957
De Bock M, Wang N, Bol M, Decrock E, Ponsaerts R, Bultynck G, Dupont G, Leybaert L (2012) Connexin 43 hemichannels contribute to cytoplasmic Ca2+ oscillations by providing a bimodal Ca2+ dependent Ca2+ entry pathway. J Biol Chem 287:12250–12266
De Vuyst E, Boengler K, Antoons G, Sipido KR, Schulz R, Leybaert L (2011) Pharmacological modulation of connexin-formed channels in cardiac pathophysiology. Br J Pharmacol 163: 469–483
De Vuyst E, Decrock E, Cabooter L, Dubyak GR, Naus CC, Evans WH, Leybaert L (2006) Intracellular calcium changes trigger connexin 32 hemichannel opening. EMBO J 25:34–44
De Vuyst E, Wang N, Decrock E, De Bock M, Vinken M, Van Moorhem M, Lai C, Culot M, Rogiers V, Cecchelli R (2009) Ca(2+) regulation of connexin 43 hemichannels in C6 glioma and glial cells. Cell Calcium 46:176–187
Duan S, Anderson CM, Keung EC, Chen Y, Swanson RA (2003) P2X7 receptor-mediated release of excitatory amino acids from astrocytes. J Neurosci 23:1320–1328
Fasciani I, Temperan A, Perez-Atencio LF, Escudero A, Martinez-Montero P, Molano J, Gomez-Hernandez JM, Paino CL, Gonzalez-Nieto D, Barrio LC (2013) Regulation of connexin hemichannel activity by membrane potential and the extracellular calcium in health and disease. Neuropharmacology 75:479–490
Fields RD, Burnstock G (2006) Purinergic signalling in neuron-glia interactions. Nat Rev Neurosci 7:423–436
Fields RD, Stevens-Graham B (2002) New insights into neuron-glia communication. Science 298:556–562
Fiori MC, Figueroa V, Zoghbi ME, Saez JC, Reuss L, Altenberg GA (2012) Permeation of calcium through purified connexin 26 hemichannels. J Biol Chem 287:40826–40834
Frayling C, Britton R, Dale N (2011) ATP-mediated glucosensing by hypothalamic tanycytes. J Physiol 589(Pt 9):2275–2286
Funk GD (2010) The 'connexin' between astrocytes, ATP and central respiratory chemoreception. J Physiol 588:4335–4337
Gidday JM (2006) Cerebral preconditioning and ischaemic tolerance. Nat Rev Neurosci 7: 437–448
Hamilton N, Vayro S, Kirchhoff F, Verkhratsky A, Robbins J, Gorecki DC, Butt AM (2008) Mechanisms of ATP- and glutamate-mediated calcium signaling in white matter astrocytes. Glia 56:734–749
Huckstepp RTR, Eason R, Sachdev A, Dale N (2010a) CO2-dependent opening of connexin 26 and related β connexins. J Physiol 588(Pt 20):3921–3931
Huckstepp RTR, Id Bihi R, Eason R, Spyer KM, Dicke N, Willecke K, Marina N, Gourine AV, Dale N (2010b) Connexin hemichannel-mediated CO2-dependent release of ATP in the medulla oblongata contributes to central respiratory chemosensitivity. J Physiol 588(20): 3901–3920

Iglesias R, Dahl G, Qiu F, Spray DC, Scemes E (2009) Pannexin 1: the molecular substrate of astrocyte "hemichannels". J Neurosci 29:7092–7097
Imura Y, Morizawa Y, Komatsu R, Shibata K, Shinozaki Y, Kasai H, Moriishi K, Moriyama Y, Koizumi S (2013) Microglia release ATP by exocytosis. Glia 61(8):1320–1330
Inoue K, Koizumi S, Tsuda M (2007) The role of nucleotides in the neuron–glia communication responsible for the brain functions. J Neurochem 102:1447–1458
Islam MR, Uramoto H, Okada T, Sabirov RZ, Okada Y (2012) Maxi-anion channel and pannexin 1 hemichannel constitute separate pathways for swelling-induced ATP release in murine L929 fibrosarcoma cells. Am J Physiol Cell Physiol 303:C924–C935
Jiang S, Yuan H, Duan L, Cao R, Gao B, Xiong Y-F, Rao Z-R (2011) Glutamate release through connexin 43 by cultured astrocytes in a stimulated hypertonicity model. Brain Res 1392:8–15
Kaneko Y, Tachikawa M, Akaogi R, Fujimoto K, Ishibashi M, Uchida Y, Couraud PO, Ohtsuki S, Hosoya K, Terasaki T (2015) Contribution of pannexin 1 and connexin 43 hemichannels to extracellular calcium-dependent transport dynamics in human blood-brain barrier endothelial cells. J Pharmacol Exp Ther 353:192–200
Kawamura M Jr, Ruskin DN, Masino SA (2010) Metabolic autocrine regulation of neurons involves cooperation among pannexin hemichannels, adenosine receptors, and KATP channels. J Neurosci 30:3886–38895
Kim JE, Kang TC (2011) The P2X7 receptor-pannexin-1 complex decreases muscarinic acetylcholine receptor-mediated seizure susceptibility in mice. J Clin Invest 121:2037–2047
Kimelberg HK, Goderie SK, Higman S, Pang S, Waniewski RA (1990) Swelling-induced release of glutamate, aspartate, and taurine from astrocyte cultures. J Neurosci 10:1583–1591
Klaassen LJ, Sun Z, Steijaert MN, Bolte P, Fahrenfort I, Sjoerdsma T, Klooster J, Claassen Y, Shields CR, Ten Eikelder HM, Janssen-Bienhold U, Zoidl G, McMahon DG, Kamermans M (2011) Synaptic transmission from horizontal cells to cones is impaired by loss of connexin hemichannels. PLoS Biol 9:e1001107
Kurtenbach S, Whyte-Fagundes P, Gelis L, Brazil E, Zoidl C, Hatt H, Shestopalov VI, Zoidl G (2014) Investigation of olfactory function in a Panx1 knock out mouse model. Front Cell Neurosci 8:266
Langlet F (2014) Tanycytes: a gateway to the metabolic hypothalamus. J Neuroendocrinol 26:753–760
Lee S, Yoon BE, Berglund K, Oh SJ, Park H, Shin HS, Augustine GJ, Lee CJ (2010) Channel-mediated tonic GABA release from glia. Science 330:790–796
Lin JH, Lou N, Kang N, Takano T, Hu F, Han X, Xu Q, Lovatt D, Torres A, Willecke K (2008) A central role of connexin 43 in hypoxic preconditioning. J Neurosci 28:681–695
Liu X, Hashimoto-Torii K, Torii M, Ding C, Rakic P (2010) Gap junctions/hemichannels modulate interkinetic nuclear migration in the forebrain precursors. J Neurosci 30:4197–4209
Liu X, Sun L, Torii M, Rakic P (2012) Connexin 43 controls the multipolar phase of neuronal migration to the cerebral cortex. Proc Natl Acad Sci USA 109:8280–8285
Mawhinney LJ, de Rivero Vaccari JP, Dale GA, Keane RW, Bramlett HM (2011) Heightened inflammasome activation is linked to age-related cognitive impairment in Fischer 344 rats. BMC Neurosci 12:123
McDougal DH, Hermann GE, Rogers RC (2013) Astrocytes in the nucleus of the solitary tract are activated by low glucose or glucoprivation: evidence for glial involvement in glucose homeostasis. Front Neurosci 7:249
Meigh L, Greenhalgh SA, Rodgers TL, Cann MJ, Roper DI, Dale N (2013) CO2 directly modulates connexin 26 by formation of carbamate bridges between subunits. Elife 2:e01213
Moore AR, Zhou WL, Sirois CL, Belinsky GS, Zecevic N, Antic SD (2014) Connexin hemichannels contribute to spontaneous electrical activity in the human fetal cortex. Proc Natl Acad Sci USA 111:E3919–E3928
Oberheim NA, Wang X, Goldman S, Nedergaard M (2006) Astrocytic complexity distinguishes the human brain. Trends Neurosci 29:547–553

Orellana JA, Froger N, Ezan P, Jiang JX, Bennett MV, Naus CC, Giaume C, Saez JC (2011) ATP and glutamate released via astroglial connexin 43 hemichannels mediate neuronal death through activation of pannexin 1 hemichannels. J Neurochem 118:826–840

Orellana JA, Saez PJ, Cortes-Campos C, Elizondo RJ, Shoji KF, Contreras-Duarte S, Figueroa V, Velarde V, Jiang JX, Nualart F, Sáez JC, García MA (2012a) Glucose increases intracellular free Ca(2+) in tanycytes via ATP released through connexin 43 hemichannels. Glia 60:53–68

Orellana JA, Saez PJ, Shoji KF, Schalper KA, Palacios-Prado N, Velarde V, Giaume C, Bennett MV, Saez JC (2009) Modulation of brain hemichannels and gap junction channels by pro-inflammatory agents and their possible role in neurodegeneration. Antioxid Redox Signal 11:369–399

Orellana JA, von Bernhardi R, Giaume C, Saez JC (2012b) Glial hemichannels and their involvement in aging and neurodegenerative diseases. Rev Neurosci 23:163–177

Paul DL, Ebihara L, Takemoto LJ, Swenson KI, Goodenough DA (1991) Connexin46, a novel lens gap junction protein, induces voltage-gated currents in nonjunctional plasma membrane of Xenopus oocytes. J Cell Biol 115(4):1077–1089

Perea G, Navarrete M, Araque A (2009) Tripartite synapses: astrocytes process and control synaptic information. Trends Neurosci 32:421–431

Ponsaerts R, De Vuyst E, Retamal M, D'Hondt C, Vermeire D, Wang N, De Smedt H, Zimmermann P, Himpens B, Vereecke J, Leybaert L, Bultynck G (2010) Intramolecular loop/tail interactions are essential for connexin 43-hemichannel activity. FASEB J 24: 4378–4395

Prochnow N, Abdulazim A, Kurtenbach S, Wildforster V, Dvoriantchikova G, Hanske J, Petrasch-Parwez E, Shestopalov VI, Dermietzel R, Manahan-Vaughan D, Zoidl G (2012) Pannexin1 stabilizes synaptic plasticity and is needed for learning. PLoS ONE 7:e51767

Rodriguez EM, Blazquez JL, Pastor FE, Pelaez B, Pena P, Peruzzo B, Amat P (2005) Hypothalamic tanycytes: a key component of brain-endocrine interaction. Int Rev Cytol 247:89–164

Rossi DJ, Oshima T, Attwell D (2000) Glutamate release in severe brain ischaemia is mainly by reversed uptake. Nature 403:316–321

Rudkouskaya A, Chernoguz A, Haskew-Layton RE, Mongin AA (2008) Two conventional protein kinase C isoforms, alpha and beta I, are involved in the ATP-induced activation of volume-regulated anion channel and glutamate release in cultured astrocytes. J Neurochem 105:2260–2270

Saez JC, Leybaert L (2014) Hunting for connexin hemichannels. FEBS Lett 588:1205–1211

Saez JC, Berthoud VM, Branes MC, Martinez AD, Beyer EC (2003) Plasma membrane channels formed by connexins: their regulation and functions. Physiol Rev 83:1359–1400

Saez JC, Schalper KA, Retamal MA, Orellana JA, Shoji KF, Bennett MV (2010) Cell membrane permeabilization via connexin hemichannels in living and dying cells. Exp Cell Res 316: 2377–2389

Sahu G, Sukumaran S, Bera AK (2014) Pannexins form gap junctions with electrophysiological and pharmacological properties distinct from connexins. Sci Rep 4:4955

Salameh A, Blanke K, Dhein S (2013) Mind the gap! Connexins and pannexins in physiology, pharmacology and disease. Front Pharmacol 4:144

Sanchez HA, Orellana JA, Verselis VK, Saez JC (2009) Metabolic inhibition increases activity of connexin-32 hemichannels permeable to Ca2+ in transfected HeLa cells. Am J Physiol Cell Physiol 297:C665–C678

Santiago MF, Veliskova J, Patel NK, Lutz SE, Caille D, Charollais A, Meda P, Scemes E (2011) Targeting pannexin1 improves seizure outcome. PLoS ONE 6:e25178

Schafer DP, Lehrman EK, Stevens B (2013) The "quad-partite" synapse: Microglia-synapse interactions in the developing and mature CNS. Glia 61:24–36

Schalper KA, Orellana JA, Berthoud VM, Saez JC (2009) Dysfunctions of the diffusional membrane pathways mediated by hemichannels in inherited and acquired human diseases. Curr Vasc Pharmacol 7:486–505

Schalper KA, Sanchez HA, Lee SC, Altenberg GA, Nathanson MH, Saez JC (2010) Connexin 43 hemichannels mediate the Ca2+ influx induced by extracellular alkalinization. Am J Physiol Cell Physiol 299:C1504–C1515

Schock SC, Leblanc D, Hakim AM, Thompson CS (2008) ATP release by way of connexin 36 hemichannels mediates ischemic tolerance in vitro. Biochem Biophys Res Commun 368:138–144

Shestopalov VI, Slepak VZ (2014) Molecular pathways of pannexin1-mediated neurotoxicity. Front Physiol 5:23

Simard M, Arcuino G, Takano T, Liu QS, Nedergaard M (2003) Signaling at the gliovascular interface. J Neurosci 23:9254–9262

Sosinsky GE, Boassa D, Dermietzel R, Duffy HS, Laird DW, MacVicar B, Naus CC, Penuela S, Scemes E, Spray DC, Thompson RJ, Zhao HB, Dahl G (2011) Pannexin channels are not gap junction hemichannels. Channels (Austin) 5:193–197

Stehberg J, Moraga-Amaro R, Salazar C, Becerra A, Echeverria C, Orellana JA, Bultynck G, Ponsaerts R, Leybaert L, Simon F, Sáez JC, Retamal MA (2012) Release of gliotransmitters through astroglial connexin 43 hemichannels is necessary for fear memory consolidation in the basolateral amygdala. FASEB J 26:3649–3657

Stout CE, Costantin JL, Naus CCG, Charles AC (2002) Intercellular calcium signaling in astrocytes via ATP release through connexin hemichannels. J Biol Chem 277(12):10482–10488

Suadicani SO, Brosnan CF, Scemes E (2006) P2X7 receptors mediate ATP release and amplification of astrocytic intercellular Ca2+ signaling. J Neurosci 26:1378–1385

Takano T, Kang J, Jaiswal JK, Simon SM, Lin JH, Yu Y, Li Y, Yang J, Dienel G, Zielke HR, Nedergaard M (2005) Receptor-mediated glutamate release from volume sensitive channels in astrocytes. Proc Natl Acad Sci USA 102:16466–16471

Takeuchi H, Jin S, Wang J, Zhang G, Kawanokuchi J, Kuno R, Sonobe Y, Mizuno T, Suzumura A (2006) Tumor necrosis factor-alpha induces neurotoxicity via glutamate release from hemichannels of activated microglia in an autocrine manner. J Biol Chem 281(30):21362–21368

Tonon MC, Lanfray D, Castel H, Vaudry H, Morin F (2013) Hypothalamic glucose-sensing: role of Glia-to-neuron signaling. Horm Metab Res 45:955–959

Tremblay ME, Lowery RL, Majewska AK (2010) Microglial interactions with synapses are modulated by visual experience. PLoS Biol 8:e1000527

Verkhratsky A, Toescu EC (2006) Neuronal-glial networks as substrate for CNS integration. J Cell Mol Med 10:826–836

Vessey DA, Li L, Kelley M (2010) Pannexin-I/P2X 7 purinergic receptor channels mediate the release of cardioprotectants induced by ischemic pre- and postconditioning. J Cardiovasc Pharmacol Ther 15:190–195

Vessey DA, Li L, Kelley M (2011) Ischemic preconditioning requires opening of pannexin-1/P2X (7) channels not only during preconditioning but again after index ischemia at full reperfusion. Mol Cell Biochem 351:77–84

Volterra A, Meldolesi J (2005) Astrocytes, from brain glue to communication elements: the revolution continues. Nat Rev Neurosci 6:626–640

Wake H, Moorhouse AJ, Miyamoto A, Nabekura J (2013) Microglia: actively surveying and shaping neuronal circuit structure and function. Trends Neurosci 36(4):209–217

Wang N, De Bock M, Decrock E, Bol M, Gadicherla A, Vinken M, Rogiers V, Bukauskas FF, Bultynck G, Leybaert L (2013) Paracrine signaling through plasma membrane hemichannels. Biochim Biophys Acta 1828:35–50

Wenker IC, Sobrinho CR, Takakura AC, Moreira TS, Mulkey DK (2012) Regulation of ventral surface CO2/H + -sensitive neurons by purinergic signalling. J Physiol 590(Pt 9):2137–2150

Williams JA, Imamura M, Fregni F (2009) Updates on the use of non-invasive brain stimulation in physical and rehabilitation medicine. J Rehabil Med 41:305–311

Ye Z-C, Wyeth MS, Baltan-Tekkok S, Ransom BR (2003) Functional hemichannels in astrocytes: a novel mechanism of glutamate release. J Neurosci 23(9):3588–3596

Zhang Q, Fukuda M, Van Bockstaele E, Pascual O, Haydon PG (2004) Synaptotagmin IV regulates glial glutamate release. Proc Natl Acad Sci USA 101:9441–9446

# Role of Astrocytes in Central Respiratory Chemoreception

**Jaime Eugenín León, María José Olivares and Sebastián Beltrán-Castillo**

**Abstract** Astrocytes perform various homeostatic functions in the nervous system beyond that of a supportive or metabolic role for neurons. A growing body of evidence indicates that astrocytes are crucial for central respiratory chemoreception. This review presents a classical overview of respiratory central chemoreception and the new evidence for astrocytes as brainstem sensors in the respiratory response to hypercapnia. We review properties of astrocytes for chemosensory function and for modulation of the respiratory network. We propose that astrocytes not only mediate between $CO_2/H^+$ levels and motor responses, but they also allow for two emergent functions: (1) Amplifying the responses of intrinsic chemosensitive neurons through feedforward signaling via gliotransmitters and; (2) Recruiting non-intrinsically chemosensitive cells thanks to volume spreading of signals (calcium waves and gliotransmitters) to regions distant from the $CO_2/H^+$ sensitive domains. Thus, astrocytes may both increase the intensity of the neuron responses at the chemosensitive sites and recruit of a greater number of respiratory neurons to participate in the response to hypercapnia.

**Keywords** Respiratory rhythm · Central chemoreception · Raphe nuclei · Locus coeruleus nuclei · Retrotrapezoid nuclei · Brainstem · Glia · Gliotransmitters · Astrocytes

### Abbreviations

| | |
|---|---|
| 5-HT | 5-hydroxytryptamine (Serotonin) |
| ACh | Acetylcholine |
| aCSF | Artificial cerebrospinal fluid |
| AMPAR | $\alpha$-amino-3-hydroxy-5-methyl-4-isoxazolepropionic acid receptor |
| ANP | Atrial natriuretic peptide |

J.E. León (✉) · M.J. Olivares · S. Beltrán-Castillo
Departamento de Biología, Universidad de Santiago de Chile (USACH), PO 9170022, Santiago, Chile
e-mail: jaime.eugenin@usach.cl; jeugenin@gmail.com

 
R. von Bernhardi (ed.), *Glial Cells in Health and Disease of the CNS*, Advances in Experimental Medicine and Biology 949,
DOI 10.1007/978-3-319-40764-7_6

| | |
|---|---|
| ATP | Adenosine triphosphate |
| CA | Carbonic anhydrase enzyme |
| CCHS | Central congenital hypoventilation syndrome |
| CNS | Central nervous system |
| CNO | Clozapine-N-oxide |
| $CO_2$ | Carbon dioxide |
| CNQX | 6-cyano-7-nitroquinoxaline-2,3-dione—competitive AMPA/kainate receptor antagonist |
| cNTS | Caudal nucleus tractus solitarius |
| CSF | Cerebrospinal fluid |
| cVLM | Caudal ventrolateral medulla |
| cVRG | Caudal ventral respiratory group |
| Cx | Connexins |
| DRC | Dorsal respiratory columns |
| EPSP | Excitatory postsynaptic potentials |
| GABA | $\gamma$-aminobutyric acid |
| GFAP | Glial fibrillary acidic protein |
| KF | Pontine Kölliker-Fuse nucleus |
| KO | Knock out |
| LC | Locus coeruleus |
| LDT | Laterodorsal tegmental nucleus |
| LPBR | Lateral parabrachial nucleus |
| LTP | Long-term potentiation |
| mRVLM | Medial portion of the rostral ventrolateral medulla |
| MS | Methionine sulfoximine |
| NK1R | Neurokinin 1 receptor |
| NMDA | N-methyl-D-aspartate |
| NMDAR | N-methyl-D-aspartate receptor |
| NO | Nitric oxide |
| NTS | Nucleus tractus solitarius |
| $PaCO_2$ | Partial arterial pressure of carbon dioxide |
| $PCO_2$ | Partial pressure of carbon dioxide |
| $PaO_2$ | Partial arterial pressure of oxygen |
| PF-LHA | Perifornical-lateral hypothalamic area |
| PNS | Peripheral nervous system |
| PPADS | Pyridoxal-phosphate-6-azophenyl-2=,4=-disulfonate |
| PPT | Pedunculopontine tegmental nucleus |
| preBötC | PreBötzinger Complex |
| ORX | Orexin |
| ORX-KO | Prepro-orexin knockout mice |
| RN | Medullary raphe nucleus |
| RPG | Respiratory pattern generator |
| RTN/pFRG | Retrotrapezoid/parafacial respiratory group |
| RVL | Nucleus reticularis rostroventrolateralis |

| | |
|---|---|
| RVLM | Rostral ventrolateral medulla |
| rVRG | Rostral ventral respiratory group |
| SERT | Serotonin transporter |
| SIDS | Sudden infant death syndrome |
| SP | Substance P |
| SSP-SAP | Saporin–substance P conjugate |
| TH | Tyrosine hydroxylase |
| TIRF | Total internal reflection fluorescence |
| TRH | Thyrotropin releasing hormone |
| TRP | Channels Transient receptor potential channels |
| TS-eEPSCs | Tractus solitaries-evoked excitatory postsynaptic currents |
| VLM | Ventrolateral medullary surface |
| VMS | Ventral medullary surface |
| VRC | Ventral respiratory columns |
| VRG | Ventral respiratory group |

## The Respiratory Network

The neural network responsible for generating the respiratory rhythm, the respiratory pattern generator (RPG), is composed of neurons preferentially discharging during inspiration or expiration and distributed along the ventral (VRC) and the dorsal (DRC) respiratory columns (Fig. 1) (Feldman et al. 2003; von Euler 1986). The RPG projects into respiratory motoneurons located at different cranial nerve nuclei (V, VII, IX, X, XII), which innervate muscles controlling airway flow and resistance. In addition, the RPG sends projections and synapses on various spinal cord motoneurons, particularly the phrenic motoneurons (C3–C5), which innervate the diaphragm muscle, and intercostal motoneurons (T1–T10), which innervate intercostal muscles. The RPG imposes on these motoneurons a synchronic and rhythmic activity responsible for generating a sequence of inspiratory, post-inspiratory, and expiratory phases observable in recordings from phrenic, abductor laryngeal, and internal intercostal nerves, respectively (Richter and Spyer 2001). The coordinated activation of these motoneurons results in a sequence of air pressure gradients commanding the inspiratory and expiratory phases of ventilation.

At the RPG, within the VRC, at least two oscillators can be recognized: at the rostral area of the VRC, the pre-inspiratory retrotrapezoid/parafacial respiratory group (RTN/pFRG), arising from *Phox2b* expressing progenitors (Guyenet and Mulkey 2010; Onimaru and Homma 2006; Onimaru et al. 2006, 2009; Stornetta et al. 2006; Wang et al. 2013; Takakura et al. 2014; Dubreuil et al. 2009b; Abbott et al. 2011), and at the caudal portions of the VRC, the inspiratory preBötzinger Complex (preBötC), which is derived from *Dbx1* progenitors and considered

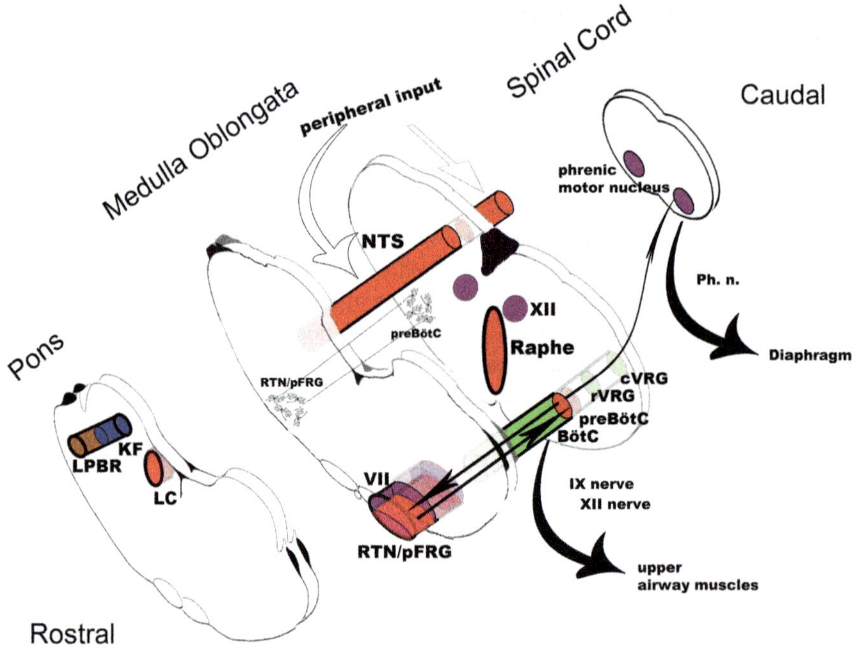


**Fig. 1** Schema of the respiratory neural network and central respiratory chemoreceptors. The dorsal respiratory column (DRC) is represented by the nucleus tractus solitarius (NTS), while the ventral respiratory column (VRC), by the retrotrapezoid/parafacial respiratory group (pFRG), the Bötzinger nucleus, the pre-Bötzinger complex (preBötC), and the rostral and caudal ventral respiratory group (rVRG and cVRG). Input and output to the central respiratory network are indicated with *white* and *black arrows*, respectively; note that the output was represented by respiratory motoneurons localized in cranial and phrenic nuclei. The phrenic nerve (Ph n.) controls the diaphragm muscle, main responsible for generating air pressure gradients during breathing. The main peripheral input is provided by vagal mechanoafferents and peripheral arterial chemoreceptors. Central chemosensitive sites (NTS, preBötc, LC, RTN, raphe) containing cells that sense changes in pH or $PCO_2$ in the interstitial or the cerebrospinal fluid of the brainstem are indicated in *red*. Note that the raphe and LC provide inputs into the respiratory network, while NTS, preBötc, and RTN belong to the respiratory network. KF, pontine Kölliker-Fuse nucleus; LC, locus coeruleus; LPBR, lateral parabrachial nucleus; VII, facial nucleus; XII, hypoglossal nucleus

essential for generating the inspiratory activity (Fig. 1) (Smith et al. 1991; Feldman et al. 2003; Gray et al. 2010).

The RPG receives input from several central nervous system (CNS) structures, including cortex, cerebellum, hypothalamus, and brainstem nuclei, and from the peripheral nervous system (PNS), including vagal mechanosensory afferents and peripheral arterial chemoreceptors (Feldman 1986; von Euler 1986). Peripheral arterial chemoreceptors (carotid and aortic bodies) sense changes in $PaO_2$, $PaCO_2$, pH, osmolarity, temperature, and flow of blood circulating through great arteries (Eyzaguirre et al. 1983). In contrast, central chemoreceptors (Fig. 1 indicated in red) are activated by changes in pH or $PCO_2$ in the interstitial or the cerebrospinal fluid of the brainstem (Nattie 1999). In fact, $CO_2$- and $H^+$-sensitive neurons are a major source of the tonic input that drives the mammalian respiratory pattern generator (Nattie 1999).

## Central Chemoreception

Central chemoreception can be defined as: "*the detection of* $CO_2$/pH *at sites within the central nervous system and the resultant effects on ventilation*" (Nattie and Li 2010), or in more systemic terms as "*the feedback process whereby changes in the brain* $CO_2$ *(or pH) bring about adaptive (homeostatic) changes in breathing to maintain arterial* $CO_2$ *(or pH) near steady-state levels*" (Funk 2010). Central chemoreception is crucial for matching breathing to physiological demands relative to $H^+$ or $CO_2$ elimination. In addition, it appears essential for generating and maintaining the respiratory rhythm (Eugenin 1995), allowing brainstem respiratory neurons to be coordinated and excitable in an optimal manner (Nattie and Li 2012). In fact, $CO_2$- and $H^+$-sensitivities are a major source of the tonic drive that sustains the activity of the RPG (Nattie 1999; Nattie and Li 2012). For example, in *en bloc* preparations from newborn opossum and mice, alkaline superfusion of the brainstem arrests the respiratory rhythm (Eugenin and Nicholls 1997; Infante et al. 2003; Eugenin et al. 2006).

## Localization of Central Chemoreceptors

Various strategies have been used to localize chemosensitive sites in the brain. Detection of c-Fos protein as a marker of neuronal activity revealed that those nuclei in which the number of c-Fos positive neurons increase after exposure to hypercapnia also contain neurons with electrophysiological responses to hypercapnic acidosis (Belegu et al. 1999; Mulkey et al. 2004; Ritucci et al. 2005; Wang and Richerson 1999; Wickstrom et al. 2002; Teppema et al. 1997). In these nuclei we can find neurons that fire in association with or in correlation with the respiratory response to hypercapnia. On the other hand, destruction (Akilesh et al. 1997; Biancardi et al. 2008; da Silva et al. 2011; Dias et al. 2007), genetic ablation (Hodges et al. 2011; Dubreuil et al. 2009b; Ramanantsoa et al. 2011), inactivation, or synaptic inhibition (Nattie and Li 2000; Curran et al. 2001) of specific nuclei reduces the ventilatory response to hypercapnia. More direct evidence of the existence of chemosensitive sites was obtained by focal acidification. Either local application of acetazolamide (Coates et al. 1993), an inhibitor of the enzyme carbonic anhydrase, or reverse microdialysis of artificial cerebrospinal fluid equilibrated with high $CO_2$ within specific CNS areas, increased ventilation (Li et al. 1999; Nattie and Li 2001, 2002a; Li and Nattie 2002; Dias et al. 2008; da Silva et al. 2010; Krause et al. 2009;

Kuwaki et al. 2010; Coates et al. 1993). Notably, as illustrated in Fig. 1 (nuclei in red), central respiratory chemosensitivity was localized to multiple sites, including such nuclei belonging to the RPG as the pFRG/RTN, preBötC, and nucleus tractus solitarius (NTS), and such nuclei or identified regions outside the RPG but projecting into as the medullary raphe (RN), locus coeruleus (LC, A6), ventrolateral medullary surface, hypothalamus, and fastigial nucleus (Ballantyne and Scheid 2001; Coates et al. 1993; Mitchell et al. 1963; Nattie 2001; Oyamada et al. 1998; Wang and Richerson 1999; Li et al. 2006, 2013; Guyenet et al. 2005; Nattie and Li 2006, 2009, 2010; Xu et al. 2001; Xu and Frazier 1995; Martino et al. 2007; Krause et al. 2009). The contribution of specific groups of cells within chemosensitive nuclei were evident from the effects of lesions of neurokinin-1 receptor expressing cells in the RTN, or serotonergic cells in the RN, or catecholaminergic cells in the LC. In all these specific lesions, the $CO_2$ response decreased by 15–30 % during both sleep and wakefulness (Nattie and Li 2008).

## Roles of the RTN and Raphe RN Neurons in Central Chemoreception

### *RTN Neurons*

RTN neurons are glutamatergic, chemosensitive, express the transcription factor Phox2b, provide excitatory projections to other sites in the central respiratory network, and when stimulated activate breathing (Mulkey et al. 2004; Onimaru et al. 2008; Stornetta et al. 2006; Wang et al. 2013; Guyenet and Mulkey 2010; Goridis et al. 2010; Dubreuil et al. 2009b).

Inhibition of RTN neurons by muscimol dialysis or their chemical (kainic acid injection) or electrical destruction reduces basal ventilation and the ventilatory responses to hypercapnia in anesthetized rats (Nattie and Li 1994). More selective lesions restricted to RTN neurons expressing the neurokinin 1 receptor (NK1R), obtained with a saporin–substance P conjugate (SSP-SAP), impairs ventilatory response to hypercapnia in rats (Nattie and Li 2002b). In anesthetized rats, elimination of at least 70 % of $Phox2b^+$ tyrosine hydroxylase negative ($TH^-$) RTN neurons is required for a significant increase of the apnea threshold, but does not affect the sensitivity of the subsequent responses to hypercapnia (Takakura et al. 2008). Allatostatin inhibition of RTN Phox2b-expressing neurons transformed with a lentiviral construct to express the G-protein-coupled Drosophila allatostatin receptor did not affect the basal respiratory activity in unanesthetized, conscious rats (Marina et al. 2010). Nevertheless, allatostatin reduced the amplitude of the phrenic nerve discharge and the $CO_2$-evoked ventilatory responses in anesthetized rats, in in situ preparations, and in conscious rats with denervated or intact peripheral chemoreceptors (by 28 and 60 %, respectively) (Marina et al. 2010; Ramanantsoa et al. 2011).

In contrast, photostimulation of RTN neurons expressing channel rhodopsin-2 under the control of the Phox2-responsive promoter PRSx8, increases ventilation in both anesthetized and conscious animals (Abbott et al. 2009, 2011; Kanbar et al. 2010; Burke et al. 2015).

The human disease called central congenital hypoventilation syndrome (CCHS) shows a spectrum of defects comparable with the ontogenic defects of the autonomic nervous system in Phox2b mutant mice (Brunet and Pattyn 2002; Pattyn et al. 1999). CCHS is a life threatening human disease characterized by hypoventilation periods or apnea during sleep and a variable reduction of ventilatory response to hypercapnia, from moderate to severe. CCHS was attributable to a mutation consisting of a polyalanine expansion in the Phox2b transcription factor (Amiel et al. 2003, 2009). Moreover, genetic generation of a knock-in mouse having the most frequent of the CCHS-mutations, the *Phox2b27Ala* allele, resulted in the selective ablation of glutamatergic neurons in the RTN and a CCHS-like phenotype. These mice showed gasping behavior, cyanosis, disruption of the respiratory chemo reflex at birth and, in contrast to human CCHS patients, they died during the first hours of postnatal life from respiratory failure (Dubreuil et al. 2008, 2009a, b; Goridis et al. 2010; Ramanantsoa et al. 2011).

## Raphe Nucleus Neurons

In brainstem slices, $CO_2/H^+$ responsive neurons can be found in the midline Raphe nucleus (RN) (Richerson 1995; Wang et al. 1998). As mentioned above, ventilation increases with focal acidification of the midline RN by microinjection of acetazolamide in anesthetized rats or by reverse microdialysis of acidified cerebrospinal fluid (CSF) in conscious rats or goats (Nattie and Li 2001; Hodges et al. 2004a, b). Inhibition of RN neurons by microdialysis of muscimol (Taylor et al. 2006), by administration of 5-hydroxytryptamine $(5\text{-HT})_{1A}$ autoreceptor agonist (8-OH-DPAT), which inhibits serotonergic neurons, or by microinjections of lidocaine or ibotenic acid significantly decreased the response to hypercapnia in piglets (Messier et al. 2002, 2004; Dreshaj et al. 1998). In the unanesthetized juvenile rat brainstem preparation perfused in situ, 5-$HT_2$ receptor antagonism with ketanserin or 5-$HT_{1A}$ autoreceptor activation with 8-OH-DPAT blunted the respiratory response (Corcoran et al. 2013). In rats, injections of a monoclonal antibody against the serotonin transporter (SERT) conjugated to saporin into the RN specifically killed serotonergic neurons, and as result decreased the average $CO_2$ response (Nattie et al. 2004). In addition, hypercapnic ventilatory response decreased by 50 % in adult knock out (KO) mice (Lmx1bf/f/p and Pet-1 knockout mice) with near complete absence of central 5-HT neurons (Hodges et al. 2008, 2011) or with absence of the 5HT transporter (Li and Nattie 2008). Egr2-null mice have, among others defects, altered serotonergic progeny, low respiratory rate, and severe apneas, dying perinatally due to respiratory insufficiency.

Selective hyperpolarization of Egr2 expressing neurons or 5HT neurons was achieved by clozapine-N-oxide (CNO) activation of the synthetic Gi/o protein–coupled receptor Di expressed selectively on 5-HT neurons using conditional intersectional genetics. Hyperpolarization of Egr2 neurons reduced the ventilatory response by 63 % (Ray et al. 2013). Hyperpolarization restricted to serotonergic neurons reduced the ventilatory chemoreflex in vivo by almost 50 % and reduced the $CO_2$-induced firing rate increase of 5HT neurons in culture (Ray et al. 2011). When Di expression was targeted to a specific subtype of 5HT neuron, the Egr2-Pet1 serotonergic subgroup was found to contribute most to the ventilatory response to hypercapnia and acidosis. Egr2-Pet1 neurons project to other chemosensory areas and show intrinsic chemosensitivity firing in response to a hypercapnic stimulus (Brust et al. 2014).

## Relative Contribution of Chemosensitive Sites to the Overall Response

Determination of the relative contribution of each chemosensory site to the full expression of chemosensitivity has been elusive. Pronounced effects after unilateral chemical or electrolytic lesion of the RTN, NTS, or RN led to the notion that each nucleus provides an essential, indispensable, and singular contribution to the full expression of central chemosensitivity (Berger and Cooney 1982; Nattie and Li 1994). However, these deleterious effects caused by lesion of chemosensitive nucleus were strongly influenced by anesthesia (Nattie and Li 2012; Nattie 2011). In fact, lesion-related impairment of the responses to systemic hypercapnia largely disappeared with recovery of consciousness (Berger and Cooney 1982). Thus, under anesthesia, destruction of the rat RTN reduced the integrated baseline activity of the phrenic nerve and the respiratory response to hypercapnia (Nattie and Li 1994). In contrast, in conscious, unanesthetized rats, similar unilateral lesions of RTN produced minor effects on baseline ventilation and the respiratory response to hypercapnia (Akilesh et al. 1997). In agreement with these results, the magnitude of the ventilatory effects evoked by acidification of chemosensitive areas using reverse microdialysis in conscious, unanesthetized animals was lower than that observed in anesthetized animals (Nattie and Li 2012; Nattie 2011). The reduction in ventilatory effects observed in conscious animals may be explained in part by an enhanced clearance of focal stimulus as a result of an increased cerebral blood flow in unanesthetized mammals.

The relative contributions of chemosensory nuclei in the conscious animal has been studied using either focal inhibition of chemosensitive sites or focal acidification by reverse microdialysis of artificial cerebrospinal fluid (aCSF) equilibrated with high $CO_2$ (Nattie and Li 2009). Assuming that the contributions of chemosensitive sites are independent, the overall respiratory response does not appear to be the result of simple additive interactions of individual contributions

(Nattie and Li 2010). Clear synergisms could be inferred, as for example, observing the ventilatory depression when RTN and caudal RN were simultaneously inhibited (Li et al. 2006). More direct evidence of this synergism was obtained with simultaneous focal acidification of the RTN and caudal RN (Dias et al. 2008). However, unrealistically complex experiments with multiple probes stimulating each chemosensory area individually or several simultaneously during wakefulness and sleep seem to be necessary to fully address this question.

Interestingly, the full expression of central chemoreception also depends on the peripheral chemoreceptor input. In fact, in unanesthetized awake dogs the ventilatory responsiveness to four progressively increasing levels of central hypercapnia depended on the degree of carotid body inhibition or stimulation with respect to basal eupneic conditions (normoxic, normocapnic carotid body perfusion). The increase in carotid body activity via carotid body perfusion with a hypoxic, normocapnic perfusate increased the ventilatory response to hypercarbia by 223 % respect basal conditions. By contrast, silencing of carotid bodies activity with hyperoxic, hypocapnic perfusate reduced the ventilatory response to hypercarbia by 81 %. This interdependence between peripheral and central chemoreception suggests that the whole system of central and peripheral chemosensory structures are functionally interrelated and integrated.

## Central Chemoreception Dependency on Functional State of the Respiratory Network

Special attention should be focused to the fact that contribution of each chemosensitive site to the overall response to hypercapnia depends on the functional state of the respiratory network. Such dependency not only may give account of the differences between conscious and anesthetized animals that are mentioned above, but also of differences in ventilation and ventilatory responses between wakefulness and sleep (Newton et al. 2014). Studies in rats with focal acidosis by reverse microdialysis along the sleep–wake cycle have shown that acidification of the RTN or the perifornical-lateral hypothalamic area (PF-LHA), where orexin neurons are found, or the caudal ventrolateral medulla (cVLM) increased ventilation predominantly in wakefulness (Li and Nattie 2002; Li et al. 2013; da Silva et al. 2010). By contrast, acidification of rostral RN increased ventilation predominantly in sleep (Nattie and Li 2001) while focal acidification of the NTS increased ventilation in both wakefulness and sleep (Nattie and Li 2002a).

Orexin neurons are good candidates to be the link between arousal state and chemoreceptive properties at the brainstem (Nattie and Li 2010, 2012). Orexin neurons are critical for generating wakefulness (Ohno and Sakurai 2008; Sakurai 2014; Alexandre et al. 2013) and controlling breathing (Nakamura et al. 2007; Li et al. 2013; Li and Nattie 2010; Dias et al. 2010; Terada et al. 2008; Dutschmann et al. 2007; Deng et al. 2007; Young et al. 2005b; Toyama et al. 2009). They are sensitive to $H^+/CO_2$ (Williams et al. 2007; Li et al. 2013; Sunanaga et al. 2009) and their firing rate is maximal during wakefulness (Lee et al. 2005) and minimal during sleep.

As mentioned above, focal acidification of the hypothalamic area containing orexin neurons increased ventilation up to 15 % only in wakefulness but not in sleep (Li et al. 2013). In prepro-orexin knockout mice (ORX-KO) basal ventilation is not affected along the sleep–wake cycle. Neither their ventilatory responses to hypercarbia during sleep period nor their ventilatory responses to hypoxia during wake–sleep cycle when compared with those in wild type mice. However, ORX-KO mice have a ventilatory response to hypercapnia reduced to the half of that in wild type mice during quiet wakefulness. The ventilatory response to hypercapnia was partially restored in ORX-KO mice administered intracerebroventricular with orexin-A or orexin–B, the two orexin subtypes derived from prepro-orexin (Deng et al. 2007).

Such results are compatible with those obtained by dialyzing the rat RTN with SB-334867, orexin receptor-1 antagonist that reduced the hyperventilation caused by hypercapnia by 30 % during wakefulness and 9 % during sleep. A much smaller effect (16 % reduction of hypercapnia-induced hyperventilation) was observed when microdialysis of SB-334867 was performed into rostral RN during wakefulness in dark period and null effect in the ventilatory chemo reflex when administered during sleep (Dias et al. 2010). In addition, almorexant, antagonist of both orexin receptor-1 and orexin receptor-2, administered orally reduced the ventilatory response to hypercapnia by 26 % only in wakefulness during the dark, active period of the diurnal cycle (Li and Nattie 2010). Then, we can conclude that projections of orexin-containing neurons to the RTN and rostral RN contribute, via orexin receptor-1, to the hypercapnic chemoreflex control during wakefulness and to a lesser extent during sleep (Dias et al. 2009). However, a possible role for orexin neurons as a "wakefulness" driver of chemosensitive properties is still uncertain.

## Astrocytes

Astrocytes are not mere intermingled cells of the CNS that outnumber neurons. As already described in Chapter "Glial Cells and Integrity of the Nervous System", they serve multiple functions: structure of the nervous tissue, trophism, metabolic support as for example the lactate shuttle, energy storage in the form of glycogen, ionic and water homeostasis, homeostasis of the synaptic environment buffering the concentration of extracellular potassium and the excess of extracellular neurotransmitters and release of gliotransmitters and neurotransmitters (most of them influencing synaptic strength, Table 1), formation and remodeling of synapses, defense against oxidative stress, scar formation, and tissue repair. Even more, astrocytes are involved in complex processes like neural network plasticity, inflammation, and neurodegeneration (Belanger et al. 2011; Grass et al. 2004; Rodriguez-Arellano et al. 2015).

**Table 1** Substances released by astrocytes

| | Neuroactive substance | Reference |
|---|---|---|
| *Neurotransmitters—neuromodulators* | | |
| Amino acids | L-glutamate | Parpura et al. (1994) and Kimelberg et al. (1990) |
| | L-aspartate | Kimelberg et al. (1990) |
| | Taurine | Kimelberg et al. (1990) |
| | D-serine | Schell et al. (1995) |
| | γ-aminobutyric acid (GABA) | Bowery et al. (1976) |
| Non-amino acids | Dopamine | Chen et al. (2005) |
| | ATP | Guthrie et al. (1999) |
| | Adenosine | Albrecht et al. (1991) |
| | Nitric oxide (NO) | Murphy et al. (1990) |
| | Met-enkephalin | Shinoda et al. (1989) |
| | Somatostatin | Mercure et al. (1996) |
| | Atrial natriuretic peptide (ANP) | Krzan et al. (2003) and Guček et al. (2012) |
| *Metabolic precursors* | | |
| | Lactate | Pellerin and Magistretti (1994) |
| | Glutamine | Yudkoff et al. (1994) |
| | α-ketoglutarate | Westergaard et al. (1994) |
| | Malate | Westergaard et al. (1994) |
| | Succinate | Westergaard et al. (1994) |
| *Growth factors* | | |
| | BDNF | Caravagna et al. (2013) |
| | NGF-β | Furukawa et al. (1986) |
| | IGF-I | Kadle et al. (1988) |
| | IGF-II | Kadle et al. (1988) |
| *Inflammatory factors* | | |
| | Prostaglandins D2, E2, I2 | Gebicke-Haerter et al. (1988) and Hartung and Toyka (1987) |
| | Thromboxane | Hartung et al. (1988) |
| | IL-1β | Corsini et al. (1996) |
| | IL-6 | Wu et al. (2005) |
| | IL-10 | Wu et al. (2005) |
| | TGF-β | Constam et al. (1992) |
| | TNF-α | Selmaj et al. (1990) |

There is a remarkable heterogeneity among astrocytes, being their phenotype largely a function of both local anatomy and regional functional demands (Oberheim et al. 2012). They are in intimate contact with most of the structures of the nervous system being largely responsible of its compartmentalization. Astrocytes send end-feet processes that enwrap blood vessels and interact with endothelial cells determining the formation of the blood brain barrier. Astrocytic end-feet processes express, among others, glucose transporters and aquaporin 4. They are involved in the cerebral neurovascular coupling regulating the microvascular flow for matching this to synaptic activity (Iadecola and Nedergaard 2007).

On the other hand, astrocytes send processes that ensheath most synapses. These perisynaptic processes express receptors for cytokines and growth factors. In addition, they express different kind of neurotransmitter receptors, transporters, and ion channels as expected of an active participant in the homeostasis of the synapse. Thus, at the synaptic compartment, astrocytes can sense the synaptic activity by means of neurotransmitter receptors activation (Araque et al. 2014), regulate the levels of neurotransmitters at the synaptic cleft influencing their recapture and release (Hamilton and Attwell 2010), modulate the synaptic transmission through gliotransmitters release, and modulate the neuron excitability by extracellular potassium buffering (Perea et al. 2014).

In hippocampus and cortex from rodent and humans, astrocytes are organized in discrete spatial domains (Oberheim et al. 2012). Each astrocyte extends its processes on a defined territory without important overlap between adjacent astrocytes. On other terms, all cellular structures in a territory (blood vessels, perikarya and synapses) interact with processes from a single astrocyte only (Oberheim et al. 2009). It is estimated that a single spatial domain for a protoplasmic astrocyte in rodent contains 20,000–120,000 synapses, while that in humans contains the extraordinary amount of 270 thousand to 2 million synapses (Oberheim et al. 2009).

A particular feature of astrocytes is that each one of them is coupled to others, through gap junction channels forming an extensive functional syncytium. In hippocampus, each astrocyte forms gap junctions with 11 others astrocytes, in average (Xu et al. 2010). This syncytium offers a route of low electrical resistance for propagation of electronic signaling and ionic currents and for cell-to-cell propagation of second messengers. This syncytium represents a huge sink for buffering the changes in potassium composition of the extracellular space. In addition, this syncytium allows the spreading of calcium waves, which, in humans reach the speed of 37–43 μm/s (Cornell-Bell et al. 1990; Oberheim et al. 2009), into neighboring astrocytes. Thus, astrocytes can be sequentially activated and recruited for performing a common task. Since each astrocytic domain represents an elementary glio-neuronal unit for monitoring the changes in activity of contiguous synapses, the existence of a functional syncytium implies the capability of influencing other astrocytic domains and the spreading of a potential astrocytic response to domains placed far away from an immediate neighborhood. This organization of highly organized and interconnected anatomical domains will allow the recruitment of distant domains, which in turn will influence a larger number of synapses within a neural network. As a consequence, a more intense, and may be, a more synchronized response will arise.

Calcium management of one astrocyte can affect many thousands of excitatory synapses nearby as shown by clamping intracellular $Ca^{2+}$ experiments. In these, clamping of calcium in individual hippocampal astrocytes is made through a whole-cell pipette while an extracellular field excitatory postsynaptic potential (EPSP) recording is done with an extracellular electrode placed either in the immediate vicinity of the clamped astrocyte or in a more distanced CA1 pyramidal cells group. Astrocytic $Ca^{2+}$ clamping blocked long term potentiation

(LTP) induced by tetanic stimulation of Schaeffer collaterals, at nearby, but not far away positions (Henneberger et al. 2010).

Astrocytes are ideally located to sense synapse activity with the perisynaptic processes and metabolic supply from blood vessels with the end feet processes. In fact, they mediate the response consisting in the modification of the local blood flow as function of synaptic or neuronal activity. It has been shown that astrocytes respond to increased neuronal activity by consuming more glucose and producing more lactate, this latter transferred into neighbor neurons as fuel during hyperactivity. As previously mentioned in Chapter "Glial Cells and Integrity of the Nervous System", this is known as the "astrocyte-neuron lactate shuttle" hypothesis (Pellerin et al. 2007).

## Astrocytes in the PreBötzinger Complex (preBötC)

The preBötC is the main generator of the inspiratory activity and a chemosensitive nucleus (Solomon 2003; Solomon et al. 2000). Fluctuations of the extracellular potassium concentrations are induced by the occurrence of rhythmic bursts of action potentials (Richter et al. 1978), which in turn are associated to fluctuation in the neurotransmitter release. Since astrocytes express $K^+$ channels (Kir4.1; KCNJ10), fluctuations in potassium concentrations generates fluctuations in the resting membrane potential, which can induce fluctuations in intracellular calcium concentration in astrocytes. Using whole-cell recordings from astrocytes and two-photon calcium imaging from rhythmic slices, none coupling between respiratory neuronal activity and astrocytic calcium signals was observed. The absence of correlation between respiratory neuronal activity and astrocytic calcium fluctuation (Schnell et al. 2011) indicates that astrocytic release of gliotransmitters is not commanding the respiratory like activity in neurons. Likely, one role of astrocytes in the preBötC is the control of extracellular levels of neurotransmitters and ions, both largely influencing the excitability of respiratory neurons.

## Astrocytes in Central Chemoreception

Over the last two decades, multiple pieces of evidence revealed that astrocytes can contribute to central chemoreception. Such contribution may be accomplished by astrocytes directly playing a role as $H^+/CO_2$ sensors or as part of the mechanisms underlying the cholinergic and glutamatergic hypothesis. Reduction in chemosensitivity of astrocytes may be involved in the pathogenesis of Rett syndrome and may explain the deficit in ventilatory responses to hypercapnia in these patients (Turovsky et al. 2015). Also, it has been proposed that astrocytes can play a modulatory role of the network in charge of the respiratory pattern generation by controlling the extracellular ion and transmitter concentrations (Neusch et al. 2006;

Szoke et al. 2006; Ballanyi et al. 2010; Erlichman and Leiter 2010). Likely, astrocytes in different chemosensitive regions also differ in their contributions to central chemoreception and the mechanisms underlying such contribution.

## Astrocyte Chemosensitivity

As illustrated in Fig. 2, several molecular mechanisms by which astrocytes detect $H^+/CO_2$ have been proposed

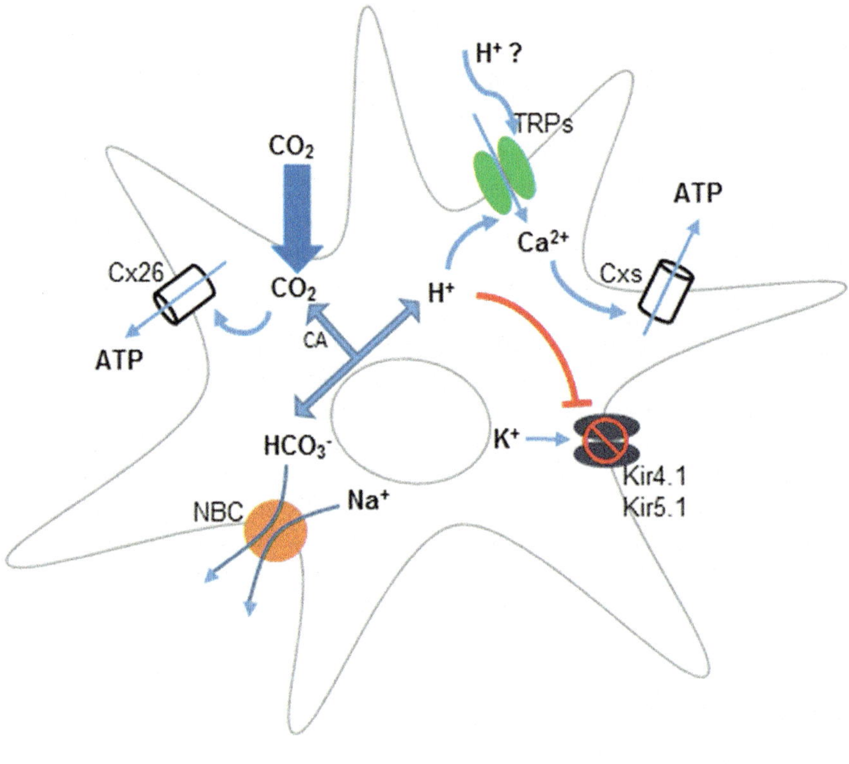


**Fig. 2** Astrocytes may sense acidosis or hypercapnia through different molecular sensors. Inwardly rectifying potassium (Kir) heteromeric channels Kir4.1–Kirk5.1 are inhibited by $CO_2$ resulting in depolarization of astrocytes (Wenker et al. 2010); carbonic anhydrase (CA) enzyme, the $Na^+$-$HCO_3^-$ cotransporter, $Na^+/H^+$ exchanger and the $Na^+$-dependent or $Na^+$-independent $Cl^-/HCO_3^-$ antiporters contribute to pH regulation (Brookes 1997; Baird et al. 1999; Makara et al. 2001; Schmitt et al. 2000; Deitmer and Rose 1996); connexins with a carbamylation motif (Cx26, Cx30, and Cx32), a site for binding $CO_2$ to induce the opening of connexin hemichannels (Meigh et al. 2013) endows cells with $CO_2$-sensitivity and the capacity for releasing ATP as a function of $PCO_2$ at constant extracellular pH (Huckstepp et al. 2010a); TRP channels endows to astrocytes with the ability for responding to hypercapnic but not isocapnic acidosis (Hirata and Oku 2010). It is possible that TRP activation could be given by extracellular acidification (Cui et al. 2011)

(1) Inwardly rectifying potassium (Kir) heteromeric channels Kir4.1–Kirk5.1. These channels contribute to the extracellular potassium regulation and are expressed in brainstem nuclei, including, among others, the LC, the ventrolateral medullary (VLM) area, the RTN, and the NTS (Wu et al. 2004). Kir4.1, and Kir5.1 channel subunits are observed in astrocytic processes contacting the pia mater, blood vessels, and synapses associated to PDZ domains containing syntrophins (Hibino et al. 2004). Depolarization of astrocytes by $CO_2$ would involve inhibition of heteromeric Kir4.1–Kir5.1 channels and contribution of $Na^+$-$HCO_3^-$ cotransporter (Wenker et al. 2010).
(2) Carbonic anhydrase enzyme and, in addition to the $Na^+$-$HCO_3^-$ cotransporter, several other transporters that contribute to pH regulation like the $Na^+/H^+$ exchanger and the $Na^+$-dependent or $Na^+$-independent $Cl^-/HCO_3^-$ antiporters (Brookes 1997; Baird et al. 1999; Makara et al. 2001; Schmitt et al. 2000; Deitmer and Rose 1996).
(3) Connexins presenting a carbamylation motif (Cx26, Cx30, and Cx32), a site for binding $CO_2$ to induce the opening of connexin hemichannels (Meigh et al. 2013) (see Chapter "Physiological Functions of Glial Cell Hemichannels" for further information on hemichannels). In particular, connexin 26 is abundantly expressed at the ventral medullary surface and its $CO_2$ sensitivity is within physiological range having a steep change in conductance centered around 40 mmHg $PCO_2$ (Huckstepp et al. 2010a, b). It is known that heterologous expression of Cx26 endows HeLa cells with $CO_2$-sensitivity and the capacity for releasing adenosine triphosphate (ATP) as a function of $PCO_2$ at constant extracellular pH (Huckstepp et al. 2010a). Accordingly, connexin hemichannel blockers reduce both the ATP release and the ventilatory response induced by hypercapnia in vivo and the ATP release induced by hypercarbia in vitro (Huckstepp et al. 2010b).
(4) Transient receptor potential (TRP) channels endows to astrocytes with the ability for responding to hypercapnic acidosis. This was assayed in enriched glia cells cultures using intracellular calcium- and pH-imaging in addition to perforated patch-clamp methods (Hirata and Oku 2010).

## Astrocyte Involvement in Respiratory Rhythm Modulation

Specific glial metabolic inhibitors have been used to evaluate astrocyte contribution to the ventilatory process. Fluorocitrate or fluoroacetate at low doses, are incorporated selectively by astrocytes and block the tricarboxylic acid (Krebs) cycle by inhibiting the enzyme aconitase. Administration of fluorocitrate into the RTN in either anesthetized mechanically ventilated or conscious adult rats increased the respiratory output (Erlichman et al. 1998; Holleran et al. 2001). This response can be explained on basis of the fluorocitrate-induced ATP and tissue pH decrease.

Inhibition of Krebs cycle reduces ATP levels, which in turn, reduces $Na^+$-$K^+$ATPase activity. Pump inactivation increases the extracellular potassium concentration and, subsequently, depolarizes, among others, chemosensitive neurons. Since chemosensitive neurons also respond to the acidification of the medium, and at the end, as overall result, the respiratory output is increased (Erlichman and Leiter 2010).

In contrast to in vivo experiments, fluoroacetate as well as methionine sulfoximine (MS), an inhibitor of glutamine synthetase, an enzyme present only in astrocytes that catalyzes the synthesis of glutamine from glutamate (see Chapter "Pharmacological Tools to Study the Role of Astrocytes in Neural Network Functions"), reduced the amplitude and frequency of the integrated inspiratory burst recorded from rhythmically active brainstem slices. At a first glance, these results suggest that astrocyte metabolic support or astrocyte functions depend on Krebs cycle and are necessary for the maintenance of the respiratory rhythm (Hulsmann 2000). In brainstem slices, evoked depolarization of the hypoglossal neurons by electrical stimulation of the ventral respiratory column (measured by optical imaging using voltage-sensitive dye) was reduced and delayed after fluoroacetate administration which is compatible with metabolic inhibition of fast synaptic transmission (Hulsmann et al. 2003). Accordingly, after fluoroacetate or MS treatment of brainstem slices, addition of glutamine restored the respiratory rhythm indicating that likely, the respiratory effects of both inhibitors were related, essentially, to impairment of the glutamate neurotransmission. In fact, fluoroacetate also impairs the astrocytic uptake of glutamate and the formation of glutamine (Swanson and Graham 1994)

In vivo administration of MS reduces basal ventilation and the ventilatory response to hypercapnia in conscious neonatal rats (Young et al. 2005a). By contrast, fluorocitrate administered into the RTN in vivo did not affect the respiratory response to acidosis or hypercapnia (Erlichman et al. 1998). Likely, the effects of fluorocitrate-induced reduction in ATP tissue pH oppose and predominate to the impairment in glutamate neurotransmission.

The hypothesis that astrocytes contribute to $H^+/CO_2$ sensitivity concatenates several steps: first, a subset of glial cells is depolarized in response to acidification (Fukuda et al. 1978; Fukuda and Honda 1975; Ritucci et al. 2005). Second, and derived from glial cell depolarization, intracellular $Ca^{2+}$ increases, which is required also for the inter-cellular propagation of calcium waves in glia (Guthrie et al. 1999); third, as a consequence of the intracellular $Ca^{2+}$ increase, ATP is released from astrocytes, likely through connexin hemichannels (Huckstepp et al. 2010b). In fact, electrochemical sensors placed at the ventral medullary surface can detect high levels of ATP ($3.8 \pm 0.9$ μM) during hypercapnia in anesthetized rats (Spyer et al. 2004; Gourine et al. 2005). Activation of glial purinoceptors by ATP can initiate self-propagating calcium waves that are proposed to influence local network excitability (Fiacco and McCarthy 2006). Finally, ATP, or other neuroactive molecules, will activate central chemoreceptor neurons such as those found in RTN/pFG (Spyer et al. 2004; Gourine et al. 2005).

ATP can act by binding to 7 subtypes of ionotropic P2X receptors (P2X1–7Rs) and eight subtypes of metabotropic P2YRs (P2Y1,2,4,6,11–14) (North 2002; Abbracchio et al. 2009).

According to this sequence of events, purinoceptor antagonists should impair the respiratory effects evoked by $CO_2$ stimulation. In fact, reduction and even abolition of ATP induced respiratory responses have been observed in vivo and in vitro (Thomas and Spyer 2000; Gourine et al. 2005; Zwicker et al. 2011; Gourine and Kasparov 2011; Gourine et al. 2010). Hypercapnia induces the release of ATP from the ventral surface of the medulla (Gourine et al. 2005). Further, application of ATP into the most rostral ATP-releasing site, corresponding likely to the retrotrapezoid nucleus, stimulated respiratory output, whereas application of ATP receptor antagonists like PPADS (pyridoxal-phosphate-6-azophenyl-2′,4′-disulfonate) to this area reduced $CO_2$ respiratory responses (Gourine et al. 2005). Hypercapnia-induced ATP-mediated excitation of the respiratory rhythm in rats is likely to involve a potent P2Y1R- activation of the preBötC (Lorier et al. 2007; Gourine et al. 2010). Such P2Y receptor-dependency has also found in vitro RTN neurons (Mulkey et al. 2006). Since pH sensitivity of RTN neurons in bicarbonate-free HEPES medium is not affected after purinergic receptor blockade with PPADS, ATP would play a role of modulator of the activity of pH-sensitive neurons, amplifying their responses to hypercapnia (Mulkey et al. 2006, 2004).

New insights of astrocyte contribution in modulating the function of respiratory neuronal circuits arise from application of molecular and electrophysiological methodology in conjunction with genetically engineered optical stimulation and $Ca^{2+}$ imaging tools. In an elegant work, Gourine et al. (2010) tested the hypothesis that rat astrocytes residing in RTN/pFRG behave as pH sensors, and trigger the respiratory response through the release of ATP. Astrocytes were genetically encoded with a $Ca^{2+}$ indicator associated to the promoter for glial fibrillary acidic protein (GFAP). They could confirm that these astrocytes, but not those from cerebral cortex, responded to physiological decreases in pH with elevations in intracellular $Ca^{2+}$ and ATP release. Accordingly, studies of vesicular fusion using total internal reflection fluorescence (TIRF) microscopy show that 35 % of astrocytes from rat brainstem in dissociated cultures respond to acidification with exocytosis of ATP-containing vesicles. Vesicles were visualized with fluorescent dyes quinacrine, an acridine derivative with very high affinity for ATP, and MANT-ATP, an ATP analogue esterified by the fluorescent methylisatoic acid. Vesicular exocytosis requires intracellular $Ca^{2+}$ signaling and was independent of autocrine ATP actions (Kasymov et al. 2013). By contrast, ATP was necessary to propagate astrocytic $Ca^{2+}$ excitation, since elimination of ATP by the ATP-hydrolyzing enzyme, apyrase, reduced importantly the $CO_2$-evoked astrocytic calcium responses (Gourine et al. 2010). In addition, ATP activating P2Y1Rs excited chemoreceptor neurons leading to the increase in the respiratory rhythm frequency (Gourine et al. 2010). Optogenetic stimulation of astrocytes expressing channelrhodopsin-2 associated under the command of GFAP promoter, resulted in a robust increase in breathing, associated to the increase in intracellular $Ca^{2+}$ in astrocytes. This optogenetic stimulation mimicked the hypercapnia and acidosis induced activation of chemoreceptor neurons via an ATP-dependent mechanism (Gourine et al. 2010). In agreement with these results, disruption of purinergic signaling decreases $CO_2$ sensitivity of RTN neurons by 25 % (Wenker et al. 2010) as well as gap junction blockers, which decrease $CO_2$-evoked ATP release in the RTN, reduced the whole-animal ventilatory response to $CO_2$ also by 25 % (Huckstepp et al. 2010b). In addition, fluorocitrate-induced depolarization of astrocytes evoked a reversible increase in firing rate of RTN neurons. This increase in neuronal firing rate was abolished by the presence of P2 receptor antagonists (PPADS or suramin) (Wenker et al. 2012) suggesting that a purinergic signaling was a mediator. Purinergic blockade also blunted the hypercapnic ventilatory response in vivo and the firing rate response of RTN neurons to hypercapnic stimulus of slices (10–15 % $CO_2$) (Wenker et al. 2012).

## Regional Differences in Contribution of Astrocytes to Central Chemoreception

As mentioned above, neurons responding to $CO_2$ with increased firing rate can be found, among other sites, at the RTN (Nattie et al. 1993a), RN (Iceman et al. 2013), and the caudal portion of the NTS (Dean et al. 1990; Nichols et al. 2009). Furthermore, focal acidification either by injection of acetazolamide within these three regions in anesthetized cats (Coates et al. 1993) or by microdialysis within these nuclei in unanesthetized awake or sleeping rats (Li et al. 1999; Nattie and Li 2001, 2002a) increases ventilation. Since RTN and RN neurons in culture have intrinsic $CO_2$-pH-sensitivities (Wang et al. 1998; Wang and Richerson 1999; Wang et al. 2013), glia would play a coadjuvant, synergic role in these chemoreceptive nuclei. There is not any study detailing the cytoarchitecture and properties of astrocytes at the different areas of the brainstem. Hitherto, the degree of cell-to-cell interconnections, the extension of astrocyte domains, and the differential expression of receptors, gliotransmitters, are mostly unknown. Since in other regions of the CNS, the population of astrocytes is heterogeneous in shapes and functions (Oberheim et al. 2012), it would not be strange that astrocytes belonging to different chemosensory nuclei at the brainstem differ in their structure and properties. Therefore, it is possible that the mechanisms through which astrocytes interact with the respiratory network at different nuclei could also be different.

As expressed before, data obtained at the RTN suggest the existence of a cascade of events triggered by hypercapnia or acidosis: depolarization of astrocytes, cytoplasmic calcium increase, ATP release, and ATP activation of respiratory neurons. It is worth to remember that this constitute the glial pathway for RTN neurons activation since RTN neurons are chemosensitive themselves. At that respect, the glial pathway appears as intensifier of the RTN neurons response to hypercapnia.

At the NTS and the RN, there is some controversial evidence pointing to the role of ATP as mediator of the response to hypercapnia. P2 receptors are expressed in the NTS and with less intensity, at the RN (Yao et al. 2000).

The administration of ATP or its analogues into the NTS, in awake rats produced cardiorespiratory responses (Antunes et al. 2005; De Paula et al. 2004). On the other hand, the injection of P2 receptor antagonists into NTS reduces the sympatho-excitatory response to peripheral chemoreflex activation (Braga et al. 2007; Boscan et al. 2002). Microinjection of ATP into the raphe magnus reduces the respiratory activity while that into the raphe pallidus increase it in anesthetized and artificially ventilated rats (Cao and Song 2007). The injection of the P2X broad-spectrum antagonist, PPADS, into the rostral medullary raphe blunted the ventilatory response to hypercapnia in conscious rats (da Silva et al. 2012), while this unaffected ventilation when injection was placed into the caudal RN of conscious rats (da Silva et al. 2012) or when it was done into raphe magnus or pallidus in anesthetized rats (Cao and Song 2007).

To test whether an astrocytic ATP-dependent mechanism was involved in central chemoreception at the RN and NTS, ATP antagonists were applied into these nuclei while chemoreflexes were evaluated in vivo as in vitro (Sobrinho et al. 2014). ATP injections into the caudal NTS (cNTS) increased cardiorespiratory activity in anesthetized rats (Sobrinho et al. 2014) confirming results obtained with the rat working heart-brainstem preparation (Antunes et al. 2005). By contrast, the injection of broad range purinergic receptor antagonists like PPADS or suramin into the cNTS did not affect basal ventilation or the ventilatory responses to changes in $CO_2/H^+$ as it does at the RTN (Sobrinho et al. 2014). In the case of RN the results were more negative, because both the injections of ATP or PPADS in anesthetized rats did not affect neither the basal ventilation nor the responsiveness to $H^+/CO_2$ (Sobrinho et al. 2014). Cell-attached NTS neurons recorded from brainstem slices increased their firing rate in response to ATP, while P2 receptors antagonists (PPDAS or suramin) did not modified NTS neurons response to hypercarbia. Likewise, the firing rate of RN neurons were not modified by ATP and their responses to changes in $PCO_2$/pH were unaffected by ATP-receptor blockade (Sobrinho et al. 2014).

Sobrinho et al. (2014) results are unexpected from previous reports indicating the existence of P2 receptors, and the respiratory-related effects of ATP agonist and antagonist injected into the NTS or RN. In fact, it is known that ATP in NTS plays a role in modulating the glutamatergic excitatory transmission as evidenced by the reduction in the amplitude of tractus solitaries-evoked excitatory postsynaptic currents (TS-eEPSCs) by purinergic antagonist (iso-PPADS). The glial cells are the source of ATP released by tractus solitarius electrical stimulation is suggested by the reduction in this TS-eEPSCs induced by the glia toxin, fluoroacetate (Accorsi-Mendonca et al. 2013). Likely, the inconsistency in results may be partly due to methodological differences, for example the use of anesthesia or the use of broad-spectrum antagonists which are weakly effective for blocking specific subset of P2 receptors. However, it remains possible that astrocytes contribute to the $CO_2/H^+$ responsiveness of cNTS and RN neurons, perhaps by an ATP-independent mechanism.

## Other Gliotransmitters

It is possible that other gliotransmitter, different to ATP, could serve as mediator in NTS or RN. A good candidate is D-serine. D-serine is a D-amino acid synthesized from L-serine by a pyridoxal 5′-phosphate-dependent serine racemase (SR) enzyme, which is present in neurons and astrocytes (Rosenberg et al. 2010; Wolosker 2011). D-serine binds with high affinity to the co-agonist (glycine) site of the N-methyl-D-aspartate (NMDA) glutamate receptor. D-serine effects have not been evaluated in the respiratory network, despite of NMDAR activation increases the respiratory frequency in vivo (Connelly et al. 1992) and in vitro (Greer et al. 1991). Preliminary data from our laboratory indicates that in *en bloc* preparations from neonatal mice, D-serine applied into the superfusion bath increases the respiratory rhythm of neonatal mice (Fig. 3).

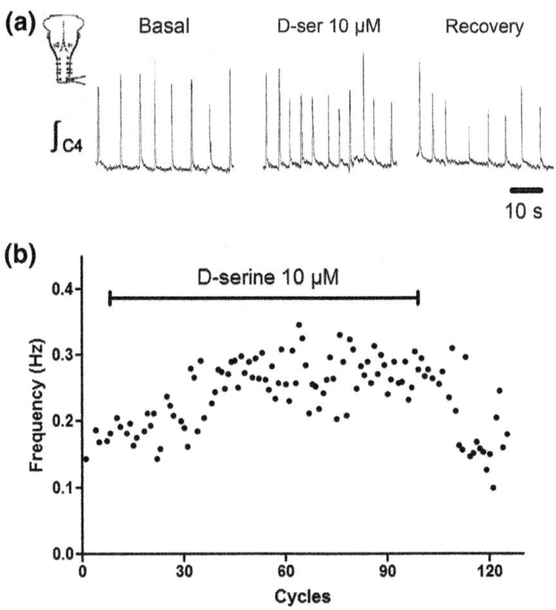


**Fig. 3** Increase of respiratory frequency induced by D-serine. **a** integrated inspiratory burst recorded from C4 ventral root in *en bloc* preparation obtained from CF1 mouse neonate at the third postnatal day before (basal), during, and after (recovery) the superfusion with aCSF containing D-serine 10 μM. **b** Instantaneous respiratory frequency measured cycle-to-cycle before, during (indicated by *horizontal bar*), and after the superfusion with aCSF containing D-serine 10 μM in the preparation from (**a**)

## Astrocytes and Cholinergic-Glutamatergic Hypothesis of Central Chemoreception

Historically, two neurotransmitters have been involved in central chemoreception, acetylcholine and glutamate, what is known as "the cholinergic and glutamate hypothesis of central chemoreception".

*Cholinergic (ACh) hypothesis*: Cholinergic neurons form part of input and output of the respiratory network. They are found at the NTS (Ruggiero et al. 1990; Armstrong et al. 1988; Gotts et al. 2015), the hypoglossal nuclei, facial nuclei, ambiguous nuclei (Kang et al. 2007), within the RN (Tatehata et al. 1987; Ruggiero et al. 1990), the nucleus reticularis rostroventrolateralis (RVL), and the ventral medullary surface (VMS); although cholinergic neurons are also detected in other localizations of the brainstem, like those in the medial portion of the rostral ventrolateral medulla (mRVLM), these would not be involved in cardiorespiratory events (Stornetta et al. 2013). The most important cholinergic inputs to the brainstem are originated from the pedunculopontine tegmental (PPT) and laterodorsal tegmental (LDT) nuclei. These inputs, as well as those provided by the serotonergic RN and the noradrenergic LC may be a clue for understanding pathogenia of respiratory dysfunctions associated to sleep–wake cycle, like sudden infant death syndrome (SIDS).

Muscarine or nicotine applied on the ventral surface of rostral and caudal medulla increase ventilation in anesthetized cats (Dev and Loeschcke 1979a, b). An endogenous cholinergic drive of the respiratory rhythm is revealed with acetylcholinesterase inhibitors (physostigmine, eserine) within rostral and caudal medulla (Dev and Loeschcke 1979a). In part, the respiratory cholinergic drive is exerted on the preBötC where activation of M3 and $\alpha 4\beta 2$ nicotinic receptors increases the frequency of the respiratory rhythm in neonatal rats and mice slices (Shao and Feldman 2005; Shao et al. 2008; Shao and Feldman 2009). A tonic cholinergic respiratory drive in the mouse *en bloc* preparation is revealed by application of atropine, a muscarinic receptor antagonist, which reduces the amplitude and frequency of the respiratory rhythm (Coddou et al. 2009).

That a cholinergic relay may be involved in central chemoreception at the surface of ventral medulla is derived from the fact that acetylcholine-sensitive areas and $H^+$- or $CO_2$-sensitive areas overlapped. In addition, application of cholinergic agonists on these sensitive areas elicits similar patterns of respiratory responses than those evoked by acidic stimulation (Loeschcke 1982; Eugenin and Nicholls 1997). Furthermore, central chemoreception and muscarinic cholinergic neurotransmission are strongly linked (Loeschcke 1982) as indicated by the brainstem distribution of muscarinic receptors (Nattie and Li 1990; Nattie et al. 1994; Mallios et al. 1995). Application of atropine to the rostral and caudal medulla decreases ventilation and, at the same time, reduces importantly the ventilatory response to $CO_2$ (Dev and Loeschcke 1979a; Nattie et al. 1989). Muscarinic blockade also reduces and, sometimes, abolishes the respiratory responses induced by $H^+$ or $CO_2$ in in vitro preparations from neonatal rats (Monteau et al. 1990), newborn opossum (Eugenin and Nicholls 1997), and neonatal mouse (Coddou et al. 2009). Microinjection of muscarinic M3 antagonist on the rostral ventrolateral medulla (RVLM) has a great efficacy for inhibiting respiratory $CO_2$-evoked response (Nattie and Li 1990). Interestingly, the arcuate nucleus, which is the human homologue of the RVLM, shows decreased muscarinic binding in SIDS infants (Kinney et al. 1995). Such probable reduction of the muscarinic binding in SIDS is compatible with the reduction of the muscarinic contribution to the chemosensory responses in *en bloc* and slices preparations from P0-P3 nicotine-exposed neonates by the prenatal-perinatal nicotine exposure (Coddou et al. 2009; Eugenin et al. 2008).

Unexpected results were obtained when muscarinic receptor knockout (KO) mice were challenged with hypercapnia (3 and 5 % CO2). M1 single KO mice showed normal, while M3 single KO mice showed reduced VT response slope to hypercapnia (Boudinot et al. 2004). Surprisingly, M1/3R or M2/4R double-KO mice showed unaltered chemosensory ventilatory responses (Boudinot et al. 2008). These results are puzzling and will require future research with conditional KO mice to evaluate muscarinic contribution to chemo reflexes in adults in absence of possible compensatory mechanisms exerted during development.

*Glutamate (Glu) hypothesis*: Excitatory glutamate neurotransmission predominates within the mammalian RPG, and the ventral surface of medulla is not an exception. Injection of glutamate into the RVLM increases ventilation in anesthetized cats (Li and Nattie 1995; Nattie and Li 1995). By contrast, microinjection of kynurenic acid, a nonselective glutamate receptor antagonist, or AP5, an NMDA receptor antagonist, or CNQX, a non-NMDA receptor antagonist, into the RVLM region decreased both the amplitude of the integrated phrenic nerve activity and the $CO_2$ sensitivity in a dose-dependent manner in anaesthetized cats (Nattie et al. 1993b). In contrast to in vivo experiments (Connelly et al. 1992), blockade of NMDARs in brainstem slices had a negligible effect on respiratory rhythm (Morgado-Valle and Feldman 2007; Greer et al. 1991), while the blockade of AMPARs completely abolished the rhythm. Similarly, NMDA receptor R1 subunit (NMDAR1) mutant mice were completely unresponsive to NMDA applications and showed a respiratory rhythm almost identical to that of controls. These results indicate that NMDA receptors are not relevant for generating the rhythm and for the development of circuits in charge of it (Funk et al. 1997). As for muscarinic receptors, the effects of glutamate antagonists have not been demonstrated to be specific for chemoreception.

Till now, acetylcholine (ACh) or glutamate (Glu) actions on chemosensitive areas are attributed to direct effects on neurons and a probable contribution of astrocytes in such responses has not been evaluated. Numerous studies demonstrate that astrocytes in different CNS regions express functional neurotransmitter receptors, which allow them to be sensitive to neurotransmitters like ACh and Glu (Perea and Araque 2010; Halassa and Haydon 2010; Ben Achour and Pascual 2010; Paixao and Klein 2010; Attwell et al. 2010; Sidoryk-Wegrzynowicz et al. 2011; Stipursky et al. 2011; Haydon and Carmignoto 2006; Erlichman et al. 2010). It is worth noting that astrocytes in the ventral respiratory group (VRG) express receptors for 5-HT, substance P (SP), and thyrotropin releasing hormone (TRH).

So, projections from chemosensitive RN neurons may modify the activation of astrocytes within the respiratory network (Hartel et al. 2009).

Astrocytes in the respiratory network respond to prevailing neuromodulators with an increase of intracellular calcium concentration (Huxtable et al. 2010; Gourine et al. 2010; Hartel et al. 2009). Besides, astrocytes are also capable of synthesizing and releasing neuro- and glio-transmitters such as ACh, Glu, ATP/adenosine, and D-serine (Haydon and Carmignoto 2006; Hamilton and Attwell 2010; Carmignoto et al. 1998; Araque et al. 2002; Hosli and Hosli 1994b; Hosli et al. 1988). So, theoretically, astrocytes may be involved in mediating or amplifying the ventilatory response to cholinergic and glutamatergic inputs by releasing gliotransmitters able of modifying the activity of the respiratory network. In addition, astrocytes can remove neurotransmitters from the synaptic cleft so they may participate in the control of the synaptic neurotransmitter concentration (Carmignoto et al. 1998; Araque et al. 2002; Hosli and Hosli 1994b; Hosli et al. 1988; Haydon and Carmignoto 2006). For example at the NTS, acidification can depolarize astrocytes by inhibition of both $K^+$ channel current and voltage-sensitive glutamate transporters (Huda et al. 2013). Therefore, as consequence of acidification at the NTS, the inhibition of this glutamate transporter, increases the levels of glutamate at the synaptic cleft affecting the excitatory synaptic transmission (Huda et al. 2013).

In the human infant, about 95 % of the arcuate nucleus neurons (corresponding to the chemosensitive RVLM in cats and rats) are glutamatergic. A large number of astrocytes in the ventral medullary surface express the vesicular glutamate transporter 2 and low levels of 5-$HT_{1A}$ and kainate (GluR5) receptors. So, it is reasonable to propose that astrocytes, which can also express muscarinic and nicotinic receptors (Gahring et al. 2004; Hosli et al. 1994; Hosli and Hosli 1994a, b), may store and release glutamate, possibly in response to stimulation by 5-HT, by ACh, or by glutamate itself (Paterson et al. 2006) affecting, in addition to the inhibition of glutamate uptake, the levels of glutamate at the synaptic cleft.

In addition, astrocytes play an essential role in glutamatergic synapses. Glutamate in the synaptic space is uptaken by astrocytes, converted by them into glutamine, and then transferred as glutamine to the presynaptic terminals for renewal of the glutamate presynaptic pool (Haydon and Carmignoto 2006). In the RPG, most of the excitatory synapses are glutamatergic; interestingly, 5 mM fluoroacetate or 0.1 mM methionine sulfoximine, both glial metabolic toxins, reduce the increase in respiratory frequency induced by ATP in brainstem slices, but they do not affect substance P evoked increase, suggesting that astrocytes contribute to the purinergic drive of the inspiratory rhythm generating network (Huxtable et al. 2010).

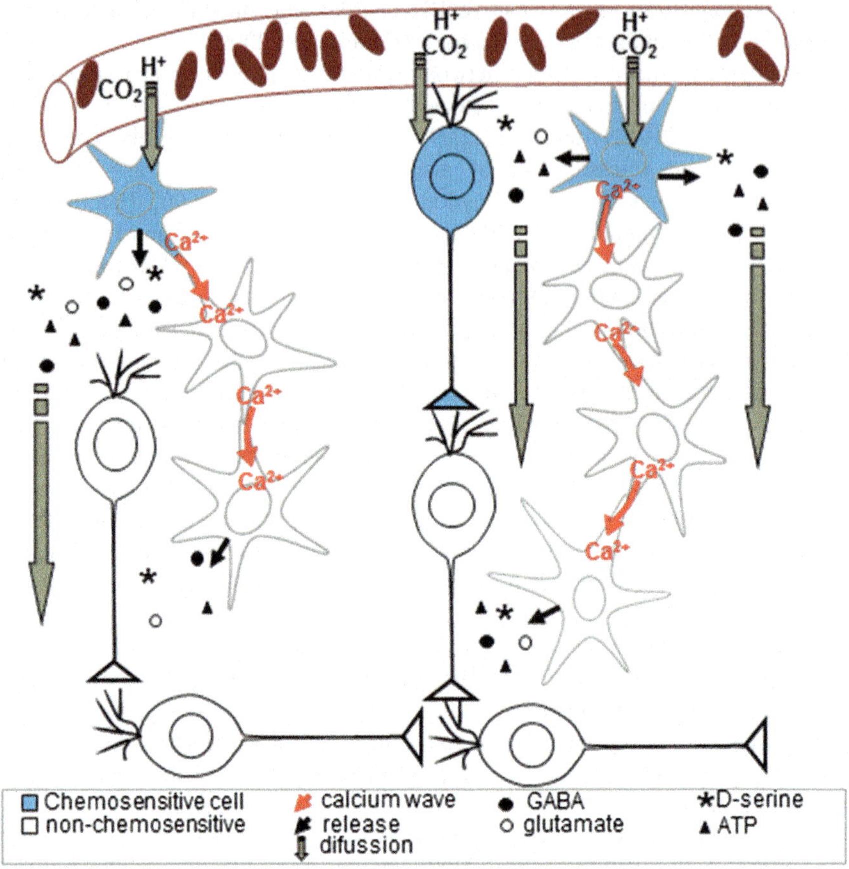


**Fig. 4** Schema of astrocyte contribution to the chemosensory response. Astrocytes enwrapping blood vessels or exposed to the CNS environment are continuously monitoring pH and $PCO_2$. As consequence of their activation (astrocyte depolarization) by acidosis or hypercapnia, there is an increase in intracellular calcium concentration, which may trigger the release, among others, of ACh, Glu, ATP, or D-serine and calcium waves that travel from astrocyte-to-astrocyte influencing the behavior of astrocyte according other astrocyte domains. Thus, the action of gliotransmitters at the local chemosensitive site may enhance the response of chemosensitive cells in the immediate environment. In addition, by volume diffusion of gliotransmitters and by activation of faraway astrocytes influenced by calcium waves more neurons of the respiratory network may be recruited

## Concluding Remarks

Astrocytes have diverse roles in modulation of the respiratory rhythm. These involve controlling neural network excitability through potassium buffering, regulation of synaptic transmitter concentrations via their synthesis, reuptake and release; in particular, at glutamatergic synapses, astrocyte is the source of glutamine, essential for replenish synaptic vesicles of glutamatergic neurons. Respect to respiratory central chemoreception, astrocytes have the ability of monitoring $PCO_2$ and pH and release gliotransmitters like ATP in the RTN, in response to changes in $CO_2$ and $H^+$ . In addition, they contribute to the regulation of the extracellular pH either by generating acidic substances derived from metabolic coupling (lactate shuttle) leading to amplification of hypercapnic stimulus or through proton buffering (transporters and channels).

On basis to the discussed properties of astrocytes (calcium waves, coupling of astrocytic domains through gap junctions, regulation of neurotransmitters and release of gliotransmitters) we propose that astrocytes may play two emergent roles in central respiratory chemoreception. A first role, as amplifiers of the responses of intrinsic chemosensitive neurons through feedforward signaling via gliotransmitters and a second role as recruiter of non-intrinsic chemosensitive cells thanks to volume spreading of signals (calcium waves and gliotransmitters) to regions far away the $CO_2/H^+$ sensitive domains (Fig. 4).

**Acknowledgments** Support from Grants Fondo Nacional de Desarrollo Científico y Tecnológico (FONDECYT) #1130874 (JE), Comisión Nacional de Ciencia y Tecnología (CONICYT) #21140669 (MJ Olivares), Comisión Nacional de Ciencia y Tecnología (CONICYT) #21120594 (S Beltrán-Castillo). DICYT-USACH (JE).

# References

Abbott SB, Stornetta RL, Fortuna MG, Depuy SD, West GH, Harris TE, Guyenet PG (2009) Photostimulation of retrotrapezoid nucleus phox2b-expressing neurons in vivo produces long-lasting activation of breathing in rats. J Neurosci 29(18):5806–5819. doi:10.1523/JNEUROSCI.1106-09.2009

Abbott SB, Stornetta RL, Coates MB, Guyenet PG (2011) Phox2b-expressing neurons of the parafacial region regulate breathing rate, inspiration, and expiration in conscious rats. J Neurosci 31(45):16410–16422. doi:10.1523/JNEUROSCI.3280-11.2011

Abbracchio MP, Burnstock G, Verkhratsky A, Zimmermann H (2009) Purinergic signalling in the nervous system: an overview. Trends Neurosci 32(1):19–29. doi:10.1016/j.tins.2008.10.001

Accorsi-Mendonca D, Zoccal DB, Bonagamba LG, Machado BH (2013) Glial cells modulate the synaptic transmission of NTS neurons sending projections to ventral medulla of Wistar rats. Physiol Rep 1(4):e00080. doi:10.1002/phy2.80

Akilesh MR, Kamper M, Li A, Nattie EE (1997) Effects of unilateral lesions of retrotrapezoid nucleus on breathing in awake rats. J Appl Physiol 82(2):469–479

Albrecht J, Simmons M, Dutton GR, Norenberg MD (1991) Aluminum chloride stimulates the release of endogenous glutamate, taurine and adenosine from cultured rat cortical astrocytes. Neurosci Lett 127(1):105–107

Alexandre C, Andermann ML, Scammell TE (2013) Control of arousal by the orexin neurons. Curr Opin Neurobiol 23(5):752–759. doi:10.1016/j.conb.2013.04.008

Amiel J, Laudier B, Attie-Bitach T, Trang H, de Pontual L, Gener B, Trochet D, Etchevers H, Ray P, Simonneau M, Vekemans M, Munnich A, Gaultier C, Lyonnet S (2003) Polyalanine expansion and frameshift mutations of the paired-like homeobox gene PHOX2B in congenital central hypoventilation syndrome. Nat Genet 33(4):459–461. doi:10.1038/ng1130

Amiel J, Dubreuil V, Ramanantsoa N, Fortin G, Gallego J, Brunet JF, Goridis C (2009) PHOX2B in respiratory control: lessons from congenital central hypoventilation syndrome and its mouse models. Respir Physiol Neurobiol 168(1–2):125–132. doi:10.1016/j.resp.2009.03.005

Antunes VR, Braga VA, Machado BH (2005) Autonomic and respiratory responses to microinjection of ATP into the intermediate or caudal nucleus tractus solitarius in the working heart-brainstem preparation of the rat. Clin Exp Pharmacol Physiol 32:467–472

Araque A, Martin ED, Perea G, Arellano JI, Buno W (2002) Synaptically released acetylcholine evokes $Ca^{2+}$ elevations in astrocytes in hippocampal slices. J Neurosci 22(7):2443–2450

Araque A, Carmignoto G, Haydon PG, Oliet SH, Robitaille R, Volterra A (2014) Gliotransmitters travel in time and space. Neuron 81(4):728–739. doi:10.1016/j.neuron.2014.02.007

Armstrong DM, Rotler A, Hersh LB, Pickel VM (1988) Localization of choline acetyltransferase in perikarya and dendrites within the nuclei of the solitary tracts. J Neurosci Res 20(3):279–290. doi:10.1002/jnr.490200302

Attwell D, Buchan AM, Charpak S, Lauritzen M, Macvicar BA, Newman EA (2010) Glial and neuronal control of brain blood flow. Nature 468(7321):232–243. doi:10.1038/nature09613

Baird NR, Orlowski J, Szabó EZ, Zaun HC, Schultheis PJ, Menon AG, Shull GE (1999) Molecular cloning, genomic organization, and functional expression of $Na^+/H^+$ exchanger isoform 5 (NHE5) from human brain. J Biol Chem 274(7):4377–4382

Ballantyne D, Scheid P (2001) Central respiratory chemosensitivity: cellular and network mechanisms. Adv Exp Med Biol 499:17–26

Ballanyi K, Panaitescu B, Ruangkittisakul A (2010) Control of breathing by "nerve glue". Sci Signal 3(147):e41. doi:10.1126/scisignal.3147pe41

Belanger M, Allaman I, Magistretti PJ (2011) Brain energy metabolism: focus on astrocyte-neuron metabolic cooperation. Cell Metab 14(6):724–738. doi:10.1016/j.cmet.2011.08.016

Belegu R, Hadziefendic S, Dreshaj IA, Haxhiu MA, Martin RJ (1999) CO2-induced c-fos expression in medullary neurons during early development. Respir Physiol 117(1):13–28

Ben Achour S, Pascual O (2010) Glia: the many ways to modulate synaptic plasticity. Neurochem Int 57(4):440–445. doi:10.1016/j.neuint.2010.02.013

Berger AJ, Cooney KA (1982) Ventilatory effects of kainic acid injection of the ventrolateral solitary nucleus. J Appl Physiol 52:131–140

Biancardi V, Bicego KC, Almeida MC, Gargaglioni LH (2008) Locus coeruleus noradrenergic neurons and $CO_2$ drive to breathing. Pflugers Arch 455(6):1119–1128. doi:10.1007/s00424-007-0338-8

Boscan P, Pickering AE, Paton JFR (2002) The nucleus of the solitary tract: an integrating station for nociceptive and cardiorespiratory afferents. Exp Physiol 87(2):259–266

Boudinot E, Yamada M, Wess J, Champagnat J, Foutz AS (2004) Ventilatory pattern and chemosensitivity in M1 and M3 muscarinic receptor knockout mice. Respir Physiol Neurobiol 139(3):237–245. doi:10.1016/j.resp.2003.10.006

Boudinot E, Champagnat J, Foutz AS (2008) M(1)/M(3) and M(2)/M(4) muscarinic receptor double-knockout mice present distinct respiratory phenotypes. Respir Physiol Neurobiol 161 (1):54–61. doi:10.1016/j.resp.2007.12.001

Bowery NG, Brown DA, Collins GG, Galvan M, Marsh S, Yamini G (1976) Indirect effects of amino-acids on sympathetic ganglion cells mediated through the release of gamma-aminobutyric acid from glial cells. Br J Pharmacol 57(1):73–91

Braga VA, Soriano RN, Braccialli AL, de Paula PM, Bonagamba LG, Paton JF, Machado BH (2007) Involvement of L-glutamate and ATP in the neurotransmission of the sympathoexcitatory component of the chemoreflex in the commissural nucleus tractus solitarii of awake rats and in the working heart-brainstem preparation. J Physiol 581(Pt 3):1129–1145. doi:10.1113/jphysiol.2007.129031

Brookes N (1997) Intracellullar pH as a regulatory signal in astrocyte metabolism. Glia 21:64–73

Brunet JF, Pattyn A (2002) Phox2 genes—from patterning to connectivity. Curr Opin Genet Dev 12(4):435–440

Brust RD, Corcoran AE, Richerson GB, Nattie E, Dymecki SM (2014) Functional and developmental identification of a molecular subtype of brain serotonergic neuron specialized to regulate breathing dynamics. Cell Rep 9(6):2152–2165. doi:10.1016/j.celrep.2014.11.027

Burke PG, Kanbar R, Viar KE, Stornetta RL, Guyenet PG (2015) Selective optogenetic stimulation of the retrotrapezoid nucleus in sleeping rats activates breathing without changing blood pressure or causing arousal or sighs. J Appl Physiol 118(12):1491–1501. doi:10.1152/japplphysiol.00164.2015 (1985)

Cao Y, Song G (2007) Purinergic modulation of respiration via medullary raphe nuclei in rats. Respir Physiol Neurobiol 155(2):114–120. doi:10.1016/j.resp.2006.04.013

Caravagna C, Soliz J, Seaborn T (2013) Brain-derived neurotrophic factor interacts with astrocytes and neurons to control respiration. Eur J Neurosci 38(9):3261–3269. doi:10.1111/ejn.12320

Carmignoto G, Pasti L, Pozzan T (1998) On the role of voltage-dependent calcium channels in calcium signaling of astrocytes in situ. J Neurosci 18(12):4637–4645

Chen X, Wang L, Zhou Y, Zheng LH, Zhou Z (2005) "Kiss-and-run" glutamate secretion in cultured and freshly isolated rat hippocampal astrocytes. J Neurosci 25(40):9236–9243. doi:10.1523/JNEUROSCI.1640-05.2005

Coates EL, Li A, Nattie EE (1993) Widespread sites of brain stem ventilatory chemoreceptors. J Appl Physiol 75(1):5–14

Coddou C, Bravo E, Eugenin J (2009) Alterations in cholinergic sensitivity of respiratory neurons induced by pre-natal nicotine: a mechanism for respiratory dysfunction in neonatal mice. Philos Trans R Soc Lond 364(1529):2527–2535

Connelly CA, Otto-Smith MR, Feldman JL (1992) Blockade of NMDA receptor-channels by MK-801 alters breathing in adult rats. Brain Res 596(1–2):99–110

Constam DB, Philipp J, Malipiero UV, ten Dijke P, Schachner M, Fontana A (1992) Differential expression of transforming growth factor-beta 1,—beta 2, and—beta 3 by glioblastoma cells, astrocytes, and microglia. J Immunol 148(5):1404–1410

Corcoran AE, Richerson GB, Harris MB (2013) Serotonergic mechanisms are necessary for central respiratory chemoresponsiveness in situ. Respir Physiol Neurobiol 186(2):214–220. doi:10.1016/j.resp.2013.02.015

Cornell-Bell AH, Finkbeiner SM, Cooper MS, Smith SJ (1990) Glutamate induces calcium waves in cultured astrocytes: long-range glial signaling. Science 247(4941):470–473

Corsini E, Dufour A, Ciusani E, Gelati M, Frigerio S, Gritti A, Cajola L, Mancardi GL, Massa G, Salmaggi A (1996) Human brain endothelial cells and astrocytes produce IL-1 beta but not IL-10. Scand J Immunol 44(5):506–511

Cui N, Zhang X, Tapedalli JS, Yu L, Gai H, Petit J, Pamulapati RT, Jin X, Jiang C (2011) Involvement of TRP channels in the $CO_2$ chemosensitivity of locus coeruleus neurons. J Neurophysiol 105:2791–2801. doi:10.1152/jn.00759.2010.-Catecholaminergic

Curran AK, Darnall RA, Filiano JJ, Li A, Nattie EE (2001) Muscimol dialysis in the rostral ventral medulla reduced the CO(2) response in awake and sleeping piglets. J Appl Physiol 90(3):971–980

da Silva GS, Li A, Nattie E (2010) High $CO_2/H^+$ dialysis in the caudal ventrolateral medulla (Loeschcke's area) increases ventilation in wakefulness. Respir Physiol Neurobiol 171(1):46–53. doi:10.1016/j.resp.2010.01.014

da Silva GS, Giusti H, Benedetti M, Dias MB, Gargaglioni LH, Branco LG, Glass ML (2011) Serotonergic neurons in the nucleus raphe obscurus contribute to interaction between central and peripheral ventilatory responses to hypercapnia. Pflugers Arch 462(3):407–418. doi:10.1007/s00424-011-0990-x

da Silva GS, Moraes DJ, Giusti H, Dias MB, Glass ML (2012) Purinergic transmission in the rostral but not caudal medullary raphe contributes to the hypercapnia-induced ventilatory response in unanesthetized rats. Respir Physiol Neurobiol 184(1):41–47. doi:10.1016/j.resp.2012.07.015

De Paula PM, Antunes VR, Bonagamba LG, Machado BH (2004) Cardiovascular responses to microinjection of ATP into the nucleus tractus solitarii of awake rats. Am J Physiol 287:R1164–R1171. doi:10.1152/ajpregu.00722.2003

Dean JB, Bayliss DA, Erickson JT, Lawing WL, Millhorn DE (1990) Depolarization and stimulation of neurons in nucleus tractus solitarii by carbon dioxide does not require chemical synaptic input. Neuroscience 36(1):207–216

Deitmer JW, Rose CR (1996) pH regulation and proton signalling by glial cells. Prog Neurobiol 48(2):73 103

Deng BS, Nakamura A, Zhang W, Yanagisawa M, Fukuda Y, Kuwaki T (2007) Contribution of orexin in hypercapnic chemoreflex: evidence from genetic and pharmacological disruption and supplementation studies in mice. J Appl Physiol 103(5):1772–1779. doi:10.1152/japplphysiol.00075.2007 (1985)

Dev NB, Loeschcke HH (1979a) A cholinergic mechanism involved in the respiratory chemosensitivity. Pflügers Arch 379:29–36

Dev NB, Loeschcke HH (1979b) Topography of the respiratory and circulatory responses to acetylcholine and nicotine on the ventral surface of the medulla oblongata. Pflügers Arch 379:19–27

Dias MB, Nucci TB, Margatho LO, Antunes-Rodrigues J, Gargaglioni LH, Branco LG (2007) Raphe magnus nucleus is involved in ventilatory but not hypothermic response to $CO_2$. J Appl Physiol 103(5):1780–1788

Dias MB, Li A, Nattie E (2008) Focal $CO_2$ dialysis in raphe obscurus does not stimulate ventilation but enhances the response to focal $CO_2$ dialysis in the retrotrapezoid nucleus. J Appl Physiol 105(1):83–90

Dias MB, Li A, Nattie EE (2009) Antagonism of orexin receptor-1 in the retrotrapezoid nucleus inhibits the ventilatory response to hypercapnia predominantly in wakefulness. J Physiol 587 (Pt 9):2059–2067. doi:10.1113/jphysiol.2008.168260

Dias MB, Li A, Nattie E (2010) The orexin receptor 1 (OX1R) in the rostral medullary raphe contributes to the hypercapnic chemoreflex in wakefulness, during the active period of the diurnal cycle. Respir Physiol Neurobiol 170(1):96–102. doi:10.1016/j.resp.2009.12.002

Dreshaj IA, Haxhiu MA, Martin RJ (1998) Role of the medullary raphe nuclei in the respiratory response to $CO_2$. Respir Physiol 111(1):15–23

Dubreuil V, Ramanantsoa N, Trochet D, Vaubourg V, Amiel J, Gallego J, Brunet JF, Goridis C (2008) A human mutation in phox2b causes lack of $CO_2$ chemosensitivity, fatal central apnea, and specific loss of parafacial neurons. Proc Natl Acad Sci USA 105(3):1069–1072. doi:10.1073/pnas.0709115105

Dubreuil V, Barhanin J, Goridis C, Brunet JF (2009a) Breathing with Phox2b. Philos Trans R Soc Lond 364:2477–2483. doi:10.1098/rstb.2009.0085

Dubreuil V, Thoby-Brisson M, Rallu M, Persson K, Pattyn A, Birchmeier C, Brunet JF, Fortin G, Goridis C (2009b) Defective respiratory rhythmogenesis and loss of central chemosensitivity in Phox2b mutants targeting retrotrapezoid nucleus neurons. J Neurosci 29(47):14836–14846. doi:10.1523/JNEUROSCI.2623-09.2009

Dutschmann M, Kron M, Morschel M, Gestreau C (2007) Activation of Orexin B receptors in the pontine Kolliker-Fuse nucleus modulates pre-inspiratory hypoglossal motor activity in rat. Respir Physiol Neurobiol 159(2):232–235

Erlichman JS, Leiter JC (2010) Glia modulation of the extracellular milieu as a factor in central $CO_2$ chemosensitivity and respiratory control. J Appl Physiol 108:1803–1811. doi:10.1152/japplphysiol.01321.2009.-We

Erlichman JS, Li A, Nattie EE (1998) Ventilatory effects of glial dysfunction in a rat brain stem chemoreceptor region. J Appl Physiol 85(5):1599–1604

Erlichman JS, Leiter JC, Gourine AV (2010) ATP, glia and central respiratory control. Respir Physiol Neurobiol 173(3):305–311. doi:10.1016/j.resp.2010.06.009

Eugenin J (1995) Generation of the respiratory rhythm: modelling the inspiratory off switch as a neural integrator. J Theor Biol 172(2):107–120

Eugenin J, Nicholls JG (1997) Chemosensory and cholinergic stimulation of fictive respiration in isolated CNS of neonatal opossum. J Physiol (London) 501(Pt 2):425–437

Eugenin J, von Bernhardi R, Muller KJ, Llona I (2006) Development and pH sensitivity of the respiratory rhythm of fetal mice in vitro. Neuroscience 141(1):223–231

Eugenin J, Otarola M, Bravo E, Coddou C, Cerpa V, Reyes-Parada M, Llona I, von Bernhardi R (2008) Prenatal to early postnatal nicotine exposure impairs central chemoreception and modifies breathing pattern in mouse neonates: a probable link to sudden infant death syndrome. J Neurosci 28(51):13907–13917

Eyzaguirre C, Fitzgerald RS, Lahiri S, Zapata P (1983) Arterial chemoreceptors. In: Shepherd JT, Abboud FM (eds) American physiological society: handbook of physiology, vol 3., The Cardiovascular SystemWilliams & Wilkins Co., Baltimore, Maryland, pp 557–621

Feldman JL (1986) Neurophysiology of breathing in mammals. In: Bloom FE (ed) Handbook of physiology, vol IV. Williams & Wilkins Co., Bethesda, Maryland, pp 463–524

Feldman JL, Mitchell GS, Nattie EE (2003) Breathing: rhythmicity, plasticity, chemosensitivity. Annu Rev Neurosci 26:239–266. doi:10.1146/annurev.neuro.26.041002.131103

Fiacco TA, McCarthy KD (2006) Astrocyte calcium elevations: properties, propagation, and effects on brain signaling. Glia 54(7):676–690. doi:10.1002/glia.20396

Fukuda Y, Honda Y (1975) pH sensitive cells at ventrolateral surface of the rat medulla oblongata. Nat New Biol 256:317–318

Fukuda Y, Honda Y, Schlaefke ME, Loeschcke HH (1978) Effect of $H^+$ on the membrane potential of silent cells in the ventral and dorsal surface layer of the rat medulla in vitro. Pflügers Arch 376:229–235

Funk GD (2010) The 'connexin' between astrocytes, ATP and central respiratory chemoreception. J Physiol 588(Pt 22):4335–4337. doi:10.1113/jphysiol.2010.200196

Funk GD, Johnson SM, Smith JC, Dong X-W, Lai J, Feldman JL (1997) Functional respiratory rhythm generating networks in neonatal mice lacking NMDAR1 gene. J Neurophysiol 78:1414–1420

Furukawa S, Furukawa Y, Satoyoshi E, Hayashi K (1986) Synthesis and secretion of nerve growth factor by mouse astroglial cells in culture. Biochem Biophys Res Commun 136(1):57–63

Gahring LC, Persiyanov K, Rogers SW (2004) Neuronal and astrocyte expression of nicotinic receptor subunit beta4 in the adult mouse brain. J Comp Neurol 468(3):322–333. doi:10.1002/cne.10942

Gebicke-Haerter PJ, Seregi A, Schobert A, Hertting G (1988) Involvement of protein kinase C in prostaglandin D(2) synthesis by cultured astrocytes. Neurochem Int 13(4):475–480

Goridis C, Dubreuil V, Thoby-Brisson M, Fortin G, Brunet JF (2010) Phox2b, congenital central hypoventilation syndrome and the control of respiration. Sem Cell Dev Biol 21(8):814–822. doi:10.1016/j.semcdb.2010.07.006

Gotts J, Atkinson L, Edwards IJ, Yanagawa Y, Deuchars SA, Deuchars J (2015) Co-expression of GAD67 and choline acetyltransferase reveals a novel neuronal phenotype in the mouse medulla oblongata. Auton Neurosci. doi:10.1016/j.autneu.2015.05.003

Gourine AV, Kasparov S (2011) Astrocytes as brain interoceptors. Exp Physiol 96(4):411–416. doi:10.1113/expphysiol.2010.053165

Gourine AV, Llaudet E, Dale N, Spyer KM (2005) ATP is a mediator of chemosensory transduction in the central nervous system. Nature 436(7047):108–111. doi:10.1038/nature03690

Gourine AV, Kasymov V, Marina N, Tang F, Figueiredo MF, Lane S, Teschemacher AG, Spyer KM, Deisseroth K, Kasparov S (2010) Astrocytes control breathing through pH-dependent release of ATP. Science 329(5991):571–575. doi:10.1126/science.1190721

Grass D, Pawlowski PG, Hirrlinger J, Papadopoulos N, Richter DW, Kirchhoff F, Hulsmann S (2004) Diversity of functional astroglial properties in the respiratory network. J Neurosci 24 (6):1358–1365. doi:10.1523/JNEUROSCI.4022-03.2004

Gray PA, Hayes JA, Ling GY, Llona I, Tupal S, Picardo MC, Ross SE, Hirata T, Corbin JG, Eugenin J, Del Negro CA (2010) Developmental origin of preBotzinger complex respiratory neurons. J Neurosci 30(44):14883–14895. doi:10.1523/JNEUROSCI.4031-10.2010

Greer JJ, Smith JC, Feldman JL (1991) Role of excitatory amino acids in the generation and transmission of respiratory drive in neonatal rat. J Physiol 437:727–749

Guček A, Vardjan N, Zorec R (2012) Exocytosis in astrocytes: transmitter release and membrane signal regulation. Neurochem Res 37(11):2351–2363. doi:10.1007/s11064-012-0773-6

Guthrie PB, Knappenberger J, Segal M, Bennett MV, Charles AC, Kater SB (1999) ATP released from astrocytes mediates glial calcium waves. J Neurosci 19(2):520–528

Guyenet PG, Mulkey DK (2010) Retrotrapezoid nucleus and parafacial respiratory group. Respir Physiol Neurobiol 173(3):244–255. doi:10.1016/j.resp.2010.02.005

Guyenet PG, Mulkey DK, Stornetta RL, Bayliss DA (2005) Regulation of ventral surface chemoreceptors by the central respiratory pattern generator. J Neurosci 25(39):8938–8947

Halassa MM, Haydon PG (2010) Integrated brain circuits: astrocytic networks modulate neuronal activity and behavior. Annu Rev Physiol 72:335–355. doi:10.1146/annurev-physiol-021909-135843

Hamilton NB, Attwell D (2010) Do astrocytes really exocytose neurotransmitters? Nat Rev 11 (4):227–238. doi:10.1038/nrn2803

Hartel K, Schnell C, Hulsmann S (2009) Astrocytic calcium signals induced by neuromodulators via functional metabotropic receptors in the ventral respiratory group of neonatal mice. Glia 57 (8):815–827. doi:10.1002/glia.20808

Hartung HP, Toyka KV (1987) Phorbol diester TPA elicits prostaglandin E release from cultured rat astrocytes. Brain Res 417(2):347–349

Hartung HP, Heininger K, Schafer B, Toyka KV (1988) Substance P stimulates release of arachidonic acid cyclooxygenation products from primary culture rat astrocytes. Ann N Y Acad Sci 540:427–429

Haydon PG, Carmignoto G (2006) Astrocyte control of synaptic transmission and neurovascular coupling. Physiol Rev 86(3):1009–1031. doi:10.1152/physrev.00049.2005

Henneberger C, Papouin T, Oliet SH, Rusakov DA (2010) Long-term potentiation depends on release of D-serine from astrocytes. Nature 463(7278):232–236. doi:10.1038/nature08673

Hibino H, Fujita A, Iwai K, Yamada M, Kurachi Y (2004) Differential assembly of inwardly rectifying K+ channel subunits, Kir4.1 and Kir5.1, in brain astrocytes. J Biol Chem 279 (42):44065–44073. doi:10.1074/jbc.M405985200

Hirata Y, Oku Y (2010) TRP channels are involved in mediating hypercapnic $Ca^{2+}$ responses in rat glia-rich medullary cultures independent of extracellular pH. Cell Calcium 48(2–3):124–132. doi:10.1016/j.ceca.2010.07.006

Hodges MR, Klum L, Leekley T, Brozoski DT, Bastasic J, Davis S, Wenninger JM, Feroah TR, Pan LG, Forster HV (2004a) Effects on breathing in awake and sleeping goats of focal acidosis in the medullary raphe. J Appl Physiol 96(5):1815–1824. doi:10.1152/japplphysiol.00992. 2003 (1985)

Hodges MR, Opansky C, Qian B, Davis S, Bonis J, Bastasic J, Leekley T, Pan LG, Forster HV (2004b) Transient attenuation of $CO_2$ sensitivity after neurotoxic lesions in the medullary raphe area of awake goats. J Appl Physiol 97(6):2236–2247

Hodges MR, Tattersall GJ, Harris MB, McEvoy SD, Richerson DN, Deneris ES, Johnson RL, Chen ZF, Richerson GB (2008) Defects in breathing and thermoregulation in mice with near-complete absence of central serotonin neurons. J Neurosci 28(10):2495–2505

Hodges MR, Best S, Richerson GB (2011) Altered ventilatory and thermoregulatory control in male and female adult Pet-1 null mice. Respir Physiol Neurobiol 177(2):133–140. doi:10.1016/j.resp.2011.03.020

Holleran J, Babbie M, Erlichman JS (2001) Ventilatory effects of impaired glial function in a brainstem chemoreceptor region in the conscious rat. J Appl Physiol 90:1539–1547

Hosli E, Hosli L (1994a) Binding of cholecystokinin, bombesin and muscarine to neurons and astrocytes in explant cultures of rat central nervous system: autoradiographic and immunohistochemical studies. Neuroscience 61(1):63–72

Hosli E, Hosli L (1994b) Colocalization of binding sites for somatostatin, muscarine and nicotine on cultured neurones of rat neocortex, cerebellum, brain stem and spinal cord: combined autoradiographic and immunohistochemical studies. Neurosci Lett 173(1–2):71–74

Hosli L, Hosli E, Della-Briotta G, Quadri L, Heuss L (1988) Action of acetylcholine, muscarine, nicotine and antagonists on the membrane potential of astrocytes in cultured rat brainstem and spinal cord. Neurosci Lett 92(2):165–170

Hosli L, Hosli E, Winter T, Stauffer S (1994) Coexistence of cholinergic and somatostatin receptors on astrocytes of rat CNS. NeuroReport 5(12):1469–1472

Huckstepp RT, Eason R, Sachdev A, Dale N (2010a) $CO_2$-dependent opening of connexin 26 and related beta connexins. J Physiol 588(Pt 20):3921–3931. doi:10.1113/jphysiol.2010.192096

Huckstepp RT, id Bihi R, Eason R, Spyer KM, Dicke N, Willecke K, Marina N, Gourine AV, Dale N (2010b) Connexin hemichannel-mediated $CO_2$-dependent release of ATP in the medulla oblongata contributes to central respiratory chemosensitivity. J Physiol 588(Pt 20):3901–3920. doi:10.1113/jphysiol.2010.192088

Huda R, McCrimmon DR, Martina M (2013) pH modulation of glial glutamate transporters regulates synaptic transmission in the nucleus of the solitary tract. J Neurophysiol 110:368–377. doi:10.1152/jn.01074.2012

Hulsmann S (2000) Metabolic coupling between glia and neurons is necessary for maintaining respiratory activity in transverse medulary slices of neonatal mouse. Eur J Neurosci 12:7

Hulsmann S, Straub H, Richter DW, Speckmann EJ (2003) Blockade of astrocyte metabolism causes delayed excitation as revealed by voltage-sensitive dyes in mouse brainstem slices. Exp Brain Res 150(1):117–121. doi:10.1007/s00221-003-1410-z

Huxtable AG, Zwicker JD, Alvares TS, Ruangkittisakul A, Fang X, Hahn LB, Posse de Chaves E, Baker GB, Ballanyi K, Funk GD (2010) Glia contribute to the purinergic modulation of inspiratory rhythm-generating networks. J Neurosci 30(11):3947–3958. doi:10.1523/JNEUROSCI.6027-09.2010

Iadecola C, Nedergaard M (2007) Glial regulation of the cerebral microvasculature. Nat Neurosci 10(11):1369–1376. doi:10.1038/nn2003

Iceman KE, Richerson GB, Harris MB (2013) Medullary serotonin neurons are $CO_2$ sensitive in situ. J Neurophysiol 110(11):2536–2544. doi:10.1152/jn.00288.2013

Infante CD, von Bernhardi R, Rovegno M, Llona I, Eugenin JL (2003) Respiratory responses to pH in the absence of pontine and dorsal medullary areas in the newborn mouse in vitro. Brain Res 984(1–2):198–205

Kadle R, Suksang C, Roberson ED, Fellows RE (1988) Identification of an insulin-like factor in astrocyte conditioned medium. Brain Res 460(1):60–67

Kanbar R, Stornetta RL, Cash DR, Lewis SJ, Guyenet PG (2010) Photostimulation of Phox2b medullary neurons activates cardiorespiratory function in conscious rats. Am J Respir Crit Care Med 182(9):1184–1194. doi:10.1164/rccm.201001-0047OC

Kang BJ, Chang DA, Mackay DD, West GH, Moreira TS, Takakura AC, Gwilt JM, Guyenet PG, Stornetta RL (2007) Central nervous system distribution of the transcription factor Phox2b in the adult rat. J Comp Neurol 503(5):627–641. doi:10.1002/cne.21409

Kasymov V, Larina O, Castaldo C, Marina N, Patrushev M, Kasparov S, Gourine AV (2013) Differential sensitivity of brainstem versus cortical astrocytes to changes in pH reveals functional regional specialization of astroglia. J Neurosci 33(2):435–441. doi:10.1523/JNEUROSCI.2813-12.2013

Kimelberg HK, Goderie SK, Higman S, Pang S, Waniewski RA (1990) Swelling-induced release of glutamate, aspartate, and taurine from astrocyte cultures. J Neurosci 10(5):1583–1591

Kinney HC, Filiano JJ, Sleeper LA, Mandell F, Valdes-Dapena M, White WF (1995) Decreased muscarinic receptor binding in the arcuate nucleus in sudden infant death syndrome. Science 269(5229):1446–1450

Krause KL, Forster HV, Davis SE, Kiner T, Bonis JM, Pan LG, Qian B (2009) Focal acidosis in the pre-Botzinger complex area of awake goats induces a mild tachypnea. J Appl Physiol 106(1):241–250. doi:10.1152/japplphysiol.90547.2008 (1985)

Krzan M, Stenovec M, Kreft M, Pangrsic T, Grilc S, Haydon PG, Zorec R (2003) Calcium-dependent exocytosis of atrial natriuretic peptide from astrocytes. J Neurosci 23(5):1580–1583

Kuwaki T, Li A, Nattie E (2010) State-dependent central chemoreception: a role of orexin. Respir Physiol Neurobiol 173(3):223–229. doi:10.1016/j.resp.2010.02.006

Lee MG, Hassani OK, Jones BE (2005) Discharge of identified orexin/hypocretin neurons across the sleep-waking cycle. J Neurosci 25(28):6716–6720. doi:10.1523/JNEUROSCI.1887-05.2005

Li A, Nattie EE (1995) Prolonged stimulation of respiration by brain stem metabotropic glutamate receptors. J Appl Physiol 79(5):1650–1656

Li A, Nattie E (2002) $CO_2$ dialysis in one chemoreceptor site, the RTN: stimulus intensity and sensitivity in the awake rat. Respir Physiol Neurobiol 133(1–2):11–22

Li A, Nattie E (2008) Serotonin transporter knockout mice have a reduced ventilatory response to hypercapnia (predominantly in males) but not to hypoxia. J Physiol 586(9):2321–2329. doi:10.1113/jphysiol.2008.152231

Li A, Nattie E (2010) Antagonism of rat orexin receptors by almorexant attenuates central chemoreception in wakefulness in the active period of the diurnal cycle. J Physiol 588(Pt 15):2935–2944. doi:10.1113/jphysiol.2010.191288

Li A, Randall M, Nattie EE (1999) CO(2) microdialysis in retrotrapezoid nucleus of the rat increases breathing in wakefulness but not in sleep. J Appl Physiol 87(3):910–919

Li A, Zhou S, Nattie E (2006) Simultaneous inhibition of caudal medullary raphe and retrotrapezoid nucleus decreases breathing and the $CO_2$ response in conscious rats. J Physiol 577(Pt 1):307–318

Li N, Li A, Nattie E (2013) Focal microdialysis of CO(2) in the perifornical-hypothalamic area increases ventilation during wakefulness but not NREM sleep. Respir Physiol Neurobiol 185 (2):349–355. doi:10.1016/j.resp.2012.09.007

Loeschcke HH (1982) Central chemosensitivity and the reaction theory. J Physiol 332:1–24

Lorier AR, Huxtable AG, Robinson DM, Lipski J, Housley GD, Funk GD (2007) P2Y1 receptor modulation of the pre-Botzinger complex inspiratory rhythm generating network in vitro. J Neurosci 27(5):993–1005. doi:10.1523/JNEUROSCI.3948-06.2007

Makara JK, Petheö GL, Tóth A, Spät A (2001) pH-sensitive inwardly rectifying chloride current in cultured rat cortical astrocytes. Glia 34:52–58

Mallios VJ, Lydic R, Baghdoyan HA (1995) Muscarinic receptor subtypes are differentially distributed across brain stem respiratory nuclei. Am J Physiol 268:L941–L949

Marina N, Abdala AP, Trapp S, Li A, Nattie EE, Hewinson J, Smith JC, Paton JF, Gourine AV (2010) Essential role of Phox2b-expressing ventrolateral brainstem neurons in the chemosensory control of inspiration and expiration. J Neurosci 30(37):12466–12473. doi:10.1523/JNEUROSCI.3141-10.2010

Martino PF, Davis S, Opansky C, Krause K, Bonis JM, Pan LG, Qian B, Forster HV (2007) The cerebellar fastigial nucleus contributes to $CO_2$-$H^+$ ventilatory sensitivity in awake goats. Respir Physiol Neurobiol 157(2–3):242–251. doi:10.1016/j.resp.2007.01.019

Meigh L, Greenhalgh SA, Rodgers TL, Cann MJ, Roper DI, Dale N (2013) CO(2)directly modulates connexin 26 by formation of carbamate bridges between subunits. Elife 2:e01213. doi:10.7554/eLife.01213

Mercure L, Tannenbaum GS, Schipper HM, Phaneuf D, Wainberg MA (1996) Expression of the somatostatin gene in human astrocytoma cell lines. Clin Diagn Lab Immunol 3(2):151–155

Messier ML, Li A, Nattie EE (2002) Muscimol inhibition of medullary raphe neurons decreases the $CO_2$ response and alters sleep in newborn piglets. Respir Physiol Neurobiol 133(3):197–214

Messier ML, Li A, Nattie EE (2004) Inhibition of medullary raphe serotonergic neurons has age-dependent effects on the $CO_2$ response in newborn piglets. J Appl Physiol 96(5):1909–1919

Mitchell RA, Loeschcke HH, Massion WH, Severinghaus JW (1963) Respiratory responses mediated through superficial chemosensitive areas on the medulla. J Appl Physiol 18(3):523–533

Monteau R, Morin D, Hilaire G (1990) Acetylcholine and central chemosensitivity: in vitro study in the newborn rat. Respir Physiol 81:241–254

Morgado-Valle C, Feldman JL (2007) NMDA receptors in preBotzinger complex neurons can drive respiratory rhythm independent of AMPA receptors. J Physiol 582(Pt 1):359–368. doi:10.1113/jphysiol.2007.130617

Mulkey DK, Stornetta RL, Weston MC, Simmons JR, Parker A, Bayliss DA, Guyenet PG (2004) Respiratory control by ventral surface chemoreceptor neurons in rats. Nat Neurosci 7 (12):1360–1369

Mulkey DK, Mistry AM, Guyenet PG, Bayliss DA (2006) Purinergic P2 receptors modulate excitability but do not mediate pH sensitivity of RTN respiratory chemoreceptors. J Neurosci 26(27):7230–7233. doi:10.1523/JNEUROSCI.1696-06.2006

Murphy S, Minor RL Jr, Welk G, Harrison DG (1990) Evidence for an astrocyte-derived vasorelaxing factor with properties similar to nitric oxide. J Neurochem 55(1):349–351
Nakamura A, Zhang W, Yanagisawa M, Fukuda Y, Kuwaki T (2007) Vigilance state-dependent attenuation of hypercapnic chemoreflex and exaggerated sleep apnea in orexin knockout mice. J Appl Physiol 102(1):241–248. doi:10.1152/japplphysiol.00679.2006 (1985)
Nattie E (1999) $CO_2$, brainstem chemoreceptors and breathing. Prog Neurobiol 59(4):299–331
Nattie EE (2001) Central chemosensitivity, sleep, and wakefulness. Respir Physiol 129(1–2):257–268
Nattie E (2011) Julius H. Comroe, Jr., distinguished lecture: central chemoreception: then ... and now. J Appl Physiol 110(1):1–8. doi:10.1152/japplphysiol.01061.2010
Nattie E, Li A (1990) Ventral medulla sites of muscarinic receptor subtypes involved in cardiorespiratory control. J Appl Physiol 69(1):33–41
Nattie EE, Li A (1994) Retrotrapezoid nucleus lesions decrease phrenic activity and $CO_2$ sensitivity in rats. Respir Physiol 97(1):63–77
Nattie EE, Li A (1995) Rat retrotrapezoid nucleus iono- and metabotropic glutamate receptors and the control of breathing. J Appl Physiol 78(1):153–163
Nattie E, Li A (2000) Muscimol dialysis in the retrotrapezoid nucleus region inhibits breathing in the awake rat. J Appl Physiol 89(1):153–162
Nattie EE, Li A (2001) $CO_2$ dialysis in the medullary raphe of the rat increases ventilation in sleep. J Appl Physiol 90(4):1247–1257
Nattie EE, Li A (2002a) $CO_2$ dialysis in nucleus tractus solitarius region of rat increases ventilation in sleep and wakefulness. J Appl Physiol 92(5):2119–2130
Nattie EE, Li A (2002b) Substance P-saporin lesion of neurons with NK1 receptors in one chemoreceptor site in rats decreases ventilation and chemosensitivity. J Physiol 544 (Pt 2):603–616
Nattie E, Li A (2006) Central chemoreception 2005: a brief review. Auton Neurosci 126–127:332–338
Nattie G, Li A (2008) Multiple central chemoreceptor sites: cell types and function in vivo. Adv Exp Med Biol 605:343–347
Nattie E, Li A (2009) Central chemoreception is a complex system function that involves multiple brain stem sites. J Appl Physiol 106(4):1464–1466. doi:10.1152/japplphysiol.00112.2008
Nattie E, Li A (2010) Central chemoreception in wakefulness and sleep: evidence for a distributed network and a role for orexin. J Appl Physiol 108(5):1417–1424. doi:10.1152/japplphysiol.01261.2009
Nattie E, Li A (2012) Central chemoreceptors: locations and functions. Compr Physiol 2(1):221–254. doi:10.1002/cphy.c100083
Nattie EE, Wood J, Mega A, Goritski W (1989) Rostral ventrolateral medulla muscarinic receptor involvement in central ventilatory chemosensitivity. J Appl Physiol 66(3):1462–1470
Nattie EE, Fung ML, Li A, St John WM (1993a) Responses of respiratory modulated and tonic units in the retrotrapezoid nucleus to $CO_2$. Respir Physiol 94(1):35–50
Nattie EE, Gdovin M, Li A (1993b) Retrotrapezoid nucleus glutamate receptors: control of $CO_2$-sensitive phrenic and sympathetic output. J Appl Physiol 74(6):2958–2968
Nattie EE, Li A, Mills J, Huang Q (1994) Retrotrapezoid nucleus muscarinic receptor subtypes localized by autoradiography. Respir Physiol 96(2–3):189–197
Nattie EE, Li A, Richerson GB, Lappi DA (2004) Medullary serotonergic neurones and adjacent neurones that express neurokinin-1 receptors are both involved in chemoreception in vivo. J Physiol 556(Pt 1):235–253. doi:10.1113/jphysiol.2003.059766
Neusch C, Papadopoulos N, Muller M, Maletzki I, Winter SM, Hirrlinger J, Handschuh M, Bahr M, Richter DW, Kirchhoff F, Hulsmann S (2006) Lack of the Kir4.1 channel subunit abolishes K+ buffering properties of astrocytes in the ventral respiratory group: impact on extracellular K+ regulation. J Neurophysiol 95(3):1843–1852. doi:10.1152/jn.00996.2005
Newton K, Malik V, Lee-Chiong T (2014) Sleep and breathing. Clin Chest Med 35(3):451–456. doi:10.1016/j.ccm.2014.06.001

Nichols NL, Wilkinson KA, Powell FL, Dean JB, Putnam RW (2009) Chronic hypoxia suppresses the $CO_2$ response of solitary complex (SC) neurons from rats. Respir Physiol Neurobiol 168 (3):272–280. doi:10.1016/j.resp.2009.07.012

North RA (2002) Molecular physiology of P2X receptors. Physiol Rev 82:1013–1067

Oberheim NA, Takano T, Han X, He W, Lin JH, Wang F, Xu Q, Wyatt JD, Pilcher W, Ojemann JG, Ransom BR, Goldman SA, Nedergaard M (2009) Uniquely hominid features of adult human astrocytes. J Neurosci 29(10):3276–3287. doi:10.1523/JNEUROSCI.4707-08.2009

Oberheim NA, Goldman SA, Nedergaard M (2012) Heterogeneity of astrocytic form and function. Methods Mol Biol 814:23–45. doi:10.1007/978-1-61779-452-0_3

Ohno K, Sakurai T (2008) Orexin neuronal circuitry: role in the regulation of sleep and wakefulness. Front Neuroendocrinol 29(1):70–87. doi:10.1016/j.yfrne.2007.08.001

Onimaru H, Homma I (2006) Point:Counterpoint: The parafacial respiratory group (pFRG)/pre-Botzinger complex (preBotC) is the primary site of respiratory rhythm generation in the mammal. Point: the PFRG is the primary site of respiratory rhythm generation in the mammal. J Appl Physiol 100(6):2094–2095

Onimaru H, Kumagawa Y, Homma I (2006) Respiration-related rhythmic activity in the rostral medulla of newborn rats. J Neurophysiol 96(1):55–61

Onimaru H, Ikeda K, Kawakami K (2008) $CO_2$-sensitive preinspiratory neurons of the parafacial respiratory group express Phox2b in the neonatal rat. J Neurosci 28(48):12845–12850. doi:10.1523/JNEUROSCI.3625-08.2008

Onimaru H, Ikeda K, Kawakami K (2009) Phox2b, RTN/pFRG neurons and respiratory rhythmogenesis. Respir Physiol Neurobiol 168(1–2):13–18. doi:10.1016/j.resp.2009.03.007

Oyamada Y, Ballantyne D, Muckenhoff K, Scheid P (1998) Respiration-modulated membrane potential and chemosensitivity of locus coeruleus neurones in the in vitro brainstem-spinal cord of the neonatal rat. J Physiol (London) 513(Pt 2):381–398

Paixao S, Klein R (2010) Neuron-astrocyte communication and synaptic plasticity. Curr Opin Neurobiol 20(4):466–473. doi:10.1016/j.conb.2010.04.008

Parpura V, Basarsky TA, Liu F, Jeftinija K, Jeftinija S, Haydon PG (1994) Glutamate-mediated astrocyte-neuron signalling. Nature 369(6483):744–747. doi:10.1038/369744a0

Paterson DS, Thompson EG, Kinney HC (2006) Serotonergic and glutamatergic neurons at the ventral medullary surface of the human infant: Observations relevant to central chemosensitivity in early human life. Auton Neurosci 124(1–2):112–124. doi:10.1016/j.autneu.2005.12.009

Pattyn A, Morin X, Cremer H, Goridis C, Brunet JF (1999) The homeobox gene Phox2b is essential for the development of autonomic neural crest derivatives. Nature 399(6734):366–370. doi:10.1038/20700

Pellerin L, Magistretti PJ (1994) Glutamate uptake into astrocytes stimulates aerobic glycolysis: a mechanism coupling neuronal activity to glucose utilization. Proc Natl Acad Sci USA 91 (22):10625–10629

Pellerin L, Bouzier-Sore AK, Aubert A, Serres S, Merle M, Costalat R, Magistretti PJ (2007) Activity-dependent regulation of energy metabolism by astrocytes: an update. Glia 55 (12):1251–1262. doi:10.1002/glia.20528

Perea G, Araque A (2010) GLIA modulates synaptic transmission. Brain Res Rev 63(1–2):93–102. doi:10.1016/j.brainresrev.2009.10.005

Perea G, Sur M, Araque A (2014) Neuron-glia networks: integral gear of brain function. Front Cell Neurosci 8:378. doi:10.3389/fncel.2014.00378

Ramanantsoa N, Hirsch MR, Thoby-Brisson M, Dubreuil V, Bouvier J, Ruffault PL, Matrot B, Fortin G, Brunet JF, Gallego J, Goridis C (2011) Breathing without $CO_2$ chemosensitivity in conditional Phox2b mutants. J Neurosci 31(36):12880–12888. doi:10.1523/JNEUROSCI.1721-11.2011

Ray RS, Corcoran AE, Brust RD, Kim JC, Richerson GB, Nattie E, Dymecki SM (2011) Impaired respiratory and body temperature control upon acute serotonergic neuron inhibition. Science 333(6042):637–642. doi:10.1126/science.1205295

Ray RS, Corcoran AE, Brust RD, Soriano LP, Nattie EE, Dymecki SM (2013) Egr2-neurons control the adult respiratory response to hypercapnia. Brain Res 1511:115–125. doi:10.1016/j.brainres.2012.12.017

Richerson GB (1995) Response to $CO_2$ of neurons in the rostral ventral medulla in vitro. J Neurophysiol 73(3):933–944

Richter DW, Spyer KM (2001) Studying rhythmogenesis of breathing: comparison of in vivo and in vitro models. Trends Neurosci 24(8):464–472

Richter DW, Camerer H, Sonnhof U (1978) Changes in extracellular potassium during the spontaneous activity of medullary respiratory neurones. Pflugers Arch 376(2):139–149

Ritucci NA, Erlichman JS, Leiter JC, Putnam RW (2005) Response of membrane potential and intracellular pH to hypercapnia in neurons and astrocytes from rat retrotrapezoid nucleus. Am J Physiol 289:R851–R861. doi:10.1152/ajpregu.00132.2005.-We

Rodriguez-Arellano JJ, Parpura V, Zorec R, Verkhratsky A (2015) Astrocytes in physiological aging and Alzheimer's disease. Neuroscience. doi:10.1016/j.neuroscience.2015.01.007

Rosenberg D, Kartvelishvily E, Shleper M, Klinker CM, Bowser MT, Wolosker H (2010) Neuronal release of D-serine: a physiological pathway controlling extracellular D-serine concentration. FASEB J 24(8):2951–2961. doi:10.1096/fj.09-147967

Ruggiero DA, Giuliano R, Anwar M, Stornetta R, Reis DJ (1990) Anatomical substrates of cholinergic-autonomic regulation in the rat. J Comp Neurol 292(1):1–53. doi:10.1002/cne.902920102

Sakurai T (2014) The role of orexin in motivated behaviours. Nat Rev 15(11):719–731. doi:10.1038/nrn3837

Schell MJ, Molliver ME, Snyder SH (1995) D-serine, an endogenous synaptic modulator: localization to astrocytes and glutamate-stimulated release. Proc Natl Acad Sci USA 92(9):3948–3952

Schmitt BM, Berger UV, Douglas RM, Bevensee MO, Hediger MA, Haddad GG, Boron WF (2000) Na/HCO3 cotransporters in rat brain: expression in glia, neurons, and choroid plexus. J Neurosci 20(18):6839–6848

Schnell C, Fresemann J, Hulsmann S (2011) Determinants of functional coupling between astrocytes and respiratory neurons in the pre-Botzinger complex. PLoS One 6(10):e26309. doi:10.1371/journal.pone.0026309

Selmaj KW, Farooq M, Norton WT, Raine CS, Brosnan CF (1990) Proliferation of astrocytes in vitro in response to cytokines. A primary role for tumor necrosis factor. J Immunol 144(1):129–135

Shao XM, Feldman JL (2005) Cholinergic neurotransmission in the preBotzinger Complex modulates excitability of inspiratory neurons and regulates respiratory rhythm. Neuroscience 130(4):1069–1081

Shao XM, Feldman JL (2009) Central cholinergic regulation of respiration: nicotinic receptors. Acta Pharmacol Sin 30(6):761–770. doi:10.1038/aps.2009.88

Shao XM, Tan W, Xiu J, Puskar N, Fonck C, Lester HA, Feldman JL (2008) Alpha4* nicotinic receptors in preBotzinger complex mediate cholinergic/nicotinic modulation of respiratory rhythm. J Neurosci 28(2):519–528. doi:10.1523/JNEUROSCI.3666-07.2008

Shinoda H, Marini AM, Cosi C, Schwartz JP (1989) Brain region and gene specificity of neuropeptide gene expression in cultured astrocytes. Science 245(4916):415–417

Sidoryk-Wegrzynowicz M, Wegrzynowicz M, Lee E, Bowman AB, Aschner M (2011) Role of astrocytes in brain function and disease. Toxicol Pathol 39(1):115–123. doi:10.1177/0192623310385254

Smith JC, Ellenberger H, Ballanyi K, Richter DW, Feldman JL (1991) Pre-Bötzinger complex: a brainstem region that may generate respiratory rhythm in mammals. Science 254(5032):726–729

Sobrinho CR, Wenker IC, Poss EM, Takakura AC, Moreira TS, Mulkey DK (2014) Purinergic signalling contributes to chemoreception in the retrotrapezoid nucleus but not the nucleus of the solitary tract or medullary raphe. J Physiol 592(Pt 6):1309–1323. doi:10.1113/jphysiol.2013.268490

Solomon IC (2003) Focal $CO_2/H^+$ alters phrenic motor output response to chemical stimulation of cat pre-Botzinger complex in vivo. J Appl Physiol 94(6):2151–2157
Solomon IC, Edelman NH, O'Neal MH 3rd (2000) CO(2)/H(+) chemoreception in the cat pre-Botzinger complex in vivo. J Appl Physiol 88(6):1996–2007
Spyer KM, Dale N, Gourine AV (2004) ATP is a key mediator of central and peripheral chemosensory transduction. Exp Physiol 89(1):53–59
Stipursky J, Romao L, Tortelli V, Neto VM, Gomes FC (2011) Neuron-glia signaling: Implications for astrocyte differentiation and synapse formation. Life Sci 89(15–16):524–531. doi:10.1016/j.lfs.2011.04.005
Stornetta RL, Moreira TS, Takakura AC, Kang BJ, Chang DA, West GH, Brunet JF, Mulkey DK, Bayliss DA, Guyenet PG (2006) Expression of Phox2b by brainstem neurons involved in chemosensory integration in the adult rat. J Neurosci 26(40):10305–10314. doi:10.1523/JNEUROSCI.2917-06.2006
Stornetta RL, Macon CJ, Nguyen TM, Coates MB, Guyenet PG (2013) Cholinergic neurons in the mouse rostral ventrolateral medulla target sensory afferent areas. Brain Struct Funct 218 (2):455–475. doi:10.1007/s00429-012-0408-3
Sunanaga J, Deng BS, Zhang W, Kanmura Y, Kuwaki T (2009) $CO_2$ activates orexin-containing neurons in mice. Respir Physiol Neurobiol 166(3):184–186. doi:10.1016/j.resp.2009.03.006
Swanson RA, Graham SH (1994) Fluorocitrate and fluoroacetate effects on astrocyte metabolism in vitro. Brain Res 664(1–2):94–100
Szoke K, Hartel K, Grass D, Hirrlinger PG, Hirrlinger J, Hulsmann S (2006) Glycine transporter 1 expression in the ventral respiratory group is restricted to protoplasmic astrocytes. Brain Res 1119(1):182–189. doi:10.1016/j.brainres.2006.08.089
Takakura AC, Moreira TS, Stornetta RL, West GH, Gwilt JM, Guyenet PG (2008) Selective lesion of retrotrapezoid Phox2b-expressing neurons raises the apnoeic threshold in rats. J Physiol 586 (Pt 12):2975–2991. doi:10.1113/jphysiol.2008.153163 jphysiol.2008.153163[pii]
Takakura AC, Barna BF, Cruz JC, Colombari E, Moreira TS (2014) Phox2b-expressing retrotrapezoid neurons and the integration of central and peripheral chemosensory control of breathing in conscious rats. Exp Physiol 99(3):571–585. doi:10.1113/expphysiol.2013.076752
Tatehata T, Shiosaka S, Wanaka A, Rao ZR, Tohyama M (1987) Immunocytochemical localization of the choline acetyltransferase containing neuron system in the rat lower brain stem. J Hirnforsch 28(6):707–716
Taylor NC, Li A, Nattie EE (2006) Ventilatory effects of muscimol microdialysis into the rostral medullary raphe region of conscious rats. Respir Physiol Neurobiol 153(3):203–216
Teppema LJ, Veening JG, Kranenburg A, Dahan A, Berkenbosch A, Olievier C (1997) Expression of c-fos in the rat brainstem after exposure to hypoxia and to normoxic and hyperoxic hypercapnia. J Comp Neurol 388(2):169–190
Terada J, Nakamura A, Zhang W, Yanagisawa M, Kuriyama T, Fukuda Y, Kuwaki T (2008) Ventilatory long-term facilitation in mice can be observed during both sleep and wake periods and depends on orexin. J Appl Physiol 104(2):499–507. doi:10.1152/japplphysiol.00919.2007 (1985)
Thomas T, Spyer KM (2000) ATP as a mediator of mammalian central $CO_2$ chemoreception. J Physiol (London) 523(Pt 2):441–447
Toyama S, Sakurai T, Tatsumi K, Kuwaki T (2009) Attenuated phrenic long-term facilitation in orexin neuron-ablated mice. Respir Physiol Neurobiol 168(3):295–302. doi:10.1016/j.resp.2009.07.025
Turovsky E, Karagiannis A, Abdala AP, Gourine AV (2015) Impaired $CO_2$ sensitivity of astrocytes in a mouse model of Rett syndrome. J Physiol 593(14):3159–3168. doi:10.1113/JP270369
von Euler C (1986) Brain stem mechanisms for generation and control of breathing pattern. In: Cherniack NS, Widdicombe JG (eds) American physiological society: handbook of physiology, vol 2., control of breathingWilliams & Wilkins Co., Baltimore, Maryland, pp 1–68

Wang W, Richerson GB (1999) Development of chemosensitivity of rat medullary raphe neurons. Neuroscience 90(3):1001–1011

Wang W, Pizzonia JH, Richerson GB (1998) Chemosensitivity of rat medullary raphe neurones in primary tissue culture. J Physiol (London) 511(Pt 2):433–450

Wang S, Shi Y, Shu S, Guyenet PG, Bayliss DA (2013) Phox2b-expressing retrotrapezoid neurons are intrinsically responsive to $H^+$ and $CO_2$. J Neurosci 33(18):7756–7761. doi:10.1523/JNEUROSCI.5550-12.2013

Wenker IC, Kreneisz O, Nishiyama A, Mulkey DK (2010) Astrocytes in the retrotrapezoid nucleus sense $H^+$ by inhibition of a Kir4.1–Kir5.1-like current and may contribute to chemoreception by a purinergic mechanism. J Neurophysiol 104(6):3042–3052. doi:10.1152/jn.00544.2010

Wenker IC, Sobrinho CR, Takakura AC, Moreira TS, Mulkey DK (2012) Regulation of ventral surface $CO_2/H^+$ -sensitive neurons by purinergic signalling. J Physiol 590(Pt 9):2137–2150. doi:10.1113/jphysiol.2012.229666

Westergaard N, Sonnewald U, Schousboe A (1994) Release of alpha-ketoglutarate, malate and succinate from cultured astrocytes: possible role in amino acid neurotransmitter homeostasis. Neurosci Lett 176(1):105–109

Wickstrom R, Hokfelt T, Lagercrantz H (2002) Development of CO(2)-response in the early newborn period in rat. Respir Physiol Neurobiol 132(2):145–158

Williams RH, Jensen LT, Verkhratsky A, Fugger L, Burdakov D (2007) Control of hypothalamic orexin neurons by acid and $CO_2$. Proc Natl Acad Sci USA 104(25):10685–10690. doi:10.1073/pnas.0702676104

Wolosker H (2011) Serine racemase and the serine shuttle between neurons and astrocytes. Biochim Biophys Acta. doi:10.1016/j.bbapap.2011.01.001

Wu J, Xu H, Shen W, Jiang C (2004) Expression and coexpression of $CO_2$-sensitive Kir channels in brainstem neurons of rats. J Membrane Biol 197:179–191. doi:10.1007/s00232-004-0652-4

Wu Z, Zhang J, Nakanishi H (2005) Leptomeningeal cells activate microglia and astrocytes to induce IL-10 production by releasing pro-inflammatory cytokines during systemic inflammation. J Neuroimmunol 167(1–2):90–98. doi:10.1016/j.jneuroim.2005.06.025

Xu F, Frazier DT (1995) Medullary respiratory neuronal activity modulated by stimulation of the fastigial nucleus of the cerebellum. Brain Res 705(1–2):53–64

Xu F, Zhang Z, Frazier DT (2001) Microinjection of acetazolamide into the fastigial nucleus augments respiratory output in the rat. J Appl Physiol 91(5):2342–2350

Xu G, Wang W, Kimelberg HK, Zhou M (2010) Electrical coupling of astrocytes in rat hippocampal slices under physiological and simulated ischemic conditions. Glia 58(4):481–493. doi:10.1002/glia.20939

Yao ST, Barden JA, Finkelstein DI, Bennett MR, Lawrence AJ (2000) Comparative study on the distribution patterns of P2X1-P2X6 receptor immunoreactivity in the brainstem of the rat and the common marmoset (Callithrix jacchus): association with catecholamine cell groups. J Comp Neurol 427:485–507

Young JK, Dreshaj IA, Wilson CG, Martin RJ, Zaidi SI, Haxhiu MA (2005a) An astrocyte toxin influences the pattern of breathing and the ventilatory response to hypercapnia in neonatal rats. Respir Physiol Neurobiol 147(1):19–30. doi:10.1016/j.resp.2005.01.009

Young JK, Wu M, Manaye KF, Kc P, Allard JS, Mack SO, Haxhiu MA (2005b) Orexin stimulates breathing via medullary and spinal pathways. J Appl Physiol 98(4):1387–1395. doi:10.1152/japplphysiol.00914.2004 (1985)

Yudkoff M, Daikhin Y, Nissim I, Pleasure D, Stern J (1994) Inhibition of astrocyte glutamine production by alpha-ketoisocaproic acid. J Neurochem 63(4):1508–1515

Zwicker JD, Rajani V, Hahn LB, Funk GD (2011) Purinergic modulation of preBotzinger complex inspiratory rhythm in rodents: the interaction between ATP and adenosine. J Physiol 589(Pt 18):4583–4600. doi:10.1113/jphysiol.2011.210930

# Purine Signaling and Microglial Wrapping

**Bernardo Castellano, Mar Bosch-Queralt, Beatriz Almolda, Nàdia Villacampa and Berta González**

**Abstract** Microglial cells are highly dynamic cells with processes continuously moving to survey the surrounding territory. Microglia possess a broad variety of surface receptors and subtle changes in their microenvironment cause microglial cell processes to extend, retract, and interact with neuronal synaptic contacts. When the nervous system is disturbed, microglia activate, proliferate, and migrate to sites of injury in response to alert signals. Released nucleotides like ATP and UTP are among the wide range of molecules promoting microglial activation and guiding their migration and phagocytic function. The increased concentration of nucleotides in the extracellular space could be involved in the microglial wrapping found around injured neurons in various pathological conditions, especially after peripheral axotomy. Microglial wrappings isolate injured neurons from synaptic inputs and facilitate the molecular dialog between endangered or injured neurons and activated microglia. Astrocytes may also participate in neuronal ensheathment. Degradation of ATP by microglial ecto-nucleotidases and the expression of various purine receptors might be decisive in regulating the function of enwrapping glial cells and in determining the fate of damaged neurons, which may die or may regenerate their axons and survive.

**Keywords** ATP · Adenosine · CD39 · 'Eat-me' signals · Neuronal degeneration · Nerve injury · Microglial migration · Phagocytosis · Purine receptors · Axotomy

## Abbreviations

| | |
|---|---|
| CNS | Central nervous system |
| PAMPs | Pathogen associated molecular patterns |
| DAMPs | Damage associated molecular patterns |
| TLRs | Toll-like receptors |

---

B. Castellano (✉) · M. Bosch-Queralt · B. Almolda · N. Villacampa · B. González
Unit of Histology, Torre M5, Department of Cell Biology, Physiology and Immunology, Institute of Neurosciences, Universitat Autònoma de Barcelona, 08193 Bellaterra, Spain
e-mail: bernardo.castellano@uab.cat



R. von Bernhardi (ed.), *Glial Cells in Health and Disease of the CNS*,
Advances in Experimental Medicine and Biology 949,
DOI 10.1007/978-3-319-40764-7_7

SAMPs Self-associated molecular patterns
TREM2 Triggering Receptor Expressed on Myeloid cells 2
PPT Perforant path transection
ECM Extracellular matrix

## "Resting" Microglia in the Healthy CNS and Their Interaction with the Microenvironment

The term "quiescent" or "resting" microglia, usually used to designate nonactivated microglia in the normal adult central nervous system (CNS), might lead one to think that these cells are in a dormant state with no apparent movement and function. However, nothing could be further from the truth. The combined use of in vivo time-lapse transcranial two-photon microscopy and transgenic mice with green fluorescent protein in resident CNS microglia has made it possible to see microglia interacting with other cortical elements (Davalos et al. 2005; Nimmerjahn et al. 2005). Microglial cells are the most dynamic cells in the healthy CNS, as their morphological changes far exceed those of both neurons (Holtmaat et al. 2008; Knott and Holtmaat 2008) and astrocytes (Hirrlinger et al. 2004). Thus, in the healthy brain, microglial cells are continuously remodeling their shape by extending and retracting their processes, surveying the local microenvironment to scan the surface of the surrounding cells and the interstitial fluid (Davalos et al. 2005; Nimmerjahn et al. 2005) (see Chapters "Glial cells and Integrity of the Nervous System" and "Microglia Function in the Normal Brain"). Under normal conditions, each microglial cell seems to be responsible for checking its own territory, and its highly dynamic processes do not overlap or enter in the territory of neighboring microglial cells. While the microglial soma and main branches remain stable in the nervous parenchyma, with few signs of movement and without any clear relationship to other cells or blood vessels, its motile processes are continuously making direct contacts with nearby neuronal cell bodies, macroglia, and blood vessels (Nimmerjahn et al. 2005; Wake et al. 2009; Tremblay et al. 2010).

Although it might appear at first glance that motility of microglial processes is random (Nimmerjahn et al. 2005), a wide range of studies indicates that microglial cells express a broad variety of surface receptors that allows them to sense subtle changes in the microenvironment (Kierdorf and Prinz 2013). In particular, in the healthy adult brain, movement of microglial processes seems to be closely related to local concentration of some neurotransmitters, neuropeptides, and neuromodulators (Pocock and Kettenmann 2007). Although not conclusive, the current data suggest that microglial motility is increased by global excitatory neurotransmission and decreased by global inhibitory neurotransmission (Nimmerjahn et al. 2005; Fontainhas et al. 2011; Eyo and Wu 2013).

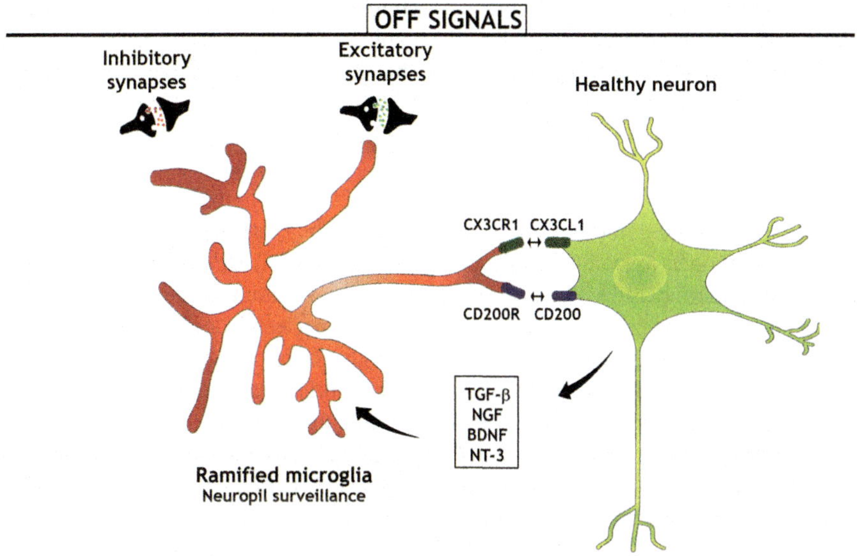


**Fig. 1** In the healthy normal brain, ramified microglia is a very dynamic cell and their processes are continuously extending and retracting, monitoring the surface of neurons and having a special predilection for excitatory synapses. Interaction between inhibitory receptors in microglia with both specific ligands in the neuronal surface and neuronal released molecules keeps microglia in a nonactivated state. These signals that are expressed constitutively are know as "Off-signals"

Furthermore, electron microscopic studies demonstrate that, under normal conditions, microglial cell processes directly contact presynaptic and postsynaptic elements and have a special predilection for excitatory synapses (Fig. 1) although the existence of microglial cell interactions with inhibitory synapses under normal physiological conditions remains yet unknown (Perry and O'Connor 2010; Siskova and Tremblay 2013). Microglial cell processes contact synapses about once per hour, remain in a close proximity to presynaptic boutons for 5 min, and then retract (Wake et al. 2009). The interactions between microglia and synapses depend on neuronal activity and, therefore, the frequency of contact declines with decreased synaptic transmission (Wake et al. 2009).

## Signaling Mechanisms Involved in Activation of Microglia

As previously discussed in Chapters "Glial cells and Integrity of the Nervous System" and "Microglia Function in the Normal Brain", microglia are activated by various changes in their microenvironment caused by acute insults and chronic disease states (Kettenmann et al. 2011; Chen et al. 2014; Gonzalez et al. 2014).

Transformation of the finely branched resting microglia into enlarged cells with short and stout processes is a hallmark of microglial cell activation (Kettenmann et al. 2011). In addition to morphological changes, microglial activation involves a stereotypical pattern of changes, including proliferation and migration to sites of injury, increased or de novo expression of cytokines and growth factors and, in some circumstances, the full transformation into phagocytes capable of clearing damaged cells and debris (Kettenmann et al. 2011). There is a wide range of molecules promoting microglial activation that can be classified as two main types: PAMPs (Pathogen associated molecular patterns) and DAMPs (damage associated molecular patterns). PAMPs warn of the presence of exogenous material, such as components of bacterial cell walls or repeats of bacterial or viral nucleic acids, whereas DAMPs warn of internal damage to the cells of the own organism and include molecules released by injured cells or modified as a consequence of tissue damage, such as oxidized lipoproteins or fragments of extracellular matrix molecules (Bianchi 2007; Matzinger 2007). Microglial cells possess a wide range of surface molecules, such as toll-like receptors (TLRs) (Lehnardt 2010), scavenger receptors (Husemann et al. 2002) and numerous cytokine and chemokine receptors, whose interaction with DAMPs and PAMPs results in a rapid activation of resting microglia to become motile effector cells (Kierdorf and Prinz 2013).

However, it would be a mistake to think that activation of microglia is a simple event; on the contrary, it is complex and includes still unidentified signaling mechanisms. In the healthy CNS, microglia exhibit a deactivated phenotype due to the interaction of inhibitory receptors ("Off receptors") in their plasma membrane, with the corresponding ligands ("Self-associated molecular patterns" or SAMPs) located on neurons and glial cells that keep microglia in a resting or nonactivated stage (Biber et al. 2007; Eyo and Wu 2013; Kierdorf and Prinz 2013) (Fig. 1). Some of the proposed inhibitory receptors in microglia are CX3CR1 and CD200R, which interact with their respective ligands, CX3CL1 (fractalkine) and CD200 on the surface of healthy neurons (Chertoff et al. 2013; Eyo and Wu 2013). Another proposed microglial inhibitory system is CD45/CD22. Recognition of CD22 on the surface of neurons by CD45 on microglia dampens microglial activation (Mott et al. 2004). Moreover, in addition to displaying membrane bound "Off-signals," neurons also release soluble Off-signals into the extracellular space, such as Transforming growth factor (TGF) $\beta$, neurotransmitters and neurotrophins including NGF, BDNF and NT-3 (Biber et al. 2007). If any of these Off-signals are lost, due to changes in the microenvironment, or are downregulated, as may occur in pathological conditions, microglial activation is triggered.

In contrast to Off-signals, which are expressed constitutively in the healthy adult brain, "On-signals" are produced on demand to initiate either a pro- or anti-inflammatory microglial activation program (Kettenmann et al. 2013) (Fig. 2). Some of the On-signals are the so-called "help-me/find-me" molecules (Marin-Teva et al. 2011; Panatier and Robitaille 2012; Xing et al. 2014). When neurons are overactive, impaired or endangered, they release these "alert" signals (Noda et al. 2013) which include nucleotides such as ATP and UTP (Sperlagh and Illes 2007); chemokines such as CCL21 and CXCL10 (Rappert et al. 2004; de Jong et al. 2005);

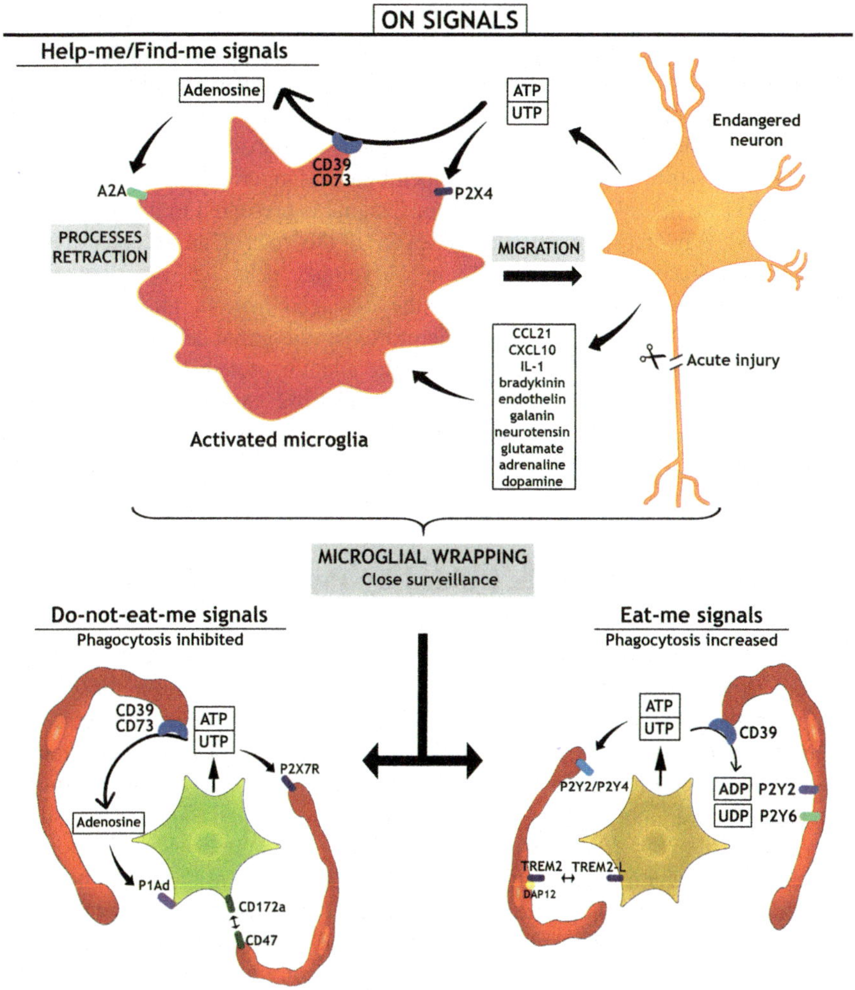


**Fig. 2** "On signals" are produced when neurons are damaged and include "Help-me"/"Find-me", "Do-not-eat-me", and "Eat-me" signals. Endangered neurons may release a wide range of alert signals (Help-me/Find-me) including nucleotides that promote microglial activation, process retraction and migration towards neuronal somata. Microglial wrapping in one hand facilitates contact-dependent neuron-microglia interactions but also isolates damaged neurons leaking nucleotides. If Do-not-eat-me signaling predominates, phagocytosis is inhibited and neurons are able to survive. If, on the contrary, Eat-me signaling prevails, an increase in the phagocytic ability of microglia takes place and damaged neurons are removed. Note the importance of the ecto-enzymes CD39 and CD73 regulating the levels of nucleotides and nucleosides in the extracellular space around injured neurons

cytokines like Interleukin (IL) 1 (Cartier et al. 2005); neuropeptides such as bradykinin (Ifuku et al. 2007), endothelin (Fleisher-Berkovich et al. 2010), galanin (Ifuku et al. 2011) and neurotensin (Martin et al. 2005); neurotransmitters such as glutamate, adrenaline, and dopamine (Farber et al. 2005; Liu et al. 2009); and cannabinoids (Walter et al. 2003) and morphine (Takayama and Ueda 2005). In response to help-me/find-me signals, microglia approach to the source of these molecules and develop either a close surveillance or a phagocytic function, depending on the presence of additional signals in the damaged neuron. If the receptor SIRP-alpha (CD172a) in the membrane of microglia interacts with the ligand CD47 on neurons, a "Do-not-eat-me" signal is presented to microglia (Biber et al. 2007; Ravichandran 2010). However, if the microglial receptor TREM2 (Triggering Receptor Expressed on Myeloid cells 2) recognizes its still-unknown-ligand on the surface of the damaged neuron, this interaction is interpreted as an "Eat-me" signal and therefore the microglial cell is able to initiate an intracellular signaling cascade, through the adaptor protein DAP12, leading to phagocytosis (Linnartz and Neumann 2013). TREM2 expression has been suggested to regulate not only phagocytic but also the migratory capacity of microglia (Melchior et al. 2010).

## Migration of Microglia Is Guided by Purinergic Signaling

Release of danger signals that act as chemoattractants at the site of damage, initiates microglial activation and stimulates migration. Time-lapse two-photon imaging demonstrates that, for example, after a small laser ablation in the cerebral cortex, all microglial cells located in the surroundings respond within minutes by enlarging and extending their processes towards the damaged site, converging and forming a spherical shaped containment around it, but without migration of the somata (Davalos et al. 2005; Nimmerjahn et al. 2005). Quick extension of microglial processes to the site of injury without significant displacement of the cell body was previously described using histological sections (Jensen et al. 1994). In this work we showed that, a few hours after a perforant path transection (PPT), microglial cells located in the inner zone of the dentate molecular layer polarize and extend their processes towards and into the denervated PP zone, and it is not until 2–3 days after PPT when microglial cell bodies move to the denervated PP zone, where they accumulate and proliferate (Jensen et al. 1994). Therefore, migration of microglial cells is probably a complex process that involves two stages: a first phase of reconnaissance and damage assessment by microglial cells processes and, if damage persists and is important enough, a second phase where the entire cell body migrates. An intense cross talk, involving the signaling mechanisms referred in the previous section, between extended microglial processes and damaged neurons and glial cells, will determine this microglial cell migration.

Purine nucleotides are among the most potent molecules involved in the migration of microglia. In fact, Davalos et al. (2005) demonstrated that ATP or ADP microinjection in the brain parenchyma was able to mimic the rapid chemotactic response of microglial processes observed following laser ablation. Moreover, lowered ATP extracellular concentration results in reduced microglial cell process movements, whereas increased ATP gradients stimulate their motility (Haynes et al. 2006).

In the healthy brain, release of ATP to the extracellular space is a common phenomenon, as this nucleotide and its derivatives act both as primary transmitter and as co-transmitter released with other neurotransmitters and peptides in many synapses. The mechanism by which intracellular ATP is released by neurons is a matter of intense debate (Cisneros-Mejorado et al. 2015), because in addition to being released by exocytosis, ATP leakage can also take place through large pores and transporters. Moreover, not only neurons but also glial cells, in particular astrocytes, can release ATP (Butt 2011; Cisneros-Mejorado et al. 2015).

Under pathological conditions when neurons are overexcited, injured or stressed in acute or chronic neurological disorders, a massive release of ATP takes place into the extracellular space (Braun et al. 1998; Melani et al. 2005). As elevated concentrations of extracellular ATP can cause cell death (Matute et al. 2007; Arbeloa et al. 2012), ATP released from endangered or dying cells may aggravate the extent of the ongoing damage. In addition, increased extracellular levels of ATP may over activate the P2X7R in neurons and trigger signaling cascades leading to neurodegeneration (Le Feuvre et al. 2003).

The concentrations of ATP, ADP, AMP and adenosine in the extracellular space are regulated by the activity of ecto-nucleotidases that are located in the plasma membrane of microglia and whose expression is dependent on the development and activation stage of these cells (Dalmau et al. 1998). One of these ecto-nucleotidases is CD39, also called Ecto-nucleoside triphosphate diphosphohydrolase 1 (E-NTPDase1), whose expression in the CNS is restricted to microglial cells and vascular endothelium (Braun et al. 2000). CD39 plays a main role hydrolysing extracellular nucleoside 5′-triphosphates to nucleoside 5′-diphosphates (NTPase enzymatic activity), as well as nucleoside 5′-diphosphates to nucleoside 5′-monophosphates (NDPase enzymatic activity). Nucleoside 5′-monophosphates are further hydrolysed to adenosine by CD73, an ecto-5′-nucleotidase also found, among other cells, in the membrane of microglia (Dalmau et al. 1998; Bulavina et al. 2013). Therefore, microglial cells could be considered as the cells responsible for the regulation of purinergic signaling in the CNS as they can control the rate, extent and timing of nucleotide degradation.

On the other hand, we should consider that microglial cells have several types of purine receptors on their surface (Ohsawa and Kohsaka 2011) whose interactions with changing concentrations of extracellular nucleotides and nucleosides (ATP/adenosine balance) may regulate microglial behavior, including process extension and retraction, microglial migration and even phagocytosis.

Purine receptors are divided into P1 (adenosine receptors) and P2 (ATP receptors). Microglia express the four subtypes of P1 receptors (A1, A3, A2A and A2B) and only some of the different subtypes of P2 receptors cloned, which are divided into ionotropic (seven subtypes: P2X1-7) and metabotropic (eight subtypes: P2Y1, −2, −4, −6, −11, −12, −13, and −14) (Kettenmann et al. 2011). Simultaneous costimulation of P1 and P2 receptors seems to be required for microglial migration (Farber et al. 2008). In particular, microglial process extension is dependent upon

ATP/ADP sensed through microglial P2Y12 receptors (Ohsawa and Kohsaka 2011), which are constitutively expressed on microglia in normal conditions (Sasaki et al. 2003) and upregulated when activated (Tozaki-Saitoh et al. 2008). P2Y12 receptors activate integrin-ß1, which accumulates in the tips of microglial processes, facilitating the adhesion of extended microglial processes with the extracellular matrix (ECM), which is a requisite for subsequent directional microglial migration (Haynes et al. 2006; Kurpius et al. 2007). Further activation of microglia, probably due to continuously elevated levels of ATP and ADP, or both (Kurpius et al. 2007), leads to upregulation of A2A and P2X4 receptors, whereas P2Y12 receptors are downregulated (Haynes et al. 2006; Orr et al. 2009). Signaling through P2X4 receptors enhances migration of microglia. As microglial activation involves increased expression of the ecto-enzymes CD39 and CD73 (causing ATP/ADP degradation), the abnormally increased levels of ATP generated by the pathological situation are gradually reduced, while the adenosine concentration increases and activates A2A receptors. Notably, adenosine causes retraction of microglial processes (Ohsawa and Kohsaka 2011). Therefore, gradually increased levels of adenosine may be the basis of microglial transformation from ramified cells into amoeboid migratory morphologies, usually found in various pathologies.

## Microglial Wrapping and Synaptic Stripping

As discussed in Chapters "Glial cells and Integrity of the Nervous System" and "Microglia Function in the Normal Brain", it has been widely reported that activated microglia migrate and accumulate near injured neurons in various pathological conditions. In addition, in certain circumstances, the somata, proximal dendrites and axons of injured neurons become ensheathed by microglia. Microglial wrapping of neuronal cell bodies is one of the most prominent features after peripheral nerve axotomy (Fig. 3). Indeed, the phenomenon of microglial wrapping has been widely described in various CNS areas in several situations involving peripheral nerve axotomy, including the facial nucleus (Moran and Graeber 2004), the hypoglossal nucleus (Sumner and Sutherland 1973; Yamada et al. 2011), the dorsal motor nucleus of the vagus nerve (Masui et al. 2002), and in the spinal cord after sciatic nerve axotomy (Gehrmann et al. 1991). Also, this phenomenon has been reported in experimental models where peripheral nerves are not affected such as hippocampal organotypic cultures after an ischemic insult (Neumann et al. 2006), in the cerebral cortex during either acute focal inflammation (Trapp et al. 2007) or following intraperitoneal LPS injection (Chen et al. 2012), and in the spinal cord after experimental autoimmune encephalomyelitis (EAE) induction (Almolda et al. 2009). Microglial wrapping occurs in parallel with a significant reduction of axosomatic synapses. It was Blinzinger and Kreutzberg (1968) who first described, following facial nerve axotomy, the displacement of presynaptic terminals from the injured motor neuron surface by the interposing of microglial pseudopods and named this phenomenon as "synaptic stripping". Although some

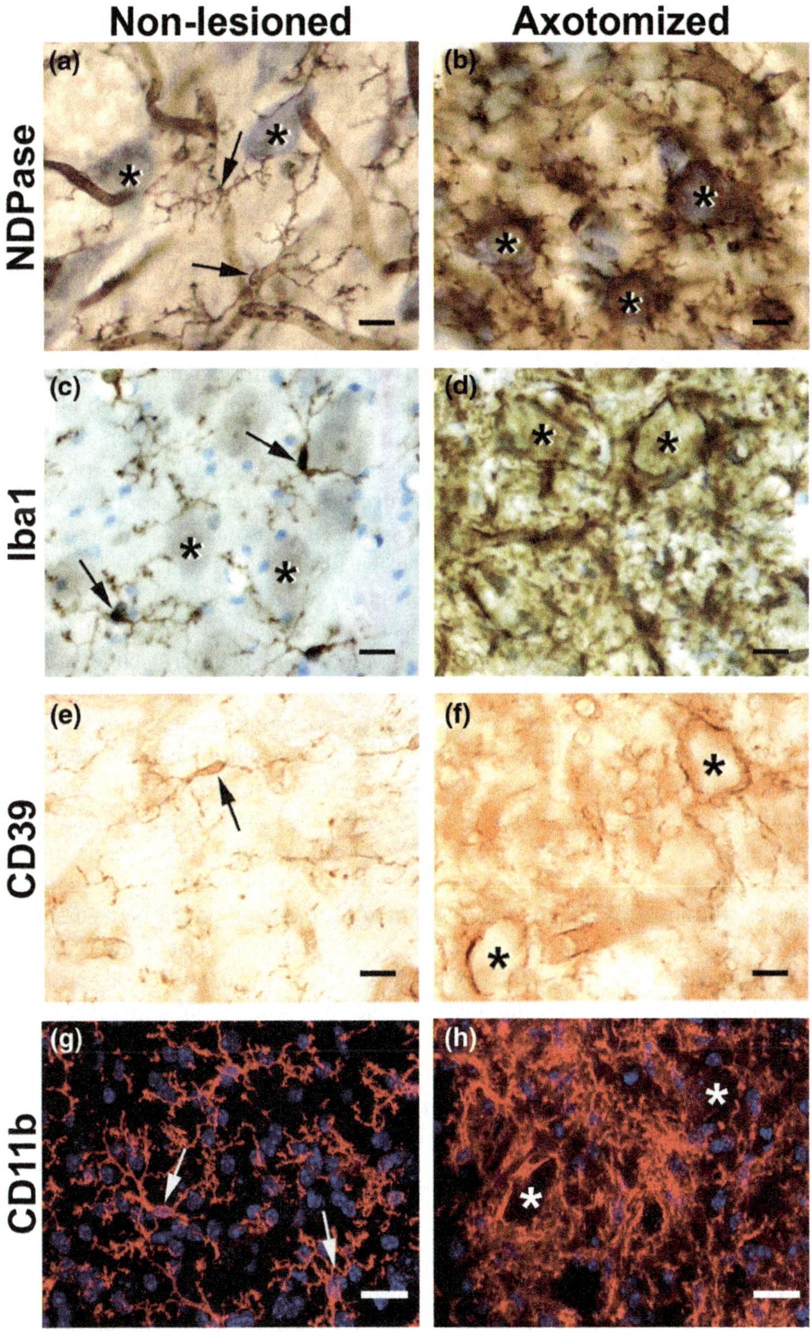


**Fig. 3** Microglial wrapping in the facial nucleus of the mouse after facial nerve axotomy. In the normal, non-lesioned facial nucleus, microglia stained with different markers including NDPase histochemistry and immunohistochemistry for Iba1, CD39 and CD11b (**a**, **c**, **e**, **g**), show a ramified morphology (*arrows*) without any particular association with neuronal motor neuron somata (asterisks). After axotomy (**b**, **d**, **f**, **h**), microglia enwrap motor neuron somata (asterisks). In (**a**)–(**d**), sections are counterstained with toluidine blue. In (**g**) and (**h**) nuclei are stained in blue with DAPI. Scale bar = 20 μm

authors claim that reactive microglia spread on the surface of motor neurons to physically disconnect synapses (Moran and Graeber 2004; Yamada et al. 2008), it is still not totally clear whether synaptic stripping is either the cause or the consequence of microglial wrapping. As microglial wrapping and synaptic stripping are associated with motor neuron regeneration, it has usually been considered to be a neuroprotective process (Kreutzberg 1996). As we will discuss below, recent studies support this neuroprotective view (Chen et al. 2014), whereas others suggest that microglial wrapping may reduce neuronal survival (Yamada et al. 2011).

Accumulating evidence indicate that synaptic stripping of either inhibitory or excitatory synapses is beneficial to damaged neurons. Since microglia wrap neuronal cell bodies and the majority of synapses terminating on projection neuron somata in the cerebral cortex are GABAergic inhibitory synapses, it has been proposed that inhibitory axosomatic synapses are preferentially stripped after focal inflammation or peripheral immune challenge (Trapp et al. 2007; Chen et al. 2012). Evidence of microglia-mediated stripping of inhibitory GABAergic presynaptic terminals from cortical neurons in adult mice has been recently confirmed by 3-D electron microscopy (Chen et al. 2014). Reduced axosomatic GABAergic innervation protects neurons against noxious insult (Hardingham et al. 2002) by increasing synchronization of neuronal firing (Woo and Lu 2006), which is critical for synaptic NMDAR-mediated neuronal survival through CREB activation and by increasing neuronal expression of anti-apoptotic and neuroprotective molecules (Hardingham and Bading 2003).

However, it is nowadays clear that microglial wrapping is not always specifically directed to disconnect inhibitory synapses because in other locations, such as the facial nucleus in the rat after nerve axotomy (Raslan et al. 2014) and the spinal cord after either intramedullary axotomy in the cat (Linda et al. 2000) or sciatic nerve transection in the rat (Arbat-Plana et al. 2015), the outcome of microglia-mediated synaptic stripping is the preferential disconnection of excitatory glutamatergic synapses. Removal of the glutamatergic input to the axotomized motor neurons is considered relevant for neuronal survival, as glutamate may exert deleterious excitotoxic effects on nerve cells (Mehta et al. 2013). In support of this possibility, blocking of the NMDA-type glutamate receptor has been reported to increase motor neuron survival after neonatal axotomy in the rat (Mentis et al. 1993). Even assuming that any changes in the synaptic input, either inhibitory or excitatory, to the lesioned neurons may reduce their stress and be beneficial for survival and repair, the question of whether microglia actively participate in this process or if instead nerve terminals simply retract from the surface of neurons remains unsolved (Linda et al. 2000).

It is generally accepted that synaptic stripping does not inevitably mean that the disconnected terminals have to be immediately engulfed by microglia, as they remain in the vicinity of ensheathed neurons and only after axotomized motor neurons regenerate their axons, synapses are restored (Navarro et al. 2007). However, some work indicates that, several weeks after nerve transection, restored synaptic inputs are not normal (Raslan et al. 2014). The usual prevalence of inhibitory over excitatory terminals seems to be shifted for surviving lesioned motor neurons in various locations (Borke et al. 1995; Linda et al. 2000; Raslan et al. 2014). Although microglia has been suggested to play a main role in regulating these synaptic rearrangements (Raslan et al. 2014), astrocytes might also be involved (Tyzack et al. 2014).

In the healthy brain, neurons, including their synapses, are generally ensheathed by fine processes of astrocytes that participate in the regulation of synapse formation, stability, and elimination. Coverage of synapses by astrocytic processes may change under various physiological conditions (Theodosis et al. 2008; Chung et al. 2013; Perez-Alvarez et al. 2014). Specifically, in the facial nerve of the mouse two weeks after axotomy, thin lamellar astrocyte processes begin to replace microglial wrapping around damaged motor neurons, and by 3 weeks they completely cover the neuron soma (Moran and Graeber 2004). Some authors have suggested that this delayed astrocyte behavior might contribute to synaptic remodeling by engulfing some disconnected presynaptic terminals (Chung et al. 2013) and promoting the rearrangement of synaptic inputs on axotomized motor neurons (Tyzack et al. 2014).

## Microglial Wrapping: Detrimental or Beneficial?

Glial wrapping, whether microglial, astroglial or both, may not only cause deafferentation, but might also facilitates contact-dependent neuron–glia interactions that prevent neuron death and promote regeneration. After facial nerve axotomy in the mouse, for example, about 65 % of axotomized neurons regenerate axons and survive, whereas 35 % of neurons degenerate. Research in our laboratory performing facial nerve axotomy on transgenic mice with astrocyte-targeted expression of either IL6 or IL10 in order to investigate how the local expression of those cytokines may affect microglial activation, showed that in addition to changes in the microglial reactivity pattern, there is an altered survival/death ratio of motor neurons (Almolda et al. 2014; Villacampa et al. 2015). Interestingly, higher motor neuron survival in IL10 transgenic mice was not associated with significant changes in microglial wrapping (Villacampa et al. 2015) although increased motor neuronal death in IL6 transgenic mice coincides with reduced microglial wrapping (Almolda et al. 2014). Moreover, ongoing studies performed on IRF8 KO mice indicate that incomplete microglial wrapping of individual axotomized motor neurons correlates with increased motor neuron death (Xie et al. 2014). In agreement with this, some evidence suggests that defects in microglia-neuron attachment after facial nerve axotomy, as occurs in microglial cathepsin deficient mice (Hao et al. 2007) and TGF$\beta$1 deficient animals (Makwana et al. 2007), might lead to more neuron death. These observations support the hypothesis that the intimate association between glial cells and neurons has a neurotrophic rather than neurotoxic function. The close physical proximity of microglia to injured neurons may facilitate the continuous supply of growth factors and other required molecules, thus supporting survival and regeneration (Trapp et al. 2007).

There is however an opposing view holding the possibility that prolonged contact of microglial cells with enwrapped neurons is detrimental (Yamada et al. 2011). Some studies have demonstrated that the survival ratio of injured motor neurons is markedly influenced by the species and the age of animals used (Moran and Graeber 2004; Kiryu-Seo et al. 2005). Interestingly, in this context, facial nerve axotomy in neonatal rats and mice kills damaged motor neurons within a week of lesion. Nevertheless, axotomized motor neurons in adult rats are able to survive, whereas in adult mice there is a slow and progressive motor neuron death after lesion (Kiryu-Seo et al. 2005). Some authors have suggested that these differences among adult rats and mice are due to differences in the ratio of microglial/astroglial wrapping (Yamada et al. 2011). If the astrocytic wrapping predominates, as found in the rat, some protective effects are exerted on axotomized motoneurons, whereas if the wrapping is mainly microglial, as observed in mouse, a slow apoptotic cell death of motor neurons might take place (Yamada et al. 2011).

Microglial wrapping may be the result of a continuous release or leakage of purine nucleotides that act as find-me signals (Fig. 2). Neuron ensheathment by activated microglia expressing ecto-nucleotidases in their plasma membrane effectively isolates damaged neurons leaking purine nucleotides and contributes to their rapid degradation to adenosine around neurons. Increasing concentrations of extracellular adenosine may develop a potentially neuroprotective function on neurons through P1 adenosine receptors (Stone 2002). In addition, adenosine can impair the phagocytic function of peripheral macrophages by binding to the P1 adenosine receptors expressed on their membrane (Hasko et al. 2007). Also, microglial phagocytosis seems to be regulated by purinergic signaling (Bulavina et al. 2013). It has been shown that activation of P1 receptors by a non-hydrolysable analog of adenosine decreases microglial phagocytosis (Bulavina et al. 2013). In the opposite way, activation of P2Y12 receptor by ADP, activation of P2Y6 by UDP and activation of P2Y2/P2Y4 receptors by UTP markedly increase microglial phagocytosis both in vitro and in vivo (Koizumi et al. 2007; Fang et al. 2009). Therefore, the increasing concentration of these nucleotides around injured neurons may be an eat-me signal for wrapping microglia. In agreement with this, CD39-deficient animals presented higher microglial phagocytic activity (Bulavina et al. 2013), suggesting that an increased concentration of extracellular ATP/ADP and UTP/UDP, due to the lack of CD39 enzymatic activity, leads to a chronic stimulation of the microglial phagocytic activity. However, other studies indicate that activation of P2X7 receptors by exposure to ATP induced inhibition of microglial phagocytic activity even if microglia are cotreated with UDP (Fang et al. 2009). Taken together, these observations suggest that a fine control of the levels of nucleosides and nucleotides in the extracellular space around injured neurons together with a fine regulation of purine receptors may be decisive to control phagocytosis and hence in determining the fate of damaged neurons wrapped by microglia.

## Concluding Remarks

The meaning of microglial wrapping around injured neurons is not completely understood. Microglial wrapping partially isolates endangered neurons from the adjacent neuropil, leading to an important deafferentation from synaptic inputs. Besides, the wide area of contact between microglia and neuronal surfaces enables an intense exchange of molecular signals between them. Injured neurons circumscribed by microglia may survive or die and their fate will depend on a plethora of signals. In this scenario, nucleosides and their phosphorylated nucleotides may play a key role, as they can be involved in regulation of apoptosis, in the synthesis and release of different trophic factors by astrocytes (Rathbone et al. 1999), in promotion of axonal growth (Heine et al. 2006), and in modulation of microglial phagocytosis (Inoue 2008). Although programmed neuronal cell death can result from axonal injury, cell regeneration and axonal outgrowth programs are also activated (Raivich and Makwana 2007; Kiryu-Seo and Kiyama 2011). The putative involvement of microglia and astroglia in the activation of these regenerative programs are still poorly understood and will be a challenge for researchers in the coming years.

**Acknowledgments** This work was supported by the Spanish Ministry of Science and Innovation BFU2011-27400 and BFU2014-55459-P.

## References

Almolda B, Costa M, Montoya M, Gonzalez B, Castellano B (2009) CD4 microglial expression correlates with spontaneous clinical improvement in the acute Lewis rat EAE model. J Neuroimmunol 209(1–2):65–80. doi:10.1016/j.jneuroim.2009.01.026 S0165-5728(09)00031-9 [pii]

Almolda B, Villacampa N, Manders P, Hidalgo J, Campbell IL, Gonzalez B, Castellano B (2014) Effects of astrocyte-targeted production of interleukin-6 in the mouse on the host response to nerve injury. Glia 62(7):1142 1161. doi:10.1002/glia.22668

Arbat-Plana A, Torres-Espin A, Navarro X, Udina E (2015) Activity dependent therapies modulate the spinal changes that motoneurons suffer after a peripheral nerve injury. Exp Neurol 263:293–305. doi:10.1016/j.expneurol.2014.10.009 S0014-4886(14)00345-8 [pii]

Arbeloa J, Perez-Samartin A, Gottlieb M, Matute C (2012) P2X7 receptor blockade prevents ATP excitotoxicity in neurons and reduces brain damage after ischemia. Neurobiol Dis 45(3): 954–961. doi:10.1016/j.nbd.2011.12.014 S0969-9961(11)00389-5 [pii]

Bianchi ME (2007) DAMPs, PAMPs and alarmins: all we need to know about danger. J Leukoc Biol 81(1):1–5. doi:10.1189/jlb.0306164 jlb.0306164 [pii]

Biber K, Neumann H, Inoue K, Boddeke HW (2007) Neuronal 'On' and 'Off' signals control microglia. Trends Neurosci 30(11):596–602. doi:10.1016/j.tins.2007.08.007 S0166-2236(07)00248-2 [pii]

Blinzinger K, Kreutzberg G (1968) Displacement of synaptic terminals from regenerating motoneurons by microglial cells. Z Zellforsch Mikrosk Anat 85(2):145–157

Borke RC, Bridwell RS, Nau ME (1995) The progression of deafferentation as a retrograde reaction to hypoglossal nerve injury. J Neurocytol 24(10):763–774

Braun N, Zhu Y, Krieglstein J, Culmsee C, Zimmermann H (1998) Upregulation of the enzyme chain hydrolyzing extracellular ATP after transient forebrain ischemia in the rat. J Neurosci 18 (13):4891–4900

Braun N, Sevigny J, Robson SC, Enjyoji K, Guckelberger O, Hammer K, Di Virgilio F, Zimmermann H (2000) Assignment of ecto-nucleoside triphosphate diphosphohydrolase-1/cd39 expression to microglia and vasculature of the brain. Eur J Neurosci 12(12):4357–4366 doi:ejn1342 [pii]

Bulavina L, Szulzewsky F, Rocha A, Krabbe G, Robson SC, Matyash V, Kettenmann H (2013) NTPDase1 activity attenuates microglial phagocytosis. Purinergic Signal 9(2):199–205. doi:10.1007/s11302-012-9339-y

Butt AM (2011) ATP: a ubiquitous gliotransmitter integrating neuron-glial networks. Semin Cell Dev Biol 22(2):205–213. doi:10.1016/j.semcdb.2011.02.023 S1084-9521(11)00036-X [pii]

Cartier L, Hartley O, Dubois-Dauphin M, Krause KH (2005) Chemokine receptors in the central nervous system: role in brain inflammation and neurodegenerative diseases. Brain Res Brain Res Rev 48(1):16–42. doi:10.1016/j.brainresrev.2004.07.021 S0165-0173(04)00116-X [pii]

Chen JL, Villa KL, Cha JW, So PT, Kubota Y, Nedivi E (2012) Clustered dynamics of inhibitory synapses and dendritic spines in the adult neocortex. Neuron 74(2):361–373. doi:10.1016/j.neuron.2012.02.030 S0896-6273(12)00267-X [pii]

Chen Z, Jalabi W, Hu W, Park HJ, Gale JT, Kidd GJ, Bernatowicz R, Gossman ZC, Chen JT, Dutta R, Trapp BD (2014) Microglial displacement of inhibitory synapses provides neuroprotection in the adult brain. Nat Commun 5:4486. doi:10.1038/ncomms5486 ncomms5486 [pii]

Chertoff M, Shrivastava K, Gonzalez B, Acarin L, Gimenez-Llort L (2013) Differential modulation of TREM2 protein during postnatal brain development in mice. PLoS One 8(8):e72083. doi:10.1371/journal.pone.0072083 PONE-D-13-15299 [pii]

Chung WS, Clarke LE, Wang GX, Stafford BK, Sher A, Chakraborty C, Joung J, Foo LC, Thompson A, Chen C, Smith SJ, Barres BA (2013) Astrocytes mediate synapse elimination through MEGF10 and MERTK pathways. Nature 504(7480):394–400. doi:10.1038/nature12776 nature12776 [pii]

Cisneros-Mejorado A, Perez-Samartin A, Gottlieb M, Matute C (2015) ATP signaling in brain: release, excitotoxicity and potential therapeutic targets. Cell Mol Neurobiol 35(1):1–6. doi:10.1007/s10571-014-0092-3

Dalmau I, Vela JM, Gonzalez B, Castellano B (1998) Expression of purine metabolism-related enzymes by microglial cells in the developing rat brain. J Comp Neurol 398(3):333–346. doi:10.1002/(SICI)1096-9861(19980831)398:3<333:AID-CNE3>3.0.CO;2-0 [pii]

Davalos D, Grutzendler J, Yang G, Kim JV, Zuo Y, Jung S, Littman DR, Dustin ML, Gan WB (2005) ATP mediates rapid microglial response to local brain injury in vivo. Nat Neurosci 8 (6):752–758. doi:10.1038/nn1472 nn1472 [pii]

de Jong EK, Dijkstra IM, Hensens M, Brouwer N, van Amerongen M, Liem RS, Boddeke HW, Biber K (2005) Vesicle-mediated transport and release of CCL21 in endangered neurons: a possible explanation for microglia activation remote from a primary lesion. J Neurosci 25(33):7548–7557. doi:10.1523/JNEUROSCI.1019-05.2005 25/33/7548 [pii]

Eyo UB, Wu LJ (2013) Bidirectional microglia-neuron communication in the healthy brain. Neural Plast 2013:456857. doi:10.1155/2013/456857

Fang KM, Yang CS, Sun SH, Tzeng SF (2009) Microglial phagocytosis attenuated by short-term exposure to exogenous ATP through P2X receptor action. J Neurochem 111(5):1225–1237. doi:10.1111/j.1471-4159.2009.06409.x JNC6409 [pii]

Farber K, Pannasch U, Kettenmann H (2005) Dopamine and noradrenaline control distinct functions in rodent microglial cells. Mol Cell Neurosci 29(1):128–138. doi:10.1016/j.mcn.2005.01.003 S1044-7431(05)00010-2 [pii]

Farber K, Markworth S, Pannasch U, Nolte C, Prinz V, Kronenberg G, Gertz K, Endres M, Bechmann I, Enjyoji K, Robson SC, Kettenmann H (2008) The ectonucleotidase cd39/ENTPDase1 modulates purinergic-mediated microglial migration. Glia 56(3):331–341. doi:10.1002/glia.20606

Fleisher-Berkovich S, Filipovich-Rimon T, Ben-Shmuel S, Hulsmann C, Kummer MP, Heneka MT (2010) Distinct modulation of microglial amyloid beta phagocytosis and migration by neuropeptides (i). J Neuroinflammation 7:61. doi:10.1186/1742-2094-7-61 1742-2094-7-61 [pii]

Fontainhas AM, Wang M, Liang KJ, Chen S, Mettu P, Damani M, Fariss RN, Li W, Wong WT (2011) Microglial morphology and dynamic behavior is regulated by ionotropic glutamatergic and GABAergic neurotransmission. PLoS One 6(1):e15973. doi:10.1371/journal.pone.0015973

Gehrmann J, Monaco S, Kreutzberg GW (1991) Spinal cord microglial cells and DRG satellite cells rapidly respond to transection of the rat sciatic nerve. Restor Neurol Neurosci 2(4):181–198. doi:10.3233/RNN-1991-245605 F2852H884540843H [pii]

Gonzalez H, Elgueta D, Montoya A, Pacheco R (2014) Neuroimmune regulation of microglial activity involved in neuroinflammation and neurodegenerative diseases. J Neuroimmunol 274(1–2):1–13. doi:10.1016/j.jneuroim.2014.07.012 S0165-5728(14)00227-6 [pii]

Hao HP, Doh-Ura K, Nakanishi H (2007) Impairment of microglial responses to facial nerve axotomy in cathepsin S-deficient mice. J Neurosci Res 85(10):2196–2206. doi:10.1002/jnr.21357

Hardingham GE, Bading H (2003) The Yin and Yang of NMDA receptor signalling. Trends Neurosci 26(2):81–89. doi:10.1016/S0166-2236(02)00040-1 S0166-2236(02)00040-1 [pii]

Hardingham GE, Fukunaga Y, Bading H (2002) Extrasynaptic NMDARs oppose synaptic NMDARs by triggering CREB shut-off and cell death pathways. Nat Neurosci 5(5):405–414. doi:10.1038/nn835 [pii]

Hasko G, Pacher P, Deitch EA, Vizi ES (2007) Shaping of monocyte and macrophage function by adenosine receptors. Pharmacol Ther 113(2):264–275. doi:10.1016/j.pharmthera.2006.08.003 S0163-7258(06)00150-1 [pii]

Haynes SE, Hollopeter G, Yang G, Kurpius D, Dailey ME, Gan WB, Julius D (2006) The P2Y12 receptor regulates microglial activation by extracellular nucleotides. Nat Neurosci 9(12):1512–1519. doi:10.1038/nn1805 nn1805 [pii]

Heine C, Heimrich B, Vogt J, Wegner A, Illes P, Franke H (2006) P2 receptor-stimulation influences axonal outgrowth in the developing hippocampus in vitro. Neuroscience 138(1):303–311. doi:10.1016/j.neuroscience.2005.11.056 S0306-4522(05)01295-9 [pii]

Hirrlinger J, Hulsmann S, Kirchhoff F (2004) Astroglial processes show spontaneous motility at active synaptic terminals in situ. Eur J Neurosci 20(8):2235–2239. doi:10.1111/j.1460-9568.2004.03689.x EJN3689 [pii]

Holtmaat A, De Paola V, Wilbrecht L, Knott GW (2008) Imaging of experience-dependent structural plasticity in the mouse neocortex in vivo. Behav Brain Res 192(1):20–25. doi:10.1016/j.bbr.2008.04.005 S0166-4328(08)00190-3 [pii]

Husemann J, Loike JD, Anankov R, Febbraio M, Silverstein SC (2002) Scavenger receptors in neurobiology and neuropathology: their role on microglia and other cells of the nervous system. Glia 40(2):195–205. doi:10.1002/glia.10148

Ifuku M, Farber K, Okuno Y, Yamakawa Y, Miyamoto T, Nolte C, Merrino VF, Kita S, Iwamoto T, Komuro I, Wang B, Cheung G, Ishikawa E, Ooboshi H, Bader M, Wada K, Kettenmann H, Noda M (2007) Bradykinin-induced microglial migration mediated by B1-bradykinin receptors depends on $Ca^{2+}$ influx via reverse-mode activity of the $Na^+/Ca^{2+}$ exchanger. J Neurosci 27(48):13065–13073. doi:10.1523/JNEUROSCI.3467-07.2007 27/48/13065 [pii]

Ifuku M, Okuno Y, Yamakawa Y, Izumi K, Seifert S, Kettenmann H, Noda M (2011) Functional importance of inositol-1,4,5-triphosphate-induced intracellular $Ca^{2+}$ mobilization in galanin-induced microglial migration. J Neurochem 117(1):61–70. doi:10.1111/j.1471-4159.2011.07176.x

Inoue K (2008) Purinergic systems in microglia. Cell Mol Life Sci 65(19):3074–3080. doi:10.1007/s00018-008-8210-3

Jensen MB, Gonzalez B, Castellano B, Zimmer J (1994) Microglial and astroglial reactions to anterograde axonal degeneration: a histochemical and immunocytochemical study of the adult rat fascia dentata after entorhinal perforant path lesions. Exp Brain Res 98(2):245–260

Kettenmann H, Hanisch UK, Noda M, Verkhratsky A (2011) Physiology of microglia. Physiol Rev 91(2):461–553. doi:10.1152/physrev.00011.2010 91/2/461 [pii]

Kettenmann H, Kirchhoff F, Verkhratsky A (2013) Microglia: new roles for the synaptic stripper. Neuron 77(1):10–18. doi:10.1016/j.neuron.2012.12.023 S0896-6273(12)01162-2 [pii]

Kierdorf K, Prinz M (2013) Factors regulating microglia activation. Front Cell Neurosci 7:44. doi:10.3389/fncel.2013.00044

Kiryu-Seo S, Kiyama H (2011) The nuclear events guiding successful nerve regeneration. Front Mol Neurosci 4:53. doi:10.3389/fnmol.2011.00053

Kiryu-Seo S, Hirayama T, Kato R, Kiyama H (2005) Noxa is a critical mediator of p53-dependent motor neuron death after nerve injury in adult mouse. J Neurosci 25(6):1442–1447. doi:10.1523/JNEUROSCI.4041-04.2005 25/6/1442 [pii]

Knott G, Holtmaat A (2008) Dendritic spine plasticity–current understanding from in vivo studies. Brain Res Rev 58(2):282–289. doi:10.1016/j.brainresrev.2008.01.002 S0165-0173(08)00004-0 [pii]

Koizumi S, Shigemoto-Mogami Y, Nasu-Tada K, Shinozaki Y, Ohsawa K, Tsuda M, Joshi BV, Jacobson KA, Kohsaka S, Inoue K (2007) UDP acting at P2Y6 receptors is a mediator of microglial phagocytosis. Nature 446(7139):1091–1095. doi:10.1038/nature05704 nature05704 [pii]

Kreutzberg GW (1996) Microglia: a sensor for pathological events in the CNS. Trends Neurosci 19(8):312–318. doi:10.1016/0166-2236(96)10049-7 [pii]

Kurpius D, Nolley EP, Dailey ME (2007) Purines induce directed migration and rapid homing of microglia to injured pyramidal neurons in developing hippocampus. Glia 55(8):873–884. doi:10.1002/glia.20509

Le Feuvre RA, Brough D, Touzani O, Rothwell NJ (2003) Role of P2X7 receptors in ischemic and excitotoxic brain injury in vivo. J Cereb Blood Flow Metab 23(3):381–384

Lehnardt S (2010) Innate immunity and neuroinflammation in the CNS: the role of microglia in Toll-like receptor-mediated neuronal injury. Glia 58(3):253–263. doi:10.1002/glia.20928

Linda H, Shupliakov O, Ornung G, Ottersen OP, Storm-Mathisen J, Risling M, Cullheim S (2000) Ultrastructural evidence for a preferential elimination of glutamate-immunoreactive synaptic terminals from spinal motoneurons after intramedullary axotomy. J Comp Neurol 425(1):10–23. doi:10.1002/1096-9861(20000911)425:1<10:AID-CNE2>3.0.CO;2-# [pii]

Linnartz B, Neumann H (2013) Microglial activatory (immunoreceptor tyrosine-based activation motif)- and inhibitory (immunoreceptor tyrosine-based inhibition motif)-signaling receptors for recognition of the neuronal glycocalyx. Glia 61(1):37–46. doi:10.1002/glia.22359

Liu GJ, Nagarajah R, Banati RB, Bennett MR (2009) Glutamate induces directed chemotaxis of microglia. Eur J Neurosci 29(6):1108–1118. doi:10.1111/j.1460-9568.2009.06659.x EJN6659 [pii]

Makwana M, Jones LL, Cuthill D, Heuer H, Bohatschek M, Hristova M, Friedrichsen S, Ormsby I, Bueringer D, Koppius A, Bauer K, Doetschman T, Raivich G (2007) Endogenous transforming growth factor beta 1 suppresses inflammation and promotes survival in adult CNS. J Neurosci 27(42):11201–11213. doi:10.1523/JNEUROSCI.2255-07.2007 27/42/11201 [pii]

Marin-Teva JL, Cuadros MA, Martin-Oliva D, Navascues J (2011) Microglia and neuronal cell death. Neuron Glia Biol 7(1):25–40. doi:10.1017/S1740925X12000014 S1740925X12000014 [pii]

Martin S, Dicou E, Vincent JP, Mazella J (2005) Neurotensin and the neurotensin receptor-3 in microglial cells. J Neurosci Res 81(3):322–326. doi:10.1002/jnr.20477

Masui K, Yamada E, Shimokawara T, Mishima K, Enomoto Y, Nakajima H, Yoshikawa T, Sakaki T, Ichijima K (2002) Expression of c-Jun N-terminal kinases after axotomy in the dorsal motor nucleus of the vagus nerve and the hypoglossal nucleus. Acta Neuropathol 104(2): 123–129. doi:10.1007/s00401-002-0519-7

Matute C, Torre I, Perez-Cerda F, Perez-Samartin A, Alberdi E, Etxebarria E, Arranz AM, Ravid R, Rodriguez-Antiguedad A, Sanchez-Gomez M, Domercq M (2007) P2X(7) receptor blockade prevents ATP excitotoxicity in oligodendrocytes and ameliorates experimental autoimmune encephalomyelitis. J Neurosci 27(35):9525–9533. doi:10.1523/JNEUROSCI.0579-07.2007 27/35/9525 [pii]

Matzinger P (2007) Friendly and dangerous signals: is the tissue in control? Nat Immunol 8(1): 11–13. doi:10.1038/ni0107-11 ni0107-11 [pii]

Mehta A, Prabhakar M, Kumar P, Deshmukh R, Sharma PL (2013) Excitotoxicity: bridge to various triggers in neurodegenerative disorders. Eur J Pharmacol 698(1–3):6–18. doi:10.1016/j.ejphar.2012.10.032 S0014-2999(12)00900-4 [pii]

Melani A, Turchi D, Vannucchi MG, Cipriani S, Gianfriddo M, Pedata F (2005) ATP extracellular concentrations are increased in the rat striatum during in vivo ischemia. Neurochem Int 47(6):442–448. doi:10.1016/j.neuint.2005.05.014 S0197-0186(05)00143-9 [pii]

Melchior B, Garcia AE, Hsiung BK, Lo KM, Doose JM, Thrash JC, Stalder AK, Staufenbiel M, Neumann H, Carson MJ (2010) Dual induction of TREM2 and tolerance-related transcript, Tmem176b, in amyloid transgenic mice: implications for vaccine-based therapies for Alzheimer's disease. ASN Neuro 2(3):e00037. doi:10.1042/AN20100010

Mentis GZ, Greensmith L, Vrbova G (1993) Motoneurons destined to die are rescued by blocking N-methyl-D-aspartate receptors by MK-801. Neuroscience 54(2):283–285. doi:10.1016/0306-4522(93)90253-C [pii]

Moran LB, Graeber MB (2004) The facial nerve axotomy model. Brain Res Brain Res Rev 44(2–3):154–178. doi:10.1016/j.brainresrev.2003.11.004 S0165017303002595 [pii]

Mott RT, Ait-Ghezala G, Town T, Mori T, Vendrame M, Zeng J, Ehrhart J, Mullan M, Tan J (2004) Neuronal expression of CD22: novel mechanism for inhibiting microglial proinflammatory cytokine production. Glia 46(4):369–379. doi:10.1002/glia.20009

Navarro X, Vivo M, Valero-Cabre A (2007) Neural plasticity after peripheral nerve injury and regeneration. Prog Neurobiol 82(4):163–201. doi:10.1016/j.pneurobio.2007.06.005 S0301-0082(07)00109-8 [pii]

Neumann J, Gunzer M, Gutzeit HO, Ullrich O, Reymann KG, Dinkel K (2006) Microglia provide neuroprotection after ischemia. FASEB J 20(6):714–716. doi:10.1096/fj.05-4882fje 05-4882fje [pii]

Nimmerjahn A, Kirchhoff F, Helmchen F (2005) Resting microglial cells are highly dynamic surveillants of brain parenchyma in vivo. Science 308(5726):1314–1318. doi:10.1126/science.1110647 1110647 [pii]

Noda M, Ifuku M, Mori Y, Verkhratsky A (2013) Calcium influx through reversed NCX controls migration of microglia. Adv Exp Med Biol 961:289–294. doi:10.1007/978-1-4614-4756-6_24

Ohsawa K, Kohsaka S (2011) Dynamic motility of microglia: purinergic modulation of microglial movement in the normal and pathological brain. Glia 59(12):1793–1799. doi:10.1002/glia.21238

Orr AG, Orr AL, Li XJ, Gross RE, Traynelis SF (2009) Adenosine A(2A) receptor mediates microglial process retraction. Nat Neurosci 12(7):872–878. doi:10.1038/nn.2341 nn.2341 [pii]

Panatier A, Robitaille R (2012) The soothing touch: microglial contact influences neuronal excitability. Dev Cell 23(6):1125–1126. doi:10.1016/j.devcel.2012.11.015 S1534-5807(12)00535-7 [pii]

Perez-Alvarez A, Navarrete M, Covelo A, Martin ED, Araque A (2014) Structural and functional plasticity of astrocyte processes and dendritic spine interactions. J Neurosci 34(38):12738–12744. doi:10.1523/JNEUROSCI.2401-14.2014 34/38/12738 [pii]

Perry VH, O'Connor V (2010) The role of microglia in synaptic stripping and synaptic degeneration: a revised perspective. ASN Neuro 2(5):e00047. doi:10.1042/AN20100024

Pocock JM, Kettenmann H (2007) Neurotransmitter receptors on microglia. Trends Neurosci 30(10):527–535. doi:10.1016/j.tins.2007.07.007 S0166-2236(07)00211-1 [pii]

Raivich G, Makwana M (2007) The making of successful axonal regeneration: genes, molecules and signal transduction pathways. Brain Res Rev 53(2):287–311. doi:10.1016/j.brainresrev.2006.09.005 S0165-0173(06)00110-X [pii]

Rappert A, Bechmann I, Pivneva T, Mahlo J, Biber K, Nolte C, Kovac AD, Gerard C, Boddeke HW, Nitsch R, Kettenmann H (2004) CXCR3-dependent microglial recruitment is essential for dendrite loss after brain lesion. J Neurosci 24(39):8500–8509. doi:10.1523/JNEUROSCI.2451-04.2004 24/39/8500 [pii]

Raslan A, Ernst P, Werle M, Thieme H, Szameit K, Finkensieper M, Guntinas-Lichius O, Irintchev A (2014) Reduced cholinergic and glutamatergic synaptic input to regenerated motoneurons after facial nerve repair in rats: potential implications for recovery of motor function. Brain Struct Funct 219(3):891–909. doi:10.1007/s00429-013-0542-6

Rathbone MP, Middlemiss PJ, Gysbers JW, Andrew C, Herman MA, Reed JK, Ciccarelli R, Di Iorio P, Caciagli F (1999) Trophic effects of purines in neurons and glial cells. Prog Neurobiol 59(6):663–690. doi:10.1016/S0301-0082(99)00017-9 [pii]

Ravichandran KS (2010) Find-me and eat-me signals in apoptotic cell clearance: progress and conundrums. J Exp Med 207(9):1807–1817. doi:10.1084/jem.20101157 jem.20101157 [pii]

Sasaki Y, Hoshi M, Akazawa C, Nakamura Y, Tsuzuki H, Inoue K, Kohsaka S (2003) Selective expression of Gi/o-coupled ATP receptor P2Y12 in microglia in rat brain. Glia 44(3):242–250. doi:10.1002/glia.10293

Siskova Z, Tremblay ME (2013) Microglia and synapse: interactions in health and neurodegeneration. Neural Plast 2013:425845. doi:10.1155/2013/425845

Sperlagh B, Illes P (2007) Purinergic modulation of microglial cell activation. Purinergic Signal 3(1–2):117–127. doi:10.1007/s11302-006-9043-x

Stone TW (2002) Purines and neuroprotection. Adv Exp Med Biol 513:249–280

Sumner BE, Sutherland FI (1973) Quantitative electron microscopy on the injured hypoglossal nucleus in the rat. J Neurocytol 2(3):315–328

Takayama N, Ueda H (2005) Morphine-induced chemotaxis and brain-derived neurotrophic factor expression in microglia. J Neurosci 25(2):430–435. doi:10.1523/JNEUROSCI.3170-04.2005 25/2/430 [pii]

Theodosis DT, Poulain DA, Oliet SH (2008) Activity-dependent structural and functional plasticity of astrocyte-neuron interactions. Physiol Rev 88(3):983–1008. doi:10.1152/physrev.00036.2007 88/3/983 [pii]

Tozaki-Saitoh H, Tsuda M, Miyata H, Ueda K, Kohsaka S, Inoue K (2008) P2Y12 receptors in spinal microglia are required for neuropathic pain after peripheral nerve injury. J Neurosci 28(19):4949–4956. doi:10.1523/JNEUROSCI.0323-08.2008 28/19/4949 [pii]

Trapp BD, Wujek JR, Criste GA, Jalabi W, Yin X, Kidd GJ, Stohlman S, Ransohoff R (2007) Evidence for synaptic stripping by cortical microglia. Glia 55(4):360–368. doi:10.1002/glia.20462

Tremblay ME, Lowery RL, Majewska AK (2010) Microglial interactions with synapses are modulated by visual experience. PLoS Biol 8(11):e1000527. doi:10.1371/journal.pbio.1000527

Tyzack GE, Sitnikov S, Barson D, Adams-Carr KL, Lau NK, Kwok JC, Zhao C, Franklin RJ, Karadottir RT, Fawcett JW, Lakatos A (2014) Astrocyte response to motor neuron injury promotes structural synaptic plasticity via STAT3-regulated TSP-1 expression. Nat Commun 5:4294. doi:10.1038/ncomms5294 ncomms5294 [pii]

Villacampa N, Almolda B, Vilella A, Campbell IL, Gonzalez B, Castellano B (2015) Astrocyte-targeted production of IL-10 induces changes in microglial reactivity and reduces motor neuron death after facial nerve axotomy. Glia. doi:10.1002/glia.22807

Wake H, Moorhouse AJ, Jinno S, Kohsaka S, Nabekura J (2009) Resting microglia directly monitor the functional state of synapses in vivo and determine the fate of ischemic terminals. J Neurosci 29(13):3974–3980. doi:10.1523/JNEUROSCI.4363-08.2009 29/13/3974 [pii]

Walter L, Franklin A, Witting A, Wade C, Xie Y, Kunos G, Mackie K, Stella N (2003) Nonpsychotropic cannabinoid receptors regulate microglial cell migration. J Neurosci 23(4):1398–1405 23/4/1398 [pii]

Woo NH, Lu B (2006) Regulation of cortical interneurons by neurotrophins: from development to cognitive disorders. Neuroscientist 12(1):43–56. doi:10.1177/1073858405284360 12/1/43 [pii]

Xie R, Villacampa N, Almolda B, González B, Castellano B, Campbell IL (2014) Interferon regulator factor (IRF) 8 regulates the microglial response to neuronal injury. Cytokine 70(1):77. doi:10.1016/j.cyto.2014.07.209

Xing C, Wang X, Cheng C, Montaner J, Mandeville E, Leung W, van Leyen K, Lok J, Lo EH (2014) Neuronal production of lipocalin-2 as a help-me signal for glial activation. Stroke 45 (7):2085–2092. doi:10.1161/STROKEAHA.114.005733 STROKEAHA.114.005733 [pii]

Yamada J, Hayashi Y, Jinno S, Wu Z, Inoue K, Kohsaka S, Nakanishi H (2008) Reduced synaptic activity precedes synaptic stripping in vagal motoneurons after axotomy. Glia 56(13): 1448–1462. doi:10.1002/glia.20711

Yamada J, Nakanishi H, Jinno S (2011) Differential involvement of perineuronal astrocytes and microglia in synaptic stripping after hypoglossal axotomy. Neuroscience 182:1–10. doi:10.1016/j.neuroscience.2011.03.030 S0306-4522(11)00288-0 [pii]

# Oligodendrocytes: Functioning in a Delicate Balance Between High Metabolic Requirements and Oxidative Damage

**Alejandro D. Roth and Marco T. Núñez**

**Abstract** The study of the metabolic interactions between myelinating glia and the axons they ensheath has blossomed into an area of research much akin to the elucidation of the role of astrocytes in tripartite synapses (Tsacopoulos and Magistretti in J Neurosci 16:877–885, 1996). Still, unlike astrocytes, rich in cytochrome-P450 and other anti-oxidative defense mechanisms (Minn et al. in Brain Res Brain Res Rev 16:65–82, 1991; Wilson in Can J Physiol Pharmacol. 75:1149–1163, 1997), oligodendrocytes can be easily damaged and are particularly sensitive to both hypoxia and oxidative stress, especially during their terminal differentiation phase and while generating myelin sheaths. In the present review, we will focus in the metabolic complexity of oligodendrocytes, particularly during the processes of differentiation and myelin deposition, and with a specific emphasis in the context of oxidative stress and the intricacies of the iron metabolism of the most iron-loaded cells of the central nervous system (CNS).

**Keywords** Myelination · Ensheathment · Reactive oxygen species (ROS) · Oligodendrocyte precursor cells (OPCs)

## A Brief Introduction to Oligodendrocyte Structure and Function

Oligodendrocytes were initially described in 1921 by Pio del Río Hortega, who demonstrated their neuroectodermal origin and association to axons (see a historical review by Pasik and Pasik 2004). Historically unlike other glial cells, oligodendrocytes have attracted a lot of attention and are probably one of the best-studied cell lineages in the CNS, reflecting the importance myelin plays in multiple diseases and in the inhibition of axonal regeneration. At the same time, and in contradiction to what would be predicted from their complex morphology and intimate

A.D. Roth (✉) · M.T. Núñez
Department of Biology, Faculty of Science, University of Chile, Santiago, Chile
e-mail: alejroth@uchile.cl



R. von Bernhardi (ed.), *Glial Cells in Health and Disease of the CNS*,
Advances in Experimental Medicine and Biology 949,
DOI 10.1007/978-3-319-40764-7_8

association to axons, oligodendrocytes have proven to be remarkably adaptable to in vitro culture, as purified precursor cells can achieve terminal differentiation in the absence of neurons or other glial cells, extending huge membrane sheets and expressing characteristic myelin markers like myelin basic protein (MBP), proteolipid protein (PLP), myelin-associated glycoprotein (MAG), and myelin oligodendrocyte glycoprotein (MOG) (Pfeiffer et al. 1993; Espinosa-Jeffrey et al. 2009).

Oligodendrocytes originate from oligodendrocyte precursor cells (OPCs) which appear late in embryonic development at different areas of the ventral neural cord and sub-ventricular zone (SVZ) (Baumann and Pham-Dinh 2001; Miller and Mi 2007). Here, we must note that in rodent telencephalon oligodendrocytes originate in three waves from increasingly dorsally localized niches and, as each OPC wave arises, it displaces the previous cells, resulting in a postnatal predominance of dorsally derived oligodendrocytes (Kessaris et al. 2006; Rowitch and Kriegstein 2010; Tomassy and Fossati 2014). As these highly motile cells (Noble et al. 1988) migrate away in response to chemoattractants and chemorepulsants (e.g., Netrin-1) (Jarjour et al. 2003), they undergo multiple rounds of division before differentiating into oligodendrocyte progenitors (pre-OLGs) which present lower motility, a higher number of cell processes, and which scan the surrounding area for axons. At this stage, pre-OLGs undergo terminal differentiation into immature oligodendrocytes, developing a large number of highly branched cell processes which, in vitro and in the absence of axons, flatten on the substrate, express characteristic protein markers, and extend large membrane sheets that recapitulate the biochemical composition of myelin, and thus are referred to as "mature oligodendrocytes." It is important to note that in vitro differentiation and maturation renders these cells highly dependent on extrinsic survival factors, reflecting their requirement on trophic factors that are provided by axons (Barres et al. 1992, 1993). In vivo, as oligodendrocytes are the last neural cell lineage to arise, myelin deposition starts shortly before birth (in rodents) and proceeds to expand in a rostral to caudal fashion in the ventral spinal cord, while expanding through the dorsal spinal cord both rostrally and caudally starting from the cervical enlargement throughout the first weeks of life (Foran and Peterson 1992). In humans, myelination is initiated close to the 17th week of gestation and continues throughout life (Tosic et al. 2002; recently reviewed in El Waly et al. 2014). It should be noted that the onset of embryological myelination marks the damage window of hypoxic-ischemic injury to the periventricular cerebral white matter (periventricular leukomalacia, between 23 and 32 weeks), which results in the most common brain damage to premature infants: cerebral palsy (Back et al. 2001; Volpe 2001) (see Chapter "Prenatal Systemic Hypoxia-Ischemia and Oligodendroglia Loss in Cerebellum" for further reading on hypoxic-ischemic injury and cerebral palsy).

Terminal oligodendrocytes maturation can be separated into three major steps: (i) Cell spreading and axon recognition; (ii) Axon–glial association and Initial wrapping; (iii) Membrane extension and myelin compaction. Once an OLG process has aligned and spiraled around the axon, it spreads laterally, covering the axon and the innermost spirals, an extensively studied event that has given rise to multiple models as to how it is achieved (see review by Bauer et al. 2009). Then, as multiple plasma membrane lamella have wrapped around and axon, a process of cytoplasm extrusion brings the inner faces of the plasma membrane close together, displacing the cytoplasm to the loops at the sides of the internode (see reviews by Pedraza et al. 2001; Ffrench-Constant et al. 2004; Simons and Trotter 2007). Through this process of compaction, the myelin sheath is divided into two structurally and biochemically different domains: compact myelin, which surrounds the internode and provides radial electrical resistance, and the non-compacted channels, cytoplasm-rich areas that connect compact myelin domains to the cell body and provide metabolic support for the compacted myelin domains and the underlying axon (see Court et al. 2008 and an extensive review by Nave 2010). It should be stressed that compact myelin lamella corresponds to a particularly lipid-enriched plasma membrane, where cholesterol (26 %), galactolipids (31 %), and plasmalogens (20 %) overshadow the overall protein content (30 %) (see Chrast et al. 2011). Here, in comparison to the rest of the CNS, unsaturated long-chain fatty acids are unusually concentrated. These lipidic characteristics lend myelin a distinct density when compared to other CNS membranes, simplifying its purification and its proteomic and lipidomic characterization (Taylor et al. 2004; Roth et al. 2006; Ishii et al. 2009; Jahn et al. 2009; Dhaunchak et al. 2010; Gopalakrishnan et al. 2013). At the same time, this highly stable lipid composition, the organization of the hydrophobic motifs, and the long half-life of myelin components have led to the interesting suggestion that myelin organization arises through an intrinsic self-organization and self-assembly into what has been compared to a lipid crystal (see Aggarwal et al. 2011 for a comprehensive discussion on this topic).

As described in Chapter "Glial Cells and Integrity of the Nervous System," myelination itself divides axons into two discrete domains: the myelin free, electrically active nodes (nodes of Ranvier) and the myelin ensheathed internodes, which separate one node from the next. At a molecular level, nodes present a characteristic concentration of voltage-gated sodium channels that are central to saltatory conduction (estimated at 1000–1500 channels/$\mu m^2$) (Rosenbluth 1976), while internodes are divided into three specific domains: paranodes, juxtaparanodes, and internodes, whose description escapes the scope of this work (interested readers should seek out Rasband et al. 1998; Poliak et al. 1999; Pedraza et al. 2001, 2009; Traka et al. 2002; Poliak and Peles 2003).

## Oligodendrocyte Metabolism and Axonal Support

Studies on myelin tend to concentrate in the acceleration of neural communication, for the concept of response time is easily conveyed and the expression "the quick and the dead" is echoed in nature by the myriad of adaptations that hasten response to stimuli by reducing the time required for an action potentials to transverse the length of an axon. Nevertheless, adaptations to increase speed may be metabolically expensive and require considerable resources, as is evident for the giant axons in squids, where the excitable membrane is depolarized and repolarized throughout the whole length of the axon, and the ion gradient must be maintained by greater action of the Na+/K+ pump. Axon myelination induces saltatory conduction of action potentials, thereby increasing conduction speed up to a hundredfold while maintaining small cross-sectional axonal diameters and reducing the areas of the excitable membrane that are depolarized during each action potential (reviewed by Hartline and Colman 2007). In this sense, the importance of axonal ensheathment is highlighted by the convergent evolution of myelin-like structures in annelids and crustacea, which do not share myelinated common ancestors with vertebrates (Xu and Terakawa 1999; Wilson and Hartline 2011). Furthermore, although most of these myelin-like structures are the result of axonal ensheathments by glial cell, recent observations in *calanoid copepods* show the construction of myelin-like structures from the axon itself, without the participation of an accompanying glial cell (Wilson and Hartline 2011). Multiple authors have put forth the idea that myelination is advantageous, being metabolically "cheap" for axons, as it minimizes the energy required to reset the axonal membrane to its chemo-electrically polarized state, to the point that some have suggested that this energy saving was the original function of myelin sheaths, and that action potential acceleration emerged later as a response to increases in depredation (Stiefel et al. 2013). Still, these ideas must be tempered by the impressive metabolic cost this places upon the myelinating glial cells, which not only must deposit and maintain huge plasma membrane extensions, but must also provide protection and metabolic support to axons (see review by Hirrlinger and Nave 2014), a condition highlighted by the plethora of demyelinating diseases (adrenoleukodystrophy, Charcot–Marie–Tooth disease, multiple sclerosis (MS), and amyotrophic lateral sclerosis), all of which have bad prognosis and present incapacitating effects, particularly when axonal damage ensues (Sauer et al. 2013) (see Chapter "Peripheral Inflammation and Demyelinating Diseases" for further reading on demyelinating diseases). In this context, considering the effects on mice lacking myelin-specific genes has led some authors to suggest that some neurodegenerative diseases could result from problems intrinsic to the myelinating cells (Popko 2003).

## Hypoxia and Oxidative Stress

Myelination requires that each oligodendrocyte first constructs and then maintains an extensive plasma membrane, for which a vast supply of both precursor molecules and oxidative substrates is required. Thus, the availability of these resources is ensured through oligodendrocyte secretion of pro-angiogenic factors [Hypoxia inducible factors, HIFs (Yuen et al. 2014)], events that underscore the involvement of oligodendrocytes in the construction of the vascular tree of the CNS. In this context, it is notable that hypoxia-mediated damage to white matter in periventricular leukomalacia occurs precisely during the early phases of CNS myelination (Back et al. 2001; Volpe 2001). While the underlying causes of periventricular leukomalacia are a matter of ongoing controversy (Dammann and Leviton 2004;

Edwards and Tan 2006; Khwaja and Volpe 2008; Volpe et al. 2011; Hu et al. 2013), it has been shown that the blood flow to white matter areas is slower when compared with gray matter areas in preterm babies (Greisen and Borch 2001), a condition that probably reflects an incomplete maturation of the vasculature (Volpe et al. 2011). White matter damage has been associated to excitotoxic-mediated death (Matute et al. 2007) and reactive oxygen species (ROS), to which OPCs are particularly sensitive as they are poor in antioxidant defenses and their maturation is inhibited by these reactive species (French et al. 2009; Volpe et al. 2011) (see Chapter "Prenatal Systemic Hypoxia-Ischemia and Oligodendroglia Loss in Cerebellum"). At the same time, areas damaged in periventricular leukomalacia present increased levels of OLG-associated ROS generating 12/15-lipoxygenase (Haynes and van Leyen 2013), and lipid peroxidation markers of oxidative damage (Back et al. 2005).

As the CNS matures, these events become less pronounced, for oligodendrocytes are better protected from oxidative stress by the presence of higher levels of Glutathione (Back et al. 1998) and by their switch from an oxidative to a glycolytic metabolism (Funfschilling et al. 2012); events that probably reflect the long half-life of myelin proteins (Toyama et al. 2013) and the metabolic support of axons by myelin sheaths (Brown et al. 2001; Morland et al. 2007; Funfschilling et al. 2012; Hirrlinger and Nave 2014). Nonetheless, as reviewed recently (van Meeteren et al. 2005), oligodendrocytes are vulnerable to oxidative stress (Smith et al. 1999), particularly in the context of autoimmune-mediated inflammatory injury, as occurs in multiple sclerosis (Guan et al. 2014), where oxidative stress markers are observed in the cerebral spinal fluid of MS patients (Sbardella et al. 2013; Mir et al. 2014). Likewise, current models of X-linked adrenoleukodystrophy have associated to the loss of peroxisomal ATP-binding cassette transporter D (ABCD1), which allows the incorporation of very long-chain fatty acids (VLCFA) into this organelle, to an increase in ROS production and generalized oxidative stress (Baes and Aubourg 2009; Baarine et al. 2012).

## *Iron Homeostasis in Oligodendrocytes—Importance and Perils*

Oligodendrocytes and iron have a love affair. Indeed, in areas acknowledged as iron-rich, namely substantia nigra, striatum, and cerebellar nuclei, the cells that stain most prominently for iron are oligodendrocytes (Hill et al. 1985; Dwork et al. 1988; Benkovic and Connor 1993; Connor and Menzies 1996; Todorich et al. 2009). Likewise, white matter, rich in oligodendrocytes, stains more strongly for iron than gray matter (LeVine and Macklin 1990). Oligodendrocytes have a particular need for iron during myelination because of iron's participation in the synthesis of cholesterol, an indispensable component of myelin membranes (Beard and Connor 2003; Saher et al. 2005; Badaracco et al. 2010), and its participation as a cofactor in a large number of enzymes involved in myelination and re-myelination (reviewed in Stephenson et al. 2014).

One of the major symptoms of iron deficiency in young children is a decline in cognitive capacity characterized by poorer cognition, decreased school achievement, and other behavior problems (Grantham-McGregor and Ani 2001; Felt et al. 2006). In a number of studies, the clinical symptoms of iron deficiency have been associated to hypomyelination (Oski et al. 1983; Kretchmer et al. 1996; Lozoff et al. 1996). Despite the evident importance of an adequate iron status for the process of myelin synthesis, particularly in regards to the enzymes involved (see review by Todorich et al. 2009), knowledge on the specific elements responsible for iron homeostasis in oligodendrocytes is incipient.

With regard to iron uptake, while the two typical routes have been reported, namely transferrin endocytosis and direct membrane uptake by iron transporters, a third and less common route has been recently been proposed: the endocytosis of the H-rich isoform of ferritin (HFn) (Fig. 1).

Besides providing iron for cell needs, transferrin (Tf) enhances oligodendrocyte progenitor differentiation, as the presence of Tf in the culture medium increases proliferation of oligodendrocyte precursors, an effect mediated at least in part by Tf receptor 1 (TfR1) (Guardia Clausi et al. 2010). Similarly, mice overexpressing Tf in the brain undergo faster oligodendrocyte lineage maturation than animals expressing regular levels of Tf (Sow et al. 2006). In in vitro experiments, Tf enhanced the proliferation rate of OPCs as apoTf added either to the culture medium or by overexpression of Tf in an immature oligodendrocyte cell line (Silvestroff et al. 2013). Likewise, Tf-induced neural stem and progenitor cells toward the oligodendroglial lineage and promoted oligodendrocyte maturation from OPCs (Silvestroff et al. 2012). Taken together, the evidence supports the notion that, besides its important role as an iron donor, Tf is a trophic factor that accelerates oligodendrocyte commitment and differentiation.

Both TfR1 and H-ferritin have been shown to be up-regulated during in vitro oligodendrocyte maturation, indicating that both proteins might contribute to the supply of iron in the process of maturation and myelination of OPCs (Li et al. 2013). Nevertheless, it appears that upon maturation, the contribution of Tf endocytosis as a mechanism of iron uptake losses relevance. Cultured OPCs express low levels of transferrin receptors (Espinosa de los Monteros and Foucaud 1987), and upon differentiation into mature oligodendrocytes, transferrin receptor protein expression becomes undetectable by Western blot (Todorich et al. 2009). Likewise, histochemical assessment of TfR1 distribution in the brain shows just a slight staining of TfR1 in mature oligodendrocytes (Giometto et al. 1990; Connor and Menzies 1995). Given the iron requirements by mature oligodendrocytes, it is likely that these cells possess an iron acquisition mechanism complementary to TfR1-mediated uptake, for, as noted, mature oligodendrocytes are the brain cells that most strikingly stain for iron.

A second form of iron incorporation into cells occurs through the direct transport across the plasma membrane by the inwards iron transporter protein DMT1. Still, in normal and Belgrade rats (which express a mutated form of DMT1 with a low

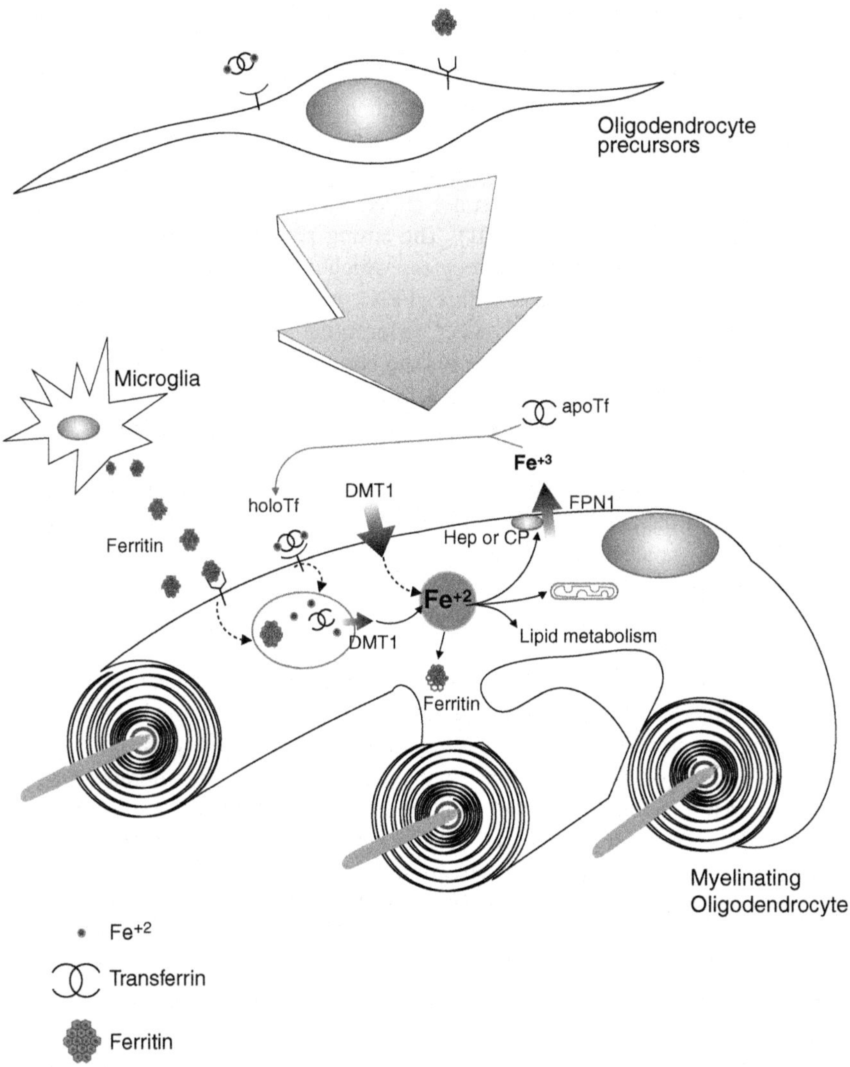


**Fig. 1** Overview of oligodendrocyte iron metabolism. Differentiating oligodendrocyte precursors accumulate iron through three different pathways: Transferrin (Tf) or ferritin endocytosis and direct membrane uptake by iron transporters (DMT1). While both endocytic-dependent pathways converge into lysosomes, where vesicular iron transporter (also DMT1) allows iron to exit to the cytoplasm, each holotransferrin (holoTf) molecule transports two iron atoms, while one ferritin molecule holds up to 4000. Intracellular iron ($Fe^{+2}$) is mobilized to mitochondria, where it participates as an enzyme cofactor in oxidative phosphorylation, or becomes bound to enzymes participating in lipid metabolism. Intracellular iron excesses are stored as intracellular ferritin or secreted by ferroportin (FPN1), a process which requires further oxidation into $Fe^{+3}$ by the membrane-associated ferroxidases Hephaestin (Hep) or Ceruloplasmin (CP). This secreted $Fe^{+3}$ is bound by apotransferrin (apoTf) resulting in endocytable holoTf

transport activity), there is a slight staining of DMT1 in oligodendrocytes at the subcortical white matter, and an overall sparse presence of DMT1 positive oligodendrocytes (Burdo et al. 2001). If this low DMT1 immunostaining pattern is a characteristic of mature oligodendrocytes, then direct iron uptake mediated by DMT1 can be predicted to be of a relative low magnitude, consistent with the need for alternative mechanisms of iron uptake. Unlike DMT1, the immunostaining for the iron export transporter Ferroportin 1 (FPN1) is relatively strong in subcortical oligodendrocytes (Burdo et al. 2001). The strong presence of FPN1 suggests an active iron efflux from oligodendrocytes, which will allow for decreased iron accumulation. For effective iron efflux, FPN1 must partner with a ferroxidase, needed to oxidize outcoming $Fe^{2+}$ to $Fe^{3+}$ prior to the binding of $Fe^{3+}$ to transferrin. Oligodendrocytes express both a membrane-bound form of the ferroxidase ceruloplasmin and hephaestin, a ferroxidase initially described in intestinal epithelia (Schulz et al. 2014), thus iron export, as a form to alleviate iron overload, is most probably at work in oligodendrocytes.

The third, and least conventional, form of iron incorporation into oligodendrocytes is through the endocytosis of the iron storage molecule ferritin in its H-rich isoform (HFn), which in its hollow center can contain up to 4500 atoms of iron (Arosio and Levi 2010). The endocytosis of HFn could turn out to be a highly efficient manner of iron uptake, for the uptake of one ferritin molecule could yield thousands of cytoplasmic iron atoms.

The evidence for the involvement of HFn endocytosis as a mechanism of iron uptake by oligodendrocytes is compelling. OPCs bind and internalize $^{125}I$-labeled HFn through a process consistent with clathrin-mediated endocytosis (Hulet et al. 2000). The iron originally present in exogenous HFn is incorporated into the intracellular iron pool, since incubation of oligodendrocyte progenitor cells with HFn for 16 h results in a decreased activity of the iron-deficit sensors: Iron regulatory proteins 1 and 2 in an amount similar to that observed after incubation for 16 h with ferric ammonium citrate (Hulet et al. 2000). Importantly, HFn endocytosis does not seem to be regulated by cell iron status, since pretreatment of the oligodendrocyte precursors for 16 h with either unlabeled HFn (Fe loading) or the iron chelator Desferal (iron depletion) did not affect the amount of $^{125}I$-HFn uptake (Hulet et al. 2000).

The search for the putative receptor for HFn in oligodendrocytes resulted in the identification of T cell immunoglobulin and mucin domain containing protein-2 (Tim-2) as a likely candidate. In contrast to an earlier study (Hulet et al. 2000), the study of Todorich et al. (2008) reported that in CG4 cells Tim-2 expression was responsive to iron, decreasing with iron loading and increasing with iron chelation.

Which is the source of the HFn that supports oligodendrocyte's needs? The answer seems to be the microglial cells, which in the CNS serve as iron stores (Erb et al. 1996; LeVine 1997; Mehlhase et al. 2006; Oshiro et al. 2008). HFn is secreted by iron-loaded microglia, while conditioned media from microglia increases the survival of oligodendrocytes. Transfecting microglia with siRNA for HFn blocks the trophic response of conditioned media on oligodendrocytes (Zhang et al. 2006). Interestingly, the activation of iron-loaded microglia with lipopolysaccharide results in the increased release of the proinflammatory cytokines tumor necrosis factor-α and interleukin-1 (see Chapters "Microglia Function in the Normal Brain" and "Purine Signaling and Microglial Wrapping"), and the decreased release of HFn (Zhang et al. 2006). These results establish a putative link between inflammation and a deficient iron supply by microglia to oligodendrocytes.

The triad of low DMT1—high FPN1—high ferritin predicts a tight management of intracellular redox-active iron. Nevertheless, the high content of iron in oligodendrocytes raises the question of how tightly is iron homeostasis regulated in these cells. Do oligodendrocytes have a tight iron homeostasis system that maintains just enough intracellular iron levels as to satisfy their metabolic needs? Alternatively, do oligodendrocytes behave like iron-starving cells with a poor homeostatic mechanism to control for iron excess? These questions wait for further experimentation.

## *Iron Homeostasis in Oligodendrocytes—Perils*

Iron is a redox-active metal that in the intracellular reductive environment and the presence of oxygen generates the highly damaging hydroxyl free radical (Nunez et al. 2012). The high iron content of oligodendrocytes may be a time bomb, as shown in neurodegenerative diseases like MS. In patients with MS, iron released by oligodendrocytes accumulates in macrophages and microglia around the rim of lesions (Hametner et al. 2013). In turn, iron-loaded macrophages and activated microglia can release redox-active iron into the extracellular space, damaging axon integrity at the border of the lesions (Stephenson et al. 2014).

In MS, iron homeostasis mechanisms in oligodendrocytes seem to be operative. A substantial decrease of iron in the normal-appearing white matter, which corresponded with disease duration, was found in chronic MS patients. This decrease of iron in oligodendrocytes and myelin was probably mediated by the upregulation of iron-exporting ferroxidases (Hametner et al. 2013). Nonetheless, iron released from dying oligodendrocytes accumulated in astrocytes and axons, leading to waves of iron liberation, which may propagate neurodegeneration together with inflammatory bursts (Hametner et al. 2013). Thus, the data indicates that MS oligodendrocytes probably have an active iron export system, and when killed, their iron content is released at the site of the lesion.

## Concluding Remarks

Oligodendrocytes and neurons operate close to the brink, for not only are these cells extremely large and ramified, requiring de novo synthesis of huge amounts of lipids and proteins, but they also extend and maintain extremely long cell processes (axons and dendrites) or enormous plasma membrane sheets. In this sense, considering the rate at which myelin proteins and membranes are produced (Barbarese and Pfeiffer 1981; Pfeiffer et al. 1993; Kramer et al. 2001), it is not surprising that oligodendrocytes are particularly susceptible to chaperone inhibition (Alcazar and Cid 2009) or endoplasmic reticulum stress derived from missense mutations of the predominant myelin proteins (Numasawa-Kuroiwa et al. 2014).

As if this were not enough, oligodendrocytes participate in the maintenance and support of the axons to which they attach. While these conditions must burden the metabolism of these cells, they have low levels of antioxidant defenses, which instead in the CNS are mostly associated with astroglial cells. This paints a picture of intrinsic instability, which is probably offset by switching from oxidative to glycolytic metabolism. Still, the fact remains that these cells present the highest iron content in the CNS. Whether the high iron content is a response to the requirements of cholesterol synthesis and other metabolic processes or to intrinsic peculiarities of the iron homeostasis in these cells is unknown. Nevertheless, the high iron content entails a great peril in cases of oligodendrocytes death, since upon its release to the surroundings it will be taken up by macrophages and microglia, which, in turn, will transform this iron accumulation into oxidative and inflammatory bursts, which can easily spiral out of control and, unlike other types of tissue damage, the consequences for neural tissues are for the most part irreversible.

**Acknowledgments** This work was made possible in part by grant PIA-CONICYT ACT1114 and a support grant 2014 from the Faculty of Science of the University of Chile.

## References

Aggarwal S, Yurlova L, Simons M (2011) Central nervous system myelin: structure, synthesis and assembly. Trends Cell Biol 21:585–593

Alcazar A, Cid C (2009) High cytotoxic sensitivity of the oligodendrocyte precursor cells to HSP90 inhibitors in cell cultures. Exp Neurol 216:511–514

Arosio P, Levi S (2010) Cytosolic and mitochondrial ferritins in the regulation of cellular iron homeostasis and oxidative damage. Biochim Biophys Acta 1800:783–792

Baarine M, Andreoletti P, Athias A, Nury T, Zarrouk A, Ragot K, Vejux A, Riedinger JM, Kattan Z, Bessede G, Trompier D, Savary S, Cherkaoui-Malki M, Lizard G (2012) Evidence of oxidative stress in very long chain fatty acid–treated oligodendrocytes and potentialization of ROS production using RNA interference-directed knockdown of ABCD1 and ACOX1 peroxisomal proteins. Neuroscience 213:1–18

Back SA, Gan X, Li Y, Rosenberg PA, Volpe JJ (1998) Maturation-dependent vulnerability of oligodendrocytes to oxidative stress-induced death caused by glutathione depletion. J Neurosci 18:6241–6253

Back SA, Luo NL, Borenstein NS, Levine JM, Volpe JJ, Kinney HC (2001) Late oligodendrocyte progenitors coincide with the developmental window of vulnerability for human perinatal white matter injury. J Neurosci 21:1302–1312

Back SA, Luo NL, Mallinson RA, O'Malley JP, Wallen LD, Frei B, Morrow JD, Petito CK, Roberts CT Jr, Murdoch GH, Montine TJ (2005) Selective vulnerability of preterm white matter to oxidative damage defined by F2-isoprostanes. Ann Neurol 58:108–120

Badaracco ME, Siri MV, Pasquini JM (2010) Oligodendrogenesis: the role of iron. BioFactors 36:98–102

Baes M, Aubourg P (2009) Peroxisomes, myelination, and axonal integrity in the CNS. Neuroscientist 15:367–379
Barbarese E, Pfeiffer SE (1981) Developmental regulation of myelin basic protein in dispersed cultures. Proc Natl Acad Sci USA 78:1953–1957
Barres BA, Hart IK, Coles HS, Burne JF, Voyvodic JT, Richardson WD, Raff MC (1992) Cell death in the oligodendrocyte lineage. J Neurobiol 23:1221–1230
Barres BA, Jacobson MD, Schmid R, Sendtner M, Raff MC (1993) Does oligodendrocyte survival depend on axons? Curr Biol 3:489–497
Bauer NG, Richter-Landsberg C, Ffrench-Constant C (2009) Role of the oligodendroglial cytoskeleton in differentiation and myelination. Glia 57:1691–1705
Baumann N, Pham-Dinh D (2001) Biology of oligodendrocyte and myelin in the mammalian central nervous system. Physiol Rev 81:871–927
Beard JL, Connor JR (2003) Iron status and neural functioning. Annu Rev Nutr 23:41–58
Benkovic SA, Connor JR (1993) Ferritin, transferrin, and iron in selected regions of the adult and aged rat brain. J Comp Neurol 338:97–113
Brown AM, Wender R, Ransom BR (2001) Metabolic substrates other than glucose support axon function in central white matter. J Neurosci Res 66:839–843
Burdo JR, Menzies SL, Simpson IA, Garrick LM, Garrick MD, Dolan KG, Haile DJ, Beard JL, Connor JR (2001) Distribution of divalent metal transporter 1 and metal transport protein 1 in the normal and Belgrade rat. J Neurosci Res 66:1198–1207
Chrast R, Saher G, Nave KA, Verheijen MH (2011) Lipid metabolism in myelinating glial cells: lessons from human inherited disorders and mouse models. J Lipid Res 52:419–434
Connor JR, Menzies SL (1995) Cellular management of iron in the brain. J Neurol Sci 134 (Suppl):33–44
Connor JR, Menzies SL (1996) Relationship of iron to oligodendrocytes and myelination. Glia 17:83–93
Court FA, Hendriks WT, Macgillavry HD, Alvarez J, van Minnen J (2008) Schwann cell to axon transfer of ribosomes: toward a novel understanding of the role of glia in the nervous system. J Neurosci 28:11024–11029
Dammann O, Leviton A (2004) Inflammatory brain damage in preterm newborns-dry numbers, wet lab, and causal inferences. Early Hum Dev 79:1–15
Dhaunchak AS, Huang JK, De Faria O Jr, Roth AD, Pedraza L, Antel JP, Bar-Or A, Colman DR (2010) A proteome map of axoglial specializations isolated and purified from human central nervous system. Glia 58:1949–1960
Dwork AJ, Schon EA, Herbert J (1988) Nonidentical distribution of transferrin and ferric iron in human brain. Neuroscience 27:333–345
Edwards AD, Tan S (2006) Perinatal infections, prematurity and brain injury. Curr Opin Pediatr 18:119–124
El Waly B, Macchi M, Cayre M, Durbec P (2014) Oligodendrogenesis in the normal and pathological central nervous system. Front Neurosci 8:145
Erb GL, Osterbur DL, LeVine SM (1996) The distribution of iron in the brain: a phylogenetic analysis using iron histochemistry. Brain Res Dev Brain Res 93:120–128
Espinosa de los Monteros A, Foucaud B (1987) Effect of iron and transferrin on pure oligodendrocytes in culture; characterization of a high-affinity transferrin receptor at different ages. Brain Res 432:123–130
Espinosa-Jeffrey A, Wakeman DR, Kim SU, Snyder EY, de Vellis J (2009) Culture system for rodent and human oligodendrocyte specification, lineage progression, and maturation. Curr Protoc Stem Cell Biol Chapter 2:Unit 2D.4
Felt BT, Beard JL, Schallert T, Shao J, Aldridge JW, Connor JR, Georgieff MK, Lozoff B (2006) Persistent neurochemical and behavioral abnormalities in adulthood despite early iron supplementation for perinatal iron deficiency anemia in rats. Behav Brain Res 171:261–270
Ffrench-Constant C, Colognato H, Franklin RJ (2004) Neuroscience. The mysteries of myelin unwrapped. Science 304:688–689

Foran DR, Peterson AC (1992) Myelin acquisition in the central nervous system of the mouse revealed by an MBP-Lac Z transgene. J Neurosci 12:4890–4897
French HM, Reid M, Mamontov P, Simmons RA, Grinspan JB (2009) Oxidative stress disrupts oligodendrocyte maturation. J Neurosci Res 87:3076–3087
Funfschilling U, Supplie LM, Mahad D, Boretius S, Saab AS, Edgar J, Brinkmann BG, Kassmann CM, Tzvetanova ID, Mobius W, Diaz F, Meijer D, Suter U, Hamprecht B, Sereda MW, Moraes CT, Frahm J, Goebbels S, Nave KA (2012) Glycolytic oligodendrocytes maintain myelin and long-term axonal integrity. Nature 485:517–521
Giometto B, Bozza F, Argentiero V, Gallo P, Pagni S, Piccinno MG, Tavolato B (1990) Transferrin receptors in rat central nervous system. An immunocytochemical study. J Neurol Sci 98:81–90
Gopalakrishnan G, Awasthi A, Belkaid W, De Faria O Jr, Liazoghli D, Colman DR, Dhaunchak AS (2013) Lipidome and proteome map of myelin membranes. J Neurosci Res 91:321–334
Grantham-McGregor S, Ani C (2001) A review of studies on the effect of iron deficiency on cognitive development in children. J Nutr 131:649S–666S; discussion 666S–668S
Greisen G, Borch K (2001) White matter injury in the preterm neonate: the role of perfusion. Dev Neurosci 23:209–212
Guan JZ, Guan WP, Maeda T, Guoqing X, GuangZhi W, Makino N (2014) Patients with multiple sclerosis show increased oxidative stress markers and somatic telomere length shortening. Mol Cell Biochem 400:183–187
Guardia Clausi M, Pasquini LA, Soto EF, Pasquini JM (2010) Apotransferrin-induced recovery after hypoxic/ischaemic injury on myelination. ASN Neuro 2:e00048
Hametner S, Wimmer I, Haider L, Pfeifenbring S, Bruck W, Lassmann H (2013) Iron and neurodegeneration in the multiple sclerosis brain. Ann Neurol 74:848–861
Hartline DK, Colman DR (2007) Rapid conduction and the evolution of giant axons and myelinated fibers. Curr Biol 17:R29–R35
Haynes RL, van Leyen K (2013) 12/15-lipoxygenase expression is increased in oligodendrocytes and microglia of periventricular leukomalacia. Dev Neurosci 35:140–154
Hill JM, Ruff MR, Weber RJ, Pert CB (1985) Transferrin receptors in rat brain: neuropeptide-like pattern and relationship to iron distribution. Proc Natl Acad Sci USA 82:4553–4557
Hirrlinger J, Nave KA (2014) Adapting brain metabolism to myelination and long-range signal transduction. Glia 62:1749–1761
Hu Y, Chen G, Wan H, Zhang Z, Zhi H, Liu W, Qian X, Chen M, Wen L, Gao F, Li J, Zhao L (2013) A rat pup model of cerebral palsy induced by prenatal inflammation and hypoxia. Neural Regen Res 8:817–824
Hulet SW, Heyliger SO, Powers S, Connor JR (2000) Oligodendrocyte progenitor cells internalize ferritin via clathrin-dependent receptor mediated endocytosis. J Neurosci Res 61:52–60
Ishii A, Dutta R, Wark GM, Hwang SI, Han DK, Trapp BD, Pfeiffer SE, Bansal R (2009) Human myelin proteome and comparative analysis with mouse myelin. Proc Natl Acad Sci USA 106:14605–14610
Jahn O, Tenzer S, Werner HB (2009) Myelin proteomics: molecular anatomy of an insulating sheath. Mol Neurobiol 40:55–72
Jarjour AA, Manitt C, Moore SW, Thompson KM, Yuh SJ, Kennedy TE (2003) Netrin-1 is a chemorepellent for oligodendrocyte precursor cells in the embryonic spinal cord. J Neurosci 23:3735–3744
Kessaris N, Fogarty M, Iannarelli P, Grist M, Wegner M, Richardson WD (2006) Competing waves of oligodendrocytes in the forebrain and postnatal elimination of an embryonic lineage. Nat Neurosci 9:173–179
Khwaja O, Volpe JJ (2008) Pathogenesis of cerebral white matter injury of prematurity. Arch Dis Child Fetal Neonatal Ed 93:F153–F161
Kramer EM, Schardt A, Nave KA (2001) Membrane traffic in myelinating oligodendrocytes. Microsc Res Tech 52:656–671

Kretchmer N, Beard JL, Carlson S (1996) The role of nutrition in the development of normal cognition. Am J Clin Nutr 63:997S–1001S

LeVine SM (1997) Iron deposits in multiple sclerosis and Alzheimer's disease brains. Brain Res 760:298–303

LeVine SM, Macklin WB (1990) Iron-enriched oligodendrocytes: a reexamination of their spatial distribution. J Neurosci Res 26:508–512

Li Y, Guan Q, Chen Y, Han H, Liu W, Nie Z (2013) Transferrin receptor and ferritin-H are developmentally regulated in oligodendrocyte lineage cells. Neural Regen Res 8:6–12

Lozoff B, Wolf AW, Jimenez E (1996) Iron-deficiency anemia and infant development: effects of extended oral iron therapy. J Pediatr 129:382–389

Matute C, Alberdi E, Domercq M, Sanchez-Gomez MV, Perez-Samartin A, Rodriguez-Antigüedad A, Perez-Cerda F (2007) Excitotoxic damage to white matter. J Anat 210:693–702

Mehlhase J, Gieche J, Widmer R, Grune T (2006) Ferritin levels in microglia depend upon activation: modulation by reactive oxygen species. Biochim Biophys Acta 1763:854–859

Miller RH, Mi S (2007) Dissecting demyelination. Nat Neurosci 10:1351–1354

Minn A, Ghersi-Egea JF, Perrin R, Leininger B, Siest G (1991) Drug metabolizing enzymes in the brain and cerebral microvessels. Brain Res Brain Res Rev 16:65–82

Mir F, Lee D, Ray H, Sadiq SA (2014) CSF isoprostane levels are a biomarker of oxidative stress in multiple sclerosis. Neurol Neuroimmunol Neuroinflamm 1:e21

Morland C, Henjum S, Iversen EG, Skrede KK, Hassel B (2007) Evidence for a higher glycolytic than oxidative metabolic activity in white matter of rat brain. Neurochem Int 50:703–709

Nave KA (2010) Myelination and the trophic support of long axons. Nat Rev Neurosci 11: 275–283

Noble M, Murray K, Stroobant P, Waterfield MD, Riddle P (1988) Platelet-derived growth factor promotes division and motility and inhibits premature differentiation of the oligodendrocyte/type-2 astrocyte progenitor cell. Nature 333:560–562

Numasawa-Kuroiwa Y, Okada Y, Shibata S, Kishi N, Akamatsu W, Shoji M, Nakanishi A, Oyama M, Osaka H, Inoue K, Takahashi K, Yamanaka S, Kosaki K, Takahashi T, Okano H (2014) Involvement of ER stress in dysmyelination of Pelizaeus-Merzbacher Disease with PLP1 missense mutations shown by iPSC-derived oligodendrocytes. Stem Cell Reports 2: 648–661

Nunez MT, Urrutia P, Mena N, Aguirre P, Tapia V, Salazar J (2012) Iron toxicity in neurodegeneration. Biometals 25:761–776

Oshiro S, Kawamura K, Zhang C, Sone T, Morioka MS, Kobayashi S, Nakajima K (2008) Microglia and astroglia prevent oxidative stress-induced neuronal cell death: implications for aceruloplasminemia. Biochim Biophys Acta 1782:109–117

Oski FA, Honig AS, Helu B, Howanitz P (1983) Effect of iron therapy on behavior performance in nonanemic, iron-deficient infants. Pediatrics 71:877–880

Pasik P, Pasik T (2004) Cajal, Achúcarro, Río Hortega, and the early exploration of Neuroglia. In: Lazzarini RA (ed) Myelin biology and disorders. Elsevier Academic Press, San Diego, pp xxiii–xli

Pedraza L, Huang JK, Colman DR (2001) Organizing principles of the axoglial apparatus. Neuron 30:335–344

Pedraza L, Huang JK, Colman D (2009) Disposition of axonal caspr with respect to glial cell membranes: implications for the process of myelination. J Neurosci Res 87:3480–3491

Pfeiffer SE, Warrington AE, Bansal R (1993) The oligodendrocyte and its many cellular processes. Trends Cell Biol 3:191–197

Poliak S, Peles E (2003) The local differentiation of myelinated axons at nodes of Ranvier. Nat Rev Neurosci 4:968–980

Poliak S, Gollan L, Martinez R, Custer A, Einheber S, Salzer JL, Trimmer JS, Shrager P, Peles E (1999) Caspr2, a new member of the neurexin superfamily, is localized at the juxtaparanodes of myelinated axons and associates with K+ channels. Neuron 24:1037–1047

Popko B (2003) Myelin: not just a conduit for conduction. Nat Genet 33:327–328

Rasband MN, Trimmer JS, Schwarz TL, Levinson SR, Ellisman MH, Schachner M, Shrager P (1998) Potassium channel distribution, clustering, and function in remyelinating rat axons. J Neurosci 18:36–47
Rosenbluth J (1976) Intramembranous particle distribution at the node of Ranvier and adjacent axolemma in myelinated axons of the frog brain. J Neurocytol 5:731–745
Roth AD, Ivanova A, Colman DR (2006) New observations on the compact myelin proteome. Neuron Glia Biol 2:15–21
Rowitch DH, Kriegstein AR (2010) Developmental genetics of vertebrate glial-cell specification. Nature 468:214–222
Saher G, Brugger B, Lappe-Siefke C, Mobius W, Tozawa R, Wehr MC, Wieland F, Ishibashi S, Nave KA (2005) High cholesterol level is essential for myelin membrane growth. Nat Neurosci 8:468–475
Sauer BM, Schmalstieg WF, Howe CL (2013) Axons are injured by antigen-specific CD8(+) T cells through a MHC class I- and granzyme B-dependent mechanism. Neurobiol Dis 59: 194–205
Sbardella E, Greco A, Stromillo ML, Prosperini L, Puopolo M, Cefaro LA, Pantano P, De Stefano N, Minghetti L, Pozzilli C (2013) Isoprostanes in clinically isolated syndrome and early multiple sclerosis as biomarkers of tissue damage and predictors of clinical course. Mult Scler 19:411–417
Schulz K, Vulpe CD, Harris LZ, David S (2014) Iron efflux from oligodendrocytes is differentially regulated in gray and white matter. J Neurosci 31:13301–13311
Silvestroff L, Franco PG, Pasquini JM (2012) ApoTransferrin: dual role on adult subventricular zone-derived neurospheres. PLoS ONE 7:e33937
Silvestroff L, Franco PG, Pasquini JM (2013) Neural and oligodendrocyte progenitor cells: transferrin effects on cell proliferation. ASN Neuro 5:e00107
Simons M, Trotter J (2007) Wrapping it up: the cell biology of myelination. Curr Opin Neurobiol 17:533–540
Smith KJ, Kapoor R, Felts PA (1999) Demyelination: the role of reactive oxygen and nitrogen species. Brain Pathol 9:69–92
Sow A, Lamant M, Bonny JM, Larvaron P, Piaud O, Lecureuil C, Fontaine I, Saleh MC, Garcia Otin AL, Renou JP, Baron B, Zakin M, Guillou F (2006) Oligodendrocyte differentiation is increased in transferrin transgenic mice. J Neurosci Res 83:403–414
Stephenson E, Nathoo N, Mahjoub Y, Dunn JF, Yong VW (2014) Iron in multiple sclerosis: roles in neurodegeneration and repair. Nat Rev Neurol 10:459–468
Stiefel KM, Torben-Nielsen B, Coggan JS (2013) Proposed evolutionary changes in the role of myelin. Front Neurosci 7:202
Taylor CM, Marta CB, Claycomb RJ, Han DK, Rasband MN, Coetzee T, Pfeiffer SE (2004) Proteomic mapping provides powerful insights into functional myelin biology. Proc Natl Acad Sci USA 101:4643–4648
Todorich B, Zhang X, Slagle-Webb B, Seaman WE, Connor JR (2008) Tim-2 is the receptor for H-ferritin on oligodendrocytes. J Neurochem 107:1495–1505
Todorich B, Pasquini JM, Garcia CI, Paez PM, Connor JR (2009) Oligodendrocytes and myelination: the role of iron. Glia 57:467–478
Tomassy GS, Fossati V (2014) How big is the myelinating orchestra? Cellular diversity within the oligodendrocyte lineage: facts and hypotheses. Front Cell Neurosci 8:201
Tosic M, Rakic S, Matthieu J, Zecevic N (2002) Identification of Golli and myelin basic proteins in human brain during early development. Glia 37:219–228
Toyama BH, Savas JN, Park SK, Harris MS, Ingolia NT, Yates JR 3rd, Hetzer MW (2013) Identification of long-lived proteins reveals exceptional stability of essential cellular structures. Cell 154:971–982
Traka M, Dupree JL, Popko B, Karagogeos D (2002) The neuronal adhesion protein TAG-1 is expressed by Schwann cells and oligodendrocytes and is localized to the juxtaparanodal region of myelinated fibers. J Neurosci 22:3016–3024

Tsacopoulos M, Magistretti PJ (1996) Metabolic coupling between glia and neurons. J Neurosci 16:877–885

van Meeteren ME, Teunissen CE, Dijkstra CD, van Tol EA (2005) Antioxidants and polyunsaturated fatty acids in multiple sclerosis. Eur J Clin Nutr 59:1347–1361

Volpe JJ (2001) Neurobiology of periventricular leukomalacia in the premature infant. Pediatr Res 50:553–562

Volpe JJ, Kinney HC, Jensen FE, Rosenberg PA (2011) The developing oligodendrocyte: key cellular target in brain injury in the premature infant. Int J Dev Neurosci 29:423–440

Wilson JX (1997) Antioxidant defense of the brain: a role for astrocytes. Can J Physiol Pharmacol 75:1149–1163

Wilson CH, Hartline DK (2011) Novel organization and development of copepod myelin. ii. nonglial origin. J Comp Neurol 519:3281–3305

Xu K, Terakawa S (1999) Fenestration nodes and the wide submyelinic space form the basis for the unusually fast impulse conduction of shrimp myelinated axons. J Exp Biol 202:1979–1989

Yuen TJ, Silbereis JC, Griveau A, Chang SM, Daneman R, Fancy SP, Zahed H, Maltepe E, Rowitch DH (2014) Oligodendrocyte-encoded HIF function couples postnatal myelination and white matter angiogenesis. Cell 158:383–396

Zhang X, Surguladze N, Slagle-Webb B, Cozzi A, Connor JR (2006) Cellular iron status influences the functional relationship between microglia and oligodendrocytes. Glia 54:795–804

# Schwann Cell and Axon: An Interlaced Unit—From Action Potential to Phenotype Expression

**Felipe A. Court and Jaime Alvarez**

**Abstract** Here we propose a model of a peripheral axon with a great deal of autonomy from its cell body—the autonomous axon—but with a substantial dependence on its ensheathing Schwann cell (SC), the axon-SC unit. We review evidence in several fields and show that (i) axons can extend sprouts and grow without the concurrence of the cell body, but regulated by SCs; (ii) axons synthesize their proteins assisted by SCs that supply them with ribosomes and, probably, with mRNAs by way of exosomes; (iii) the molecular organization of the axoplasm, i.e., its phenotype, is regulated by the SC, as illustrated by the axonal microtubular content, which is down-regulated by the SC; and (iv) the axon has a program for self-destruction that is boosted by the SC. The main novelty of this model axon-SC unit is that it breaks with the notion that all proteins of the nerve cell are specified by its own nucleus. The notion of a collaborative specification of the axoplasm by more than one nucleus, which we present here, opens a new dimension in the understanding of the nervous system in health and disease and is also a frame of reference to understand other tissues or cell associations.

**Keywords** Axoplasm · Exosomes · mRNA · Microtubular density · Wallerian degeneration

F.A. Court (✉)
Center for Integrative Biology, Universidad Mayor, Santiago, Chile
e-mail: fcourt@bio.puc.cl

F.A. Court
FONDAP Center for Geroscience, Brain Health and Metabolism, Santiago, Chile

F.A. Court · J. Alvarez
Millenium Nucleus for Regenerative Biology, Pontificia Universidad Católica de Chile, Santiago 8331150, Chile
e-mail: jaime.alvarez24@gmail.com

 
R. von Bernhardi (ed.), *Glial Cells in Health and Disease of the CNS*,
Advances in Experimental Medicine and Biology 949,
DOI 10.1007/978-3-319-40764-7_9

## Early Model

A decade ago, we proposed a model for the reciprocal interaction between Schwann cells (SC) and axon at the cell biology level, which configured an inextricable functional unit (Alvarez 2001; Court and Alvarez 2005). In this article, we will present the contribution of our group to the understanding of this unit. The main features of this model are, on the side of the axon, (i) the axon contains a sprouting or growth program fully present in it; (ii) it contains a destruction program; (iii) the axoplasm synthesizes its intrinsic proteins; and (iv) the axon regulates the SC, in particular, the axon induces its differentiation. On the side of the SC, its role for the axon changes radically from the differentiated to the dedifferentiated condition. While the differentiated SC (v) regulates the phenotype of the axon, and (vi) represses the growth program, the proliferating or dedifferentiated SC (vii) promotes growth (Fig. 1). The supporting evidence will be summarized below.

## Sprouting Program of Axons and Its Repression by Schwann Cells

It is well known that adult axons have a rather conserved anatomy along the space axis (trajectory) and along the time axis. Perturbations of axons, e.g., severance of fibers, result in a destruction of the distal domain and, at the proximal domain, the

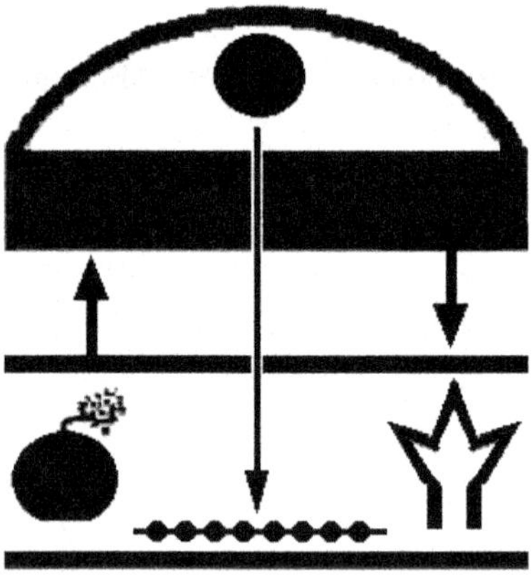

**Fig. 1** Model of the axon–Schwann cell unit. Parallel bars represent the membrane of the axon; inside, the bomb is the destruction program, the trident/fork is the sprouting program, and the beaded string is machinery for protein synthesis. The *upper* structure is the Schwann cell. *Thick arrows* pointing *up* and *down* indicate intercellular regulation between axon and Schwann cell. The *thin* transcellular *arrow* spanning from Schwann cell to axon indicates transfer of macromolecular component bypassing the extracellular space. Present status of our model adds the transcellular transfer to the previous one (Alvarez et al. 2000; Court and Alvarez 2005)

severed end of axons develops a regenerative response. We will examine first the regenerative response. Cajal (1928) with the optic microscope and later Zelena et al. (1968), with the electron microscope, showed that severed axons a few hours after injury developed membranous extensions at their cut ends, which suggested that axons had the ability to start a remodeling/regenerative response without direct involvement of the cell body. Later, with the pervading notion that all proteins of the axon originated in the neuronal cell body, it was assumed the cell body developed a regenerative response triggered by the arrest of ongoing retrograde signals, or by new signals coming from the lesion. The observed delay between the lesion—usually a crush—and the beginning of the axonal elongation—1 or 2 days post crush—was considered the time required for the processes linking the lesion to the regrowth response developed in the cell body (Bisby and Keen 1985; McQuarrie and Jacob 1991). We contested this view when we found that a nerve injected with inhibitors of proteases, such as aprotinin, leupeptin, APP with the Kunitz insert, and others, resulted in a local sprouting by otherwise uninterrupted axons. At the same time, SCs proliferated (Fig. 2) (Alvarez et al. 1992, 1995; Moreno et al. 1996). We reasoned that the axon started to sprout at the treated zone

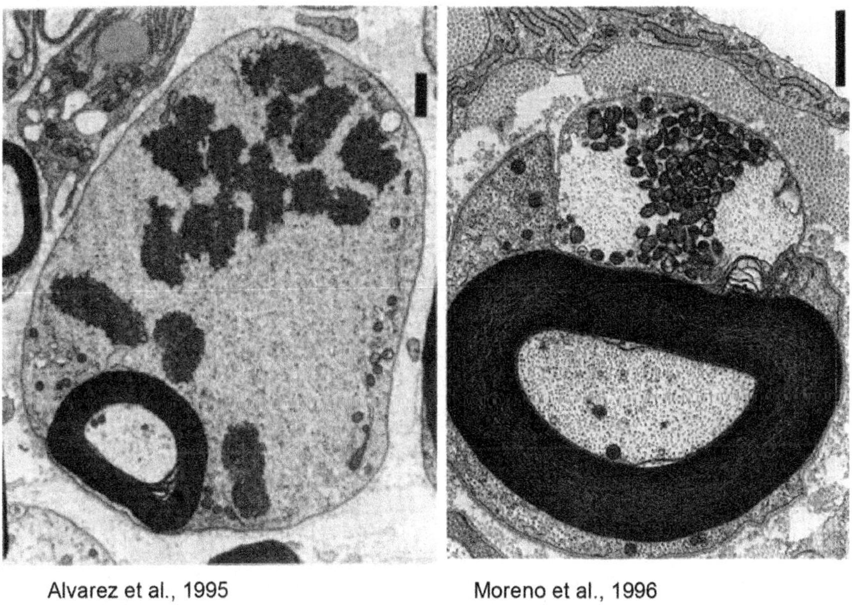


**Fig. 2** Sprouting response of intact axons and dedifferentiation of Schwann cells. Sural nerves of rats treated with intraneural injections of aprotinin, a serine antiprotease. *Left panel*, the Schwann cell presents a normal axon encased in a normal myelin, but is proliferating as shown by the chromosomes floating in the cytoplasm (*dark masses*). *Right panel*, the Schwann cell is surrounded by its basal lamina, the myelin looks normal and encases an axon while a second axonal profile loaded with mitochondria (*dark bodies*) is located between the myelin sheath and the basal lamina. This axonal profile is a sprout of the encased axon. Proliferation of Schwann cells and sprouting of axons precede the breakdown of myelin

of its own accord, i.e., the cell body could not cause the initial extension of these sprouts. From these observations, we surmised that (i) the axon contained a sprouting program and (ii) repressed by the differentiated but not by the proliferating SC.

In the light of the hypothesis that the differentiated SC suppresses the axonal sprouting program, we re-examined the delay of regeneration. Instead of retrograde signals from the lesion to the cell body, we assumed that the delay of regeneration was due to the time taken by the SC of the distal stump to proliferate, a response secondary to the degeneration of the severed axon. In other words, we conjectured that the delay was caused by the time taken for a change in the territory to be invaded by regrowing axons instead of signals going from lesion to cell body. In this scenario, we predicted that the delay of regeneration should be reduced by treating the distal stump appropriately. When we treated the sciatic nerve locally with aprotinin to induce SC proliferation and crushed the nerve in this region, the delay was obliterated. Moreover, when the distal stump was frozen to kill all cells, the delay was obliterated as well (Fig. 3) (Court and Alvarez 2000; Tapia et al. 1995). These observations supported the hypothesis that local mechanisms, probably involving the SC of the distal stump, were repressing the axonal growth. As a consequence, the role of the cell body in the initial stages of axonal regrowth needed re-evaluation.

To further explore the repression of the growth program of axons by SCs, we took advantage of the inability of axons of Wld$^s$ mice to regenerate. In this strain, severed fibers do not degenerate for several weeks, and surviving axons do not

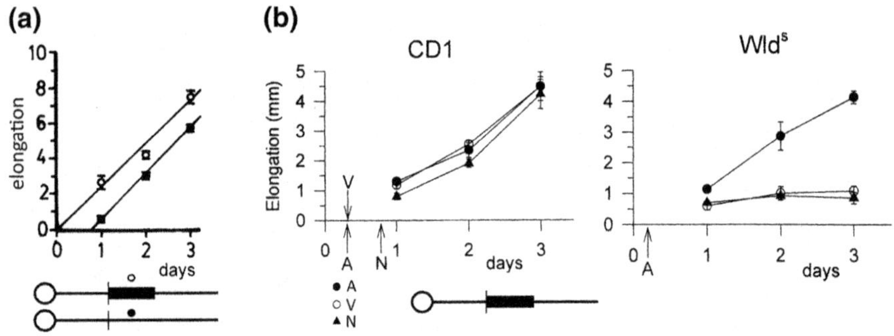


**Fig. 3** Delay of regeneration is locally controlled. **a** Rat sural nerves. Ordinates, elongation of axons in mm, post crush; pinch test. Abscissas, time post crush, in days. Diagram: One side was frozen over a span of 4–5 mm (*thick black line*) (*open circle*) and the other was manipulated mechanically (*filled circle*), before the crush (*vertical line*). The treatment obliterates the delay but does not change the rate of growth (modified from Tapia et al. 1995). **b** CD1 albino mice, and Wld$^s$ mice. Regeneration after 3 days of pretreatment with *A*, actinomycin D; *V*, vehicle. *N*, naive, for comparison. Diagram shows the infiltrated region (*thick line*) and the crush (*vertical line*). *Vertical arrows* pointing to abscissas indicate the intercepts of the regression lines. Notice that CD1 nerves reduce the delay of regeneration after pretreatment with *A* or *V*, and that Wld$^s$ nerves, which do not spontaneously regenerate, do so after actinomycin D pretreatment, with a short delay (modified from Court and Alvarez 2000)

regenerate into the distal stump either. We surmised that SCs of the distal stump—the territory to be invaded—were impairing the ingrowth of regenerating axons. In $Wld^s$, SCs were destroyed in a restricted zone of the sciatic nerve with actinomycin D, and the crush performed in that region. This treatment normalized the elongation of regenerating axons, which otherwise do not regrow (Fig. 3) (Benavides and Alvarez 1998; Court and Alvarez 2000). Finally, to rule out the involvement of the cell body in the sprouting response of axons, we used the distal domain of severed nerves of $Wld^s$ mice. The sciatic nerve was cut and the central stump eliminated. Several mm of the distal stump were crushed to kill all cells. Therefore, the blind end of surviving $Wld^s$ axons disconnected from their cell bodies was next to an acellular nerve domain. This domain was invaded by sprouts regrowing from axons isolated from their cell bodies (Fig. 4) (Iñiguez and Alvarez 1999). From these observations we conclude that axons start sprouting as soon as the ongoing repression is removed, a process that neither requires interruption of axons nor involvement of the cell body, and that the differentiated SC represses the axonal growth. Since the freezing of the SC unleashed the sprouting program of axons with no delay, we surmise that the repression is the result of an active process requiring a functional SC; in particular, the repression is not due to components of the extracellular matrix or myelin products. More generally, this view emphasizes the

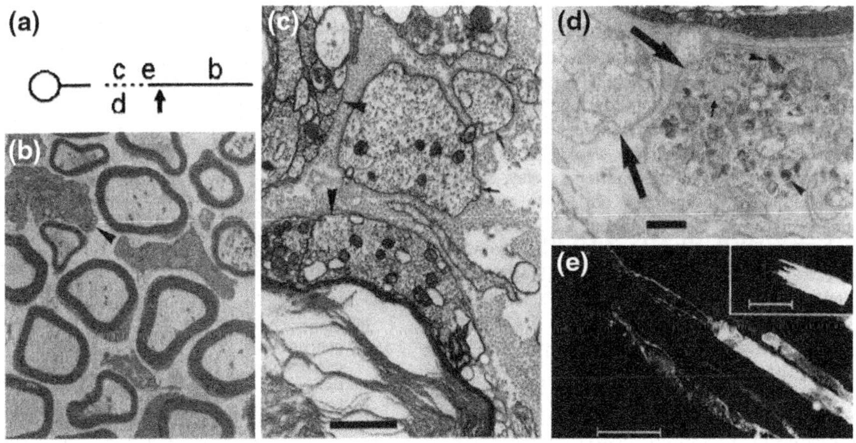


**Fig. 4** Cell bodies are unnecessary for axons to grow. **a** Diagram illustrates the $Wld^s$ preparation in vivo. A segment of sciatic nerve was removed, the distal segment received an extensive crush, 4–5 mm (*dotted line*), and the preparation was examined a few days later. Letters in the diagram correspond to those of micrographs. **b** The surviving distal nerve looks normal, which is a characteristic of this strain. **c** The crushed domain presents axonal profiles (*arrowheads*) amidst cell debris and cytoplasmic remnants (*arrows*). **d** the surviving nerve was cut 3 days post lesion (*arrow* in diagram of panel **a**) and exposed to horseradish peroxidase. The reaction product is seen in the axonal profile within membranous structures (*arrows*). **e** The fixed nerve was cut (*arrow* in diagram of panel **a**) and exposed to DiI. Dissociated fibers; the blind ends of surviving axons exhibit sprouts emerging from them (*thin threads*). *Inset*, whole nerve at the junction between surviving fibers and crushed region (from Iñiguez and Alvarez 1999)

presence of ongoing local regulatory mechanisms, what we have called the autonomous axon, diminishing the importance of the cell body as the "mastermind" of the neuron. In other words, the axon has a repertoire of behaviors regulated on a local basis that are not under the direct command of the cell body.

In invertebrates, an even more pushed set of observations has been reported for axons separated from their cell bodies. They survive unusually long periods, initiate regrowth, and may fuse with the blind central segments; axons deprived of their cell body can regenerate to make synaptic contact with a postsynaptic apparatus, intact axons divested of their ensheathing glial cells extend sprouts, to mention a few (Bittner and Mann 1976; Hoy et al. 1967; Krasne and Lee 1977; Mason and Muller 1982; Masuda-Nakagawa et al. 1993). These observations suggest that these properties of nerve fibers appeared way back in evolution, and that vertebrates have conserved a reduced set compared with invertebrates.

Let us focus now on the SC. In mature nerve fibers, the SC is in a post-mitotic condition. When the axon dies, its widow SCs proliferate, implying that the axon was preventing this proliferation. However, the SC enters the cell cycle when extracellular proteases are inhibited, despite the fact that the axon remains alive and functional. Therefore, the cascade that represses the cell cycle of SCs involves at one stage the activity of an extracellular protease. The group of Shubayev (Chattopadhyay and Shubayev 2009; Liu et al. 2010) has identified the matrix metalloproteinase 9 (MMP-9) as a candidate to carry on this control of the SC. Moreover, proliferation of SCs has been shown to be associated with the neuregulin cascade (Hayworth et al. 2006), as also with an increase of the mRNA coding for the zinc finger protein Zipro1 (Ellerton et al. 2008), which they suggest to be involved in the activation of SCs.

## Regulation of the Phenotype of the Axon by the Schwann Cell: Microtubular Content

As mentioned in Chapter "Glial cells and Integrity of the Nervous System," it is well known that the glial cell regulates some features of the axonal phenotype. For example, the caliber of the axon increases upon myelination in cultured peripheral neurons (Windebank et al. 1985). In the central nervous system, optic nerve axons acquire myelin past the lamina cribrosa, and at the same time increase in caliber (Hernandez et al. 1989). In peripheral myelinated fibers, Yokota (1984) reported the occurrence of intercalated unmyelinated segments where the axon was thinner. These observations point to a local and ongoing control of the anatomy of the axon by the glial cell. In this paragraph we will focus on axonal microtubules. In vertebrates, the microtubular content of axons is an extremely constant feature. The density (microtuble/μm$^2$ of cross-sectional area of axon) exhibits an inverse correlation with the caliber, but for a given caliber, the density is the same, whether the axon is motor, sensory, or autonomic, myelinated or unmyelinated, central or peripheral, in adulthood, during regeneration or development, whether the animal species is minute as a lizard or large as a cow, in wasted or well-nourished animals, warm- or cold-blooded, whether the terminal field of the axon is extensive or very restricted, and even if the caliber varies along its course, the microtubular density of the axon accords with the local caliber (Alvarez et al. 1982; Faundez et al. 1990; Friede and Samorajski 1970; Pannese et al. 1984b; Vergara et al. 1991; Fadic and Alvarez 1986; Alvarez and Zarour 1983; Espejo and Alvarez 1986; Saitua and Alvarez 1989; Pannese et al. 1988; Smith 1973). This tight correlation, though, changes abruptly in the ventral and dorsal roots, where the microtubular density is one half that observed in a peripheral or central axon of the same caliber (Fig. 5a) (Lopez and Alvarez 1990; Pannese et al. 1984a; Fadic et al. 1985; Saitua and Alvarez 1989).

What makes this structural feature change in this manner? Motor and sensory axons are continuous from spinal cord, through roots, to the peripheral nerve. Their cell bodies alone cannot specify an axonal microtubular density for the spinal cord domain, reduce it by half in the root, and resume the original microtubular density in the peripheral nerve. A local cue seems necessary. The nodosal ganglion of the vagal nerve lies in the neck, so the central branches of these neurons are in the periphery before becoming root fibers inside the skull. The central branches of the nodosal sensory neurons share the phenotype of sensory radicular fibers and have a low microtubular density compared to peripheral branches of the same caliber. When nodosal central branches regenerate along their anatomical course, the microtubular density remains unchanged, but when they regenerate along the hypoglossal peripheral nerve, the original low microtubular density increases to match the higher density typical of peripheral axons (Fig. 5b) (Serra and Alvarez 1989). In brief, the molecular architecture of the axon does not accord with its being a central branch but with the local environment.

The best candidate to provide this local regulation was the SC. To explore this conjecture, the SC was crippled with actinomycin D, which blocks transcription and as a consequence the cell runs down. The drug was applied to a short 4-mm span of nerve for a few days with a sleeve. Under the electron microscope, SCs debris was seen throughout while axons were conserved albeit with abundant microtubules. In unmyelinated fibers, the inverse correlation between microtubular density and axon caliber changed to a constant density across the range of cross-sectional areas. The density corresponded to the highest value observed in the normal nerve, that of the smallest axons. In brief, in the normal condition, the larger the encased axon is, the greater the Schwann cell down regulates its microtubular density (Fig. 5c). A few mm away from either side of the sleeve, axons and SCs appeared as usual (Bustos et al. 1991). We highlight that the molecular make-up of the axon varies from one point to the next, depending on local cues. In this scenario, the microtubular density of axons appears to arise from an internal axonal program whose default set point is very high but is down-regulated by an external agent—we surmise it is the glial cell —and this repression is more effective the larger is the axon. In this view, the SC of

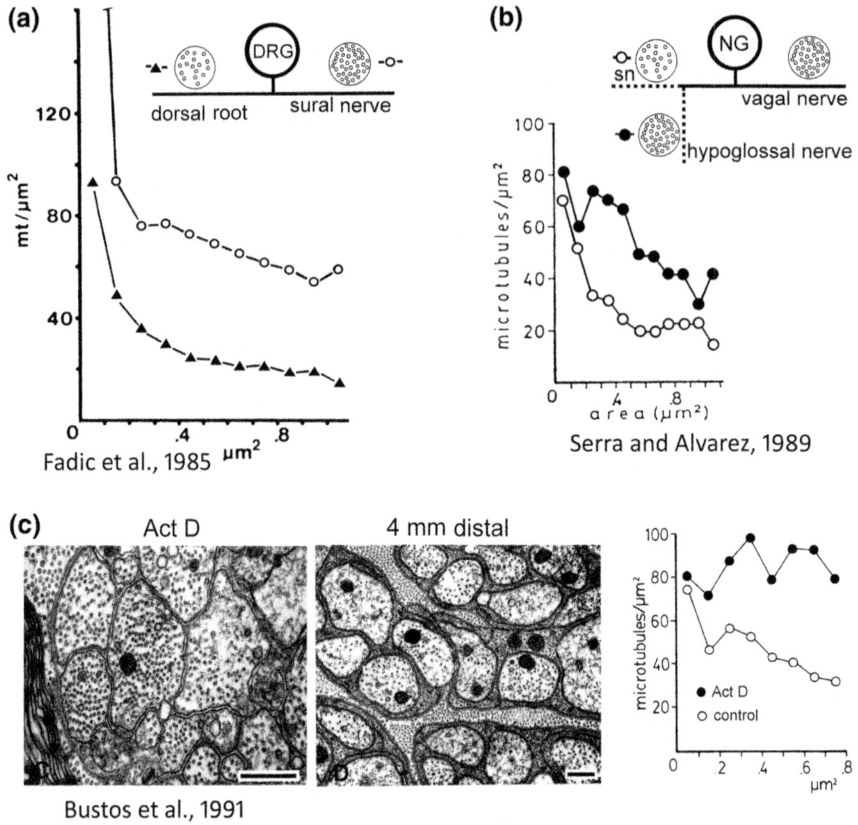


**Fig. 5** Control of the axonal microtubular content by the Schwann cell. **a** Rat. Diagram shows the preparation and corresponding symbols. Microtubular density (μt/μm$^2$) of peripheral and radicular unmyelinated axons as a function of the cross-sectional area of axons. Notice that for each caliber, the density is about one half in the root compared to that of the peripheral branch. **b** Cat. Diagram shows the nodose ganglion (*NG*), supranodosal (*sn*) and vagal nerves. When *sn* regenerated along its anatomical course, microtubular densities were low, i.e., similar to intact *sn* or radicular fibers. When *sn* regenerated along the hypoglossal nerve, microtubular densities increased to the level of peripheral fibers. **c** Sural nerve of rat. A 4-mm cuff loaded with actinomycin D was placed around the nerve. A week later, at the cuff, the nerve exhibited no healthy Schwann or resident cells, while axons were filled with microtubules (*Act D*), whereas 4-mm distal to the treated segment, the nerve appeared normal (4 mm distal). The *graph* shows that the crippling of resident cells abrogated the inverse correlation between caliber of axon and microtubular density, which was replaced by constant values for all calibers, corresponding to the highest density of control axons

the root must repress this axonal microtubular program even more than its peripheral sister does.

What is the biological significance of the constant microtubular density of axons and its drastic reduction at the roots? We do not have an answer. The microtubular density of axons follows an architectural feature of axons, their caliber. On the other hand, other features of the axons, e.g., length, transport demand, axoplamic mass, and size of the terminal, field do not correlate with the microtubular density. The mechanism by which the SC regulates this axonal feature remains conjectural (see below).

## Destruction Program of the Axon and Its Regulation by the Schwann Cell

A number of agents trigger axonal destruction, including those associated with genes, toxins, and mechanical damage to neurons. By far, a crush lesion to a nerve trunk is the most used perturbation to trigger nerve fiber destruction, also referred to as Wallerian degeneration. In Wallerian degeneration, axonal mechanisms lead to the activation of axoplasmic proteases. Generation of reactive oxygen species (ROS), the activation of the mitochondrial permeability transition pore (mPTP), and calcium rise in the axoplasm are instrumental for the activation of caspases resulting in the destruction of the axoplasm (Court and Coleman 2012). Cultured neurons free of other cell types can lose neurites in a manner resembling thus Wallerian degeneration (Barrientos et al. 2011; Villegas et al. 2014). This similarity has led to the notion that axonal destruction following disconnection with the cell body is essentially an internal program of axons. In Wallerian degeneration, the SC participates in the clearing of axonal remnants—a late stage of degeneration—and in its own remodeling.

In Wallerian degeneration, the myelin sheath breaks into ovoids at the Schmidt-Lanterman incisures, 2–3 days post lesion. We are currently studying the role of SCs in the early stages of Wallerian degeneration, and are focused on the process of ovoid formation by SCs as a possible promoter of axonal degeneration. Preliminary results suggest that the pharmacological impairment of ovoid formation delays the destruction of the axoplasm (Catenaccio and Court, unpublished), suggesting that early stages of axonal degeneration are executed by the SC in a non-cell autonomous mechanism, followed by a later axonal destruction stage involving mitochondrial dysfunction and calpain activation.

## Synthesis of Protein in Axons and Its Dependence from Schwann Cells

When we proposed our model of the axon-SC unit over a decade ago (Alvarez 2001; Court and Alvarez 2005), the model axon synthesized its own proteins. This view broke with the current notion at the time that the cell body supplied the bulk of axoplasmic proteins by a slow transport mechanism. Although evidence for axoplasmic protein synthesis has a long history (Edström 1966; Koenig 1984; Alvarez et al. 2000), only recently has become an established notion (Lin and Holt 2008), making the slow transport as supplier of most axoplasmic proteins a seriously flawed model (Court and Alvarez 2011). Mature axons have a small complement of ribosomes (Kun et al. 2007; Court et al. 2008), so small that they went unnoticed by electron microscopists (Palay and Palade 1955) despite the fact that micrographs did record polyribosomes (see Fig. 5-2 in Peters et al. 1991; Court and Alvarez 2011). This low ribosomal content accords with the low rate of amino acid incorporation into axoplasmic proteins, 1–4 % that of the cell body per unit of volume, as reported for the Mauthner neuron of the goldfish. However, owing to the large volume of the axoplasm, its contribution to the proteins of the neuron as a whole could be greater than that of the cell body (Alvarez and Benech 1983).

What came as a surprise is that a population of ribosomes in axons originated in the SC. In severed sciatic nerve of Wld$^{s}$ mice, the distal surviving axons exhibit an enormous amount of ribosomes as assessed with the electron microscope. These ribosomes could not come from the cell body because of the anatomical discontinuity of axons. In Wld$^{s}$ mice, SC infected with a lentivirus coding for a tagged ribosomal protein expressed the tagged protein in their cytoplasm, which co-localized with other ribosomal markers. This co-localization was also observed in the axoplasm after a crush (Fig. 6) (Court et al. 2008) and during axonal regeneration (Court et al. 2011). These observations support the idea that SCs supply ribosomes to axons. In the axoplasm, ribosomes were seen singly and as polyribosomes, the morphological correlate of several ribosomes translating an mRNA. Ribosomes were also seen densely packed inside multimembrane vesicles and in broken vesicles. These morphological structures suggest that ribosomes are transferred by way of vesicles (see below).

Our finding that ribosomes are transferred from one cell to its neighbor breaks with the notion that all components of a cell originate in the same cell, which is seldom stated explicitly but always present in the back of the mind. It is unlikely that this transfer mechanism appeared during evolution only in the axon-SC unit. Therefore, it is important to bear in mind such a transcellular tranfer when standard mechanisms appear to be insufficient to account for a set of results.

## Further Developments of Axonal Protein Synthesis and Its Support by Schwann Cells

Transfer of proteins between cells is an old notion. In the nervous system protein transfer from glia-to-axon has been previously proposed in the squid giant axon. Recently, intercellular transfer of proteins and RNA by means of extracellular vesicles has been described as a novel mechanism for cell-to-cell communication. In the 80s, experimental evidence indicated that during reticulocyte maturation, vesicles were released to the extracellular space (Harding et al. 1983). These vesicles, named exosomes, were produced in the endosomal compartment, moved into

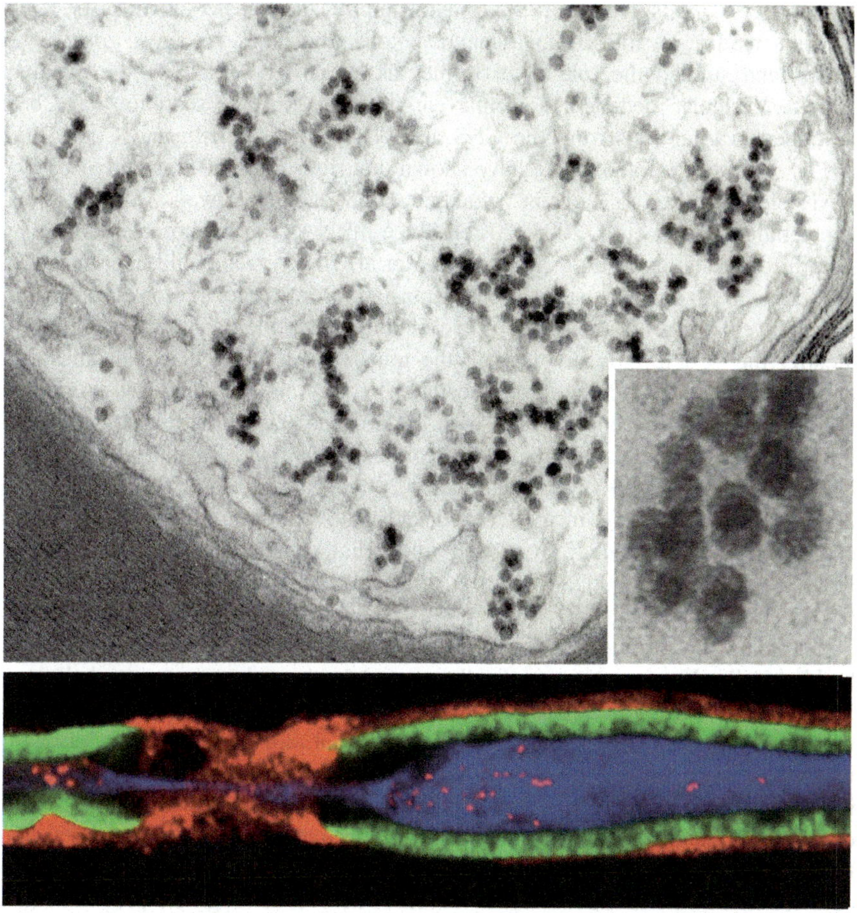

**Fig. 6** Transfer of ribosomes from Schwann cell to axon. Wld[s] sciatic nerve axons severed for a week. In this strain, severed axons survive for several weeks. *Upper panel*, electron micrograph; myelin and axoplasm are conserved; the dark particles in it are ribosomes (*inset*). *Lower panel*, immunostained teased fiber; *red*, ribosomal marker; *green*, P0, a myelin glycoprotein; *blue*, neurofilament. Notice the *red puncta* in an ocean of neurofilaments. Since axons were interrupted, this large amount of ribosomes presumably appeared in the axoplasm after its disconnection from the cell body; hence, these ribosomes did not originate in the neuronal cell body (modified from Court and Alvarez 2011)

multivesicular endosomes and secreted after fusion of these multivesicular bodies with the plasma membrane (Théry et al. 2002). A second class of vesicles, produced by evagination of the plasma membrane, were named microvesicles (Cocucci et al. 2009; Théry et al. 2009). Nowadays, exosome secretion has been described in vitro and in vivo. Physiological body fluids (plasma, urine, CSF, milk, and saliva) contain exosomes secreted by different cell types, and their molecular cargoes are rather specific. This field is currently under intense investigation to detect early biomarkers for pathological conditions (Russo et al. 2012; Skog et al. 2008).

Transfer of ribosomes from SC-to-axon implies the transfer of mRNAs, as the subunits require an mRNA to assemble into a ribosome. We have gone a step farther and studied the role of exosomes as a transfer vehicle in the SC–axon system. Vesicular-mediated molecular cargoes and exosome secretion have been described in glial cells from the CNS and PNS (Lopez-Verrilli and Court 2012, 2013). As described above, we have shown that ribosomes are transferred by way of vesicles from SCs to axons in vivo after axonal damage as well as during axonal regeneration (Court et al. 2008, 2011). Since mRNAs can be stored in a dormant state in the distal axon until needed (Yoo et al. 2010), SCs could supply mRNA transcript to axons for storage and translation in response to acute stimuli (e.g., nerve damage) or the stimulus itself could trigger the transfer. In our quest to identify the vesicle type involved in the transfer, we found that exosomes containing proteins and mRNAs were secreted by SCs and taken up by axons both in vitro and in vivo (Lopez-Verrilli et al. 2013). It is known that exosomes deliver mRNAs that can be translated in the recipient cells and also can deliver micro RNAs (miRNA) (Baj-Krzyworzeka et al. 2006; Valadi et al. 2007; Skog et al. 2008). In fact, elongation factors needed for mRNA translation have been found in exosomes from oligodendrocytes and microglial cells (Valenzuela et al. 2012; Kramer-Albers et al. 2007; Rigaud et al. 2008), as well as from Schwann cells (Picou et al., unpublished).

Summing up, through exosomes SCs might provide an efficient, specific, and highly localized support to axons for their maintenance and responses to challenges, including regenerative responses. Secreted vesicles interact specifically with the target cell (Cocucci et al. 2009; Lopez-Verrilli et al. 2013), supplying a variety of macromolecules; this allows SCs to regulate axonal functions without immediate involvement of the neuronal cell body. Our preliminary studies reveal SC-derived exosomes contain mRNA for neuronal-specific proteins. Distal to a nerve lesion, mRNA for neurofilament, a transcript for a neuron-specific intermediate filament, is detected in SCs including SCs of intact sciatic nerves, although the protein itself is absent from SCs (Roberson et al. 1992; Fabrizi et al. 1997). These findings were initially not considered relevant, since axons were believed unable to synthesize proteins, but that long-held dogma has been shown to be untenable (Alvarez et al. 2000; Twiss and Fainzilber 2009; Jung et al. 2012). Therefore, Schwann cells appear to supply ribosomes to axons and a set of mRNAs to be translated in the axoplasm.

## SC Response After Nerve Injury and Vesicular Transfer During Axonal Regeneration

Axonal regeneration is a subject of intense investigation, as it is one of the underlying processes of functional recovery after nervous system damage (Rigaud et al. 2008; Wang and Sun 2010). In the mammalian nervous system, the regenerative capabilities of neurons show regional variability. In the PNS, axons efficiently regenerate after nerve damage, a process that is supported by SCs (Chen et al. 2007). In contrast, in the CNS the regenerative capability is poor due to both neuronal intrinsic limitations and glial responses that restrict regeneration (Filbin 2003; Blesch and Tuszynski 2009).

In the PNS, axonal degeneration due to trauma, toxic agents, or genetic mutations triggers the dedifferentiation of Schwann cells, which support axonal regeneration in contrast to differentiated SCs that repress axonal growth (vide supra). In fact, dedifferentiated SCs proliferate, upregulate regeneration-associated genes, and secrete trophic factors that promote axonal growth (Jessen and Mirsky 2008). In addition, SCs play an active role in removing axonal and myelin debris and secrete cytokines and chemokines that recruit immune cells to the injured region, which further eliminate myelin debris (Glenn and Talbot 2013). This leads to an efficient removal of myelin-associated proteins that are inhibitory for regeneration. At a later stage, dedifferentiated SCs align under the basal lamina forming endoneurial tubes, which are efficient substrates for the regrowing axon. Together, these tissue changes generate a highly supportive environment for axonal regeneration in the PNS (Chen et al. 2007).

In contrast, the response of glial cells to injury in the CNS (oligodendrocytes and astrocytes) generates a mechanical and molecular barrier which greatly limits regeneration (Yiu and He 2006). SCs have been used as a tool for coaxing CNS axons to regenerate. In fact, optic nerve axons regrow through a sciatic nerve implant as a conduit to synapse onto their CNS targets (Aguayo et al. 1990). Subsequent work has largely focused on the extracellular matrix and growth factors secreted by Schwann cells. Recently, we showed that SC exosomes, which contain mRNA and miRNA, are selectively internalized by axons, increase neurite growth substantially (Fig. 7), and greatly enhance axonal elongation in vitro and in vivo (Lopez-Verrilli et al. 2013). The main novelty of the mechanism we found is that a complex array of macromolecules—the cargo of exosomes—simultaneously affects a broad spectrum of processes of the targeted axon.

The use of exosomes in patients is emerging, which adds a clinical relevance to this field. For example, SC-derived exosomes can be used by exploiting their endogenous pro-regenerative activity or by loading them with specific transcript or proteins using modified glial cells (Zhang et al. 2010; Schmitte et al. 2010); neuronal-targeted exosomes obtained in vitro have been electroporated with specific siRNAs, and after intravenous injection, they specifically knock-down their target gene in brain neurons (Alvarez-Erviti et al. 2011). Thus vesicle-mediated delivery of biological agents to specific targets appears to be a potentially valuable tool in clinical medicine.

Intercellular regulatory mechanisms are generally thought to rely on such agents as neurotransmitters, hormones, trophic factors, cytokines, and adhesion molecules, which trigger signaling cascades in the target cell, even modulating its genetic programs. But now in the nerve, composed of glial cells and their associated axons, we have unveiled a regulatory mechanism that goes one step further. A cell, by delivering RNAs, specifies the protein architecture of the recipient cell. More

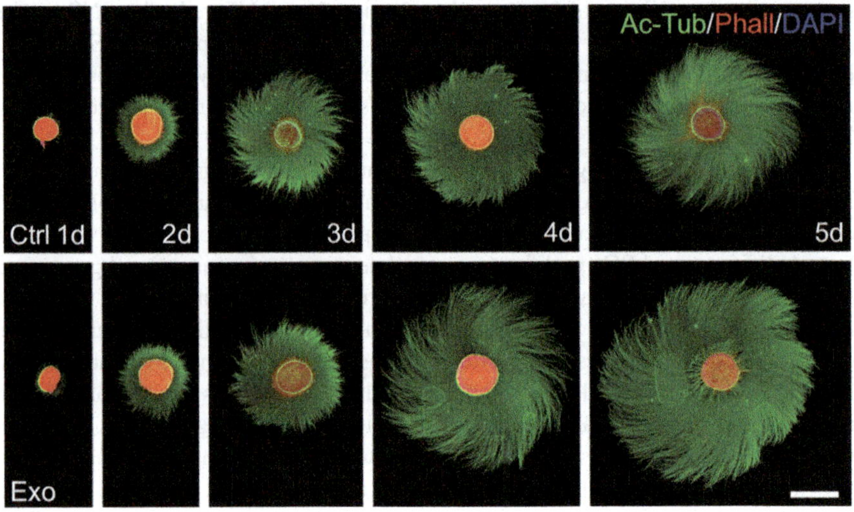


**Fig. 7** SC-derived exosomes stimulate axonal growth. Sensory neurons from dorsal root ganglia (DRGs) explanted in vitro, after 1 day, SC-derived exosomes or vehicle solution (PBS) were supplemented in a daily basis. Micrograph shown axonal growth of DRGs during 5 days. DRGs are stained for acetylated tubulin (Ac-Tub, *green*), phalloidin rhodamine (Phall, *red*) and nuclei with DAPI (*blue*). Impressively, at 5 days, exosome treatment increased the regeneration rate compared to vehicle-treated explants (mean ± SEM, 41 ± 4 vs. 26 ± 3 $mm^2$/day compared to vehicle, *$p < 0.05$, linear regression) suggesting that SC exosomes stimulate axonal regeneration in vitro. Scale bar, 50 μm (modified from Lopez-Verrilli et al. 2013)

generally, the phenotype of one cell is completed by the genome of a second cell. This mode of interaction provides a new dimension to the understanding of intercellular regulation; we foresee that several phenomena of the nervous system still poorly understood will be recast under this new light. For example, synapses are exquisite associations involving various cell types. In the motor endplate, the mutual regulation enlists at least three main leads, presynaptic axon, muscle fiber, and terminal Schwann cell, each one with an elaborated part (Hayworth et al. 2006; Kang et al. 2014; Li and Thompson 2011; Smith 1973), adding to the complexity of the system. It is likely that the mechanisms operating in the Schwann cell–axon system presented here are extensive to synapses.

## Perspectives and Concluding Remarks

The axon–glia unit proposed over a decade ago is now better understood. Axons have internal programs under local control, and the organization of their cytoskeleton and their ability to sprout, to grow, and to destroy themselves, as well as other programs, are controlled by their associated glial cells. Moreover, Schwann cells supply axons with ribosomes and mRNA to synthesize their proteins. Therefore, the molecular architecture of axons relies also on the genome of their ensheathing glial cells. Since proteins underlie nearly every conceivable cellular function, both the axon as a structure as well as its repertoire of functions are the collaborative efforts of the neuronal nucleus and a string of associated glial cells. The discovery of this mechanism opens a wide avenue of research. The old idea, still prevalent, that the cell body supplies all axoplasmic proteins, has to be revised; this notion was put forth to explain a model in which axons are unable to synthesize proteins, which does not apply to real axons.

SCs also control ongoing functions of axons. They regulate the axonal cytoskeleton, repress or promote the ability of axons to grow, and promote the destruction of doomed axons—processes which are important in nerve repair. The axon–glia unit requires that our understanding of the axon includes its inextricable association with its supporting cell. This also implies that overt malfunction of a nerve fiber, which we may call its "clinical scream," will be always uttered by the axon even if the primary alteration is seated in the glial cell, as the latter is clinically mute. This model is particularly relevant to the understanding of axonopathies, as the primary disturbance may be seated in the inconspicuous glial cell while the boisterous axon is altered indirectly. We foresee that our model will be rewarding in the study of neurological disorders in general.

Finally, a comment on vesicular-mediated intercellular transfer is discussed. We surmise that the transfer of RNAs via exosomes (or other extracellular vesicles) is a general mechanism for fine tuning between partner cells in order to optimize their coupling and operation. By extension, the transfer of RNAs plays a pivotal role in the organization of tissues.

## References

Aguayo AJ, Bray GM, Rasminsky M, Zwimpfer T, Carter D, Vidal-Sanz M (1990) Synaptic connections made by axons regenerating in the central nervous system of adult mammals. J Exp Biol 153:199–224

Alvarez J (2001) The autonomous axon: a model based on local synthesis of proteins. Biol Res 34 (2):103–109

Alvarez J, Benech CR (1983) Axoplasmic incorporation of amino acids in a myelinated fiber exceeds that of its soma: a radioautographic study. Exp Neurol 82(1):25–42

Alvarez J, Zarour J (1983) Microtubules in short and in long axons of the same caliber: implications for the maintenance of the neuron. Exp Neurol 79(1):283–286

Alvarez J, Arredondo F, Espejo F, Williams V (1982) Regulation of axonal microtubules: effect of sympathetic hyperactivity elicited by reserpine. Neuroscience 7(10):2551–2559

Alvarez J, Moreno RD, Llanos O, Inestrosa NC, Brandan E, Colby T, Esch FS (1992) Axonal sprouting induced in the sciatic nerve by the amyloid precursor protein (APP) and other antiproteases. Neurosci Lett 144(1–2):130–134

Alvarez J, Moreno RD, Inestrosa NC (1995) Mitosis of Schwann cells and demyelination are induced by the amyloid precursor protein and other protease inhibitors in the rat sciatic nerve. Eur J Neurosci 7(1):152–159

Alvarez J, Giuditta A, Koenig E (2000) Protein synthesis in axons and terminals: significance for maintenance, plasticity and regulation of phenotype. With a critique of slow transport theory. Prog Neurobiol 62:1–62

Alvarez-Erviti L, Seow Y, Yin H, Betts C, Lakhal S, Wood MJA (2011) Delivery of siRNA to the mouse brain by systemic injection of targeted exosomes. Nat Biotechnol 29(4):341–345

Baj-Krzyworzeka M, Szatanek R, Weglarczyk K, Baran J, Urbanowicz B, Brański P, Ratajczak MZ, Zembala M (2006) Tumour-derived microvesicles carry several surface determinants and mRNA of tumour cells and transfer some of these determinants to monocytes. Cancer Immunol Immunother 55:808–818

Barrientos SA, Martinez NW, Yoo S, Jara JS, Zamorano S, Hetz C, Twiss JL, Alvarez J, Court FA (2011) Axonal degeneration is mediated by the mitochondrial permeability transition pore. J Neurosci 31:359–370

Benavides E, Alvarez J (1998) Peripheral axons of Wld$^s$ mice, which regenerate after a delay of several weeks, do so readily when transcription is inhibited in the distal stump. Neurosci Lett 258(2):77–80

Bisby MA, Keen P (1985) The effect of a conditioning lesion on the regeneration rate of peripheral nerve axons containing substance P. Brain Res 336(2):201–206

Bittner GD, Mann DW (1976) Differential survival of isolated portions of crayfish axons. Cell Tissue Res 169(3):301–311

Blesch A, Tuszynski MH (2009) Spinal cord injury: plasticity, regeneration and the challenge of translational drug development. Trends Neurosci 32(1):41–47

Bustos J, Vial JD, Faundez V, Alvarez J (1991) Axons sprout and microtubules increase after local inhibition of RNA synthesis, and microtubules decrease after inhibition of protein synthesis: a morphometric study of rat sural nerves. Eur J Neurosci 3(11):1123–1133

Cajal SR (1928) Degeneration and regeneration of the nervous system. Oxford University Press

Chattopadhyay S, Shubayev V (2009) MMP-9 controls Schwann cell proliferation and phenotypic remodeling via IGF-1 and ErbB receptor-mediated activation of MEK/ERK pathway. Glia 57 (12):1316–1325

Chen Z-L, Yu W-M, Strickland S (2007) Peripheral regeneration. Annu Rev Neurosci 30:209–233

Cocucci E, Racchetti G, Meldolesi J (2009) Shedding microvesicles: artefacts no more. Trends Cell Biol 19(2):43–51

Court F, Alvarez J (2000) Nerve regeneration in Wld(s) mice is normalized by actinomycin D. Brain Res 867:1–8

Court FA, Alvarez J (2005) Local regulation of the axonal phenotype, a case of merotrophism. Biol Res 38:365–374

Court FA, Alvarez J (2011) Slow axoplasmic transport under scrutiny. Biol Res 44:283–293

Court FA, Coleman MP (2012) Mitochondria as a central sensor for axonal degenerative stimuli. Trends Neurosci 35:364–372

Court FA, Hendriks WTJ, Macgillavry HD, Alvarez J, van Minnen J (2008) Schwann cell to axon transfer of ribosomes: toward a novel understanding of the role of glia in the nervous system. J Neurosci 28:11024–11029

Court FA, Midha R, Cisterna BA, Grochmal J, Shakhbazau A, Hendriks WT, van Minnen J (2011) Morphological evidence for a transport of ribosomes from Schwann cells to regenerating axons. Glia 59:1529–1539

Edström A (1966) Amino acid incorporation in isolated Mauthner nerve fibre of goldfish. J Neurochem 13:315–321

Ellerton EL, Thompson WJ, Rimer M (2008) Induction of zinc-finger proliferation 1 expression in non-myelinating Schwann cells after denervation. Neuroscience 153(4):975–985. doi:10.1016/j.neuroscience.2008.02.078

Espejo F, Alvarez J (1986) Microtubules and calibers in normal and regenerating axons of the sural nerve of the rat. J Comp Neurol 250(1):65–72. doi:10.1002/cne.902500106

Fabrizi C, Kelly BM, Gillespie CS, Schlaepfer WW, Scherer SS, Brophy PJ (1997) Transient expression of the neurofilament proteins NF-L and NF-M by Schwann cells is regulated by axonal contact. J Neurosci Res 50(2):291–299

Fadic R, Alvarez J (1986) Calibers and microtubules of sympathetic axons are not subject to trophic control by the preganglionic nerve. Exp Neurol 94(1):237–240
Fadic R, Vergara J, Alvarez J (1985) Microtubules and caliber of central and peripheral processes of sensory axons. J Comp Neurol 236(2):258–264. doi:10.1002/cne.902360209
Faundez V, Cordero ME, Rosso P, Alvarez J (1990) Calibers and microtubules of nerve fibers: differential effect of undernutrition in developing and adult rats. Brain Res 509(2):198–204
Filbin MT (2003) Myelin-associated inhibitors of axonal regeneration in the adult mammalian CNS. Nat Rev Neurosci 4:703–713
Friede RL, Samorajski T (1970) Axon caliber related to neurofilaments and microtubules in sciatic nerve fibers of rats and mice. Anat Rec 167(4):379–387. doi:10.1002/ar.1091670402
Glenn TD, Talbot WS (2013) Signals regulating myelination in peripheral nerves and the Schwann cell response to injury. Curr Opin Neurobiol 23:1041–1048
Harding C, Heuser J, Stahl P (1983) Receptor-mediated endocytosis of transferrin and recycling of the transferrin receptor in rat reticulocytes. J Cell Biol 97(2):329–339
Hayworth CR, Moody SE, Chodosh LA, Krieg P, Rimer M, Thompson WJ (2006) Induction of neuregulin signaling in mouse schwann cells in vivo mimics responses to denervation. J Neurosci 26(25):6873–6884. doi:10.1523/JNEUROSCI.1086-06.2006
Hernandez C, Blackburn E, Alvarez J (1989) Calibre and microtubule content of the non-medullated and myelinated domains of optic nerve axons of rats. Eur J Neurosci 1 (6):654–658
Hoy RR, Bittner GD, Kennedy D (1967) Regeneration in crustacean motoneurons: evidence for axonal fusion. Science 156(3772):251–252
Iñiguez A, Alvarez J (1999) Isolated axons of $Wld^s$ mice regrow centralward. Neurosci Lett 268:57–114
Jessen KR, Mirsky R (2008) Negative regulation of myelination: relevance for development, injury, and demyelinating disease. Glia 56(14):1552–1565
Jung H, Yoon BC, Holt CE (2012) Axonal mRNA localization and local protein synthesis in nervous system assembly, maintenance and repair. Nat Rev Neurosci 13:308–324
Kang H, Tian L, Mikesh M, Lichtman JW, Thompson WJ (2014) Terminal Schwann cells participate in neuromuscular synapse remodeling during reinnervation following nerve injury. J Neurosci 34(18):6323–6333. doi:10.1523/JNEUROSCI.4673-13.2014
Koenig E (1984) Local synthesis of axonal protein. In: Lajtha A (ed) Handbook of neurochemistry, vol 7. Plenum, New York, pp 315–340
Kramer-Albers EM, Bretz N, Tenzer S, Winterstein C, Mobius W, Berger H, Nave KA, Schild H, Trotter J (2007) Oligodendrocytes secrete exosomes containing major myelin and stress-protective proteins: trophic support for axons? Proteomics Clin Appl 1(11): 1446–1461. doi:10.1002/prca.200700522
Krasne FB, Lee SH (1977) Regenerating afferents establish synapses with a target neuron that lacks its cell body. Science 198(4316):517–519
Kun A, Otero L, Sotelo-Silveira JR, Sotelo JR (2007) Ribosomal distributions in axons of mammalian myelinated fibers. J Neurosci Res 85(10):2087–2098. doi:10.1002/jnr.21340
Li Y, Thompson WJ (2011) Nerve terminal growth remodels neuromuscular synapses in mice following regeneration of the postsynaptic muscle fiber. J Neurosci 31(37):13191–13203. doi:10.1523/JNEUROSCI.2953-11.2011
Lin AC, Holt CE (2008) Function and regulation of local axonal translation. Curr Opin Neurobiol 18:60–68
Liu H, Kim Y, Chattopadhyay S, Shubayev I, Dolkas J, Shubayev VI (2010) Matrix metalloproteinase inhibition enhances the rate of nerve regeneration in vivo by promoting dedifferentiation and mitosis of supporting schwann cells. J Neuropathol Exp Neurol 69 (4):386–395. doi:10.1097/NEN.0b013e3181d68d12
Lopez JM, Alvarez J (1990) The microtubular pattern changes at the spinal cord-root junction and reverts at the root-peripheral nerve junction in sensory and motor fibres of the rat. Eur J Neurosci 2(10):873–878

Lopez-Verrilli MA, Court FA (2012) Transfer of vesicles from schwann cells to axons: a novel mechanism of communication in the peripheral nervous system. Front Physiol 3:205

Lopez-Verrilli MA, Court FA (2013) Exosomes: mediators of communication in eukaryotes. Biol Res 46:5–11

Lopez-Verrilli MA, Picou F, Court FA (2013) Schwann cell-derived exosomes enhance axonal regeneration in the peripheral nervous system. Glia 61:1795–1806

Mason A, Muller KJ (1982) Axon segments sprout at both ends: tracking growth with fluorescent D-peptides. Nature 296(5858):655–657

Masuda-Nakagawa LM, Muller KJ, Nicholls JG (1993) Axonal sprouting and laminin appearance after destruction of glial sheaths. Proc Natl Acad Sci USA 90(11):4966–4970

McQuarrie IG, Jacob JM (1991) Conditioning nerve crush accelerates cytoskeletal protein transport in sprouts that form after a subsequent crush. J Comp Neurol 305(1):139–147. doi:10.1002/cne.903050113

Moreno RD, Inestrosa NC, Culwell AR, Alvarez J (1996) Sprouting and abnormal contacts of nonmedullated axons, and deposition of extracellular material induced by the amyloid precursor protein (APP) and other protease inhibitors. Brain Res 718:13–24

Palay SL, Palade GE (1955) The fine structure of neurons. J Biophys Biochem Cytol 1(1):69–88

Pannese E, Ledda M, Arcidiacono G, Rigamonti L, Procacci P (1984a) A comparison of the density of microtubules in the central and peripheral axonal branches of the pseudounipolar neurons of lizard spinal ganglia. Anat Rec 208(4):595–605. doi:10.1002/ar.1092080415

Pannese E, Procacci P, Ledda M, Arcidiacono G, Rigamonti L (1984b) A quantitative study of microtubules in motor and sensory axons. Acta Anat 118(4):193–200

Pannese E, Ledda M, Matsuda S (1988) Nerve fibres with myelinated and unmyelinated portions in dorsal spinal roots. J Neurocytol 17(5):693–700

Peters A, Palay S, Webster HdF (1991) Fine structure of the nervous system: neurons and their supporting cells. Oxford Press, New York

Rigaud M, Gemes G, Barabas M-E, Chernoff DI, Abram SE, Stucky CL, Hogan QH (2008) Species and strain differences in rodent sciatic nerve anatomy: implications for studies of neuropathic pain. Pain 136:188–201

Roberson MD, Toews AD, Goodrum JF, Morell P (1992) Neurofilament and tubulin mRNA expression in Schwann cells. J Neurosci Res 33(1):156–162

Russo F, Di Bella S, Nigita G, Macca V, Laganà A, Giugno R, Pulvirenti A, Ferro A (2012) miRandola: extracellular circulating microRNAs database. PLoS ONE 7:e47786. doi:10.1371/journal.pone.0047786

Saitua F, Alvarez J (1989) Microtubular packing varies along the course of motor and sensory axons: possible regulation of microtubules by environmental cues. Neurosci Lett 104(3):249–252

Schmitte R, Tipold A, Stein VM, Schenk H, Flieshardt C, Grothe C, Haastert K (2010) Genetically modified canine Schwann cells–In vitro and in vivo evaluation of their suitability for peripheral nerve tissue engineering. J Neurosci Methods 186(2):202–208. doi:10.1016/j.jneumeth.2009.11.023

Serra M, Alvarez J (1989) On the asymmetry of the primary branching of vagal sensory axons: possible role of the supporting tissue. J Comp Neurol 284(1):108–118. doi:10.1002/cne.902840108

Skog J, Würdinger T, van Rijn S, Meijer DH, Gainche L, Sena-Esteves M, Curry WT, Carter BS, Krichevsky AM, Breakefield XO (2008) Glioblastoma microvesicles transport RNA and proteins that promote tumour growth and provide diagnostic biomarkers. Nat Cell Biol 10:1470–1476

Smith RS (1973) Microtubule and neurofilament densities in amphibian spinal root nerve fibers: relationship to axoplasmic transport. Can J Physiol Pharmacol 51(11):798–806

Tapia M, Inestrosa NC, Alvarez J (1995) Early axonal regeneration: repression by Schwann cells and a protease? Exp Neurol 131(1):124–132

Théry C, Zitvogel L, Amigorena S (2002) Exosomes: composition, biogenesis and function. Nat Rev Immunol 2:569–579

Théry C, Ostrowski M, Segura E (2009) Membrane vesicles as conveyors of immune responses. Nat Rev Immunol 9:581–593

Twiss J, Fainzilber M (2009) Ribosomes in axons - scrounging from the neighbors? Trends Cell Biol 19:236–243

Valadi H, Ekström K, Bossios A, Sjöstrand M, Lee JJ, Lötvall JO (2007) Exosome-mediated transfer of mRNAs and microRNAs is a novel mechanism of genetic exchange between cells. Nat Cell Biol 9(6):654–659

Valenzuela V, Collyer E, Armentano D, Parsons GB, Court FA, Hetz C (2012) Activation of the unfolded protein response enhances motor recovery after spinal cord injury. Cell Death Dis 3: e272. doi:10.1038/cddis.2012.8

Vergara J, Serra M, Saitua F, Iturriaga R, Alvarez J (1991) Axonal microtubules: comparative anatomy in vertebrates, including man. J Submicrosc Cytol Pathol 23(3):357–363

Villegas R, Martinez NW, Lillo J, Pihan P, Hernandez D, Twiss JL, Court FA (2014) Calcium release from intra-axonal endoplasmic reticulum leads to axon degeneration through mitochondrial dysfunction. J Neurosci 34(21):7179–7189. doi:10.1523/JNEUROSCI.4784-13.2014

Wang D, Sun T (2010) Neural plasticity and functional recovery of human central nervous system with special reference to spinal cord injury. Spinal Cord 49:486–492

Windebank AJ, Wood P, Bunge RP, Dyck PJ (1985) Myelination determines the caliber of dorsal root ganglion neurons in culture. J Neurosci 5(6):1563–1569

Yiu G, He Z (2006) Glial inhibition of CNS axon regeneration. Nat Rev Neurosci 7:617–627

Yokota R (1984) Occurrence of long non-myelinated axonal segments intercalated in myelinated, presumably sensory axons: electron microscopic observations in the dog atrial endocardium. J Neurocytol 13(1):127–143

Yoo S, van Niekerk EA, Merianda TT, Twiss JL (2010) Dynamics of axonal mRNA transport and implications for peripheral nerve regeneration. Exp Neurol 223(1):19–27. doi:10.1016/j.expneurol.2009.08.011

Zelena J, Lubinska L, Gutmann E (1968) Accumulation of organelles at the ends of interrupted axons. Z Mikrosk Anat Forsch 91(2):200–219

Zhang J, Zhao F, Wu G, Li Y, Jin X (2010) Functional and histological improvement of the injured spinal cord following transplantation of Schwann cells transfected with NRG1 gene. Anat Rec 293(11):1933–1946. doi:10.1002/ar.21223

# Part II
# Glial Cell Dysfunction and Pathological Implications

# Age-Dependent Changes in the Activation and Regulation of Microglia

**Francisca Cornejo and Rommy von Bernhardi**

**Abstract** As we age, a large number of physiological and molecular changes affect the normal functioning of cells, tissues, and the organism as a whole. One of the main changes is the establishment of a state of systemic inflammatory activation, which has been termed "inflamm-aging"; a mild chronic inflammation of the aging organism that reduces the ability to generate an efficient response against stressor stimuli. As any other system, the nervous system undergoes these aging-related changes; the neuroinflammatory state depends mainly on the dysregulated activation of microglia, the innate immune cells of the central nervous system (CNS) and the principal producers of reactive oxygen species. As the brain ages, microglia acquire a phenotype that is increasingly inflammatory and cytotoxic, generating a hostile environment for neurons. There is mounting evidence that this process facilitates development of neurodegenerative diseases, for which the greatest risk factor is age. In this chapter, we will review key aging-associated changes occurring in the central nervous system, focusing primarily on the changes that occur in aging microglia, the inflammatory and oxidative stressful environment they establish, and their impaired regulation. In addition, we will discuss the effects of aged microglia on neuronal function and their participation in the development of neurodegenerative pathologies such as Parkinson's and Alzheimer's diseases.

**Keywords** Aging · Cytokines · Microglia · Neuroinflammation · Neurodegenerative diseases · Oxidative stress

F. Cornejo · R. von Bernhardi (✉)
Departamento de Neurología, Escuela de Medicina, Pontificia Universidad Católica de Chile, Marcoleta 391, Santiago, Chile
e-mail: rvonb@med.puc.cl

F. Cornejo
e-mail: flcornej@uc.cl

 
R. von Bernhardi (ed.), *Glial Cells in Health and Disease of the CNS*,
Advances in Experimental Medicine and Biology 949,
DOI 10.1007/978-3-319-40764-7_10

## Introduction

Aging is a functional decline that affects all living organisms in a time-dependent fashion. It is characterized by several cellular and molecular hallmarks, such as genomic instability, epigenetic modifications, loss of protein homeostasis, mitochondrial dysfunction, and cellular senescence among others (López-Otín et al. 2013).

Aging concerns society and the world's economy, particularly because the human lifespan has drastically extended without reduced disease load (Vaupel 2010). As it is common for many aging cell systems, the brain goes through a certain loss-of-function process during aging. This has led to scientific research focused on several neurodegenerative diseases with aging as their main risk factor, such as Alzheimer's disease (AD) and Parkinson's diseases (PD) (Friedrich 2014).

It has been widely reported that brain aging results in several cognitive changes: information processing and long-term memory show a mild reduction, the speed of processing becomes slower, and both the working memory and inhibitory functions played by the brain appear to be to some extent impaired (Denise et al. 1996; Hultsch 1998; Park et al. 2002). However, there are also some reports suggesting that some older adults keep their cognitive capabilities and they may have even better memory function than young individuals (Gefen et al. 2015). The Baltimore longitudinal study of aging of more than 2000 individuals followed for a decade, showed a broad interindividual variability in cognitive decline with age (Shock et al. 1984). Cognitive changes are due to age-related physiological and molecular changes in the brain that depend mostly on inflammatory process and oxidative stress induced as the CNS ages (Berr et al. 2000; reviewed in von Bernhardi et al. 2015b). Many of those changes have microglia as the main responsible effectors, leading to several alterations that favor neurodegenerative diseases (Dröge and Schipper 2007; von Bernhardi et al. 2010, 2015a; Tsurumi and Li 2012; Tichauer et al. 2014).

## Main Age-Related Changes in the CNS

Several phenotypic changes are reported to occur as our brains age. The Baltimore longitudinal study of aging, in which non-demented subjects of 65–85 years old were evaluated with regular Magnetic Resonance Imaging analysis (Driscoll et al. 2009), revealed regional changes in brain volume. There are structural changes, including shrinking of nuclei and cortical areas such as the caudate nucleus, cerebellum, hippocampus, and prefrontal cortex (Salat et al. 2004; Raz et al. 2005). Age-related differences in white matter integrity accompany a conspicuous reduction in frontal areas of the brain, establishing an anterior-posterior gradient (Head et al. 2004), and gray matter of prefrontal, parietal, and temporal cortices shrinks (Courchesne et al. 2000; Good et al. 2001; Ge et al. 2002). There are also characteristic metabolic changes associated with aging: glucose metabolism declines in the prefrontal, anterior cingulate, ventral and dorso-lateral cortex, the medial prefrontal and the pre-central areas (Zuendorf et al. 2003; Kalpouzos et al. 2009; Hsieh et al. 2012). Age-dependent anatomical changes of the brain affect cognitive skills (Faith and Naftali 2003; Rosen et al. 2003; Rodrigue and Raz 2004), and as glucose metabolism declines, so do synaptic density and structural integrity (Kochunov et al. 2009).

Increasing evidence supports the "free-radical theory of aging" (Harman 1956), which proposes that an organism's lifespan depends on its regulation of antioxidants. In terms of brain function, age-related memory impairment is correlated with a decreased level of antioxidant moieties in plasma (Perrig et al. 1997; Perkins et al. 1999; Berr et al. 2000). Also, glutathione, the most abundant endogenous antioxidant, is reduced in several regions of aged rodent brains, including the hippocampus (Calabrese et al. 2004; Balu et al. 2005; Zhu et al. 2006). In addition, some studies have shown that the increased oxidative stress is due to reduced antioxidant enzyme activity in aged brains, such as glutathione peroxidase (Rodrigues Siqueira et al. 2005), catalase (Tian et al. 1998) and Mn- and Cu, Zn-superoxide dismutase (Gupta et al. 1991; Navarro et al. 2004). Dysregulation of the homeostatic oxidative response would induce oxidative protein damage in the aged brain (Fig. 1), impairing its function, which would support the theory that age-induced oxidative stress (von Bernhardi and Eugenin 2012) is associated with the impairment of cognitive skills (Forster et al. 1996).

Oxidative stress, associated with the accumulation of reactive oxygen species (ROS), increases as organisms age. The brain is particularly susceptible to ROS-induced damage, given that it demands very high amounts of oxygen and has low levels of antioxidants molecules compared with other tissues (Perluigi et al. 2014). The major source of cellular ROS in non-macrophage cells is mitochondria. ROS are generated as a consequence of oxidative phosphorylation in the mitochondrial inner membrane (Balaban et al. 2005). It has been shown that mitochondrial integrity is reduced with age due to inefficient electron transport, reducing generation of energy rich molecules (ATP) and favoring the formation of reactive oxidants (Shigenaga et al. 1994). The mitochondrial DNA (mtDNA) is the first target; mtDNA is particularly vulnerable to oxidative damage because of the proximity of mtDNA to the main ROS source and the absence of histones to protect it, causing mtDNA to have a higher mutation rate than nuclear DNA (Richter et al. 1988). Mutations in mtDNA can affect the expression of cytochromes, altering the electron transport chain, which in turn would increase ROS production, favoring a vicious circle leading to oxidative stress (Fig. 1) (Dröge and Schipper 2007), destabilizing the mitochondrial membrane and promoting cell death.

At a molecular level, microarray studies have shown age-related changes in genome-wide gene expression in the brain. Changes are most conspicuous after age 70. The principal genes that are downregulated are those associated with cellular processes related to memory and learning, calcium signaling, vesicle-mediated protein transport, and mitochondrial function. At the same time, genes related to

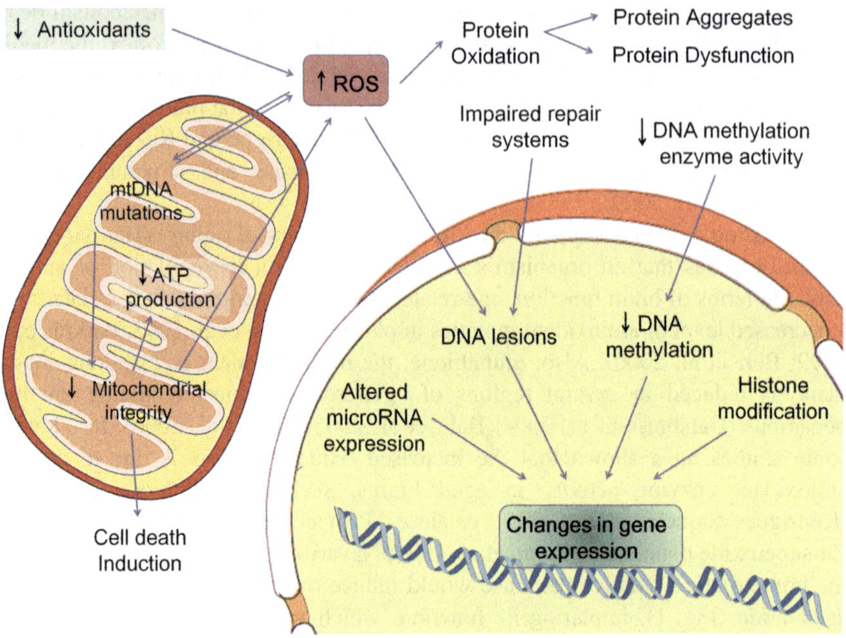


**Fig. 1** Cellular changes induced in the aged brain. Schematic representation of the main alterations induced by age in brain cells, and their consequences for ROS production, cell viability, and gene expression. Aged brain cells have reduced antioxidant mechanisms and increased intracellular ROS levels that affect mtDNA structure, inducing mutations and leading to a reduction in mitochondrial integrity, which in turn induce more ROS production. ROS accumulation and reduced activity of DNA-repair systems in aged cells damage DNA. In addition, aged cells have altered microRNA expression, several histone modifications, and reduced DNA methylation enzyme activity, shifting the methylation state of some promoter genes. These age-induced cell alterations produce changes in gene expression and increase the oxidative stress on the brain

antioxidant functions, DNA repair and the inflammatory response are upregulated (Lu et al. 2004; Fraser et al. 2005; Erraji-Benchekroun et al. 2005).

Some of the changes in gene expression could depend on DNA damage or mutations in DNA-repair genes. In fact, ROS have deleterious effects on DNA structure and protein function (Fig. 1), among others (Fraga et al. 1990). The most common oxidative DNA lesions observed in the aged brain are single base modifications of DNA, which accumulate in the promoter region of most genes that are downregulated during aging (Lu et al. 2004). In addition, there is also evidence showing that DNA-repair systems become impaired with aging (Fig. 1), which directly affects age-induced functional decline. There are several mutations in DNA-repair genes that induce accelerated aging phenotypes in mice and humans, characterized by organism deterioration and neurodegeneration, commonly known as progeroid syndromes (Lombard et al. 2005).

Increasing evidence show that age-related changes in gene expression could also occur due to epigenetic effects, such as histone modifications and DNA methylation (Gravina and Vijg 2010). It has long been known that aged tissues, including in brain, heart, and spleen, show a global methylation reduction associated with changes in the activity of DNA methylation enzymes (Vanyushin et al. 1973). These changes are the basis of a novel epigenetic clock that can be used to measure tissue age from its DNA methylation (Horvath 2013). Furthermore, several age-specific modifications have been observed on histones, such as methylation (Thakur and Kanungo 1981; Wang et al. 2010) and acetylation (Pruitt et al. 2006), together with changes in the expression of some micro-RNAs also observed in aged tissues (Fig. 1) (Cencioni et al. 2013). In addition, changes in DNA methylation and posttranslational histone modifications cause alterations in normal chromatin structure, inducing global heterochromatin loss with aging, which has been established as the main cause of the deleterious processes observed in aged tissues (Tsurumi and Li 2012).

## Inflammation in the Aging Brain

As observed in the aging organism as a whole, the aging brain is also characterized by the presence of a chronic mild inflammation, which reduces neuronal dendritic and axonal branching, synapse density, dendritic spines, and presynaptic markers (Yankner et al. 2008).

The term "inflamm-aging" was coined in 2000 (Franceschi et al. 2000) to describe the reduced capability of the organism to deal with stressor stimuli and the progression to a more inflammatory state as individuals age. It is postulated that the phenomenon is induced by the continuous stress and antigenic load to which the organism is subjected throughout life, and that the persistent exposure to this inflammatory condition could predispose the organism to develop several age-related diseases, such as atherosclerosis, AD, PD, osteoporosis, and diabetes (Franceschi et al. 2000). Age-related immune alterations, known as immunosenescence (Larbi et al. 2008), might also be induced by chronic mild inflammation. It has been shown that changes in gene expression related to inflammation and immune response (Lee et al. 2000; de Magalhães et al. 2009), increased plasma levels of inflammatory cytokines (Singh and Newman 2011), and the activation of inflammatory intracellular pathways such as NFκB occur with aging (Helenius et al. 1996).

NFκB is the main modulator of immune-related gene expression, and its relation to aging has been a recent focus of research. Genetic blockade of NFκB causes reversion of the aged phenotype and shifts the gene expression pattern of aged mice to that of young mice (Adler et al. 2007). Accordingly, blocking expression of a NFκB subunit acting as a repressor of inflammatory gene transcription, induces a progressive low-grade inflammation, and as a consequence, mice undergo premature aging (Jurk et al. 2014). Furthermore, there is evidence that appears to link aging directly with inflammation. Rap1, a protein that is part of the mammalian telomeric complex, acts as a potent modulator of NFκB activity by establishing a macromolecular complex with IκB kinases (IKKs), driving NFκB transcriptional activity (Teo et al. 2010). The evidence suggests that as we age, telomere shortening could increase free Rap1 in the cytoplasm, which in turn could increase IKKs activity, increasing the transcriptional activity of NFκB. This sequence might explain the link between aging and "inflamm-aging" related changes (Arka Subhra and Vinay 2010) with telomere shortening having a direct effect on the NFκB inflammatory pathway (Teo et al. 2010). Moreover, microglia, as one of the few CNS cells with significant mitotic potential, are susceptible to telomere shortening and have low telomerase activity (Flanary and Streit 2004).

On the other hand, increased levels of TGFβ have been reported in the brains of aged individuals (Bye et al. 2001; Werry et al. 2010). Non-neuronal cells appear to be responsible for the increased production of TGFβ, since TGFβ transcripts are reduced in aged neurons (de Sampaio e Spohr et al. 2002). In the brain, TGFβ favors cell survival, is neuroprotective (Dhandapani and Brann 2003), and reduces activation of microglia and their production of oxidative agents (Herrera-Molina and von Bernhardi 2005; Abutbul et al. 2012).

There are age-related changes in TGFβ signaling at several levels. As the response to inflammatory stimulation appears to become more oxidative and potentially more cytotoxic in aged animals (Tichauer et al. 2014), there is an increased secretion of TGFβ, but a reduced activation of its canonical signaling pathway, Smad. Both age and chronic inflammation reduce the activation of the Smad3 pathway in mice hippocampus (Tichauer et al. 2014). The activation of the Smad pathway in young animals could depend on the induction of TGFβ1 by inflammatory stimulation (Wynne et al. 2010) and the induction of Smad3 by the activation of MAPK1 (Ross et al. 2007). In adult mice, increased basal levels of TGFβ1 (Colangelo et al. 2002; Lukiw 2004) maintain elevated Smad3, which becomes unresponsive to new inflammatory stimulation. Increased levels of TGFβ1 with an inhibited Smad signaling can result in an impaired regulation of inflammatory activation by TGFβ1 (Schmierer and Hill 2007). Other changes depend on the interaction of TGFβ with other inflammatory mediators or their transcription factors, such as IFNγ and NFκB, or on regulatory components, such as MKP-1 (von Bernhardi et al. 2015b). Furthermore, considering that the non-Smad TGFβ1 pathways MAPKs and PI3K, are not abolished in aged mice, inhibition of Smad could abolish the regulatory anti-inflammatory effect of TGFβ1 on inflammation, facilitating the cytotoxic activation of glia (von Bernhardi et al. 2015a).

## Age-Induced Changes in Microglial Cells

The main mediators of inflammatory responses in the CNS are microglia, the resident immune cell of the brain, which are responsible for the detection of disturbances and orchestrate the innate immune response after any event. As previously discussed in Chapters "Glial Cells and Integrity of the Nervous System", "Microglia Function in the Normal Brain" and "Purine Signaling and Microglial Wrapping", the microglial cell immune response is activated after neuronal damage by the release of ATP, growth factors, and cytokines, by changes in the extracellular ionic balance, and by the absence of molecules normally released by healthy neurons, such as neurotransmitters (Hanisch and Kettenmann 2007; Pocock and Kettenmann 2007; Block et al. 2007). Initially, it was proposed that the microglial cells fluctuate between two extreme phenotypes: the classically activated phenotype, M1, which appears to be more inflammatory, and the alternatively activated phenotype, M2, a "regulatory" state (Cherry et al. 2014). However, recent studies have shown that microglia can be activated in several ways, with a number of adaptive responses to different stimuli (Hanisch and Kettenmann 2007). When microglia detect pathogen-associated molecular patterns (PAMPs) or danger-associated molecular patterns (DAMPs) through their pattern-recognition receptors (PRRs), they shift their activity state to a more "reactive" phenotype, which is pro-inflammatory (Lucin and Wyss-Coray 2009). Normally, Toll-like receptors (TLRs) play an important role in PAMPs and DAMPs recognition, initiating the immune response mediated by microglia (Janeway and Medzhitov 2002). The intracellular signaling of TLRs activates NFκB and transcription of numerous pro-inflammatory genes encoding cytokines, chemokines, complement proteins, enzymes, adhesion molecules, and immune receptors such as the major histocompatibility complex II (MHCII) and complement receptor 3 (CD11b) (Nguyen et al. 2002; Hanisch and Kettenmann 2007; Rozovsky et al. 1998).

In the absence of an inflammatory stimulus, microglia are actively searching for pathogenic signals (Nimmerjahn et al. 2005), whereby they have been considered to be in a "surveilling" instead of "resting" state (Hanisch and Kettenmann 2007), having reduced nitric oxide production and increased anti-inflammatory cytokine release (Mantovani et al. 2004).

As our brain ages, microglia become more activated and irregularly distributed in several cortical and subcortical areas (Schuitemaker et al. 2012; Tremblay et al. 2012; von Bernhardi et al. 2015b), with diverse cell morphologies and decreased process length and complexity (Fig. 2a, b), resulting in a smaller volume of parenchyma being probed by a single cell, and appearing more granular due to increased phagocytic inclusions (Tremblay et al. 2012; Vaughan and Peters 1974). In addition, microglial cell process movement speed, and the dynamics of the cell's response to tissue injury decline with age (Hefendehl et al. 2014). Age-induced changes in microglia phenotype suggest that, rather than being in an over-activated state, aged microglia have a reduced ability to develop a normal immune response, becoming dysregulated (Sheng et al. 1998; von Bernhardi 2007; von Bernhardi et al. 2015a).

It has been reported that aged microglia show a higher proliferation rate in response to nerve injury in murine models, suggesting a less-regulated proliferative response under pathological conditions than that of young microglia (Conde and Streit 2006; Tremblay et al. 2012). However, there is some agreement about the effect of aging alone (non-pathological aging) on microglial cell replication: there is

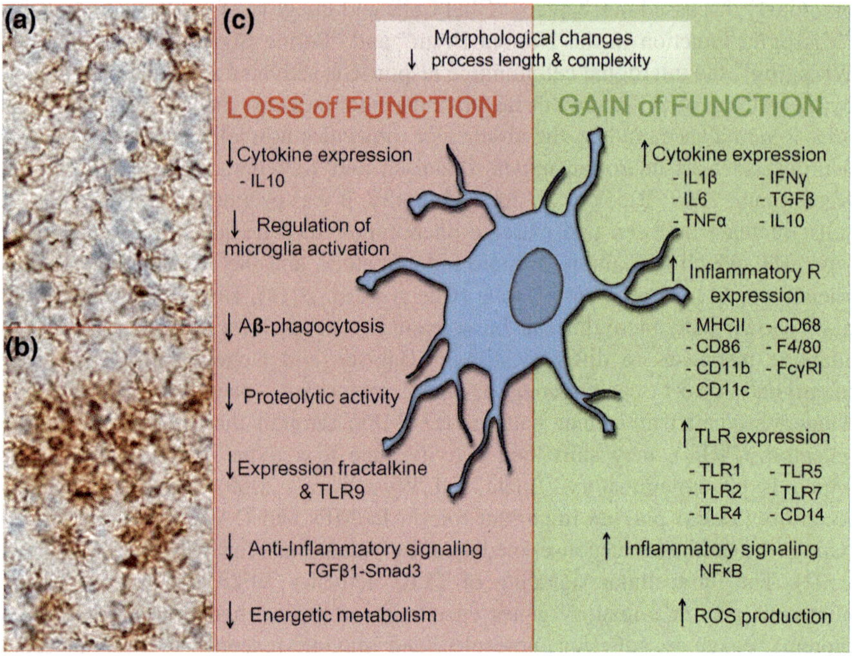


**Fig. 2** Microglia morphological, physiological and molecular changes in the aged brain. Microglial cells of young (**a**) and old (**b**) mice were labeled in hippocampal cryosections for Iba1, a microglia identity marker, to compare their morphological features. **a** Microglia have large and ramified processes in 2 month old mice, whereas **b** microglia from 18 month old mice have a big cell body with shorter processes. **c** Schematic representation of functions lost and gained by aged microglia. Some of the main functions lost involve impaired processes integrity, reduced expression of the fractalkine and TLR9 inflammatory mediators, and impaired proteolytic and phagocytic activity. On the other hand, several inflammatory markers are up regulated as microglia age, such as cytokines, TLRs and several other membrane receptors involving the gain of inflammatory features

no significant increase in the number of microglia as the brain ages (VanGuilder et al. 2011), owing to an age-induced reduction in microglia replication. Decreased replication implies a reduced cell turnover that could result in having fewer healthy microglia and more aged and dysfunctional cells in aged brains (Mosher and Wyss-Coray 2014).

As summarized in Fig. 2c, aged mice have a transcriptional profile characteristic of activated microglia (Godbout et al. 2005) with an increased expression of mRNA for the inflammatory cytokines Tumor Necrosis Factor α (TNFα), Interleukin 1β (IL1β), Interleukin 6 (IL6), and Interferon γ (IFNγ), a decrease in the anti-inflammatory cytokines Interleukin 10 (IL10) and Transforming Growth Factor β (TGFβ) (Sierra et al. 2007; Vaughan and Peters 1974; Frank et al. 2006), and increased inflammatory receptors MHCII (Henry et al. 2009) and CD86 (Frank et al. 2006), suggesting a shift to a more inflammatory phenotype.

Recently, an RNA-seq study has shown that several activation markers can either be upregulated or downregulated in aged microglia. Whereas, 31 % of genes involved in the "sensing" activity of microglia are downregulated, 13 % are significantly upregulated with age; where 81 % of deregulated genes encode for proteins involved in sensing endogenous ligands and 62 % of the upregulated genes are related to sensing infectious microbial ligands (Hickman et al. 2013). Because sensing is one of the main functions of microglia, these changes in the microglial "sensome" induced by age will alter its homeostatic functions, perhaps inducing a pathological microglial phenotype, less protective and more cytotoxic (Yan et al. 2014; von Bernhardi et al. 2010, 2011).

Some epigenetic changes have also been observed in aged microglia. The histone H3K27me3 demethylase Jumonji domain 3 (Jmjd3), which is essential for microglial activation, is reduced in the midbrain of aged mice, resulting in an elevated level of histone methylation, exacerbating the reactive microglia inflammatory response (Tang et al. 2014). Also, it has been recently shown that aged microglia CpG sites on the IL1β promoter are hypomethylated, inducing high IL1β expression (Cho et al. 2015). Both epigenetic alterations in aged microglia seem to contribute to the development of an activated microglial phenotype and the associated neuroinflammation observed in the aged brain.

Aged microglia also have membrane markers (Fig. 2c) that in young cells are either poorly expressed, such as TLR1, TLR2, TLR4, TLR5, TLR7, and CD14 (Letiembre et al. 2007), or even absent, such as MHCII, CD11b, CD14 (Perry et al. 1993). Some of the markers are increased in certain regions of the brain; CD11b, CD68, CD11c, F4/80, and FcγRI increases in aged microglia in both the white matter and caudal areas and CD11c in the white matter (Hart et al. 2012). In contrast, the fractalkine receptor (CX3CR1), critical for microglia migration and activation, and TLR9 have a reduced expression in aged microglia (Fig. 2c) (Wynne et al. 2010; Letiembre et al. 2007).

Class A scavenger receptor (SR-A) has recently been proposed as another target for microglial activation (Cornejo and von Bernhardi 2013). SR-A is a PRR and has an important role in the phagocytosis of Aβ and other anionic molecules. Thus, SR-A inhibition could increase Aβ burden in the brain of AD patients, potentially promoting neurotoxic effects and disease progression (Frenkel et al. 2013). Nevertheless, SR-A also modulates glial cell activation (Murgas et al. 2014). It has been reported that treatment with SR-A antagonists appears to improve the phenotypic features of AD (Handattu et al. 2009) by reducing activation of microglia (Handattu et al. 2013). These pieces of evidence have led to the idea that the modulation of SR-A activity could be a potent mediator of glial activation, such that SR-A inhibition could reduce key neuroinflammatory effects secondary to the microglial cell dysfunction observed in aging.

Cytokine release is also altered in aged microglia (von Bernhardi et al. 2015b), with increased release of IL1α (Sheng et al. 1998), IL6, and TNFα (Njie et al. 2012). Aged microglia challenged with an inflammatory stimulus release more IL1β than do young cells (Combrinck et al. 2002; Cunningham et al. 2005; Sierra et al. 2007; Henry et al. 2009).

## Effects of Aged Microglia on Neuronal Activity

Several investigators report that the neuroimmune response impairs memory and cognition, since inflammation due to illness or injury impairs cognitive and memory tasks (Hudetz et al. 2009; Selnes et al. 2003; Shapira-Lichter et al. 2008). In agreement with in vivo studies, in vitro work shows that conditioned medium derived from LPS-stimulated microglia reduces the number of synapses in neuronal cultures (Moraes et al. 2014), with the inflammatory cytokine IL1β being the main cytokine responsible for the regulation of synaptic activation and long-term potentiation (LTP) (Bellinger et al. 1993; Cunningham et al. 1996; Moraes et al. 2014).

In aged rats, several inflammatory cytokines are upregulated in the hippocampus, accompanied by progressive deficits in LTP (Griffin et al. 2006). In fact, direct administration of IL1β to the CNS impairs memory consolidation (Rachal Pugh et al. 2001), and inhibition of IL1 receptor activity in the hippocampus promotes short-term and long-term memory retention (Depino et al. 2004). It has been proposed that the effect of IL1β could involve downregulation of BDNF (Barrientos et al. 2003, 2004) or reduction of glutamate release in the hippocampus (Gonzalez et al. 2013). However, it has also been shown that IL1β acutely impairs neurogenesis (Goshen et al. 2008; Wu et al. 2012; Zunszain et al. 2012). Thus, IL1β-induced neuroinflammation would both modulate synapses and impair neurogenesis in the hippocampus.

IFNγ is also upregulated in the aged hippocampus and induces microglial activation when injected in rat hippocampus, diminishing LTP (Maher et al. 2006). This demonstrates that inflammatory cytokines released by microglia subjected to aging-induced changes have a role both in neuronal homeostasis and in the modulation of synapse integrity and neuronal function. This identifies activated microglia as important participants in the genesis and progression of neurological diseases, as previously mentioned in Chapter "Microglia Function in the Normal Brain".

## Aged Microglia in the Development of Neurodegenerative Diseases

Microglia have been linked to the pathology and disease progression of several neurodegenerative disorders, including prion diseases (Perry et al. 2002), multiple sclerosis (Takeuchi et al. 2006), amyotrophic lateral sclerosis (McGeer and McGeer 2002), Huntington's disease (Sapp et al. 2001), Pick's disease (Schofield et al. 2003), and HIV-associated dementia (Sopper et al. 1996). Similarly, age-induced microglial dysfunction appears to have an important role in the onset of age-related neurodegenerative diseases, such as PD and AD (von Bernhardi 2007).

PD is a disease characterized by the loss of the ascending nigrostriatal dopaminergic projections, resulting in motor dysfunctions such as rigidity, tremor, slowness, difficulty to initiate movements, and loss of balance. One main cause of dopaminergic neuron loss is activated microglia. Activated microglia are observed associated with degenerating dopaminergic neurons in the *substantia nigra* (McGeer et al. 1988) and in the hippocampus, trans-entorhinal cortex, cingulate cortex, and temporal cortex of PD patients (Imamura et al. 2003). Microglial cells are activated by α-synuclein aggregation, involving NADPH oxidase, leading to persistent and progressive nigral neurodegeneration in PD (Zhang et al. 2005). This selective loss of dopaminergic neurons in the nigrostriatal pathway might also depend on the high density of microglia that exists in this region (Kim et al. 2000). It has been reported that IFNγ-mediated death of dopaminergic neurons is observed only in the presence of microglia (Mount et al. 2007), with TNFα also having a role in PD progression (McCoy et al. 2006; Sriram et al. 2002).

As already discussed, one of the main changes associated with CNS aging is aberrant activation of microglia, generating cells that are incapable of developing a normal immune response (von Bernhardi 2007). In AD, one of those abnormal responses of aged microglia is their loss of the ability to deal with amyloid beta peptide (Aβ) aggregates (Fig. 2c), so that aged microglia internalize less Aβ (Floden and Combs 2011; Njie et al. 2012) and have less capacity to process it (Hickman et al. 2008; Mawuenyega et al. 2010; Nixon et al. 2001). Studies in AD mouse models show that aged microglia have fewer scavenger receptors and Aβ-degrading enzymes, in addition to increased inflammatory cytokines, as already mentioned. IL1β and TNFα appear to be the main agents responsible for the reduced expression of scavenger receptors (Hickman et al. 2008), leading to the idea that the increased levels of inflammatory cytokines observed in aging down-regulate the expression of genes involved in Aβ clearance and promote Aβ accumulation. This Aβ accumulation not only induces neurotoxicity, but also affects neuronal function, inhibiting LTP in aged brains (Lynch et al. 2007). Additionally, Aβ accumulation induces activation of microglia and the expression of the glycoprotein of human histocompatibility complex HLA-DR in the vicinity of neuritic plaques (McGeer et al. 1987; Rogers et al. 1988; Xiang et al. 2006), increases iNOS (Dheen et al. 2005), produces NO and superoxide anion (Ii et al. 1996; Qin et al. 2002), increases expression of TNFα (Meda et al. 1995), leads to microglial cell dystrophy, and exacerbates telomere shortening induced by aging in microglial cells (Flanary et al. 2007). These observations imply that as microglia get older, their Aβ-phagocytic activity declines. This facilitates the accumulation of Aβ aggregates in the aged brain, which in turn leads microglia to a more reactive phenotype, establishing a vicious circle leading to neurodegenerative changes. In fact, injection of Aβ is neurotoxic only in aged individuals, but not when injected in the brains of young monkeys, suggesting that Aβ-induced neurodegeneration is a pathological response of the aged brain (Geula et al. 1998).

It has been reported that Aβ reactivity and phagocytic activity of microglia are regulated by astrocytes, which attenuate their cytotoxic activation (DeWitt et al. 1998; von Bernhardi and Ramirez 2001). However, this regulation is not observed when primed microglia are exposed to Aβ (von Bernhardi and Eugenín 2004). Compared with nonactivated microglia, primed microglia show increased cytotoxicity, Aβ precursor protein (APP) synthesis and Aβ aggregation, as well as impairment of the uptake and degradation of Aβ (Rogers et al. 2002; Ramírez et al. 2008; von Bernhardi 2007).

TGFβ1 secreted by hippocampal neurons and astrocytes has been identified as an important modulator cytokine, attenuating the release of pro-inflammatory mediators (Chen et al. 2002; Mittaud et al. 2002; Herrera-Molina and von Bernhardi 2005) and promoting microglia-mediated Aβ phagocytosis and degradation (Wyss-Coray et al. 2001). We have recently shown that these regulatory effects are mediated by a Smad3-dependent mechanism (Tichauer and von Bernhardi 2012), the TGFβ1 signaling pathway that is inhibited in aged brain, as discussed above. Interestingly, the Smad signaling pathway is also impaired in the brains of AD patients and mouse models for AD, even though TGFβ1 levels are elevated in the cerebrospinal fluid of these patients (Blobe et al. 2000). Thus, impairment of Smad signaling appears to result in Aβ accumulation, Aβ-induced neurodegeneration, and neurofibrillary tangle formation (Tesseur et al. 2006; Ueberham et al. 2006), suggesting it could play a role in the genesis of AD, and eventually be considered as a therapeutic target.

## Concluding Remarks

As we age, several morphological and metabolic changes occur in our brain. Many of those changes are a consequence of alterations in the normal functioning of brain cells and how they deal with stressor stimuli, leading to genetic and epigenetic changes. Here we have discussed two theories that aim to explain age-induced molecular changes: the "free-radical theory of aging," in which the mediator for aging is ROS, and the "inflamm-aging theory" that proposes that cumulative inflammatory signals direct age-related impairment. The latter is particularly striking because it directly associates aging and telomere shortening with inflammatory activation through Rap1 and IKK interaction.

Microglia are the cells most prone to ROS damage and telomere shortening in our brains, and both theories finger microglia as the main cell type affected by aging. Aged microglia show a dysregulated inflammatory response associated with several changes in cytokines and PRR expression compared with young cells. These changes not only affect immune homeostasis of the brain but also neural function and cognitive skills.

Among the molecular signaling pathways affected by aging in microglia, the TGFβ pathway appears to be one of the most relevant, because it modulates microglial cell reactivity and activation state (Herrera-Molina and von Bernhardi 2005; Boche et al. 2006; von Bernhardi et al. 2010). Study of the TGFβ pathway as a possible target for age-related diseases should reveal new approaches to promote the well-being of elderly people, especially those affected by neurodegenerative pathologies such as AD.

**Acknowledgments** This work was supported by grant FONDECYT 1131025 and fellowship CONICYT 21120013.

## References


Abutbul S, Shapiro J, Szaingurten-Solodkin I, Levy N, Carmy Y, Baron R, Jung S, Monsonego A (2012) TGF-β signaling through SMAD2/3 induces the quiescent microglial phenotype within the CNS environment. Glia 60(7):1160–1171. doi:10.1002/glia.22343

Adler A, Sinha S, Kawahara T, Zhang J, Segal E, Chang H (2007) Motif module map reveals enforcement of aging by continual NF-κB activity. Genes Dev 21(24):3244–3257. doi:10.1101/gad.1588507

Arka Subhra G, Vinay T (2010) Telomeres and inflammation: rap1 joins the ends? Cell Cycle 9. doi:10.4161/cc.9.19.13383

Balaban R, Nemoto S, Finkel T (2005) Mitochondria, oxidants, and aging. Cell 120(4):483–495. doi:10.1016/j.cell.2005.02.001

Balu M, Sangeetha P, Murali G, Panneerselvam C (2005) Age-related oxidative protein damages in central nervous system of rats: modulatory role of grape seed extract. Int J Dev Neurosci 23 (6):501–507. doi:10.1016/j.ijdevneu.2005.06.001

Barrientos R, Sprunger D, Campeau S, Higgins E, Watkins L, Rudy J, Maier S (2003) Brain-derived neurotrophic factor mRNA downregulation produced by social isolation is blocked by intrahippocampal interleukin-1 receptor antagonist. Neuroscience 121(4):847–853. doi:10.1016/S0306-4522(03)00564-5

Barrientos R, Sprunger D, Campeau S, Watkins L, Rudy J, Maier S (2004) BDNF mRNA expression in rat hippocampus following contextual learning is blocked by intrahippocampal IL-1beta administration. J Neuroimmunol 155(1–2):119–126. doi:10.1016/j.jneuroim.2004.06.009

Bellinger F, Madamba S, Siggins G (1993) Interleukin 1 beta inhibits synaptic strength and long-term potentiation in the rat CA1 hippocampus. Brain Res 628(1–2):227–234. doi:10.1016/0006-8993(93)90959-Q

Berr C, Balansard B, Arnaud J, Roussel A, Alpérovitch A (2000) Cognitive decline is associated with systemic oxidative stress: the EVA study. Etude du Vieillissement Artériel. J Am Geriatr Soc 48(10):1285–1291

Blobe G, Schiemann W, Lodish H (2000) Role of transforming growth factor beta in human disease. New Engl J Med 342(18):1350–1358. doi:10.1056/NEJM200005043421807

Block M, Zecca L, Hong J-S (2007) Microglia-mediated neurotoxicity: uncovering the molecular mechanisms. Nat Rev Neurosci 8(1):57–69. doi:10.1038/nrn2038

Boche D, Cunningham C, Docagne F, Scott H, Perry V (2006) TGFbeta1 regulates the inflammatory response during chronic neurodegeneration. Neurobiol Dis 22(3):638–650. doi:10.1016/j.nbd.2006.01.004

Bye N, Zieba M, Wreford N, Nichols N (2001) Resistance of the dentate gyrus to induced apoptosis during ageing is associated with increases in transforming growth factor-beta1 messenger RNA. Neuroscience 105(4):853–862. doi:10.1016/S0306-4522(01)00236-6

Calabrese V, Scapagnini G, Ravagna A, Colombrita C, Spadaro F, Butterfield D, Giuffrida Stella A (2004) Increased expression of heat shock proteins in rat brain during aging: relationship with mitochondrial function and glutathione redox state. Mech Ageing Dev 125 (4):325–335. doi:10.1016/j.mad.2004.01.003

Cencioni C, Spallotta F, Martelli F, Valente S, Mai A, Zeiher A, Gaetano C (2013) Oxidative stress and epigenetic regulation in ageing and age-related diseases. Int J Mol Sci 14 (9):17643–17663. doi:10.3390/ijms140917643

Chen S, Luo D, Streit W, Harrison J (2002) TGF-beta1 upregulates CX3CR1 expression and inhibits fractalkine-stimulated signaling in rat microglia. J Neuroimmunol 133(1–2):46–55

Cherry J, Olschowka J, O'Banion M (2014) Neuroinflammation and M2 microglia: the good, the bad, and the inflamed. J Neuroinflammation 11:98. doi:10.1186/1742-2094-11-98

Cho S-H, Chen J, Sayed F, Ward M, Gao F, Nguyen T, Krabbe G, Sohn P, Lo I, Minami S, Devidze N, Zhou Y, Coppola G, Gan L (2015) SIRT1 deficiency in microglia contributes to cognitive decline in aging and neurodegeneration via epigenetic regulation of IL-1β. J Neurosci 35(2):807–818. doi:10.1523/JNEUROSCI.2939-14.2015

Colangelo V, Schurr J, Ball MJ, Pelaez RP, Bazan NG, Lukiw WJ (2002) Gene expression profiling of 12633 genes in Alzheimer hippocampal CA1: transcription and neurotrophic factor down-regulation and up-regulation of apoptotic and pro-inflammatory signaling. J Neurosci Res 70(3):462–473. doi:10.1002/jnr.10351

Combrinck M, Perry V, Cunningham C (2002) Peripheral infection evokes exaggerated sickness behaviour in pre-clinical murine prion disease. Neuroscience 112(1):7–11. doi:10.1016/S0306-4522(02)00030-1

Conde J, Streit W (2006) Effect of aging on the microglial response to peripheral nerve injury. Neurobiol Aging 27(10):1451–1461. doi:10.1016/j.neurobiolaging.2005.07.012

Cornejo F, von Bernhardi R (2013) Role of scavenger receptors in glia-mediated neuroinflammatory response associated with Alzheimer's disease. Mediators Inflamm 2013:895651. doi:10.1155/2013/895651

Courchesne E, Chisum H, Townsend J, Cowles A, Covington J, Egaas B, Harwood M, Hinds S, Press G (2000) Normal brain development and aging: quantitative analysis at in vivo MR imaging in healthy volunteers. Radiology 216(3):672–682. doi:10.1148/radiology.216.3.r00au37672

Cunningham A, Murray C, O'Neill L, Lynch M, O'Connor J (1996) Interleukin-1 beta (IL-1 beta) and tumour necrosis factor (TNF) inhibit long-term potentiation in the rat dentate gyrus in vitro. Neurosci Lett 203(1):17–20

Cunningham C, Wilcockson D, Campion S, Lunnon K, Perry V (2005) Central and systemic endotoxin challenges exacerbate the local inflammatory response and increase neuronal death during chronic neurodegeneration. J Neurosci 25(40):9275–9284. doi:10.1523/JNEUROSCI.2614-05.2005

de Magalhães J, Curado J, Church G (2009) Meta-analysis of age-related gene expression profiles identifies common signatures of aging. Bioinformatics 25(7):875–881. doi:10.1093/bioinformatics/btp073

de Sampaio e Spohr T, Martinez R, da Silva E, Neto V, Gomes F (2002) Neuro-glia interaction effects on GFAP gene: a novel role for transforming growth factor-beta1. Eur J Neurosci 16 (11):2059–2069. doi:10.1046/j.1460-9568.2002.02283.x

Denise CP, Anderson DS, Gary L, Julie LE et al (1996) Mediators of long-term memory performance across the life span. Psychol Aging 11. doi:10.1037/0882-7974.11.4.621

Depino A, Alonso M, Ferrari C, del Rey A, Anthony D, Besedovsky H, Medina J, Pitossi F (2004) Learning modulation by endogenous hippocampal IL-1: blockade of endogenous IL-1 facilitates memory formation. Hippocampus 14(4):526–535. doi:10.1002/hipo.10164

DeWitt D, Perry G, Cohen M, Doller C, Silver J (1998) Astrocytes regulate microglial phagocytosis of senile plaque cores of Alzheimer's disease. Exp Neurol 149(2):329–340. doi:10.1006/exnr.1997.6738

Dhandapani K, Brann D (2003) Transforming growth factor-beta: a neuroprotective factor in cerebral ischemia. Cell Biochem Biophys 39(1):13–22. doi:10.1385/CBB:39:1:13

Dheen S, Jun Y, Yan Z, Tay S, Ling E (2005) Retinoic acid inhibits expression of TNF-alpha and iNOS in activated rat microglia. Glia 50(1):21–31. doi:10.1002/glia.20153

Driscoll I, Davatzikos C, An Y, Wu X, Shen D, Kraut M, Resnick S (2009) Longitudinal pattern of regional brain volume change differentiates normal aging from MCI. Neurology 72 (22):1906–1913. doi:10.1212/WNL.0b013e3181a82634

Dröge W, Schipper H (2007) Oxidative stress and aberrant signaling in aging and cognitive decline. Aging Cell 6(3):361–370. doi:10.1111/j.1474-9726.2007.00294.x

Erraji-Benchekroun L, Underwood M, Arango V, Galfalvy H, Pavlidis P, Smyrniotopoulos P, Mann J, Sibille E (2005) Molecular aging in human prefrontal cortex is selective and continuous throughout adult life. Biol Psychiatry 57(5):549–558. doi:10.1016/j.biopsych.2004.10.034

Faith MG-D, Naftali R (2003) Neuroanatomical correlates of selected executive functions in middle-aged and older adults: a prospective MRI study. Neuropsychologia 41. doi:10.1016/S0028-3932(03)00129-5

Flanary B, Streit W (2004) Progressive telomere shortening occurs in cultured rat microglia, but not astrocytes. Glia 45(1):75–88. doi:10.1002/glia.10301

Flanary B, Sammons N, Nguyen C, Walker D, Streit W (2007) Evidence that aging and amyloid promote microglial cell senescence. Rejuvenation Res 10(1):61–74. doi:10.1089/rej.2006.9096

Floden A, Combs C (2011) Microglia demonstrate age-dependent interaction with amyloid-β fibrils. J Alzheimers Dis 25(2):279–293. doi:10.3233/JAD-2011-101014

Forster M, Dubey A, Dawson K, Stutts W, Lal H, Sohal R (1996) Age-related losses of cognitive function and motor skills in mice are associated with oxidative protein damage in the brain. Proc Nat Acad Sci USA 93(10):4765–4769. doi:10.1073/pnas.93.10.4765

Fraga C, Shigenaga M, Park J, Degan P, Ames B (1990) Oxidative damage to DNA during aging: 8-hydroxy-2′-deoxyguanosine in rat organ DNA and urine. Proc Nat Acad Sci USA 87 (12):4533–4537. doi:10.1073/pnas.87.12.4533

Franceschi C, Bonafè M, Valensin S, Olivieri F, De Luca M, Ottaviani E, De Benedictis G (2000) Inflamm-aging. An evolutionary perspective on immunosenescence. Ann N Y Acad Sci 908:244–254. doi:10.1111/j.1749-6632.2000.tb06651.x

Frank M, Barrientos R, Biedenkapp J, Rudy J, Watkins L, Maier S (2006) mRNA up-regulation of MHC II and pivotal pro-inflammatory genes in normal brain aging. Neurobiol Aging 27 (5):717–722. doi:10.1016/j.neurobiolaging.2005.03.013

Fraser H, Khaitovich P, Plotkin J, Pääbo S, Eisen M (2005) Aging and gene expression in the primate brain. PLoS Biol 3(9). doi:10.1371/journal.pbio.0030274

Frenkel D, Wilkinson K, Zhao L, Hickman S, Means T, Puckett L, Farfara D, Kingery N, Weiner H, El Khoury J (2013) Scara1 deficiency impairs clearance of soluble amyloid-β by mononuclear phagocytes and accelerates Alzheimer's-like disease progression. Nat Commun 4:2030. doi:10.1038/ncomms3030

Friedrich MJ (2014) Researchers probe the aging brain in health and disease. JAMA 311(3):231–232. doi:10.1001/jama.2013.284609

Ge Y, Grossman R, Babb J, Rabin M, Mannon L, Kolson D (2002) Age-related total gray matter and white matter changes in normal adult brain. Part I: volumetric MR imaging analysis. AJNR Am J Neuroradiol 23(8):1327–1333

Gefen T, Peterson M, Papastefan S, Martersteck A, Whitney K, Rademaker A, Bigio E, Weintraub S, Rogalski E, Mesulam MM, Geula C (2015) Morphometric and histologic substrates of cingulate integrity in elders with exceptional memory capacity. J Neurosci 35 (4):1781–1791. doi:10.1523/JNEUROSCI.2998-14.2015

Geula C, Wu C, Saroff D, Lorenzo A, Yuan M, Yankner B (1998) Aging renders the brain vulnerable to amyloid beta-protein neurotoxicity. Nat Med 4(7):827–831. doi:10.1038/nm0798-827

Godbout J, Chen J, Abraham J, Richwine A, Berg B, Kelley K, Johnson R (2005) Exaggerated neuroinflammation and sickness behavior in aged mice following activation of the peripheral innate immune system. FASEB J 19(10):1329–1331. doi:10.1096/fj.05-3776fje

Gonzalez P, Machado I, Vilcaes A, Caruso C, Roth G, Schiöth H, Lasaga M, Scimonelli T (2013) Molecular mechanisms involved in interleukin 1-beta (IL-1β)-induced memory impairment. Modulation by alpha-melanocyte-stimulating hormone (α-MSH). Brain Behav Immun 34:141–150. doi:10.1016/j.bbi.2013.08.007

Good C, Johnsrude I, Ashburner J, Henson R, Friston K, Frackowiak R (2001) A voxel-based morphometric study of ageing in 465 normal adult human brains. NeuroImage 14(1 Pt 1):21–36. doi:10.1006/nimg.2001.0786

Goshen I, Kreisel T, Ben-Menachem-Zidon O, Licht T, Weidenfeld J, Ben-Hur T, Yirmiya R (2008) Brain interleukin-1 mediates chronic stress-induced depression in mice via adrenocortical activation and hippocampal neurogenesis suppression. Mol Psychiatry 13(7):717–728. doi:10.1038/sj.mp.4002055

Gravina S, Vijg J (2010) Epigenetic factors in aging and longevity. Pflugers Arch 459(2):247–258. doi:10.1007/s00424-009-0730-7

Griffin R, Nally R, Nolan Y, McCartney Y, Linden J, Lynch M (2006) The age-related attenuation in long-term potentiation is associated with microglial activation. J Neurochem 99(4):1263–1272. doi:10.1111/j.1471-4159.2006.04165.x

Gupta A, Hasan M, Chander R, Kapoor N (1991) Age-related elevation of lipid peroxidation products: diminution of superoxide dismutase activity in the central nervous system of rats. Gerontology 37(6):305–309. doi:10.1159/000213277

Handattu S, Garber D, Monroe C, van Groen T, Kadish I, Nayyar G, Cao D, Palgunachari M, Li L, Anantharamaiah G (2009) Oral apolipoprotein A-I mimetic peptide improves cognitive function and reduces amyloid burden in a mouse model of Alzheimer's disease. Neurobiol Dis 34(3):525–534. doi:10.1016/j.nbd.2009.03.007

Handattu S, Monroe C, Nayyar G (2013) In vivo and in vitro effects of an apolipoprotein E mimetic peptide on amyloid-β pathology. J Alzheimers Dis 36(2):335–347

Hanisch U-K, Kettenmann H (2007) Microglia: active sensor and versatile effector cells in the normal and pathologic brain. Nat Neurosci 10(11):1387–1394. doi:10.1038/nn1997

Harman D (1956) Aging: a theory based on free radical and radiation chemistry. J Gerontol 11 (3):298–300. doi:10.1016/0921-8734(92)90030-S

Hart A, Wyttenbach A, Perry V, Teeling J (2012) Age related changes in microglial phenotype vary between CNS regions: grey versus white matter differences. Brain Behav Immun 26 (5):754–765. doi:10.1016/j.bbi.2011.11.006

Head D, Buckner R, Shimony J, Williams L, Akbudak E, Conturo T, McAvoy M, Morris J, Snyder A (2004) Differential vulnerability of anterior white matter in nondemented aging with minimal acceleration in dementia of the Alzheimer type: evidence from diffusion tensor imaging. Cereb Cortex 14(4):410–423. doi:10.1093/cercor/bhh003

Hefendehl J, Neher J, Sühs R, Kohsaka S, Skodras A, Jucker M (2014) Homeostatic and injury-induced microglia behavior in the aging brain. Aging Cell 13(1):60–69. doi:10.1111/acel.12149

Helenius M, Hänninen M, Lehtinen S, Salminen A (1996) Changes associated with aging and replicative senescence in the regulation of transcription factor nuclear factor-κB. Biochem J 318(Pt 2):603–608

Henry C, Huang Y, Wynne A, Godbout J (2009) Peripheral lipopolysaccharide (LPS) challenge promotes microglial hyperactivity in aged mice that is associated with exaggerated induction of both pro-inflammatory IL-1beta and anti-inflammatory IL-10 cytokines. Brain Behav Immun 23(3):309–317. doi:10.1016/j.bbi.2008.09.002

Herrera-Molina R, von Bernhardi R (2005) Transforming growth factor-beta 1 produced by hippocampal cells modulates microglial reactivity in culture. Neurobiol Dis 19(1–2):229–236. doi:10.1016/j.nbd.2005.01.003

Hickman S, Allison E, El Khoury J (2008) Microglial dysfunction and defective beta-amyloid clearance pathways in aging Alzheimer's disease mice. J Neurosci 28(33):8354–8360. doi:10.1523/JNEUROSCI.0616-08.2008

Hickman S, Kingery N, Ohsumi T, Borowsky M, L-C Wang, Means T, El Khoury J (2013) The microglial sensome revealed by direct RNA sequencing. Nat Neurosci 16(12):1896–1905. doi:10.1038/nn.3554

Horvath S (2013) DNA methylation age of human tissues and cell types. Genome Biol 14(10). doi:10.1186/gb-2013-14-10-r115

Hsieh T-C, Lin W-Y, Ding H-J, Sun S-S, Wu Y-C, Yen K-Y, Kao C-H (2012) Sex- and age-related differences in brain FDG metabolism of healthy adults: an SPM analysis. J Neuroimaging 22(1):21–27. doi:10.1111/j.1552-6569.2010.00543.x

Hudetz J, Iqbal Z, Gandhi S, Patterson K, Byrne A, Hudetz A, Pagel P, Warltier D (2009) Ketamine attenuates post-operative cognitive dysfunction after cardiac surgery. Acta Anaesthesiol Scand 53(7):864–872. doi:10.1111/j.1399-6576.2009.01978.x

Hultsch D (1998) Memory change in the aged. Cambridge University Press

Ii M, Sunamoto M, Ohnishi K, Ichimori Y (1996) beta-Amyloid protein-dependent nitric oxide production from microglial cells and neurotoxicity. Brain Res 720(1–2):93–100. doi:10.1016/0006-8993(96)00156-4

Imamura K, Hishikawa N, Sawada M, Nagatsu T, Yoshida M, Hashizume Y (2003) Distribution of major histocompatibility complex class II-positive microglia and cytokine profile of Parkinson's disease brains. Acta Neuropathol 106(6):518–526. doi:10.1007/s00401-003-0766-2

Janeway C, Medzhitov R (2002) Innate immune recognition. Annu Rev Immunol 20:197–216. doi:10.1146/annurev.immunol.20.083001.084359

Jurk D, Wilson C, Passos J, Oakley F, Correia-Melo C, Greaves L, Saretzki G, Fox C, Lawless C, Anderson R, Hewitt G, Pender S, Fullard N, Nelson G, Mann J, van de Sluis B, Mann D, von Zglinicki T (2014) Chronic inflammation induces telomere dysfunction and accelerates ageing in mice. Nat Commun 2:4172. doi:10.1038/ncomms5172

Kalpouzos G, Chételat G, Baron J-C, Landeau B, Mevel K, Godeau C, Barré L, Constans J-M, Viader F, Eustache F, Desgranges B (2009) Voxel-based mapping of brain gray matter volume and glucose metabolism profiles in normal aging. Neurobiol Aging 30(1):112–124. doi:10.1016/j.neurobiolaging.2007.05.019

Kim W, Mohney R, Wilson B, Jeohn G, Liu B, Hong J (2000) Regional difference in susceptibility to lipopolysaccharide-induced neurotoxicity in the rat brain: role of microglia. J Neurosci 20 (16):6309–6316

Kochunov P, Ramage A, Lancaster J, Robin D, Narayana S, Coyle T, Royall D, Fox P (2009) Loss of cerebral white matter structural integrity tracks the gray matter metabolic decline in normal aging. NeuroImage 45(1):17–28. doi:10.1016/j.neuroimage.2008.11.010

Larbi A, Franceschi C, Mazzatti D, Solana R, Wikby A, Pawelec G (2008) Aging of the immune system as a prognostic factor for human longevity. Physiology (Bethesda) 23:64–74. doi:10.1152/physiol.00040.2007

Lee C, Weindruch R, Prolla T (2000) Gene-expression profile of the ageing brain in mice. Nat Genet 25(3):294–297. doi:10.1038/77046

Letiembre M, Hao W, Liu Y, Walter S, Mihaljevic I, Rivest S, Hartmann T, Fassbender K (2007) Innate immune receptor expression in normal brain aging. Neuroscience 146(1):248–254. doi:10.1016/j.neuroscience.2007.01.004

Lombard D, Chua K, Mostoslavsky R, Franco S, Gostissa M, Alt F (2005) DNA repair, genome stability, and aging. Cell 120(4):497–512. doi:10.1016/j.cell.2005.01.028

López-Otín C, Blasco M, Partridge L, Serrano M, Kroemer G (2013) The hallmarks of aging. Cell 153(6):1194–1217. doi:10.1016/j.cell.2013.05.039

Lu T, Pan Y, Kao S-Y, Li C, Kohane I, Chan J, Yankner B (2004) Gene regulation and DNA damage in the ageing human brain. Nature 429(6994):883–891. doi:10.1038/nature02661

Lucin K, Wyss-Coray T (2009) Immune activation in brain aging and neurodegeneration: too much or too little? Neuron 64(1):110–122. doi:10.1016/j.neuron.2009.08.039

Lukiw WJ (2004) Gene expression profiling in fetal, aged, and Alzheimer hippocampus: a continuum of stress-related signaling. Neurochem Res 29(6):1287–1297

Lynch A, Loane D, Minogue A, Clarke R, Kilroy D, Nally R, Roche O, O'Connell F, Lynch M (2007) Eicosapentaenoic acid confers neuroprotection in the amyloid-beta challenged aged hippocampus. Neurobiol Aging 28(6):845–855. doi:10.1016/j.neurobiolaging.2006.04.006

Maher F, Clarke R, Kelly A, Nally R, Lynch M (2006) Interaction between interferon gamma and insulin-like growth factor-1 in hippocampus impacts on the ability of rats to sustain long-term potentiation. J Neurochem 96(6):1560–1571. doi:10.1111/j.1471-4159.2006.03664.x

Mantovani A, Sica A, Sozzani S, Allavena P, Vecchi A, Locati M (2004) The chemokine system in diverse forms of macrophage activation and polarization. Trends Immunol 25(12):677–686. doi:10.1016/j.it.2004.09.015

Mawuenyega K, Sigurdson W, Ovod V, Munsell L, Kasten T, Morris J, Yarasheski K, Bateman R (2010) Decreased clearance of CNS beta-amyloid in Alzheimer's disease. Science 330 (6012):1774. doi:10.1126/science.1197623

McCoy M, Martinez T, Ruhn K, Szymkowski D, Smith C, Botterman B, Tansey K, Tansey M (2006) Blocking soluble tumor necrosis factor signaling with dominant-negative tumor necrosis factor inhibitor attenuates loss of dopaminergic neurons in models of Parkinson's disease. J Neurosci 26(37):9365–9375. doi:10.1523/JNEUROSCI.1504-06.2006

McGeer P, McGeer E (2002) Inflammatory processes in amyotrophic lateral sclerosis. Muscle Nerve 26(4):459–470. doi:10.1002/mus.10191

McGeer P, Itagaki S, Tago H, McGeer E (1987) Reactive microglia in patients with senile dementia of the Alzheimer type are positive for the histocompatibility glycoprotein HLA-DR. Neurosci Lett 79(1–2):195–200. doi:10.1016/0304-3940(87)90696-3

McGeer P, Itagaki S, Boyes B, McGeer E (1988) Reactive microglia are positive for HLA-DR in the substantia nigra of Parkinson's and Alzheimer's disease brains. Neurology 38(8):1285–1291

Meda L, Cassatella M, Szendrei G, Otvos L, Baron P, Villalba M, Ferrari D, Rossi F (1995) Activation of microglial cells by beta-amyloid protein and interferon-gamma. Nature 374 (6523):647–650. doi:10.1038/374647a0

Mittaud P, Labourdette G, Zingg H, Guenot-Di Scala D (2002) Neurons modulate oxytocin receptor expression in rat cultured astrocytes: involvement of TGF-beta and membrane components. Glia 37(2):169–177. doi:10.1002/glia.10029

Moraes C, Santos G, Spohr T, D'Avila J, Lima F, Benjamim C, Bozza F, Gomes F (2014) Activated Microglia-Induced Deficits in Excitatory Synapses Through IL-1β: Implications for Cognitive Impairment in Sepsis. Mol Neurobiol. doi:10.1007/s12035-014-8868-5

Mosher K, Wyss-Coray T (2014) Microglial dysfunction in brain aging and Alzheimer's disease. Biochem Pharmacol 88(4):594–604. doi:10.1016/j.bcp.2014.01.008

Mount M, Lira A, Grimes D, Smith P, Faucher S, Slack R, Anisman H, Hayley S, Park D (2007) Involvement of interferon-gamma in microglial-mediated loss of dopaminergic neurons. J Neurosci 27(12):3328–3337. doi:10.1523/JNEUROSCI.5321-06.2007

Murgas P, Cornejo F, Merino G, von Bernhardi R (2014) SR-A regulates the inflammatory activation of astrocytes. Neurotox Res 25(1):68–80. doi:10.1007/s12640-013-9432-1

Navarro A, Gomez C, López-Cepero J, Boveris A (2004) Beneficial effects of moderate exercise on mice aging: survival, behavior, oxidative stress, and mitochondrial electron transfer. Am J Physiol Regul Integr Comp Physiol 286(3):11. doi:10.1152/ajpregu.00208.2003

Nguyen M, Julien J-P, Rivest S (2002) Innate immunity: the missing link in neuroprotection and neurodegeneration? Nature Rev Neurosci 3(3):216–227. doi:10.1038/nrn752

Nimmerjahn A, Kirchhoff F, Helmchen F (2005) Resting microglial cells are highly dynamic surveillants of brain parenchyma in vivo. Science 308(5726):1314–1318. doi:10.1126/science.1110647

Nixon R, Mathews P, Cataldo A (2001) The neuronal endosomal-lysosomal system in Alzheimer's disease. J Alzheimers Dis 3(1):97–107

Njie E, Boelen E, Stassen F, Steinbusch H, Borchelt D, Streit W (2012) Ex vivo cultures of microglia from young and aged rodent brain reveal age-related changes in microglial function. Neurobiol Aging 33(1):195.e1–195.e12. doi:10.1016/j.neurobiolaging.2010.05.008

Park D, Lautenschlager G, Hedden T, Davidson N, Smith A, Smith P (2002) Models of visuospatial and verbal memory across the adult life span. Psychol Aging 17(2):299–320

Perkins A, Hendrie H, Callahan C, Gao S, Unverzagt F, Xu Y, Hall K, Hui S (1999) Association of antioxidants with memory in a multiethnic elderly sample using the Third National Health and Nutrition Examination Survey. Am J Epidemiol 150(1):37–44. doi:10.1093/oxfordjournals.aje.a009915

Perluigi M, Swomley A, Butterfield D (2014) Redox proteomics and the dynamic molecular landscape of the aging brain. Ageing Res Rev 13:75–89. doi:10.1016/j.arr.2013.12.005

Perrig W, Perrig P, Stähelin H (1997) The relation between antioxidants and memory performance in the old and very old. J Am Geriatr Soc 45(6):718–724

Perry V, Matyszak M, Fearn S (1993) Altered antigen expression of microglia in the aged rodent CNS. Glia 7(1):60–67. doi:10.1002/glia.440070111

Perry V, Cunningham C, Boche D (2002) Atypical inflammation in the central nervous system in prion disease. Curr Opin Neurol 15(3):349–354. doi:10.1097/00019052-200206000-00020

Pocock J, Kettenmann H (2007) Neurotransmitter receptors on microglia. Trends Neurosci 30 (10):527–535. doi:10.1016/j.tins.2007.07.007

Pruitt K, Zinn R, Ohm J, McGarvey K, Kang S-HL, Watkins D, Herman J, Baylin S (2006) Inhibition of SIRT1 reactivates silenced cancer genes without loss of promoter DNA hypermethylation. PLoS Genet 2(3). doi:10.1371/journal.pgen.0020040

Qin L, Liu Y, Cooper C, Liu B, Wilson B, Hong J-S (2002) Microglia enhance beta-amyloid peptide-induced toxicity in cortical and mesencephalic neurons by producing reactive oxygen species. J Neurochem 83(4):973–983. doi:10.1046/j.1471-4159.2002.01210.x

Rachal Pugh C, Fleshner M, Watkins L, Maier S, Rudy J (2001) The immune system and memory consolidation: a role for the cytokine IL-1β. Neurosci Biobehav Rev 25(1):29–41

Ramírez G, Rey S, von Bernhardi R (2008) Proinflammatory stimuli are needed for induction of microglial cell-mediated AβPP_{244-C} and Aβ-neurotoxicity in hippocampal cultures. J Alzheimers Dis 15(1):45–59

Raz N, Lindenberger U, Rodrigue K, Kennedy K, Head D, Williamson A, Dahle C, Gerstorf D, Acker J (2005) Regional brain changes in aging healthy adults: general trends, individual differences and modifiers. Cereb Cortex 15(11):1676–1689. doi:10.1093/cercor/bhi044

Richter C, Park J, Ames B (1988) Normal oxidative damage to mitochondrial and nuclear DNA is extensive. Proc Nat Acad Sci USA 85(17):6465–6467. doi:10.1073/pnas.85.17.6465

Rodrigue K, Raz N (2004) Shrinkage of the entorhinal cortex over five years predicts memory performance in healthy adults. J Neurosci 24(4):956–963. doi:10.1523/JNEUROSCI.4166-03.2004

Rodrigues Siqueira I, Fochesatto C, da Silva Torres I, Dalmaz C, Alexandre Netto C (2005) Aging affects oxidative state in hippocampus, hypothalamus and adrenal glands of Wistar rats. Life Sci 78(3):271–278. doi:10.1016/j.lfs.2005.04.044

Rogers J, Luber-Narod J, Styren S, Civin W (1988) Expression of immune system-associated antigens by cells of the human central nervous system: relationship to the pathology of Alzheimer's disease. Neurobiol Aging 9(4):339–349. doi:10.1016/S0197-4580(88)80079-4

Rogers J, Strohmeyer R, Kovelowski C, Li R (2002) Microglia and inflammatory mechanisms in the clearance of amyloid beta peptide. Glia 40(2):260–269. doi:10.1002/glia.10153

Rosen A, Prull M, Gabrieli J, Stoub T, O'Hara R, Friedman L, Yesavage J, deToledo-Morrell L (2003) Differential associations between entorhinal and hippocampal volumes and memory performance in older adults. Behav Neurosci 117(6):1150–1160. doi:10.1037/0735-7044.117.6.1150

Ross KR, Corey DA, Dunn JM, Kelley TJ (2007) SMAD3 expression is regulated by mitogen-activated protein kinase kinase-1 in epithelial and smooth muscle cells. Cell Signal 19 (5):923–931. doi:10.1016/j.cellsig.2006.11.008

Rozovsky I, Finch C, Morgan T (1998) Age-related activation of microglia and astrocytes: in vitro studies show persistent phenotypes of aging, increased proliferation, and resistance to down-regulation. Neurobiol Aging 19(1):97–103. doi:10.1016/S0197-4580(97)00169-3

Salat D, Buckner R, Snyder A, Greve D, Desikan R, Busa E, Morris J, Dale A, Fischl B (2004) Thinning of the cerebral cortex in aging. Cereb Cortex 14(7):721–730. doi:10.1093/cercor/bhh032

Sapp E, Kegel K, Aronin N, Hashikawa T, Uchiyama Y, Tohyama K, Bhide P, Vonsattel J, DiFiglia M (2001) Early and progressive accumulation of reactive microglia in the Huntington disease brain. J Neuropathol Exp Neurol 60(2):161–172

Schmierer B, Hill CS (2007) TGFbeta-SMAD signal transduction: molecular specificity and functional flexibility. Nat Rev Mol Cell Biol 8(12):970–982. doi:10.1038/nrm2297

Schofield E, Kersaitis C, Shepherd C, Kril J, Halliday G (2003) Severity of gliosis in Pick's disease and frontotemporal lobar degeneration: tau-positive glia differentiate these disorders. Brain 126(Pt 4):827–840. doi:10.1093/brain/awg085

Schuitemaker A, van der Doef T, Boellaard R, van der Flier W, Yaqub M, Windhorst A, Barkhof F, Jonker C, Kloet R, Lammertsma A, Scheltens P, van Berckel B (2012) Microglial activation in healthy aging. Neurobiol Aging 33(6):1067–1072. doi:10.1016/j.neurobiolaging. 2010.09.016

Selnes O, Grega M, Borowicz Jr L, Royall R, McKhann G, Baumgartner W (2003) Cognitive changes with coronary artery disease: a prospective study of coronary artery bypass graft patients and nonsurgical controls. Ann Thorac Surg 75(5):1377–1386

Shapira-Lichter I, Beilin B, Ofek K, Bessler H, Gruberger M, Shavit Y, Seror D, Grinevich G, Posner E, Reichenberg A, Soreq H, Yirmiya R (2008) Cytokines and cholinergic signals co-modulate surgical stress-induced changes in mood and memory. Brain Behav Immun 22 (3):388–398. doi:10.1016/j.bbi.2007.09.006

Sheng J, Mrak R, Griffin W (1998) Enlarged and phagocytic, but not primed, interleukin-1 alpha-immunoreactive microglia increase with age in normal human brain. Acta Neuropathol 95(3):229–234

Shigenaga M, Hagen T, Ames B (1994) Oxidative damage and mitochondrial decay in aging. Proc Nat Acad Sci USA 91(23):10771–10778. doi:10.1073/pnas.91.23.10771

Shock N, Greulich R, Costa P, Andres R, Lakatta E, Arenberg D, Tobin J (1984) Normal human aging: the baltimore longitudinal study of aging. US Department of Health and Human Services, Baltimore, USA

Sierra A, Gottfried-Blackmore A, McEwen B, Bulloch K (2007) Microglia derived from aging mice exhibit an altered inflammatory profile. Glia 55(4):412–424. doi:10.1002/glia.20468

Singh T, Newman A (2011) Inflammatory markers in population studies of aging. Ageing Res Rev 10(3):319–329. doi:10.1016/j.arr.2010.11.002

Sopper S, Demuth M, Stahl-Hennig C, Hunsmann G, Plesker R, Coulibaly C, Czub S, Ceska M, Koutsilieri E, Riederer P, Brinkmann R, Katz M, ter Meulen V (1996) The effect of simian immunodeficiency virus infection in vitro and in vivo on the cytokine production of isolated microglia and peripheral macrophages from rhesus monkey. Virology 220(2):320–329. doi:10. 1006/viro.1996.0320

Sriram K, Matheson J, Benkovic S, Miller D, Luster M, O'Callaghan J (2002) Mice deficient in TNF receptors are protected against dopaminergic neurotoxicity: implications for Parkinson's disease. FASEB J 16(11):1474–1476. doi:10.1096/fj.02-0216fje

Takeuchi H, Wang J, Kawanokuchi J, Mitsuma N, Mizuno T, Suzumura A (2006) Interferon-gamma induces microglial-activation-induced cell death: a hypothetical mechanism of relapse and remission in multiple sclerosis. Neurobiol Dis 22(1):33–39. doi:10.1016/j.nbd. 2005.09.014

Tang Y, Li T, Li J, Yang J, Liu H, Zhang X, Le W (2014) Jmjd3 is essential for the epigenetic modulation of microglia phenotypes in the immune pathogenesis of Parkinson's disease. Cell Death Differ 21(3):369–380. doi:10.1038/cdd.2013.159

Teo H, Ghosh S, Luesch H, Ghosh A, Wong E, Malik N, Orth A, de Jesus P, Perry A, Oliver J, Tran N, Speiser L, Wong M, Saez E, Schultz P, Chanda S, Verma I, Tergaonkar V (2010) Telomere-independent Rap1 is an IKK adaptor and regulates NF-κB-dependent gene expression. Nat Cell Biol 12(8):758–767. doi:10.1038/ncb2080

Tesseur I, Zou K, Esposito L, Bard F, Berber E, Can J, Lin A, Crews L, Tremblay P, Mathews P, Mucke L, Masliah E, Wyss-Coray T (2006) Deficiency in neuronal TGF-beta signaling promotes neurodegeneration and Alzheimer's pathology. J Clin Invest 116(11):3060–3069. doi:10.1172/JCI27341

Thakur M, Kanungo M (1981) Methylation of chromosomal proteins and DNA of rat brain and its modulation by estradiol and calcium during aging. Exp Geront 16(4):331–336. doi:10.1016/ 0531-5565(81)90052-8

Tian L, Cai Q, Wei H (1998) Alterations of antioxidant enzymes and oxidative damage to macromolecules in different organs of rats during aging. Free Radic Biol Med 24(9):1477–1484

Tichauer J, von Bernhardi R (2012) Transforming growth factor-β stimulates β amyloid uptake by microglia through Smad3-dependent mechanisms. J Neurosci Res 90(10):1970–1980. doi:10. 1002/jnr.23082

Tichauer J, Flores B, Soler B, Eugenín-von Bernhardi L, Ramírez G, von Bernhardi R (2014) Age-dependent changes on TGFβ1 Smad3 pathway modify the pattern of microglial cell activation. Brain Behav Immun 37:187–196. doi:10.1016/j.bbi.2013.12.018

Tremblay M-È, Zettel M, Ison J, Allen P, Majewska A (2012) Effects of aging and sensory loss on glial cells in mouse visual and auditory cortices. Glia 60(4):541–558. doi:10.1002/glia.22287

Tsurumi A, Li W (2012) Global heterochromatin loss: a unifying theory of aging? Epigenetics 7 (7):680–688. doi:10.4161/epi.20540

Ueberham U, Ueberham E, Gruschka H, Arendt T (2006) Altered subcellular location of phosphorylated Smads in Alzheimer's disease. Eur J Neurosci 24(8):2327–2334. doi:10.1111/j.1460-9568.2006.05109.x

VanGuilder H, Bixler G, Brucklacher R, Farley J, Yan H, Warrington J, Sonntag W, Freeman W (2011) Concurrent hippocampal induction of MHC II pathway components and glial activation with advanced aging is not correlated with cognitive impairment. J Neuroinflammation 8:138. doi:10.1186/1742-2094-8-138

Vanyushin B, Nemirovsky L, Klimenko V, Vasiliev V, Belozersky A (1973) The 5-methylcytosine in DNA of rats. Tissue and age specificity and the changes induced by hydrocortisone and other agents. Gerontologia 19(3):138–152

Vaughan D, Peters A (1974) Neuroglial cells in the cerebral cortex of rats from young adulthood to old age: an electron microscope study. J Neurocytol 3(4):405–429. doi:10.1007/BF01098730

Vaupel JW (2010) Biodemography of human ageing. Nature. doi:10.1038/nature08984

von Bernhardi R (2007) Glial cell dysregulation: a new perspective on Alzheimer disease. Neurotox Res 12(4):215–232. doi:10.1007/BF03033906

von Bernhardi R, Eugenín J (2004) Microglial reactivity to beta-amyloid is modulated by astrocytes and proinflammatory factors. Brain Res 1025(1–2):186–193. doi:10.1016/j.brainres.2004.07.084

von Bernhardi R, Eugenin J (2012) Alzheimer's disease: redox dysregulation as a common denominator for diverse pathogenic mechanisms. Antioxid Redox Signal 16(9):974–1031. doi:10.1089/ars.2011.4082

von Bernhardi R, Ramirez G (2001) Microglia-astrocyte interaction in Alzheimer's disease: friends or foes for the nervous system? Biol Res 34(2):123–128. doi:10.4067/S0716-97602001000200017

von Bernhardi R, Tichauer J, Eugenín J (2010) Aging-dependent changes of microglial cells and their relevance for neurodegenerative disorders. J Neurochem 112(5):1099–1114. doi:10.1111/j.1471-4159.2009.06537.x

von Bernhardi R, Tichauer J, Eugenin-von Bernhardi L (2011) Proliferating culture of aged microglia for the study of neurodegenerative diseases. J Neurosci Methods 202(1):65–69. doi:10.1016/j.jneumeth.2011.08.027

von Bernhardi R, Cornejo F, Parada G, Eugenín J (2015a) Role of TGFβ signaling in the pathogenesis of Alzheimer's disease. Front Cell Neurosci 9:426. doi:10.3389/fncel.2015.00426

von Bernhardi R, Eugenin-von Bernhardi L, Eugenin J (2015b) Microglial cell dysregulation in brain aging and neurodegeneration. Front Aging Neurosci 7:124. doi:10.3389/fnagi.2015.00124

Wang C, Tsai S, Yew T, Kwan Y, Ngai S (2010) Identification of histone methylation multiplicities patterns in the brain of senescence-accelerated prone mouse 8. Biogerontology 11 (1):87–102. doi:10.1007/s10522-009-9231-5

Werry E, Enjeti S, Halliday G, Sachdev P, Double K (2010) Effect of age on proliferation-regulating factors in human adult neurogenic regions. J Neurochem 115 (4):956–964. doi:10.1111/j.1471-4159.2010.06992.x

Wu M, Hein A, Moravan M, Shaftel S, Olschowka J, O'Banion M (2012) Adult murine hippocampal neurogenesis is inhibited by sustained IL-1β and not rescued by voluntary running. Brain Behav Immun 26(2):292–300. doi:10.1016/j.bbi.2011.09.012

Wynne A, Henry C, Huang Y, Cleland A, Godbout J (2010) Protracted downregulation of CX3CR1 on microglia of aged mice after lipopolysaccharide challenge. Brain Behav Immun 24(7):1190–1201. doi:10.1016/j.bbi.2010.05.011

Wyss-Coray T, Lin C, Yan F, Yu G, Rohde M, McConlogue L, Masliah E, Mucke L (2001) TGF-β1 promotes microglial amyloid-β clearance and reduces plaque burden in transgenic mice. Nat Med 7(5):612–618. doi:10.1038/87945

Xiang Z, Haroutunian V, Ho L, Purohit D, Pasinetti G (2006) Microglia activation in the brain as inflammatory biomarker of Alzheimer's disease neuropathology and clinical dementia. Dis Markers 22(1–2):95–102. doi:10.1155/2006/276239

Yan J, Zhang H, Yin Y, Li J, Tang Y, Purkayastha S, Li L, Cai D (2014) Obesity- and aging-induced excess of central transforming growth factor-β potentiates diabetic development via an RNA stress response. Nat Med 20(9):1001–1008. doi:10.1038/nm.3616

Yankner B, Lu T, Loerch P (2008) The aging brain. Annu Rev Pathol 3:41–66. doi:10.1146/annurev.pathmechdis.2.010506.092044

Zhang W, Wang T, Pei Z, Miller D, Wu X, Block M, Wilson B, Zhang W, Zhou Y, Hong J-S, Zhang J (2005) Aggregated alpha-synuclein activates microglia: a process leading to disease progression in Parkinson's disease. FASEB J 19(6):533–542. doi:10.1096/fj.04-2751com

Zhu Y, Carvey P, Ling Z (2006) Age-related changes in glutathione and glutathione-related enzymes in rat brain. Brain Res 1090(1):35–44. doi:10.1016/j.brainres.2006.03.063

Zuendorf G, Kerrouche N, Herholz K, Baron J-C (2003) Efficient principal component analysis for multivariate 3D voxel-based mapping of brain functional imaging data sets as applied to FDG-PET and normal aging. Hum Brain Mapp 18(1):13–21. doi:10.1002/hbm.10069

Zunszain P, Anacker C, Cattaneo A, Choudhury S, Musaelyan K, Myint A, Thuret S, Price J, Pariante C (2012) Interleukin-1β: a new regulator of the kynurenine pathway affecting human hippocampal neurogenesis. Neuropsychopharmacology 37(4):939–949. doi:10.1038/npp.2011.277

# Astrocyte Dysfunction in Developmental Neurometabolic Diseases

**Silvia Olivera-Bravo, Eugenia Isasi, Anabel Fernández, Gabriela Casanova, Juan Carlos Rosillo and Luigi Barbeito**

**Abstract** Astrocytes play crucial roles in maintaining brain homeostasis and in orchestrating neural development, all through tightly coordinated steps that cooperate to maintain the balance needed for normal development. Here, we review the alterations in astrocyte functions that contribute to a variety of developmental neurometabolic disorders and provide additional data on the predominant role of astrocyte dysfunction in the neurometabolic neurodegenerative disease glutaric acidemia type I. Finally, we describe some of the therapeutical approaches directed to neurometabolic diseases and discuss if astrocytes can be possible therapeutic targets for treating these disorders.

**Keywords** Astrocytes · Astrocyte dysfunction · Neurodevelopmental diseases · Neurodegeneration · Myelin damage

S. Olivera-Bravo (✉) · E. Isasi
Cellular and Molecular Neurobiology, Instituto de Investigaciones Biológicas Clemente Estable (IIBCE), Montevideo, Uruguay
e-mail: solivera@iibce.edu.uy

E. Isasi
e-mail: eisasi@iibce.edu.uy

A. Fernández
Neuroscience Division, IIBCE and Comparative Neuroanatomy-Associated Unit to the School of Sciences, UDELAR, Montevideo, Uruguay
e-mail: afernandez@iibce.edu.uy

G. Casanova
UMET, School of Sciences, UDELAR, Montevideo, Uruguay
e-mail: casanova@fcien.edu.uy

J.C. Rosillo
Neuroscience Division IIBCE, UDELAR, Montevideo, Uruguay
e-mail: jrosillo@iibce.edu.uy

L. Barbeito
Neurodegeneration Laboratory, Institut Pasteur Montevideo, Montevideo, Uruguay
e-mail: barbeito@pasteur.edu.uy

 
R. von Bernhardi (ed.), *Glial Cells in Health and Disease of the CNS*,
Advances in Experimental Medicine and Biology 949,
DOI 10.1007/978-3-319-40764-7_11

**Abbreviations**

| | |
|---|---|
| ALS | Amyotrophic Lateral Sclerosis |
| AQP4 | Aquaporin 4 |
| BBB | Blood–Brain Barrier |
| GA | Glutaric Acid |
| GA-I | Glutaric Acidemia Type I |
| GCDH | Glutaryl CoA Dehydrogenase |
| GDNF | Glial-Derived Neurotrophic Factor |
| GFAP | Glial Fibrillary Acidic Protein |
| Icv | Intracerebroventricular |
| IEM | Inborn Errors of Metabolism |
| MAPK | Mitogen-Activated Protein Kinases |
| MLC | Megalencephalic Leukoencephalopathy with Subcortical Cysts |
| NPC | Niemann-Pick type C Disease |
| NVU | Neurovascular Unit |
| PC | Pyruvate Carboxylase |
| VWM | Vanishing White Matter |

## Introduction to Astrocyte Functions

As previously discussed in Chapters "Glial Cells and Integrity of the Nervous System," "Pharmacological Tools to Study the Role of Astrocytes in Neural Network Functions," and "Role of Astrocytes in Central Respiratory Chemoreception," astrocytes are indispensable for neuronal survival and activity. They play many key regulatory functions in brain homeostasis, including glutamate uptake that is vital for maintaining a physiological balance and protecting against neurotoxicity (Rothstein et al. 1996). Astrocytes are the only cells in the CNS that store and process glycogen. They thus contribute to influx of glucose and energy intermediates into neurons, and are also a significant source of neurotrophins and antioxidant defenses (Maragakis and Rothstein 2006). Moreover, astrocytes are the nearly unique neural cells that express glutamine synthase, the only brain source of endogenous glutamine, which is the major amino moiety donor for glucose precursors and nitrogenated compounds and the preferred metabolite for inter-organ transport and temporary storage of nitrogen (Haberle et al. 2005, 2011). Astrocytes also participate in blood–brain barrier formation and maintenance (Abbott et al. 2006) and have a unique role in the clearance of brain solutes along veins (Iliff et al. 2013). Mediated by a gap junction-coupled network that allows the direct cytoplasmic passage of ions and small molecules through significant distances, astrocytes communicate with each other and modulate the activity of adjacent cells (Maragakis and Rothstein 2006; Verkhratsky et al. 2012, 2015). In addition, astrocytes have receptors to a wide range of neurotransmitters, peptides, hormones, and cytokines that regulate their own functions and also influence neurons, oligodendrocytes, and microglia (Maragakis and Rothstein 2006; Verkhratsky et al. 2012, 2015). Although astrocytes are not electrically excitable, they are capable of detecting and modulating neuronal activity by releasing neuroactive substances (glutamate, ATP, D-serine) that can feed back onto presynaptic terminals or stimulate postsynaptic neurons (Araque 2008; Verkhratsky et al. 2012; Pekny and Pekna 2014). Moreover, as each astrocyte domain covers up to $2 \times 10^6$ synapses, and human astrocytes enhanced learning and memory skills in transplanted rodents (Han et al. 2013), important roles in integrating and processing complex cognitive have been proposed for astrocytes (Araque 2008; Han et al. 2013).

Finally, in damaging conditions, astrocytes can suffer early alterations that elicit downstream mechanisms that have the potential to irreversibly damage neurons and oligodendrocytes. Among the most important astrocyte alterations are included the decreased expression of glutamate transporters and subsequent delayed clearance of synaptic glutamate, impaired glutamate-glutamine cycle, depletion of glutamine and neurotransmitter precursors, reductions in glutathione levels as well as induced expression of nitric oxide synthase and pro-inflammatory cytokines (Maragakis and Rothstein 2006; De Keyser et al. 2008; Verkhratsky et al. 2012, 2015; Pekny and Pekna 2014).

## Astrocyte Dysfunction and Development: A Focus on Inborn Errors of Metabolism

Astrocytes play crucial roles during development by actively contributing to the finely coordinated steps required to establish functional neural networks (Clarke and Barres 2013; Chung et al. 2013; Sloan and Barres 2014). Moreover, normal synaptic development and pruning requires this tightly regulated timing and communication between astrocytes and neurons (Helmuth 2001; Eroglu et al. 2009; Dodla et al. 2010). In turn, astrogenesis depends on the signaling of newly formed neurons (Barnabe-Heider et al. 2005). In this context of mutual influence it has been proposed that alterations of astrocyte differentiation or function may irreversibly unbalance the coordinated developmental events, thereby producing neurodevelopmental disorders. Moreover, as most glial cells arise from clonal divisions of early differentiated astrocytes (Ge et al. 2012), if damage occurs early during astrogenesis or in the first astrocytes, it might be sustained and amplified by the clonal descendants, thus perpetuating astrocyte dysfunction (Sloan and Barres 2014) which can further account for neurodevelopmental diseases. In this regard, we will review as astrocyte dysfunction affects neurons and brain development in some inborn errors of metabolism (IEM) with a further focus on glutaric acidemia I, an IEM belonging to the group of neurometabolic diseases because of the very predominant CNS damage over the rest of the body.

## Astrocyte Dysfunction-Associated IEMs: Impact on Neuron Survival

The term inborn errors of metabolism (IEMs) encompasses an enormous group of rare disorders (more than 750) that in common affect about 1:2000 newborns. IEMs are produced by genetic mutations that alter biochemical pathways producing an excess of a damaging substance or deficiency of a normal biochemical compound, both causing toxic effects in the brain and/or the rest of the body (Scriver et al. 1995). Brain and neurons are significantly affected either by a direct toxicity or indirectly by a primary glial cell dysfunction. In the IEMs in which biochemical pathways are primarily altered in astrocytes instead of neurons, astrocyte damage elicit downstream mechanisms such as glutamate receptor-mediated excitotoxicity, lactic acidosis, energetic deprivation, oxidative/nitrosative stress, or neuroinflammation that acting alone or together have the potential to kill neurons (Maragakis and Rothstein 2006; De Keyser et al. 2008; Verkhratsky et al. 2012, 2015; Pekny and Pekna 2014). Furthermore, the imbalance of critical and exclusive astrocytic enzymatic routes such as the glutamine synthase may trigger different neurometabolic diseases associated to either excess or lack of the enzymatic product. In this regard, hepatic encephalopathy (HE) is an IEM characterized by mild to very severe neuropsychiatric manifestations that are attributed to glutamine overload produced by the activity of astrocytic glutamine synthase in response to the brain increased uptake of ammonia produced by liver failure (Butterworth 2010). Excessive brain glutamine causes significant astrocyte swelling that in turn triggers a complex signaling cascade which relies on NMDA receptor activation, elevation of intracellular calcium, and glutamate exocytosis, all together resulting in a self-amplifying signaling loop that causes excitotoxicity and increased oxidative stress that trigger neuron death for one side and sustain astrocyte activation for the other (De Keyser et al. 2008; Butterworth 2010).

Conversely to what is observed in HE, the loss of function of glutamine synthase produces the congenital glutamine synthase deficiency, a very rare IEM that usually causes neonatal death associated with severe brain malformations, including abnormal gyration and white matter lesions. Glutamine starvation impairs both the synthesis of nitrogenated compounds and the nitrogen storage and transport among organs. It is believed that the combination of all of these distorted events produces a systemic collapse which is the main responsible for unviable brain malformations (Haberle et al. 2005, 2011).

Lack of pyruvate carboxylase activity (PC), another metabolic pathway confined to astrocytes, causes the rare autosomal recessive IEM known as PC deficiency, whose neonatal form is characterized by multi-organ metabolic imbalance, lactic acidemia, and significant neurological dysfunction. PC catalyzes the conversion of pyruvate into oxaloacetate which is crucial for the replenishment of citric acid cycle intermediates, gluconeogenesis, synthesis of glycogen, maintenance of the antioxidant glutathione system, and anaplerotic support to neurons (Robinson et al. 1984; Garcia-Cazorla et al. 2006). As the main pathological findings correlate with impaired astrocytic anaplerosis, it has been proposed that loss of PC leads to generalized hyperammonemia and death in the first few months of life. Likely, this happens because of the impaired astrocyte ammonia buffering and support to microvascular morphogenesis and myelination (Robinson et al. 1984; García-Cazorla et al. 2006). Moreover, as PC is involved in the synthesis of myelin lipids (De Keyser et al. 2008), an altered astrocyte–oligodendrocyte communication may underlie the paucity of myelin and white matter lesions observed in patients.

Niemann-Pick type C disease (NPC) is an autosomal recessive lipid-storage IEM characterized by progressive neurodegeneration, hepatosplenomegaly, and general organ dysfunction. It may be caused by mutations in the NPC-1 gene which in brain is predominantly present in the astrocytic processes closely associated to the nerve terminals (Patel et al. 1999). In NPC animal models, astrocytes become activated, and show decreased gap junctional communication and increased hemichannel activity that might predispose the surrounding neurons to death by either neuroinflammation and/or reduced astrocyte mediated-spatial buffering (Saez et al. 2013). In accordance with a crucial astrocyte role in the disease, rescuing NPC1 expression in astrocytes delays neuronal loss and prolongs life span in $NPC1^{-/-}$ mice; and simultaneous NPC1 recovery in neurons and astrocytes exhibited better protection than the sole recovery in neurons (Borbon et al. 2012).

Aceruloplasminemia is another IEM linked to primary astrocyte dysfunction with neuron death as a secondary phenomenon. Its neuropathological hallmarks include excessive iron deposition, astrocyte perivascular endfeet deformation, and progressive neuronal loss (Oide et al. 2006; Miyajima 2015). All defects are attributed to the lack of astrocytic ceruloplasmin ferroxidase activity which mediates the ferrous ion oxidation needed for transferrin-dependent iron efflux. Impaired ferroxidase activity leads to prominent redox-active iron accumulation in astrocytes that causes significant endfeet deformation, lipid peroxidation, and hydroxyl radical formation as well as impaired transferrin/iron shuttle to neurons. Thus, neurons will die either by increased oxidative stress, decreased astrocyte protection, and/or iron starvation (Oide et al. 2006; De Keyser et al. 2008; Miyajima 2015).

## Astrocyte Dysfunctions and Leukodystrophies: Effects on Myelin Formation

The term leukodystrophy comprises all of the diseases genetically determined as white matter disorders characterized by abnormal myelin formation (Hagemann et al. 2009; van der Knaap et al. 2006). Astrocyte disturbed functions may also be the leading cause of several leukodystrophies. Alexander disease, a paradigmatic leukodystrophy, clearly manifests as a primary astrocyte dysfunction that compromises myelin development and integrity. It is considered a primary astrogliopathy with myelin alterations as secondary damage (Messing et al. 2012). The only pathological causes known in Alexander disease are the several mutations in the glial acidic fibrillary protein (GFAP) gene that lead to a toxic gain of function. Astrocytes but no other cells show the defining features of the disease that include hypertrophic cell bodies with GFAP-positive processes and characteristic intracytoplasmic multiprotein aggregates named Rosenthal fibers. As myelin failure and oligodendrocyte loss correlate with increased presence of astrocytic Rosenthal fibers, it is proposed that an altered pattern of the astrocyte signals that promote myelination is the main pathological underlying mechanism (De Keyser et al. 2008; Hagemann et al. 2009; Liem and Messing 2009; Messing et al. 2012) (see Chapter "Oligodendrocytes: Functioning in a Delicate Balance Between High Metabolic Requirements and Oxidative Damage" for further reading on oligodendrocytes and myelination).

Astrocytes also play a central role in the pathogenesis of megalencephalic leukoencephalopathy with subcortical cysts (MLC), which commonly affects infants, lead to progressive spasticity and ataxia, and exhibit myelin splitting and intra-myelin vacuole formation as characteristic features (De Keyser et al. 2008; Ridder et al. 2011). The disease is directly linked to mutations in the MLC1 gene that is mainly expressed in astrocyte perivascular endfeet (Boor et al. 2005); and defective astrocyte MLC1 is enough to elicit myelin damage (Duarri et al. 2011). These findings lead to the proposition that disturbances in astrocytic MLC1 may irreversibly affect myelin by impeding the transport of essential molecules for oligodendrocytes or myelin formation (Duarri et al. 2011). Another paradigmatic leukodystrophy associated with astrocyte dysfunction is the vanishing white matter (VWM) disease that is caused by alterations in the genes that encode for the subunits of the eukaryotic initiation factor eIF2B that is expressed in astrocytes and oligodendrocytes (van der Knaap et al. 2006; Bugiani et al. 2011). VWM is predominantly manifested in early childhood, and usually presents with a rapid clinical decline triggered by stress-related events that may lead to death (van der Knaap et al. 2006). Brain patient samples from patients show astrocytes with an increased proliferation, immature phenotype, abnormal composition of intermediate filament network, and metabolic stress (Bugiani et al. 2011). Loss of white matter may be at least partially explained by excessive astrocyte proliferation at the expense of oligodendrocytes that may occur due to aberrant specification of glial progenitors, as has been shown in astrocytosis linked to myelin alterations found in children born prematurely or surviving cerebral ischemia (Bain et al. 2010).

## A Novel Mechanism: Astrocyte Dysfunction Sustaining the Pathogenesis of the Neurometabolic IEM GA-I

In spite of abundant evidence directly linking abnormal astrocyte pathways to neurodevelopmental disorders, the participation of astrocyte dysfunction in complex multifactorial IEMs that selectively affect the brain is much less known. Among these particular IEMs, usually named neurometabolic diseases because the almost exclusive CNS damage over the rest of the body (Scriver et al. 1995), organic acidemias and acidurias comprise a predominant group that includes all abnormal amino acid catabolism pathways. Although most acidemias are characterized by an acute life-threatening stage, some of them present a significant and progressive neurodegenerative component that cannot be directly linked to the accumulation or lack of the metabolites transiently produced by the altered pathways. Moreover, as has been shown for glutaric acidemia type I (GA-I), that is, among the most prevalent organic acidemias/acidurias, the accumulated metabolites may not be directly toxic to neurons (Freudenberg et al. 2004; Olivera-Bravo et al. 2011; Jafari et al. 2013), so the underlying mechanisms remain mostly unknown.

GA-I is characterized by extensive acute striatal and progressive cortical neurodegeneration together with white matter and vascular abnormalities, causing permanent motor and cognitive sequels. The disease is caused by a loss of function of glutaryl-CoA dehydrogenase (GCDH), a mitochondrial enzyme involved in the catabolism of lysine, hydroxylysine, and tryptophan. It results in accumulated millimolar concentrations of glutaric acid (GA) and related metabolites in fluids and tissues (Goodman et al. 1977; Strauss et al. 2003; Funk et al. 2005). In spite of having a strong and progressive neurodegenerative profile, even high concentrations of GA-I metabolites did not directly account for neuron death (Freudenberg et al. 2004); and the $GCDH^{-/-}$ mice characterized by permanent high levels of GA-I metabolites do not suffer any neurological deficit (Koeller et al. 2002, 2004). Therefore, based on previous data that showed an altered glutamate uptake in astrocytes exposed to GA (Magni et al. 2009), we propose that instead of neurons, astrocytes are the earliest cell targets of GA-I accumulated metabolites. We have found that astrocytes, but not neurons or oligodendrocytes, respond to pathophysiological concentrations of GA by increasing their proliferation that is mediated by activation of the MEK/ERK pathway, augmenting oxidative stress, and significantly decreasing both their mitochondrial potential and glutathione production (Olivera-Bravo et al. 2008, 2011). Astrocyte increased proliferation was also found when newborn rat pups were injected intracerebroventricularly (icv) with 5 mM GA. Remarkably, astrocytes pretreated with GA were toxic to co-cultured neurons even many days after removing the acid, and GA-injected animals presented delayed striatal neuronal death that became significant only many days after the peak of astrocytosis (Olivera-Bravo et al. 2011). Astrocyte response to GA also preceded delayed oligodendrocyte differentiation, which resulted in a significant and progressive demyelination. Interestingly, striatal axon bundles from GA-injected animals exhibited not only significantly low levels of myelin, but also significant decreases in expression of myelin-associated glycoprotein and myelin-binding protein, suggesting significant white matter damage (Olivera-Bravo et al. 2014). Transmission electron microscopy of the striatum of GA-injected animals showed oligodendrocytes with significant endoplasmic reticulum stress, altered myelination, and thinner axons (Olivera-Bravo et al. 2014), all resembling the progressive myelin alterations reported in affected patients (Bähr et al. 2002; Funk et al. 2005). At the ages analyzed, GA-induced myelination failure was restricted to the basal ganglia, the most vulnerable brain area in GA-I patients

**Fig. 1** Striatal damage elicited upon astrocyte response to GA. **a** Sustained astrocytosis and axonal bundle shrinkage. Newborn rat pups received an icv injection of 1 μg/g GA or vehicle and were sacrificed 30 days postinjection (DPI). At that time, GA-injected rats showed increased number of S100β astrocytes, which present enlarged bodies and stronger processes (*white arrows*) together with reduced phosphorylated neurofilaments (pNF) axonal bundle areas. *Inset* shows the region analyzed at all conditions and ages. **b** Delayed neuronal death. Pictures show an increased labeling of the marker of degenerating neurons, Fluoro-Jade C (FJC), in GA-injected rats. The *inset* shows absence of significant signal in 21 DPI controls when neuron death became significant in GA-injected animals. **c** Altered neurovascular unit (NVU) in GA-injected animals. Platelet-derived growth factor β-receptor (PDGFRβ) labeled positive pericyte bodies (*) and processes contacting blood vessels (*white arrows*) in 30 DPI vehicle-injected animals. In age-matched GA-injected rats, PDGFRβ immunoreactivity was confined to blood vessels and there appeared abundant Evans Blue positive cells, indicating abnormal permeation through the blood–brain barrier. **d** Quantitation of alterations observed in GA-injected animals. Charts show a significant and early astrocytosis that precedes increasing neuron damage and decreasing expression of NVU components such as pericytes and astrocyte endfeet that were recognized with PDGFRβ and aquaporin 4 (AQP4), respectively. All data are related to corresponding age-matched controls. Calibration: 40 μm in (**a**), 100 μm in (**b**), 20 μm in (**c**). *Asterisks* indicate $p < 0.05$

(Goodman et al. 1977; Strauss et al. 2003, 2007; Funk et al. 2005). On the other hand, GA did not induce acute damage to oligodendrocyte precursors in neonatal pups suggesting the delayed oligodendrocyte cytopathology found several days after a single exposure to GA was produced by an indirect neurotoxic mechanism instead of a direct action. Recently, we have found that the perinatal icv administration of GA caused significant long-lasting alterations in the permeability of the maturing brain–blood barrier (BBB) to substrates of low molecular weight, such as Evans Blue and a significant reduction in the expression of markers of critical components of the neurovascular unit (NVU), including astrocyte podocytes, pericytes, and basal lamina. These findings were statistically significant in the striatum, and a similar trend was described in parietal cortex (Isasi et al. 2014) (Fig. 1), confirming the striatal vulnerability to GA-I metabolites. In summary, our data indicate that a transient increase in GA levels in the CNS triggers an early but long-lasting astrocyte dysfunction that leads to delayed myelination, altered BBB permeability, and NVU failure together with significant loss of striatal neurons. The single exposure to GA not only caused an acute astrocyte response but also a sustained altered phenotype that seems to elicit the sequential events that account for most of the GA-I neurological features (Fig. 2). Thus, GA-induced neurotoxicity in our hands appears to require a complex cellular interplay that, after an initial trigger elicited by toxic levels of GA, can perpetuate itself in ensuing weeks, leading to progressive damage that resembles the alterations seen in patients independently on the occurrence of encephalopathic crises (Bähr et al. 2002; Funk et al. 2005). On the other hand, whereas the genetic model of GA-I expected to reproduce the human disease fails to develop neuronal loss (Koeller et al. 2002, 2004), even after overload with high amounts of lysine to boost endogenous GA production (Seminotti et al. 2012), our icv GA injection model moderately reproduces the encephalopathic crisis, striatal neuronal loss, myelination, and BBB defects reported in patients.

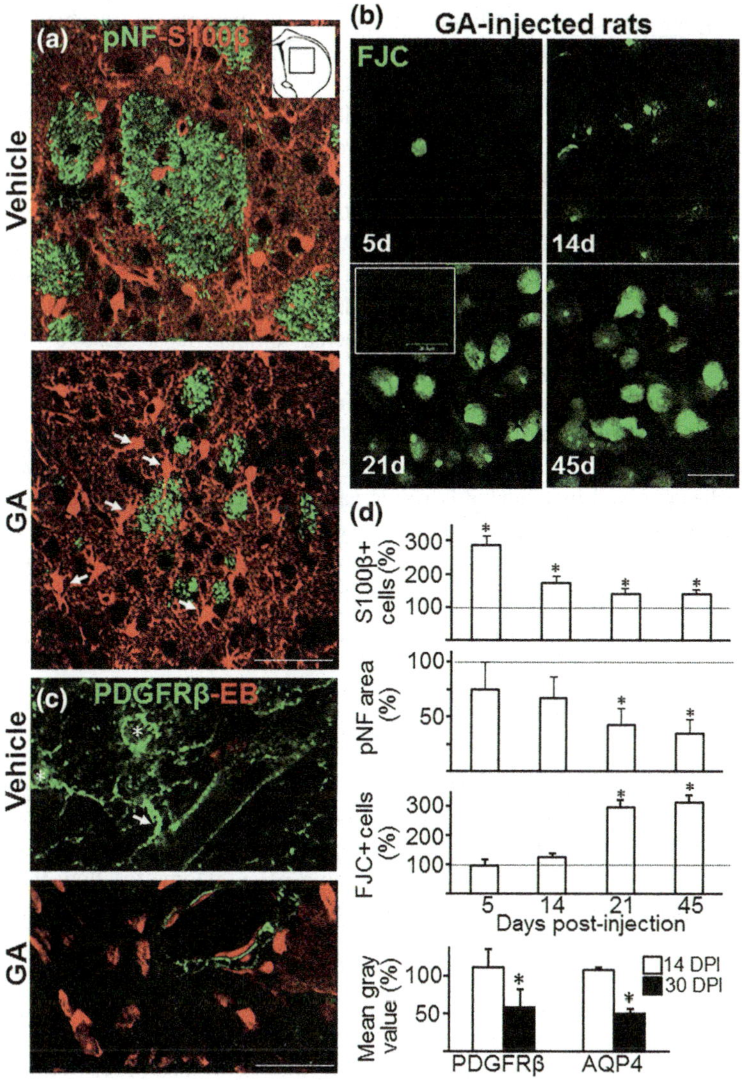


Our evidence also supports a crucial role of astrocytes in GA-I. Astrocytes are preferentially vulnerable because they can actively take up GA (Magni et al. 2009), which leads to mitochondrial dysfunction, increased proliferation, and oxidative stress, all causing a poorly differentiated phenotype with high S100β and low GFAP expression (Olivera-Bravo et al. 2008, 2011). Such inability of astrocytes to fully differentiate may critically compromise their supportive functions, as proposed in Fig. 2. Accordingly, astrocytes from GCDH$^{-/-}$ mice, which suffer permanently elevated levels of GA-I metabolites, cannot offer normal anaplerotic support to neurons (Lamp et al. 2011), and hyperammonemia caused by selective vulnerability of astrocytes to GA-I metabolites is a leading neurotoxic mechanism in a three-dimensional embryo cultures that also contain neurons and oligodendrocytes (Jafari et al. 2013).

On the other hand, as multiple trophic factors produced by astrocytes are required for the survival of developing oligodendrocytes (Barres et al. 1993; Nash et al. 2011), even small changes in astrocytic function may be detrimental to oligodendrocyte survival and lead to defective myelination, as has been shown in Alexander disease (Liem and Messing 2009; Messing et al. 2012). Regarding BBB and NVU failures, it has been demonstrated that astrocytes are key players in BBB maturation and stabilization (Abbott et al. 2006). The progressive decrease of AQP4 close to blood vessels in GA-injected animals is suggestive of altered astrocyte differentiation and polarization at the NVU and probably of a deficiency of AQP4 in the neurovascular coupling. Moreover, defective laminin and pericytes, both influencing AQP4 polarization, indicate altered communication among NVU components long after GA injection (Isasi et al. 2014). In summary, the "toxic" process has evolved independently of increased GA levels but is tightly dependent on astrocyte dysfunction. Although these events should be studied in GA-I patients, astrocytes must be taken into account not only as a main player in GA-I pathogenesis but also when exploring therapies for this up to now incurable disease.

## Current Therapeutic Approaches For IEMs

Early diagnosis and systemic treatments for IEMs and neurometabolic disorders in particular have in many cases improved quality of life and increased life expectancy. Perinatal or gestational diagnosis pursues disease detection before the occurrence of precipitating crises or worsening of symptoms to allow the preservation of the systemic equilibrium and neurological performance. Disease management at this stage is directed to obtain some amounts of lacking products or avoid the accumulation of toxic compounds that becomes the organism prone to suffer encephalopathic crises (Ruiz Pons et al. 2007). Unfortunately, perinatal screening is available for very few IEMs and most patients are often diagnosed when some symptoms become evident or after the occurrence of crises that usually cause an acute life-threatening stage where the CNS is exposed to huge amounts of toxic compounds. At this time, emergency management looks for the restoration of general homeostasis including the rescue of glucose, oxygen, and pH levels. For example, glucose supplementation provided adequate preferred energetic substrates, and reduced both the demand for alternate substrates and the amino acid turnover in the liver all favoring the return of systemic parameters close to normal values (Zinnanti et al. 2007). The acute life-threatening stage produced during encephalopathic crises is followed by a sub-acute progressive neurodegenerative disorder that currently lacks effective medical treatment. In this period, neurological symptoms become more pronounced and worsening is attributed to the progressive destruction of motor, mental, and/or perceptual abilities, including loss of function

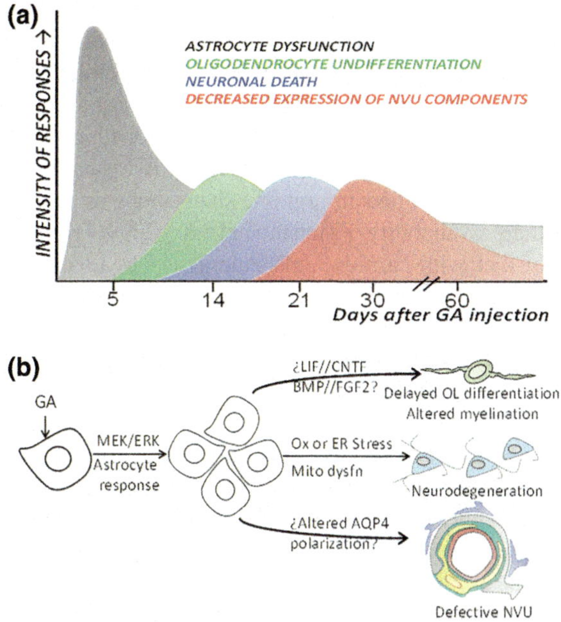


**Fig. 2** Temporal course of astrocyte response to GA and a gliocentric hypothesis for GA-I. **a** The *scheme* shows the early and long-lasting astrocyte response to GA, which was followed by delayed oligodendrocyte differentiation, which in turn preceded significant neuronal death. Significant decreases in the expression of the NVU components PDGFRβ, AQP4, and laminin close to blood vessels occurred later. **b** Proposed pathological events that occur after perinatal exposure to GA and are mediated by the consequent astrocyte dysfunction, rather than direct effects on oligodendrocytes, neurons, and NVU components. We hypothesized that astrocyte dysfunction may trigger (i) delayed oligodendrocyte (OL) differentiation and altered myelination, probably by a yet unknown imbalance of the astrocyte signals that influence myelination; (ii) astrocyte oxidative stress and mitochondrial dysfunction may account for the progressive neuron death observed; (iii) altered AQP4 polarization or imbalanced astrocyte–endothelial and/or astrocyte–pericyte communication may explain defective NVU composition and increased permeability of blood–brain barrier observed in GA-injected animals. NVU colors: *red* (endothelial cells), *cyan* (basal lamina), *yellow* (pericytes), *gray* (astrocyte endfeet), *blue* (neuronal processes). Abbreviations: *BMP* bone morphogenetic protein; *CNTF* ciliary neurotrophic factor; *ERK* extracellular signal-regulated kinases; *FGF2* fibroblast growth factor 2; *LIF* leukemia inhibitor factor; *MEK* kinase of ERK; Ox/ER stress, oxidative and endoplasmic reticulum stress

and seizures that commonly leads to early death. At this stage, all of the actions are directed to impede the amplification of damage and the recurrence of crises and catabolic states that will produce major sequels. Dietary treatment is the most important tool to avoid recurrence and worsening. It includes all of the manipulations needed to avoid the accumulation of toxic metabolites either by potentiating or inhibiting remaining enzyme action, using alternate substrates, inhibiting its precursors, or administering cofactors that favor enzyme activation such as cobalamin, thiamine, biotin, and riboflavin (Ruiz Pons et al. 2007). In some diseases such as hepatic encephalopathy, the dietary management aimed at reducing nitrogenous load is accompanied with non-absorbed agents (rifamixin) or enhancement of ammonia removal, all attempting to reduce the levels of blood ammonia and other gut-derived toxins (Butterworth 2010). In GA-I if dietary management starts in the newborn period—before the occurrence of encephalopathic crises—most of the patients remain asymptomatic (Kölker et al. 2012). GA-I usual diets are lysine-free, tryptophan-reduced, and fortified in arginine that competes with lysine to reduce enzymatic substrate availability (Zinnanti et al. 2007; Kölker et al. 2012). A combination of diet with carnitine and emergency treatment has been demonstrated to be effective in preventing neurological disease when the GA-I patient is thought to be at risk (Kölker et al. 2012).

In addition to diet management, as oxidative stress may contribute to the pathophysiology of many IEMs, antioxidant treatment is a common complementary therapeutic approach. It is directed to reduce the brain vulnerability against the attack of oxygen and nitrogen reactive species generated by abnormal metabolites or subsequent mitochondrial dysfunction. In this regard, antioxidants such as ascorbic acid, N-acetylcysteine, folic acid, lipoic acid, α-tocopherol, and L-carnitine prevented learning/memory deficits and convulsions in animal models of organic acidemias (Ribas et al. 2014) or hepatic encephalopathy (Butterworth 2010). In aceruloplasminemia, the direct attack of iron-mediated lipid peroxidation and oxidative stress with a systemic iron chelation alone or together with oral zinc sulfate administration ameliorated the neurological symptoms (Miyajima 2015).

Symptomatic therapies are usually employed in neurometabolic diseases. For example, antiepileptic drugs sometimes are used to keep patients free of seizures for prolonged periods. Tricyclic antidepressants or CNS stimulants have been shown to ameliorate cataplexy, whereas dystonia and tremor are usually treated with anticholinergic drugs or gamma-aminobutyric acid derivatives when dystonia is advanced (Ruiz Pons et al. 2007).

In summary, in spite of some advances in treatment and development of special diets, IEMs remain as yet incurable diseases with scarce disease-specific therapies and only general approaches that try to preserve overall systems avoiding recurrence of encephalopathic crises.

## Could Astrocytes Be Therapeutic Targets in IEMs and Developmental Neurometabolic Diseases in Particular?

The fact that astrocytes accomplish both adaptive and pathological functions makes targeting them difficult. The astrogliosis observed in most neurodevelopmental or neurological diseases is a good example of this problem, whereas glutamate uptake is beneficial because of its antioxidant effects and associated release of neuroprotective agents; reactive astrocytes release inflammatory cytokines, produce reactive oxygen species, and impede axonal growth (Maragakis and Rothstein 2006; Verkhratsky et al. 2012, 2015; Pekny and Pekna 2014). Another complication in targeting astrocytes is the fact that many of the molecules that astrocyte secrete may be detrimental during some phases of damage, yet advantageous during others. For example, vascular endothelial growth factor and matrix metalloproteinase inhibitors acutely increase edema and stroke volume, but many days later both are critical for angiogenesis and recovery (Zhao and Rempe 2010). In spite of these problems, some neuroprotective attempts have been focused on astrocytes. It has been shown that astrocyte-derived neuroprotective factors, such as glial-derived neurotrophic factor (GDNF), reduce neuronal death in animal models of Parkinson Disease, ALS, and stroke (Rappold and Tieu 2010; Vargas and Johnson 2010; Zhao and Rempe 2010). However, side effects and conflicting results in human trials have diminished the enthusiasm for GDNF as a potential treatment (Rappold and Tieu 2010). Similarly, erythropoietin, a neuroprotective molecule released by astrocytes that reduces stroke volume in animal models showed promising results in phase 2, but in phase 3 of clinical trials did not show efficacy (Ehrenreich et al. 2009). The potentiation of astrocyte glutamate uptake by increasing the expression of glutamate transporters has also been explored with promising effects in ALS animal models (Vargas and Johnson 2010), but results in humans are still lacking. Arundic acid reduces the production of S-100β, a calcium-binding protein proposed as indicator of CNS damage, in activated cultured astrocytes (Vargas and Johnson 2010) but was not effective for the treatment of stroke (Zhao and Rempe 2010).

Fewer approaches focused on astrocytes have been made in neurodevelopmental disorders, probably because we are far from having a complete knowledge of astrocyte roles in these diseases. In hepatic encephalopathy, targeting of astrocytic glutamine synthase is proposed as a potential treatment to avoid brain ammonia accumulation (Brusilow et al. 2010), but more results are needed. Compounds that had the ability to reduce GFAP expression alone or together with antioxidant effects or enhanced glutamate astrocytic uptake were employed to ameliorate the GFAP alterations found as the main pathological features of Alexander disease. Whether these findings obtained in cellular models can be easily translated into in vivo treatments, and eventually to humans, remains to be seen (Cho et al. 2010; Messing et al. 2012). Our experience in GA-I—that we propose as another primary astrogliopathy—indicates that abrogating the initial astrocyte response to GA with neuroprotective iron porphyrins (Wu et al. 2003) or MAPK inhibitors offers significant neuroprotection in animal models (Olivera-Bravo et al. 2008). Although we do not know the astrocyte pathways that are targeted by these compounds, the early astrocyte response to GA could offer a potential therapeutic window for a selective focus on astrocytes before the occurrence of significant myelin or neuronal damage. Furthermore, as most IEMs are diagnosed after the whole CNS is damaged, it makes unlikely that a pharmacological strategy focused on a single cell type can be successful.

## Concluding Remarks

The diversity and number of IEMs that have already been associated with astrocyte dysfunction are remarkable. However, more significant research is needed to understand what types of astrocyte dysfunction or how astrocytes themselves contribute to the pathogenesis of each of these conditions. During development, adequate neurogenesis and gliogenesis represent only the initial steps for nervous system formation. Furthermore, a close neuron–astrocyte correlation is needed to allow proper development and CNS function. Therefore, no therapeutic intervention or pharmacological manipulation of astrocytes can be assumed solely to affect astrocytes. Instead, each therapeutic approach should target astrocytes in concert with neurons, other glial cells, and the vasculature, because all together will sculpt the response to each strategy and finally determine its efficacy. Therefore, it is necessary to consider the complex cell interactions in the search for efficacious treatments for neurodevelopmental diseases, which represents an enormous challenge.

## References

Abbott NJ, Ronnback L, Hansson E (2006) Astrocyte-endothelial interactions at the blood-brain barrier. Nat Rev Neurosci 7:41–53

Araque A (2008) Astrocytes process synaptic information. Neuron Glia Biol 4:3–10

Bähr O, Mader I, Zschocke J et al (2002) Adult onset glutaric aciduria type I presenting with a leukoencephalopathy. Neurology 59(11):1802–1804

Bain JM, Ziegler A, Yang Z et al (2010) TGFß1 stimulates the over-production of white matter astrocytes from precursors of the "brain marrow" in a rodent model of neonatal encephalopathy. PLoS ONE 5(3):e9567

Barnabe-Heider F, Wasylnka JA, Fernandes KJL et al (2005) Evidence that embryonic neurons regulate the onset of cortical gliogenesis via cardiotrophin-1. Neuron 48:253–265

Barres BA, Schmid R, Sendnter M et al (1993) Multiple extracellular signals are required for long-term oligodendrocyte survival. Development 118(1):283–295

Boor PK, de Groot K, Waisfisz Q et al (2005) MLC1: a novel protein in distal astroglial processes. J Neuropathol Exp Neurol 64:412–419

Borbon I, Totenhagen J, Fiorenza MT et al (2012) Niemann-Pick C1 mice, a model of "juvenile Alzheimer's disease", with normal gene expression in neurons and fibrillary astrocytes show long term survival and delayed neurodegeneration. J Alzheimers Dis 30:875–887

Brusilow SW, Koehler RC, Traystman RJ et al (2010) Astrocyte glutamine synthetase: importance in hyperammonemic syndromes and potential target for therapy. Neurotherapeutics 7(4): 452–470

Bugiani M, Boor I, van Kollenburg B et al (2011) Defective glial maturation in vanishing white matter disease. J Neuropathol Exp Neurol 70:69–82

Butterworth RF (2010) Altered glial-neuronal crosstalk: cornerstone in the pathogenesis of hepatic encephalopathy. Neurochem Int 57:383–388

Cho W, Brenner M, Peters N et al (2010) Drug screening to identify suppressors of GFAP expression. Hum Mol Genet 19:3169–3178

Chung W-S, Clarke LE, Wang GX et al (2013) Astrocytes mediate synapse elimination through MEGF10 and MERTK pathways. Nature 504:394–400

Clarke LE, Barres BA (2013) Emerging roles of astrocytes in neural circuit development. Nat Rev Neurosci 14:311–321

De Keyser J, Mostert JP, Koch MW (2008) Dysfunctional astrocytes as key players in the pathogenesis of central nervous system disorders. J Neurol Sci 267:3–16

Dodla MC, Mumaw J, Stice SL (2010) Role of astrocytes, soluble factors, cells adhesion molecules and neurotrophins in functional synapse formation: implications for human embryonic stem cell derived neurons. Curr Stem Cell Res Ther 5:251–260

Duarri A, Lopez de Heredia M, Capdevila-Nortes X et al (2011) Knockdown of MLC1 in primary astrocytes causes cell vacuolation: a MLC disease cell model. Neurobiol Dis 43:228–238

Ehrenreich H, Weissenborn K, Prange H et al (2009) Recombinant human erythropoietin in the treatment of acute ischemic stroke. Stroke 40(12):e647–656

Eroglu C, Allen NJ, Susman MW et al (2009) Gabapentin receptor alpha 2 delta-1 is a neuronal thrombospondin receptor responsible for excitatory CNS synaptogenesis. Cell 139:380–392

Freudenberg F, Lukacs Z, Ullrich K (2004) 3-Hydroxyglutaric acid fails to affect the viability of primary neuronal rat cells. Neurobiol Dis 16(3):581–584

Funk CB, Prasad AN, Frosk P et al (2005) Neuropathological, biochemical and molecular findings in a glutaric acidemia type 1 cohort. Brain 128:711–722

Garcia-Cazorla A, Rabier D, Touati G et al (2006) Pyruvate carboxylase deficiency: metabolic characteristics and new neurological aspects. Ann Neurol 59:121–127

Ge W-P, Miyawaki A, Gage FH et al (2012) Local generation of glia is a major astrocyte source in postnatal cortex. Nature 484:376–380

Goodman SI, Norenberg MD, Shikes RH et al (1977) Glutaric aciduria: biochemical and morphologic considerations. J Pediatr 90:746–750

Haberle J, Gorg B, Rutsch F et al (2005) Congenital glutamine deficiency with glutamine synthetase mutations. N Engl J Med 353:1926–1933

Haberle J, Shahbeck N, Ibrahim K et al (2011) Natural course of glutamine synthetase deficiency in a 3 year old patient. Mol Genet Metab 103(1):89–91

Hagemann TL, Boelens WC, Wawrousek EF et al (2009) Suppression of GFAP toxicity by αβ-crystallin in mouse models of Alexander disease. Hum Mol Genet 18(7):1190–1199

Han X, Chen M, Wang F et al (2013) Forebrain engraftment by human glial progenitor cells enhances synaptic plasticity and learning in adult mice. Cell Stem Cell 12:342–353

Helmuth L (2001) Glia tell neurons to build synapses. Science 291:569–570

Iliff JJ, Lee H, Yu M et al (2013) Brain-wide pathway for waste clearance captured by contrast-enhanced MRI. J Clin Investig 123:1299–1309

Isasi E, Barbeito L, Olivera-Bravo S (2014) Increased blood–brain barrier permeability and alterations in perivascular astrocytes and pericytes induced by intracisternal glutaric acid. Fluids Barriers CNS 11:15

Jafari P, Braissant O, Zavadakova P et al (2013) Ammonium accumulation and cell death in a rat 3D brain cell model of glutaric aciduria type I. PLoS ONE 8(1):e53735

Koeller DM, Woontner M, Crnic LS et al (2002) Biochemical, pathologic and behavioral analysis of a mouse model of glutaricacidemia type I. Hum Mol Genet 11:347–357

Koeller DM, Sauer S, Wajner M et al (2004) Animal models for glutaryl-CoA dehydrogenase deficiency. J Inherit Metab Dis 27:813–818

Kölker S, Boy SP, Heringer J et al (2012) Complementary dietary treatment using lysine-free, arginine-fortified amino acid supplements in glutaric aciduria type I—a decade of experience. Mol Genet Metab 107(1–2):72–80

Lamp J, Keyser B, Koeller DM et al (2011) Glutaric aciduria type 1 metabolites impair the succinate transport from astrocytic to neuronal cells. J Biol Chem 286:17777–17784

Liem RK, Messing A (2009) Dysfunctions of neuronal and glial intermediate filaments in disease. J Clin Investig 119:1814–1824

Magni DV, Furian AF, Oliveira MS et al (2009) Kinetic characterization of l-[(3)H]glutamate uptake inhibition and increase oxidative damage induced by glutaric acid in striatal synaptosomes of rats. Int J Dev Neurosci 27:65–72

Maragakis NJ, Rothstein JD (2006) Mechanisms of disease: astrocytes in neurodegenerative disease. Nat Clin Pract Neurol 2(12):679–689
Messing A, Brenner M, Feany MB et al (2012) Alexander disease. J Neurosci 32(15):5017–5032
Miyajima H (2015) Aceruloplasminemia. Neuropathology 35:83–90
Nash B, Thomson CE, Linington C et al (2011) Functional duality of astrocytes in myelination. J Neurosci 31:13028–31308
Oide T, Yoshida K, Kaneko K et al (2006) Iron overload and antioxidative role of perivascular astrocytes in aceruloplasminemia. Neuropathol Appl Neurobiol 32:170–176
Olivera-Bravo S, Fernández A, Latini A et al (2008) Astrocytic proliferation and mitochondrial dysfunction induced by accumulated glutaric acidemia I (GAI) metabolites: possible implications for GAI pathogenesis. Neurobiol Dis 32:528–534
Olivera-Bravo S, Fernández A, Sarlabós MN et al (2011) Neonatal astrocyte damage is sufficient to trigger progressive striatal degeneration in a rat model of Glutaric Acidemia-I. PLoS ONE 6: e20831–20840
Olivera-Bravo S, Isasi E, Fernández A et al (2014) White matter injury induced by perinatal exposure to glutaric acid. Neurotox Res 25:381–391
Patel SC, Suresh S, Kumar U et al (1999) Localization of Niemann-Pick C1 protein in astrocytes: implications for neuronal degeneration in Niemann-Pick type C disease. Proc Natl Acad Sci USA 96:1657–1662
Pekny M, Pekna M (2014) Astrocyte reactivity and reactive astrogliosis: costs and benefits. Physiol Rev 94(4):1077–1098
Rappold PM, Tieu K (2010) Astrocytes and therapeutics for Parkinson's disease. Neurotherapeutics 7(4):413–423
Ribas GZ, Vargas CR, Wajner M (2014) L-carnitine supplementation as a potential antioxidant therapy for inherited neurometabolic disorders. Gene 533:469–476
Ridder MC, Boor I, Lodder JC et al (2011) Megalencephalic leucoencephalopathy with cysts: defect in chloride currents and cell volume regulation. Brain 134:3342–3354
Robinson BH, Oei J, Sherwood WG et al (1984) The molecular basis for the two different clinical presentations of classical pyruvate carboxylase deficiency. Am J Hum Genet 36:283–294
Rothstein JD, Dykes-Hoberg M, Pardo CA et al (1996) Knockout of glutamate transporters reveals a major role for astroglial transport in excitotoxicity and clearance of glutamate. Neuron 16:675–686
Ruiz Pons M, Sanchez-Valverde Visus F, Dalmau Serra J et al (2007) Nutritional treatment of inborn errors of metabolism, 1st edn. Ergon, Madrid
Saez JP, Orellana JA, Vega-Riveros N et al (2013) Disruption in connexin-based communication is associated with intracellular $Ca^{2+}$ signal alterations in astrocytes from Niemann-Pick Type C mice. PLoS ONE 8(8):e71361
Scriver CR, Beaudet AL, Sly WS, Valle D (1995) The metabolic and molecular bases of inherited disease, 7th edn. McGraw-Hill, New York
Seminotti B, da Rosa MS, Fernandes CG et al (2012) Induction of oxidative stress in brain of glutaryl-CoA dehydrogenase deficient mice by acute lysine administration. Mol Genet Metab 106:31–38
Sloan SA, Barres BA (2014) Mechanisms of astrocyte development and their contributions to neurodevelopmental disorders. Curr Opin Neurobiol 27:75–81
Strauss KA, Puffenberger EG, Robinson DL et al (2003) Type I glutaric aciduria, part 1: natural history of 77 patients. Am J Med Gen Part C Semin Med Gen 121C:38–52
Strauss KA, Lazovic J, Wintermark M et al (2007) Multimodal imaging of striatal degeneration in Amish patients with glutaryl-CoA dehydrogenase deficiency. Brain 130:1905–1920
van der Knaap MS, Pronk JC, Scheper GC (2006) Vanishing white matter disease. Lancet Neurol 5:413–423
Vargas MR, Johnson JA (2010) Astrogliosis in amyotrophic lateral sclerosis: role and therapeutic potential of astrocytes. Neurotherapeutics 7(4):471–481
Verkhratsky A, Sofroniew MV, Messing A et al (2012) Neurological diseases as primary gliopathies: a reassessment of neurocentrism. ASN Neuro 5:4(3)

Verkhratsky A, Nedergaard M, Hertz L (2015) Why are astrocytes important? Neurochem Res 40 (2):389–401
Wu AS, Kiaei M, Aguirre N et al (2003) Iron porphyrin treatment extends survival in a transgenic animal model of amyotrophic lateral sclerosis. J Neurochem 85:142–150
Zhao Y, Rempe DA (2010) Targeting astrocytes for stroke therapy. Neurotherapeutics 7 (4):439–451
Zinnanti WJ, Lazovic J, Housman C et al (2007) Mechanism of age-dependent susceptibility and novel treatment strategy in glutaric acidemia type I. J Clin Invest 117:3258–3270

# Microglia in Cancer: For Good or for Bad?

**Anna Carolina Carvalho da Fonseca, Rackele Amaral, Celina Garcia, Luiz Henrique Geraldo, Diana Matias and Flavia Regina Souza Lima**

**Abstract** Glioblastoma is a malignant tumor of astrocytic origin that is highly invasive, proliferative and angiogenic. Despite current advances in multimodal therapies, such as surgery, radio- and chemotherapy, the outcome for patients with glioblastoma is nearly always fatal. The glioblastoma microenvironment has a tremendous influence over the tumor growth and spread. Microglia and macrophages are abundant cells in the tumor mass. Increasing evidence indicates that glioblastoma recruits these cell populations and signals in a way that microglia and macrophages are subverted to promote tumor progression. In this chapter, we discuss some aspects of the interaction between microglia and glioblastoma, consequences of this interaction for tumor progression and the possibility of microglial cells being used as therapeutic vectors, which opens up new alternatives for the development of GBM therapies targeting microglia.

**Keywords** Microglia · Glioblastoma · Central nervous system · Therapy

---

A.C.C. da Fonseca · R. Amaral · C. Garcia · L.H. Geraldo · D. Matias · F.R.S. Lima (✉)
Instituto de Ciências Biomédicas, Centro de Ciências da Saúde,
Universidade Federal do Rio de Janeiro, Avenida Carlos Chagas Filho,
373, Rio de Janeiro, RJ 21949-590, Brazil
e-mail: flima@icb.ufrj.br

A.C.C. da Fonseca
e-mail: fonseca.anna@gmail.com

R. Amaral
e-mail: rackele@icb.ufrj.br

C. Garcia
e-mail: celinagarcia@icb.ufrj.br

L.H. Geraldo
e-mail: lh_geraldo@icb.ufrj.br

D. Matias
e-mail: dimtias@gmail.com



R. von Bernhardi (ed.), *Glial Cells in Health and Disease of the CNS*,
Advances in Experimental Medicine and Biology 949,
DOI 10.1007/978-3-319-40764-7_12

**Abbreviations and Acronyms**

| | |
|---|---|
| CNS | Central nervous system |
| IL | Interleukin |
| TNF | Tumor necrosis factor |
| GBM | Glioblastoma |
| MMP | Matrix metalloproteinase |
| VEGF | Vascular endothelial growth factor |
| DLL4 | Delta-like ligand 4 |
| NO | Nitric oxide |
| MCP-1 (CCL2) | Macrophage chemoattractant protein 1 |
| CCR2 | CCL2 receptor |
| TNFR1 | TNF receptor 1 |
| IkBa | Nuclear factor of kappa light polypeptide gene enhancer in B-cells inhibitor, alpha |
| NF-kB | Nuclear factor of kappa light polypeptide gene enhancer in B-cells |
| uPA | Urokinase-type plasminogen activator |
| STI1 | Stress-inducible protein 1 |
| Hsp | Heat shock protein |
| MG CM | Microglial conditioned medium |
| MYD88/TLR8 | Myeloid Differentiation Primary Response 88/Toll-like receptor 8 |
| TLR | Toll-like receptor |
| MT1-MMP | Membrane-type-1 MMP |
| GDNF | Glial cell-line-derived neurotrophic factor |
| GFRa | GDNF receptor |
| EGFR | Epidermal growth factor receptor |
| EGF | Epidermal growth factor |
| poly [I:C] | Polyinosinic-polycytidylic acid |
| TRAIL | TNF-related apoptosis inducing ligand |
| mTOR | Mammalian target of rapamycin |
| iNOS | Inducible nitric oxide synthase |
| MIF | Macrophage migration inhibitory factor |
| STAT | Signal transducer and activator of transcription |
| RAGE | Receptor for Advanced Glycation End products |
| S100B | S100 calcium binding protein B |
| CpG-ODN | Oligodeoxynucleotides containing CpG motifs |
| TROY/TNFRSF19 | Tumor necrosis factor receptor of mouse embryo |
| GSC | Glioma stem cell |
| MRI | Magnetic resonance imaging |
| GCV | Gancyclovir |

## Tumor Microenvironment and the Microglia

The central nervous system (CNS) is composed of several cell types, including neurons, astrocytes, oligodendrocytes, ependymal cells and microglia. Each cell type has distinct and essential roles for the optimal functioning of the CNS. As discussed in Chapters "Glial Cells and Integrity of the Nervous System," "Microglia Function in the Normal Brain" and "Purine Signaling and Microglial Wrapping," microglia are the resident immune cells in the CNS, but they are increasingly recognized to play diverse roles. Their embryonic origin is mesodermal, unlike other CNS cells, which have ectodermal origin. It appears that microglia progenitors come from the yolk sac early in development (Ginhoux et al. 2010). Phagocytosis of microorganisms, antigen presentation to lymphocytes, phagocytosis of cell debris, transient or aberrant axons and apoptotic cells during development, and secretion of neurotrophic factors are some of microglia functions in CNS (Vilhardt 2005; Mallat et al. 2005; Lima et al. 2010). Following lesions, microglia become active and assume an amoeboid phenotype and a high metabolic rate, synthesizing and secreting several cytokines, such as interleukin (IL)6, IL1$\beta$ and tumor necrosis factor $\alpha$ (TNF$\alpha$) (Vilhardt 2005; Yang et al. 2010).

Among all CNS pathologies, one of the deadliest is glioblastoma (GBM). This malignant tumor of astrocytic origin is highly invasive, proliferative and angiogenic. Its invasive nature explains the high recurrence even after surgical resection (Lima et al. 2012). Survival is commonly about 14 months despite all efforts (Stupp et al. 2005). The GBM microenvironment has a tremendous influence over the tumor growth and spread. In a still not completely defined way, GBM subverts cells to act in its favor. Astrocytes were shown to have an increase in MMP-2 (matrix metalloproteinase-2) expression and to convert pro-MMP-2 to active form only in the presence of glioma cells (Le et al. 2003; Gagliano et al. 2009), suggesting a pro-tumor role of astrocytes. Besides, GBM cells produce VEGF (vascular endothelial growth factor) and DLL4 (delta-like ligand 4), which stimulate the angiogenesis that sustains tumor survival and growth (Bao et al. 2006; Li et al. 2007).

Microglia and macrophages are abundant cells in the tumor mass. GBM recruits these cell populations (Fig. 1) and signals in a way that microglia and macrophages are subverted to promote tumor progression. Moreover, GBM establishes an immunosuppressed niche, favoring even more its survival, growth, and spread (da Fonseca and Badie 2013). A current and important topic regarding these glioma-associated microglia and macrophages is the existence of two phenotypically distinct cell populations classified as M1 and M2. M1 macrophages are classically activated, developing an anti-tumor response through the activation of the immune system and production of reactive oxygen species, nitric oxide (NO) and proinflammatory cytokines, such as TNF. M2 macrophages are alternatively activated, performing immunosuppressive roles, such as release of IL10 and tumor promotion, as well as inducing metastatic processes by promoting angiogenesis and extracellular matrix degradation (Sica et al. 2008; Yang et al. 2010;

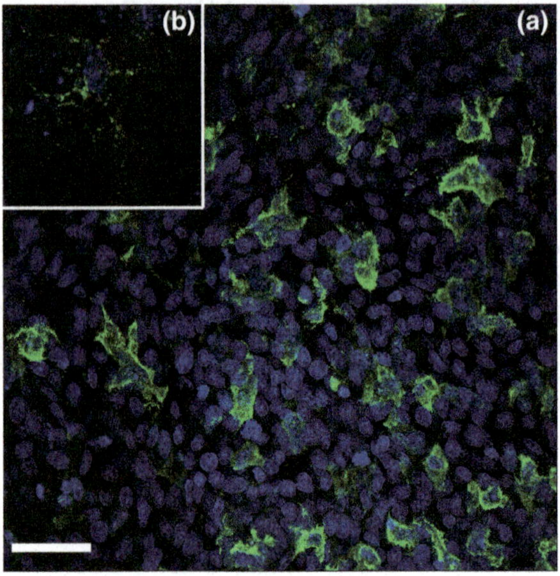


**Fig. 1** Xenotransplanted tumor produced from human GBM cells injected into the caudate putamen of mouse brain. In this study, we used the human tumor cell line GBM95, established in our lab (Faria et al. 2006). After 15 days, the brains were perfused with fixative 4 % paraformaldehyde, cut into slices and immunostained with rabbit anti-mouse Iba1 (*green*) antibody (Confocal Microscope/Leica TCS-SP5), a marker of microglia/macrophages cells. Nuclei were stained with DAPI (*blue*). **a** Tumor mass. **b** Contralateral hemisphere, without tumor cells. Note the presence of activated microglia/macrophages in the tumor mass (**a**) and ramified resident microglia cells in the contralateral hemisphere (**b**). *Bar* 40 μm. This study was approved by the Ethics Committee of the Health Sciences Center at the Federal University of Rio de Janeiro (Protocol no. DAHEICB 015) and by the Brazilian Ministry of Health Ethics Committee (CONEP no. 2340). The "Principles of laboratory animal care" (NIH publication no. 85-23, revised 1996) guidelines as well as The Code of Ethics of EU Directive 2010/63/EU were strictly followed for all experiments

Albesiano et al. 2010; Herrera et al. 2013). Indeed, the escape of tumor cells from the immune system has been related to a change of M1 to M2 phenotype during tumor progression (Schmieder et al. 2012). Similar to macrophages, glioma-associated microglia also present the M2 phenotype (Komohara et al. 2008; Fonseca et al. 2012; Gabrusiewicz et al. 2011).

## Microglia-Glioblastoma Interaction

Malignant gliomas, particularly GBMs, the most aggressive astrocytoma, contain high levels of microglia infiltrates; about 30 % of tumor mass is composed of glioma-associated microglia and macrophages which has led to the hypothesis that microglia may have a role in GBM immunology (Badie and Schartner 2000; da Fonseca and Badie 2013; Yang et al. 2010). Indeed, evidence strongly suggests that microglia contribute to the immunosuppressive environment of GBMs and may promote tumor growth (Schartner et al. 2005; Yi et al. 2011). In this context, the accumulation of microglia in GBM tissue is due to local production of growth factors and chemoattractants, such as CCL2, by GBM cells (Prat et al. 2000). CCL2, also recognized as MCP-1 (macrophage chemoattractant protein 1), is one of the most highly expressed chemokines in many CNS injuries and exerts its biological function by binding to its high affinity receptor CCR2, which is expressed by microglia, astrocytes and brain microvascular endothelial cells (Yao and Tsirka 2014). The interaction between CCL2 and CCR2 triggers IL6 release by microglia, which is associated with GBM aggressiveness (Li et al. 2010; Rolhion et al. 2001). IL6 is implicated in many aspects of tumorigenesis, and it is identified as a growth factor for glioma stem cells (Wang et al. 2009). In addition, IL6 has been found to increase microglia production of MMP-2, facilitating tumor migration (Li et al. 2010). Markovic et al. (2005), using cultured brain slices where microglia were previously depleted with clodronate-filled liposomes, showed that injected glioma cells had decreased infiltrative capacity compared with glioma cells injected into control slices, possibly because of the decrease in MMP-2 levels, produced by microglia. Moreover, Platten et al. (2003), using a rat model of intracerebral glioma cell line implant, demonstrated that the glioma cell line that recruited more microglia resulted in a larger tumor mass, and they attributed this effect to CCL2/CCR2 pathways. Another important factor during GBM progression is TNFα, a proinflammatory cytokine widely secreted by microglia (Rivest 2009), which stimulates the secretion of several molecules including CCL2, IL6, IL1β, and NO (Allan and Rothwell 2001; D'Mello et al. 2009; Nadeau and Rivest 2000). On TNF receptor 1 (TNFR1) activation, IκBa, a protein that blocks NFκB signaling in resting cells, gets phosphorylated and degraded, leading to p65/p50 nuclear translocation and transcriptional activation of NFκB target genes, including TNFα itself (Baker et al. 2008; Tchoghandjian et al. 2013). The constitutively activated NFκB has been associated with invasive behavior and malignancy of GBM (Raychaudhuri et al. 2007; Tsunoda et al. 2005). NFκB activation triggers transcriptional activation of pro-migratory genes, like CXC chemokines, urokinase-type plasminogen activator (uPA) and matrix metalloproteinases, contributing to invasiveness of GBM (Tchoghandjian et al. 2013; Wu and Zhou 2010).

In addition to the factors released in the tumor microenvironment, proteins constitutively expressed by microglia may be implicated in GBM maintenance. Stress-inducible protein 1 (STI1) is a 66 kDa protein described as a co-chaperone that binds to both Hsp70 and Hsp90 and regulates their activities (Chen and Smith 1998; Song and Masison 2005). Our group has shown that STI1 released by microglia promotes tumor proliferation, modulates MMP-9 activity and stimulates the migration of human GBM cell lines in vitro (Fonseca et al. 2012). We demonstrated that microglial conditioned medium (MG CM) stimulated proliferation and migration of the GBM cell lines, and this effect was reversed when the anti-STI1 antibody was added to the MG CM or STI1 was removed by immunodepletion from the MG CM. Furthermore, the addition of STI1 antibody to MG CM significantly decreased MMP-9 activity (Fonseca et al. 2012). These data suggest that STI1 is an important factor for glioma progression. Using a glioma model of intracranial and subcutaneous implant of GL261, a murine glioma cell line, Fonseca et al. (2012) have also shown that STI1 expression increased with tumor progression, and was also upregulated in glioma-associated microglia and macrophages and infiltrating lymphocytes. On the other hand, STI1 expression did not significantly change in circulating leukocytes, and even decreased in leukocytes that infiltrated tumors propagated in the subcutaneous tissue. Therefore, these results demonstrated that STI1 expression is modulated by the brain tumor microenvironment, and for the first time correlated STI1 expression and glioma progression (Carvalho da Fonseca et al. 2014). Altogether, we conclude that microglia-GBM interaction determine the degree of GBM invasion, opening the way for the development of new therapeutic approaches.

Microglial cells are substantial producers of MMPs and inducers of GBM invasiveness (Alves et al. 2011; Hu et al. 2014). Hu et al. (2014) showed that MMP-9 levels were expressed by Iba1$^{+}$ cells in human tumor samples, indicating that glioma-associated microglia were responsible for MMP-9 local production. Also, they demonstrated that GBM cells released soluble factors that induced the MMP-9 expression in glioma-associated microglia via Myeloid Differentiation Primary Response 88/Toll-like receptor 8 (MYD88/TLR8) signaling pathway. Interestingly, in this study, when microglial cells were treated with their inhibitor minocycline, levels of MMP-9 and TLR2 (Toll-like receptor 2) in glioma-associated microglia decreased, and consequently tumor invasion declined (Hu et al. 2014). A previous study had shown that GBM cells induced the expression of membrane-type-1 MMP (MT1-MMP) in glioma-associated microglia, promoting tumor invasion via the TLR2 signaling pathway. In this work, GL261 murine glioma cells were injected into TLR2 knockout mice, resulting in smaller tumors and reduced MT1-MMP levels in glioma-associated microglia (Vinnakota et al. 2013). Thus, it seems important to explore the role of MMPs and TLRs in microglial cells, since they can stimulate tumor progression.

Another factor that contributes to microglial recruitment during tumor progression is the glial cell-line-derived neurotrophic factor (GDNF). GDNF is a neurotrophic factor involved in dopaminergic neuronal survival, but it can also contribute to tumor progression. High levels of GDNF have been observed in GBM cells (Ng et al. 2009; Wiesenhofer et al. 2000); however, little is known about the correlation of GDNF and the attraction of microglia during tumor progression. Ku et al. (2013) have recently demonstrated that GDNF is expressed in GBM cells and plays an important role in microglia recruitment during tumor progression They showed that microglia expressed both GDNF receptors, GFRa-1 and GFRa-2, and then they used transwell assays to understand how GDNF could modulate the migration of microglia. For this, they used GBM cell conditioned medium depleted of GDNF, by using a specific shRNA, and observed that there was reduced microglia migration. In addition, they injected GBM cells not expressing GDNF into mouse brain and, after 2 weeks, observed lower levels of Iba1$^{+}$ microglia infiltration and a reduced tumor size. So, GDNF is indeed an important factor expressed by GBM cells for microglia attraction (Ku et al. 2013).

Epidermal growth factor receptor (EGFR) also plays a significant role during GBM invasion and aggressiveness (Ohgaki and Kleihues 2007; Sangar et al. 2014). It is possible that microglia cells stimulate GBM invasion through the EGFR signaling (Nolte et al. 1997). In particular, Coniglio et al. (2012) demonstrated in vitro that microglia secreted EGF, which, in turn, activated the EGFR on GBM cells and consequently induced tumor migration.

## Change in the Microglial Profile May Be Useful

Studies by Penfield (1925) led to the suggestion that microglial cells fight tumors (for review see Charles et al. 2011; Li and Graeber 2012). However, other investigators continue to believe that microglial cells behave as expected for macrophages and are able to promote antigen presentation, release of cytokines and phagocytosis even in the presence of GBM.

Kren et al. (2010) observed the expression of HLA-G and HLA-E, immune-modulatory nonclassical molecules with anti-tumor activity, by glioma-associated microglia and macrophages in most cases of human GBM of 26 samples analyzed. The role of these molecules in GBM is not well described. Even so, their observations go against the hypothesis that microglial cells and macrophages may be attracted to the tumor site and promote tumor invasion through inhibition of the cytolysis by NK-cell and T-cell, once they consider that the detected expression of HLA-G and HLA-E in glioma-associated microglia and macrophages indicates a role in immune functions (Kren et al. 2010). In this sense, an in vitro study showed that the conditioned culture medium of microglia promoted apoptotic cell death of glioma cells; when microglial cells were previously treated with LPS or IFN$\gamma$ (Interferon $\gamma$), this effect was more pronounced. Proteomic analysis was used to identify the secreted proteins, and several cathepsin proteases were found to be expressed, especially cathepsin B, as was NO, suggesting a microglial role in tumor cytotoxicity (Hwang et al. 2009). Despite these observations, much clinical evidence and many in vitro studies indicate that microglia and macrophages that infiltrate the brain tumor have pro-tumor functions, promoting cell growth and migration (Li and Graeber 2012; Alves et al. 2011; da Fonseca and Badie 2013). On the other hand, after treatment with polyinosinic-polycytidylic acid (poly [I:C]), an agonist for Toll-like receptor 3, glioma-associated microglia obtained from patients with GBM started to secrete toxic soluble factors when cocultured with different GBM cell lines. This was also true when they used the supernatant of glioma-associated microglia previously stimulated with poly (I:C). Interestingly, these factors had toxic effects only on tumor cells, since astrocytes and neurons cultures were not affected (Kees et al. 2012). Thus, Kees et al. (2012) demonstrated that it is possible to change the behavior of microglial cells from a tumor-supporting role to a tumor-suppressing function after poly (I:C) exposure. In other words, switching the M2 profile described for glioma-associated microglia, to an M1 profile led to gaining anti-tumor activities. In the same way, Chiu et al. (2011) proposed that in vitro microglial anti-tumor functions could be reestablished with treatment with IL12. Indeed, after IL12 stimulation, microglia increased the levels of TRAIL (TNF-related apoptosis inducing ligand) releasing and phagocytic activity.

In addition, Lisi et al. (2014) demonstrated that the inhibition of mTOR (mammalian target of rapamycin), which is activated in gliomas by many deregulated pathways, polarizes glioma-activated microglia to an M1 profile conferring cytotoxic functions upon microglial cells and preventing the M2 state that is involved in tumor establishment. In fact, iNOS was increased followed by a decrease in IL10 gene expression after treatment with rapamycin and its analog RAD001 in microglia (Lisi et al. 2014).

A recent in vivo study showed that MIF (macrophage migration inhibitory factor) is highly expressed on glioma cells, whereas its receptor, CD74, is expressed only in glioma-associate microglia. In this study, GBM cells and glioma-associate microglia were isolated from primary human tumors. A higher level of CD74-positive glioma-associate microglia was associated with increased patient survival, representing a positive prognostic parameter associated with the anti-tumor M1 profile (Zeiner et al. 2014).

In face of all these studies, we can conclude that microglia are a potential therapeutic target for the treatment of GBM. Certainly, the more we know about the interaction between microglia and GBM cells, the more we will know about the tumor biology. Data showing microglial cells promote tumors are substantial, especially from studies mentioned above that have attempted to manipulate the activation state of microglia, rescuing their M1 profile instead of M2 profile to fight against tumors.

## Possible Therapies Using Microglia as a Therapeutic Target in the Fight Against Cancer

Currently, the state of the art therapy for GBM consists of surgical resection of the tumor, followed by chemotherapy with Temozolomide and radiotherapy. Despite the aggressive therapy, median survival remains only 14.6 months after diagnosis (Stupp et al. 2005). Many preclinical and clinical studies are trying to improve patient survival, but without success. Some strategies focus on the interaction of the tumor cells with the microenvironment; we will discuss some of the strategies that are being currently developed focusing on the microglia-glioma interaction.

The STAT 3 (signal transducer and activator of transcription 3) pathway is constitutively expressed in high-grade gliomas (Yu et al. 2007; Weissenberger et al. 2004) and has already been implicated in GBM pathogenesis, progression and immune evasion (Takeda et al. 1999; O'Farrell et al. 1998; Lang et al. 2002). STAT

3 is also upregulated in microglial cells under glioma influence (Zhang et al. 2009a), at least in part by interaction between glioma S100B and IL-6 with microglial RAGE (Receptor for Advanced Glycation End products) and IL6R (Bromberg and Wang 2009; Zhang et al. 2009b). The activation of this pathway inhibits macrophage activation and reduces expression of co-stimulatory molecules necessary for antigen presentation by naive T-cells; it increases the secretion of immunomodulatory cytokines IL6 and IL10, while reducing lymphocyte-stimulating cytokines (IL2, IL4, IL12 and IL15) (Cheng et al. 2003; Walker et al. 2003; Hussain et al. 2007). Consistent with this, STAT 3 inhibition in glioma cells using siRNAs reverses the cytokine expression profile, leading to microglia/macrophage activation and tumor growth inhibition in a mouse model (Zhang et al. 2009a). Pharmacological agents such as small STAT 3 inhibitors that penetrate the CNS have an anti-proliferative and proapoptotic effect on glioma cell lines (Takeda et al. 1999; O'Farrell et al. 1998; Lang et al. 2002; Iwamaru et al. 2007). Apart from that, these agents are capable of reversing the immune tolerant microenvironment by activating microglial cells, through production of lymphocyte-stimulating cytokines (IL2, IL4, IL12, IL15 and CXCL10) and upregulation of co-stimulatory molecules (CD80 and CD86), and also by stimulating T-cell proliferation and inducing a $T_h1$-response (Iwamaru et al. 2007; Cheng et al. 2003; Hussain et al. 2007). There are on-going phase I and II clinical trials with STAT 3 and its more important activator IL6 for several malignancies for which this pathway is important, such as head and neck cancer, multiple myeloma and prostate cancer (Sansone and Bromberg 2012). These trials might be translated into a new therapeutic option for malignant gliomas, where this pathway was also recently shown to be important.

The role of the immune system in the treatment of the CNS tumors gained prominence with clinical trials using oligodeoxynucleotides containing CpG motifs (CpG-ODN) for GBM patients. CpG-ODNs are strongly immunostimulating agents, activating both innate and specific immunity. Biological effects of CpG-ODN are mediated by Toll-like receptor 9 (TLR9) (Klinman 2004; Krieg 2004), mainly expressed by B-lymphocytes and plasmacytoid dendritic cells in humans, and also by microglial and glioma cells (Ribes et al. 2010; El Andaloussi et al. 2006). In pre-clinical models, local treatment with CpG-ODN injections, either alone or combined with radiation therapy, reduced tumor size, with no toxicity to brain parenchyma (Carpentier et al. 2000; Auf et al. 2001; Meng et al. 2005). It was shown that tumor rejection was due not only to direct toxicity in tumor cells, but also to modulation of microglia/macrophages and induction of a $T_h1$ response (El Andaloussi et al. 2006; Carpentier et al. 2000; Auf et al. 2001). This new therapy was so promising in preclinical models that it was rapidly translated into phase I and II clinical trials. After a promising phase I study, with the few side effects limited to transient worsening of neurological condition and fever (Carpentier et al. 2006), phase II trials presented at the 2009 American Society of Clincal Oncology (ASCO) annual meeting showed only modest activity in the 6-month progression-free survival (PFS) of the cohort, with only a few cases showing radiological response (Carpentier et al. 2010; Ursu et al. 2009). This trial did not define the clinical or molecular characteristics of patients who might have benefitted from this trial, and this therapy was not continued.

Apart from these innovative strategies, old medications with newly discovered functions are also being investigated. Some glioma drugs reduce tumor growth in preclinical models by modulating microglial activity. The first, minocycline, a semi-synthetic broad spectrum tetracycline antibiotic described as capable of counteracting microglial activation into a proinflammatory phenotype by p38-MAPK inhibition (Suk 2004), reduces tumor growth in vitro and in vivo by inhibiting microglial MT1-MMP expression (Markovic et al. 2011). Another is propentofylline, an atypical synthetic methylxanthine with CNS glial modulating and anti-inflammatory actions (Si et al. 1996, 1998), described as capable of decreasing tumor growth in preclinical GBM models by a direct effect on microglial cells (and not in tumor-infiltrating macrophages) through tumor necrosis factor receptor of mouse embryo (TROY/TNFRSF19) inhibition (Jacobs et al. 2012a, b). And, more recently, Sarkar et al. (2014) demonstrated that microglial cells derived from non-glioma human subjects can markedly reduce the sphere-forming capacity of glioma stem cells (GSCs) by inducing cell-cycle arrest, reducing proliferation and inducing differentiation, most likely through IL-8 and MCP-1. Apart from that, Amphotericin B (a polyene antifungal drug) stimulates glioma-associated microglia through TLR signaling, reducing GSC survival and sphere-forming capacity in a manner resembling the action of microglia from healthy subjects. Daily treatment of mice harboring intracranial GSCs with non-toxic doses of Amphotericin B also substantially prolongs mouse survival (Sarkar et al. 2014).

Aside from pharmacological approaches, strategies have also been developed using microglia as vehicles for gene therapy in conjunction with MRI tracking. Ribot et al. (2007) labeled microglial cells with MRI contrast agents to ascertain that the injected cells were migrating to the tumor mass. The labeled cells were also transfected with a thymidine kinase suicide gene, which causes cell death after administration of its substrate, gancyclovir (GCV). This system is suitable because it induces a bystander effect: first, monophosphorylated GCV passes through intercellular gap junctions and thereby triggers the death of cells that have not been transduced; second, apoptotic bodies released by dead cells are taken up by adjacent viable cells which then die, amplifying this phenomenon (Caruso et al. 1993; Qiao et al. 2000; Burrows et al. 2002). Thus, a small quantity of enzyme and a low level of transduction are sufficient to cause tumor regression (Spencer 2000) under pharmacological control, since intracellular signaling occurs only if GCV is administered. The investigators demonstrated that the injected microglial cells are attracted to the tumor mass and that suicide gene activation with GCV reduces tumor growth and prolongs survival in a preclinical model of human GBM in nude mice (Ribot et al. 2007; Caruso et al. 1993; Qiao et al. 2000; Ribot et al. 2011).

Recent knowledge of microglia's effects on the microenvironment of malignant gliomas has led to the discovery of several pathways that are promising therapeutic targets and a new prognostic molecular marker. This time, instead of a mutation or protein expression in tumor cells, the prognostic marker is a polymorphism in a microglial chemokine receptor gene associated with cell migration: the

CX3CR1-I249 allele. This allele variant is associated with prolonged mean survival of GBM patients (23.5 v 14.1 months; $P < 0.0001$) and with reduced tumor infiltration by microglia (Rodero et al. 2008). For the first time, a microglial marker has been characterized as an independent, favorable prognostic factor and might be useful in predicting survival in GBM patients.

Since the phase III trials of Temozolomide in 2005, many options have been studied for the treatment of malignant gliomas without much success, including anti-angiogenic therapy with Bevacizumab (Avastin®) (Chinot et al. 2014) and such other chemotherapy regimens as Procarbazine, Lomustine (CCNU), and Vincristine (PCV) (Brada et al. 2010). Also, many strategies that seemed promising in preclinical trials, such as the CpG-ODNs, are rather disappointing in clinical trials. Thus, understanding of the biology of the CNS tumors and of the microenvironment's influence on tumor progression is becoming increasingly important for developing new therapeutic strategies for this deadly disease.

## Concluding Remarks

GBMs are the most aggressive tumors of astrocytic lineage. Despite significant progress in cancer research, which has led to the development of more effective therapies for some types of solid tumors, there is no effective treatment for GBM. In this chapter, we discussed some relevant properties of microglia in contact with GBM. Better understanding of the interactions between the tumor and its microenvironment, particularly microglial cells, is important for combating GBM. In this sense, the development of new therapies targeting the microglia or the factors produced by them that are specifically related to tumor progression may be an effective alternative.

## References

Albesiano E, Han JE, Lim M (2010) Mechanisms of local immunoresistance in glioma. Neurosurg Clin N Am 21(1):17–29. doi:10.1016/j.nec.2009.08.008

Allan SM, Rothwell NJ (2001) Cytokines and acute neurodegeneration. Nat Rev Neurosci 2 (10):734–744. doi:10.1038/35094583

Alves TR, Lima FR, Kahn SA, Lobo D, Dubois LG, Soletti R, Borges H, Neto VM (2011) Glioblastoma cells: a heterogeneous and fatal tumor interacting with the parenchyma. Life Sci 89(15–16):532–539. doi:10.1016/j.lfs.2011.04.022

Auf G, Carpentier AF, Chen L, Le Clanche C, Delattre JY (2001) Implication of macrophages in tumor rejection induced by CpG-oligodeoxynucleotides without antigen. Clin Cancer Res 7 (11):3540–3543

Badie B, Schartner JM (2000) Flow cytometric characterization of tumor-associated macrophages in experimental gliomas. Neurosurgery 46(4):957–961; discussion 961–952

Baker BJ, Qin H, Benveniste EN (2008) Molecular basis of oncostatin M-induced SOCS-3 expression in astrocytes. Glia 56(11):1250–1262. doi:10.1002/glia.20694

Bao S, Wu Q, Sathornsumetee S, Hao Y, Li Z, Hjelmeland AB, Shi Q, McLendon RE, Bigner DD, Rich JN (2006) Stem cell-like glioma cells promote tumor angiogenesis through vascular endothelial growth factor. Cancer Res 66(16):7843–7848. doi:10.1158/0008-5472.CAN-06-1010

Brada M, Stenning S, Gabe R, Thompson LC, Levy D, Rampling R, Erridge S, Saran F, Gattamaneni R, Hopkins K, Beall S, Collins VP, Lee SM (2010) Temozolomide versus procarbazine, lomustine, and vincristine in recurrent high-grade glioma. J Clin Oncol 28 (30):4601–4608

Bromberg J, Wang TC (2009) Inflammation and cancer: IL-6 and STAT3 complete the link. Cancer Cell 15(2):79–80

Burrows FJ, Gore M, Smiley WR, Kanemitsu MY, Jolly DJ, Read SB, Nicholas T, Kruse CA (2002) Purified herpes simplex virus thymidine kinase retroviral particles: III. Characterization of bystander killing mechanisms in transfected tumor cells. Cancer Gene Ther 9(1):87–95

Carpentier AF, Xie J, Mokhtari K, Delattre JY (2000) Successful treatment of intracranial gliomas in rat by oligodeoxynucleotides containing CpG motifs. Clin Cancer Res 6(6):2469–2473

Carpentier A, Laigle-Donadey F, Zohar S, Capelle L, Behin A, Tibi A, Martin-Duverneuil N, Sanson M, Lacomblez L, Taillibert S, Puybasset L, Van Effenterre R, Delattre JY, Carpentier AF (2006) Phase 1 trial of a CpG oligodeoxynucleotide for patients with recurrent glioblastoma. Neuro Oncol 8(1):60–66

Carpentier A, Metellus P, Ursu R, Zohar S, Lafitte F, Barrié M, Meng Y, Richard M, Parizot C, Laigle-Donadey F, Gorochov G, Psimaras D, Sanson M, Tibi A, Chinot O, Carpentier AF (2010) Intracerebral administration of CpG oligonucleotide for patients with recurrent glioblastoma: a phase II study. Neuro Oncol 12(4):401–408

Caruso M, Panis Y, Gagandeep S, Houssin D, Salzmann JL, Klatzmann D (1993) Regression of established macroscopic liver metastases after in situ transduction of a suicide gene. Proc Natl Acad Sci USA 90(15):7024–7028

Carvalho da Fonseca AC, Wang H, Fan H, Chen X, Zhang I, Zhang L, Lima FR, Badie B (2014) Increased expression of stress inducible protein 1 in glioma-associated microglia/macrophages. J Neuroimmunol 274(1–2):71–77. doi:10.1016/j.jneuroim.2014.06.021

Charles NA, Holland EC, Gilbertson R, Glass R, Kettenmann H (2011) The brain tumor microenvironment. Glia 59(8):1169–1180. doi:10.1002/glia.21136

Chen S, Smith DF (1998) Hop as an adaptor in the heat shock protein 70 (Hsp70) and hsp90 chaperone machinery. J Biol Chem 273(52):35194–35200

Cheng F, Wang HW, Cuenca A, Huang M, Ghansah T, Brayer J, Kerr WG, Takeda K, Akira S, Schoenberger SP, Yu H, Jove R, Sotomayor EM (2003) A critical role for Stat3 signaling in immune tolerance. Immunity 19(3):425–436

Chinot OL, Wick W, Mason W, Henriksson R, Saran F, Nishikawa R, Carpentier AF, Hoang-Xuan K, Kavan P, Cernea D, Brandes AA, Hilton M, Abrey L, Cloughesy T (2014) Bevacizumab plus radiotherapy-temozolomide for newly diagnosed glioblastoma. N Engl J Med 370(8):709–722

Chiu TL, Peng CW, Wang MJ (2011) Enhanced anti-glioblastoma activity of microglia by AAV2-mediated IL-12 through TRAIL and phagocytosis in vitro. Oncol Rep 25(5): 1373–1380. doi:10.3892/or.2011.1213

Coniglio SJ, Eugenin E, Dobrenis K, Stanley ER, West BL, Symons MH, Segall JE (2012) Microglial stimulation of glioblastoma invasion involves epidermal growth factor receptor (EGFR) and colony stimulating factor 1 receptor (CSF-1R) signaling. Mol Med 18:519–527. doi:10.2119/molmed.2011.00217

da Fonseca AC, Badie B (2013) Microglia and macrophages in malignant gliomas: recent discoveries and implications for promising therapies. Clin Dev Immunol 2013:264124. doi:10.1155/2013/264124

D'Mello C, Le T, Swain MG (2009) Cerebral microglia recruit monocytes into the brain in response to tumor necrosis factorα signaling during peripheral organ inflammation. J Neurosci 29(7):2089–2102. doi:10.1523/JNEUROSCI.3567-08.2009

El Andaloussi A, Sonabend AM, Han Y, Lesniak MS (2006) Stimulation of TLR9 with CpG ODN enhances apoptosis of glioma and prolongs the survival of mice with experimental brain tumors. Glia 54(6):526–535

Faria J, Romao L, Martins S, Alves T, Mendes FA, de Faria GP, Hollanda R, Takiya C, Chimelli L, Morandi V, de Souza JM, Abreu JG, Moura Neto V (2006) Interactive properties of human glioblastoma cells with brain neurons in culture and neuronal modulation of glial laminin organization. Differentiation 74(9–10):562–572. doi:10.1111/j.1432-0436.2006.00090.x

Fonseca AC, Romao L, Amaral RF, Assad Kahn S, Lobo D, Martins S, Marcondes de Souza J, Moura-Neto V, Lima FR (2012) Microglial stress inducible protein 1 promotes proliferation and migration in human glioblastoma cells. Neuroscience 200:130–141. doi:10.1016/j.neuroscience.2011.10.025

Gabrusiewicz K, Ellert-Miklaszewska A, Lipko M, Sielska M, Frankowska M, Kaminska B (2011) Characteristics of the alternative phenotype of microglia/macrophages and its modulation in experimental gliomas. PLoS ONE 6(8):e23902. doi:10.1371/journal.pone.0023902

Gagliano N, Costa F, Cossetti C, Pettinari L, Bassi R, Chiriva-Internati M, Cobos E, Gioia M, Pluchino S (2009) Glioma-astrocyte interaction modifies the astrocyte phenotype in a co-culture experimental model. Oncol Rep 22(6):1349–1356

Ginhoux F, Greter M, Leboeuf M, Nandi S, See P, Gokhan S, Mehler MF, Conway SJ, Ng LG, Stanley ER, Samokhvalov IM, Merad M (2010) Fate mapping analysis reveals that adult microglia derive from primitive macrophages. Science 330(6005):841–845. doi:10.1126/science.1194637

Herrera M, Herrera A, Dominguez G, Silva J, Garcia V, Garcia JM, Gomez I, Soldevilla B, Munoz C, Provencio M, Campos-Martin Y, Garcia de Herreros A, Casal I, Bonilla F, Pena C (2013) Cancer-associated fibroblast and M2 macrophage markers together predict outcome in colorectal cancer patients. Cancer Sci 104(4):437–444. doi:10.1111/cas.12096

Hu F, Ku MC, Markovic D, Od AD, Lehnardt S, Synowitz M, Wolf SA, Kettenmann H (2014) Glioma-associated microglial MMP9 expression is upregulated by TLR2 signaling and sensitive to minocycline. Int J Cancer 135 (11):2569–2578. doi:10.1002/ijc.28908

Hussain SF, Kong LY, Jordan J, Conrad C, Madden T, Fokt I, Priebe W, Heimberger AB (2007) A novel small molecule inhibitor of signal transducers and activators of transcription 3 reverses immune tolerance in malignant glioma patients. Cancer Res 67(20):9630–9636

Hwang SY, Yoo BC, Jung JW, Oh ES, Hwang JS, Shin JA, Kim SY, Cha SH, Han IO (2009) Induction of glioma apoptosis by microglia-secreted molecules: the role of nitric oxide and cathepsin B. Biochim Biophys Acta 1793(11):1656–1668. doi:10.1016/j.bbamcr.2009.08.011

Iwamaru A, Szymanski S, Iwado E, Aoki H, Yokoyama T, Fokt I, Hess K, Conrad C, Madden T, Sawaya R, Kondo S, Priebe W, Kondo Y (2007) A novel inhibitor of the STAT3 pathway induces apoptosis in malignant glioma cells both in vitro and in vivo. Oncogene 26(17):2435–2444

Jacobs VL, Landry RP, Liu Y, Romero-Sandoval EA, De Leo JA (2012a) Propentofylline decreases tumor growth in a rodent model of glioblastoma multiforme by a direct mechanism on microglia. Neuro Oncol 14(2):119–131

Jacobs VL, Liu Y, De Leo JA (2012b) Propentofylline targets TROY, a novel microglial signaling pathway. PLoS ONE 7(5):e37955

Kees T, Lohr J, Noack J, Mora R, Gdynia G, Todt G, Ernst A, Radlwimmer B, Falk CS, Herold-Mende C, Regnier-Vigouroux A (2012) Microglia isolated from patients with glioma gain antitumor activities on poly (I:C) stimulation. Neuro Oncol 14(1):64–78. doi:10.1093/neuonc/nor182

Klinman DM (2004) Immunotherapeutic uses of CpG oligodeoxynucleotides. Nat Rev Immunol 4 (4):249–258

Komohara Y, Ohnishi K, Kuratsu J, Takeya M (2008) Possible involvement of the M2 anti-inflammatory macrophage phenotype in growth of human gliomas. J Pathol 216(1):15–24. doi:10.1002/path.2370

Kren L, Muckova K, Lzicarova E, Sova M, Vybihal V, Svoboda T, Fadrus P, Smrcka M, Slaby O, Lakomy R, Vanhara P, Krenova Z, Michalek J (2010) Production of immune-modulatory nonclassical molecules HLA-G and HLA-E by tumor infiltrating ameboid microglia/macrophages in glioblastomas: a role in innate immunity? J Neuroimmunol 220(1–2):131–135. doi:10.1016/j.jneuroim.2010.01.014

Krieg AM (2004) Antitumor applications of stimulating toll-like receptor 9 with CpG oligodeoxynucleotides. Curr Oncol Rep 6(2):88–95

Ku MC, Wolf SA, Respondek D, Matyash V, Pohlmann A, Waiczies S, Waiczies H, Niendorf T, Synowitz M, Glass R, Kettenmann H (2013) GDNF mediates glioblastoma-induced microglia attraction but not astrogliosis. Acta Neuropathol 125(4):609–620. doi:10.1007/s00401-013-1079-8

Lang R, Patel D, Morris JJ, Rutschman RL, Murray PJ (2002) Shaping gene expression in activated and resting primary macrophages by IL-10. J Immunol 169(5):2253–2263

Le DM, Besson A, Fogg DK, Choi KS, Waisman DM, Goodyer CG, Rewcastle B, Yong VW (2003) Exploitation of astrocytes by glioma cells to facilitate invasiveness: a mechanism involving matrix metalloproteinase-2 and the urokinase-type plasminogen activator-plasmin cascade. J Neurosci 23(10):4034–4043

Li W, Graeber MB (2012) The molecular profile of microglia under the influence of glioma. Neuro Oncol 14(8):958–978. doi:10.1093/neuonc/nos116

Li JL, Sainson RC, Shi W, Leek R, Harrington LS, Preusser M, Biswas S, Turley H, Heikamp E, Hainfellner JA, Harris AL (2007) Delta-like 4 Notch ligand regulates tumor angiogenesis, improves tumor vascular function, and promotes tumor growth in vivo. Cancer Res 67 (23):11244–11253. doi:10.1158/0008-5472.CAN-07-0969

Li R, Li G, Deng L, Liu Q, Dai J, Shen J, Zhang J (2010) IL-6 augments the invasiveness of U87MG human glioblastoma multiforme cells via up-regulation of MMP-2 and fascin-1. Oncol Rep 23(6):1553–1559

Lima FRS, da Fonseca ACC, Faria GP, Dubois LGF, Alves TR, Faria J, Moura Neto V (2010) The origin of microglia and the development of the brain. In: Ulrich H (ed) Perspectives of stem cells: from tools for studying mechanisms of neuronal differentiation towards therapy. Springer, Netherlands. doi:10.1007/978-90-481-3375-8_12

Lima FR, Kahn SA, Soletti RC, Biasoli D, Alves T, da Fonseca AC, Garcia C, Romao L, Brito J, Holanda-Afonso R, Faria J, Borges H, Moura-Neto V (2012) Glioblastoma: therapeutic challenges, what lies ahead. Biochim Biophys Acta 1826(2):338–349. doi:10.1016/j.bbcan.2012.05.004

Lisi L, Laudati E, Navarra P, Dello Russo C (2014) The mTOR kinase inhibitors polarize glioma-activated microglia to express a M1 phenotype. J Neuroinflammation 11:125. doi:10.1186/1742-2094-11-125

Mallat M, Marin-Teva JL, Cheret C (2005) Phagocytosis in the developing CNS: more than clearing the corpses. Curr Opin Neurobiol 15(1):101–107. doi:10.1016/j.conb.2005.01.006

Markovic DS, Glass R, Synowitz M, Rooijen N, Kettenmann H (2005) Microglia stimulate the invasiveness of glioma cells by increasing the activity of metalloprotease-2. J Neuropathol Exp Neurol 64(9):754–762

Markovic DS, Vinnakota K, van Rooijen N, Kiwit J, Synowitz M, Glass R, Kettenmann H (2011) Minocycline reduces glioma expansion and invasion by attenuating microglial MT1-MMP expression. Brain Behav Immun 25(4):5

Meng Y, Carpentier AF, Chen L, Boisserie G, Simon JM, Mazeron JJ, Delattre JY (2005) Successful combination of local CpG-ODN and radiotherapy in malignant glioma. Int J Cancer 116(6):992–997

Nadeau S, Rivest S (2000) Role of microglial-derived tumor necrosis factor in mediating CD14 transcription and nuclear factor κ B activity in the brain during endotoxemia. J Neurosci 20 (9):3456–3468

Ng WH, Wan GQ, Peng ZN, Too HP (2009) Glial cell-line derived neurotrophic factor (GDNF) family of ligands confer chemoresistance in a ligand-specific fashion in malignant gliomas. J Clin Neurosci 16(3):427–436. doi:10.1016/j.jocn.2008.06.002

Nolte C, Kirchhoff F, Kettenmann H (1997) Epidermal growth factor is a motility factor for microglial cells in vitro: evidence for EGF receptor expression. Eur J Neurosci 9(8):1690–1698
O'Farrell AM, Liu Y, Moore KW, Mui AL (1998) IL-10 inhibits macrophage activation and proliferation by distinct signaling mechanisms: evidence for Stat-3-dependent and -independent pathways. EMBO J 17(4):1006–1018
Ohgaki H, Kleihues P (2007) Genetic pathways to primary and secondary glioblastoma. Am J Pathol 170(5):1445–1453. doi:10.2353/ajpath.2007.070011
Penfield W (1925) Microglia and the process of phagocytosis in gliomas. Am J Pathol 1(1):77–90
Platten M, Kretz A, Naumann U, Aulwurm S, Egashira K, Isenmann S, Weller M (2003) Monocyte chemoattractant protein-1 increases microglial infiltration and aggressiveness of gliomas. Ann Neurol 54(3):388–392. doi:10.1002/ana.10679
Prat E, Baron P, Meda L, Scarpini E, Galimberti D, Ardolino G, Catania A, Scarlato G (2000) The human astrocytoma cell line U373MG produces monocyte chemotactic protein (MCP)-1 upon stimulation with beta-amyloid protein. Neurosci Lett 283(3):177–180
Qiao J, Black ME, Caruso M (2000) Enhanced ganciclovir killing and bystander effect of human tumor cells transduced with a retroviral vector carrying a herpes simplex virus thymidine kinase gene mutant. Hum Gene Ther 11(11):1569–1576
Raychaudhuri B, Han Y, Lu T, Vogelbaum MA (2007) Aberrant constitutive activation of nuclear factor κB in glioblastoma multiforme drives invasive phenotype. J Neurooncol 85(1):39–47. doi:10.1007/s11060-007-9390-7
Ribes S, Ebert S, Regen T, Agarwal A, Tauber SC, Czesnik D, Spreer A, Bunkowski S, Eiffert H, Hanisch UK, Hammerschmidt S, Nau R (2010) Toll-like receptor stimulation enhances phagocytosis and intracellular killing of nonencapsulated and encapsulated streptococcus pneumoniae by murine microglia. Infect Immun 78(2):865–871
Ribot E, Bouzier-Sore AK, Bouchaud V, Miraux S, Delville MH, Franconi JM, Voisin P (2007) Microglia used as vehicles for both inducible thymidine kinase gene therapy and MRI contrast agents for glioma therapy. Cancer Gene Ther 14(8):724–737
Ribot EJ, Miraux S, Konsman JP, Bouchaud V, Pourtau L, Delville MH, Franconi JM, Thiaudière E, Voisin PJ (2011) In vivo MR tracking of therapeutic microglia to a human glioma model. NMR Biomed 24(10):1361–1368
Rivest S (2009) Regulation of innate immune responses in the brain. Nat Rev Immunol 9(6):429–439. doi:10.1038/nri2565
Rodero M, Marie Y, Coudert M, Blondet E, Mokhtari K, Rousseau A, Raoul W, Carpentier C, Sennlaub F, Deterre P, Delattre JY, Debré P, Sanson M, Combadière C (2008) Polymorphism in the microglial cell-mobilizing CX3CR1 gene is associated with survival in patients withglioblastoma. J Clin Oncol 26(36):5957–5964
Rolhion C, Penault-Llorca F, Kemeny JL, Lemaire JJ, Jullien C, Labit-Bouvier C, Finat-Duclos F, Verrelle P (2001) Interleukin-6 overexpression as a marker of malignancy in human gliomas. J Neurosurg 94(1):97–101. doi:10.3171/jns.2001.94.1.0097
Sangar V, Funk CC, Kusebauch U, Campbell DS, Moritz RL, Price ND (2014) Quantitative proteomic analysis reveals effects of EGFR on invasion-promoting proteins secreted by glioblastoma cells. Mol Cell Proteomics 13(10):2618–2631. doi:10.1074/mcp.M114.040428
Sansone P, Bromberg J (2012) Targeting the interleukin-6/Jak/stat pathway in human malignancies. J Clin Oncol 30(9):10
Sarkar S, Döring A, Zemp FJ, Silva C, Lun X, Wang X, Kelly J, Hader W, Hamilton M, Mercier P, Dunn JF, Kinniburgh D, van Rooijen N, Robbins S, Forsyth P, Cairncross G, Weiss S, Yong VW (2014) Therapeutic activation of macrophages and microglia to suppress brain tumor-initiating cells. Nat Neurosci 17(1):46–55
Schartner JM, Hagar AR, Van Handel M, Zhang L, Nadkarni N, Badie B (2005) Impaired capacity for upregulation of MHC class II in tumor-associated microglia. Glia 51(4):279–285. doi:10.1002/glia.20201
Schmieder A, Michel J, Schonhaar K, Goerdt S, Schledzewski K (2012) Differentiation and gene expression profile of tumor-associated macrophages. Semin Cancer Biol 22(4):289–297. doi:10.1016/j.semcancer.2012.02.002

Si QS, Nakamura Y, Schubert P, Rudolphi K, Kataoka K (1996) Adenosine and propentofylline inhibit the proliferation of cultured microglial cells. Exp Neurol 137(2):345–349

Si Q, Nakamura Y, Ogata T, Kataoka K, Schubert P (1998) Differential regulation of microglial activation by propentofylline via cAMP signaling. Brain Res 812(1–2):97–104

Sica A, Larghi P, Mancino A, Rubino L, Porta C, Totaro MG, Rimoldi M, Biswas SK, Allavena P, Mantovani A (2008) Macrophage polarization in tumour progression. Semin Cancer Biol 18 (5):349–355. doi:10.1016/j.semcancer.2008.03.004

Song Y, Masison DC (2005) Independent regulation of Hsp70 and Hsp90 chaperones by Hsp70/Hsp90-organizing protein Sti1 (Hop1). J Biol Chem 280(40):34178–34185. doi:10.1074/jbc.M505420200

Spencer DM (2000) Developments in suicide genes for preclinical and clinical applications. Curr Opin Mol Ther 2(4):433–440

Stupp R, Mason WP, van den Bent MJ, Weller M, Fisher B, Taphoorn MJ, Belanger K, Brandes AA, Marosi C, Bogdahn U, Curschmann J, Janzer RC, Ludwin SK, Gorlia T, Allgeier A, Lacombe D, Cairncross JG, Eisenhauer E, Mirimanoff RO, European Organisation for Research and Treatment of Cancer Brain Tumor and Radiotherapy Groups, National Cancer Institute of Canada Clinical Trials Group (2005) Radiotherapy plus concomitant and adjuvant temozolomide for glioblastoma. N Engl J Med 352 (10):987–996

Suk K (2004) Minocycline suppresses hypoxic activation of rodent microglia in culture. Neurosci Lett 366(2):167–171

Takeda K, Clausen BE, Kaisho T, Tsujimura T, Terada N, Förster I, Akira S (1999) Enhanced Th1 activity and development of chronic enterocolitis in mice devoid of Stat3 in macrophages and neutrophils. Immunity 10(1):39–49

Tchoghandjian A, Jennewein C, Eckhardt I, Rajalingam K, Fulda S (2013) Identification of non-canonical NF-κB signaling as a critical mediator of Smac mimetic-stimulated migration and invasion of glioblastoma cells. Cell Death Dis 4:e564. doi:10.1038/cddis.2013.70

Tsunoda K, Kitange G, Anda T, Shabani HK, Kaminogo M, Shibata S, Nagata I (2005) Expression of the constitutively activated RelA/NF-κB in human astrocytic tumors and the in vitro implication in the regulation of urokinase-type plasminogen activator, migration, and invasion. Brain Tumor Pathol 22(2):79–87. doi:10.1007/s10014-005-0186-1

Ursu R, Carpentier A, Metellus P, Barrie M, Meng Y, Laigle-Donadey F, Tibi A, Chinot O, Carpentier AF (2009) Phase II trial of intracerebral administration of CpG oligonucleotide for patients with recurrent glioblastoma. J Clin Oncol 27:15S

Vilhardt F (2005) Microglia: phagocyte and glia cell. Int J Biochem Cell Biol 37(1):17–21. doi:10.1016/j.biocel.2004.06.010

Vinnakota K, Hu F, Ku MC, Georgieva PB, Szulzewsky F, Pohlmann A, Waiczies S, Waiczies H, Niendorf T, Lehnardt S, Hanisch UK, Synowitz M, Markovic D, Wolf SA, Glass R, Kettenmann H (2013) Toll-like receptor 2 mediates microglia/brain macrophage MT1-MMP expression and glioma expansion. Neuro Oncol 15(11):1457–1468. doi:10.1093/neuonc/not115

Walker PR, Calzascia T, de Tribolet N, Dietrich PY (2003) T-cell immune responses in the brain and their relevance for cerebral malignancies. Brain Res Brain Res Rev 42(2):97–122

Wang H, Lathia JD, Wu Q, Wang J, Li Z, Heddleston JM, Eyler CE, Elderbroom J, Gallagher J, Schuschu J, MacSwords J, Cao Y, McLendon RE, Wang XF, Hjelmeland AB, Rich JN (2009) Targeting interleukin 6 signaling suppresses glioma stem cell survival and tumor growth. Stem Cells 27(10):2393–2404. doi:10.1002/stem.188

Weissenberger J, Loeffler S, Kappeler A, Kopf M, Lukes A, Afanasieva TA, Aguzzi A, Weis J (2004) IL-6 is required for glioma development in a mouse model. Oncogene 23(19):3308–3316

Wiesenhofer B, Stockhammer G, Kostron H, Maier H, Hinterhuber H, Humpel C (2000) Glial cell line-derived neurotrophic factor (GDNF) and its receptor (GFR-alpha 1) are strongly expressed in human gliomas. Acta Neuropathol 99(2):131–137

Wu Y, Zhou BP (2010) TNF-α/NF-κB/Snail pathway in cancer cell migration and invasion. Br J Cancer 102(4):639–644. doi:10.1038/sj.bjc.6605530

Yang I, Han SJ, Kaur G, Crane C, Parsa AT (2010) The role of microglia in central nervous system immunity and glioma immunology. J Clin Neurosci 17(1):6–10. doi:10.1016/j.jocn.2009.05.006

Yao Y, Tsirka SE (2014) Monocyte chemoattractant protein-1 and the blood-brain barrier. Cell Mol Life Sci 71(4):683–697. doi:10.1007/s00018-013-1459-1

Yi D, Hua TX, Lin HY, Kui CL, Ning LX, Wang ZZ (2011) Antitumor treatment efficacy by targeting epidermal growth factor receptor and vascular endothelial growth factor receptor-2 in an orthotopic human glioblastoma model. J Neurooncol 104(1):93–101. doi:10.1007/s11060-010-0479-z

Yu H, Kortylewski M, Pardoll D (2007) Crosstalk between cancer and immune cells: role of STAT3 in the tumour microenvironment. Nat Rev Immunol 7(1):41–51

Zeiner PS, Preusse C, Blank AE, Zachskorn C, Baumgarten P, Caspary L, Braczynski AK, Weissenberger J, Bratzke H, Reiss S, Pennarz S, Winkelmann R, Senft C, Plate KH, Wischhusen J, Stenzel W, Harter PN, Mittelbronn M (2014) MIF receptor CD74 is restricted to microglia/macrophages, associated with a M1-polarized immune milieu, and prolonged patient survival in gliomas. Brain Pathol 25(4):491–504. doi:10.1111/bpa.12194

Zhang L, Alizadeh D, Van Handel M, Kortylewski M, Yu H, Badie B (2009a) Stat3 inhibition activates tumor macrophages and abrogates glioma growth in mice. Glia 57(13):1458–1467

Zhang L, Liu W, Alizadeh D, Zhao D, Farrukh O, Lin J, Badie SA, Badie B (2009b) S100B attenuates microglia activation in gliomas: possible role of STAT3 pathway. Glia 59(3):486–498

# Peripheral Inflammation and Demyelinating Diseases

**Verónica Murta and Carina Ferrari**

**Abstract** In recent decades, several neurodegenerative diseases have been shown to be exacerbated by systemic inflammatory processes. There is a wide range of literature that demonstrates a clear but complex relationship between the central nervous system (CNS) and the immunological system, both under *naïve* or pathological conditions. In diseased brains, peripheral inflammation can transform "primed" microglia into an "active" state, which can trigger stronger pathological responses. Demyelinating diseases are a group of neurodegenerative diseases characterized by inflammatory lesions associated with demyelination, which in turn induces axonal damage, neurodegeneration, and progressive loss of function. Among them, the most important are multiple sclerosis (MS) and neuromyelitis optica (NMO). In this review, we will analyze the effect of specific peripheral inflammatory stimuli in the progression of demyelinating diseases and discuss their animal models. In most cases, peripheral immune stimuli are exacerbating.

**Keywords** Demyelinating diseases · Systemic inflammation · Microglia · Multiple sclerosis · Neuromyelitis optica · Experimental autoimmune encephalomyelitis

## Abbreviations and Acronyms

| | |
|---|---|
| AQP4 | Aquaporin-4 |
| BBB | Blood–brain barrier |
| CCL2 | Chemokine CC motif ligand 2 |
| CCR2 | Chemokine CC motif receptor 2 |
| CD | Cluster of differentiation |

V. Murta
Laboratorio de Neuropatología Molecular, Instituto de Biología
Celular y Neurociencias, Universidad de Buenos Aires, Buenos Aires, Argentina
e-mail: vmurta.fmed@gmail.com

C. Ferrari (✉)
Instituto de Ciencias Básicas y Medicina Experimental,
Instituto Universitario del Hospital Italiano, Buenos Aires, Argentina
e-mail: carina.ferrari@hospitalitaliano.org.ar; carinaferrari@gmail.com



R. von Bernhardi (ed.), *Glial Cells in Health and Disease of the CNS*,
Advances in Experimental Medicine and Biology 949,
DOI 10.1007/978-3-319-40764-7_13

| | |
|---|---|
| CNS | Central nervous system |
| CSF | Cerebrospinal fluid |
| CXCR2 | CXC motif chemokine receptor type 2 |
| EAE | Experimental autoimmune encephalomyelitis |
| GC | Glucocorticoids |
| HPA | Hypothalamic-Pituitary-Adrenal |
| HPG | Hypothalamic-Pituitary-Gonadal |
| IFN | Interferons |
| IgG | Immunoglobulin G |
| IL | Interleukin |
| iNOS | Inducible nitric oxide synthase |
| MHC | Major histocompatibility complex |
| MS | Multiple sclerosis |
| NMO | Neuromyelitis optica |
| PMN | Polymorphonuclear |
| PPMS | Primary progressive MS |
| RRMS | Relapsing remitting multiple sclerosis |
| SGK1 | Serum glucocorticoid kinase 1 |
| SPMS | Secondary progressive multiple sclerosis |
| TGF-β | Transforming growth factor beta |
| Th | T helper |
| TLR | Toll-like receptors |
| TNF-α | Tumor necrosis factor α |
| WBC | White blood cells |

## Peripheral Inflammation and Neurodegenerative Diseases

Inflammation can be viewed as one of the primary responses of the immune system to infections or body injury. Systemic inflammation is associated with several chronic diseases, including obesity, type 2 diabetes, atherosclerosis, liver disease, and cancer (reviewed in Wilson et al. 2010; Fung et al. 2012). Additionally, it may also be associated with an acute stimulus, such as infection, surgery, and acute organ injury (Ottani et al. 2009). Systemic inflammatory stimuli that circulate in the blood may induce the synthesis of cytokines in the central nervous system (CNS) (Besedovsky and del Rey 1996; Pitossi et al. 1997; Combrinck et al. 2002; Dantzer et al. 1998, 2008; Londono and Cadavid 2010). In a diseased brain, this production of proinflammatory molecules exacerbates ongoing brain damage in several neurodegenerative diseases, such as Alzheimer's disease, multiple sclerosis (MS), Parkinson's disease, prion disease, and stroke (Perry et al. 2002; Cunningham et al. 2005a, b; McColl et al. 2007; Palin et al. 2008; Ferrari and Tarelli 2011; Murta and Ferrari 2013). In this review, we will discuss the influence of specific systemic proinflammatory stimuli on different demyelinating diseases and animal models, and the role of several cells and molecules in this phenomenon.

## *Microglia as a Mediator of Systemic Inflammation and Neurodegenerative Diseases*

Microglia are the resident immune cells of the CNS; their main role is monitoring the local environment and triggering an immune response after specific stimuli in the nervous tissue. As discussed in Chapters "Glial cells and Integrity of the Nervous System", "Microglia function in the normal brain", and "Purine Signaling and Microglial Wrapping", microglia activation is characterized by morphological and physiological changes such as secretion of proinflammatory and anti-inflammatory cytokines. Therefore, microglia can exert either cytotoxic or repairing actions, and these are referred as the M1-like and M2-like responses (Samad et al. 2001).

**Resting microglia** have a ramified morphology and represent a more quiescent basal state of this cell type. Systemic infections or mild central neurodegenerative processes can activate and prime the resting microglia. **Priming of microglia** precedes a further neurotoxic activation, which a secondary inflammatory stimulus can transform into an "**active**" **state** (Samad et al. 2001; Cunningham et al. 2005b; McColl et al. 2007). Microglia activation to an M1 phenotype increases neurotoxicity and, therefore, contributes to neurodegeneration through the release of free radicals such as superoxide radicals and nitric oxide (through the action of inducible nitric oxide synthase, iNOS) (Minghetti et al. 1999; Czlonkowska et al. 2002; Arimoto and Bing 2003), and immunomodulatory cytokines such as interleukin (IL) 1β, tumor necrosis factor α (TNFα), IL6, IL8, IL12, IL15, and IL10 (Kim and de Vellis 2005; Dilger and Johnson 2008; Henry et al. 2009). Therefore, ongoing inflammatory degenerative processes can be accelerated by systemic inflammation through the stimulation of "primed" microglial cells toward a more aggressive state, which in turn exacerbates damage in the nervous tissue (Fig. 1).

## *Communication Between the Periphery and the CNS*

The brain used to be considered an "immune-privileged" organ isolated from the peripheral immune system. Nowadays, it is well known that a bidirectional pathway between the brain and the peripheral immune system exists.

Circulating cytokines and other inflammatory molecules can affect the brain through several routes, mainly through the neural or humoral pathways. The neural pathway is mainly related to the transmission of peripheral inflammatory signals

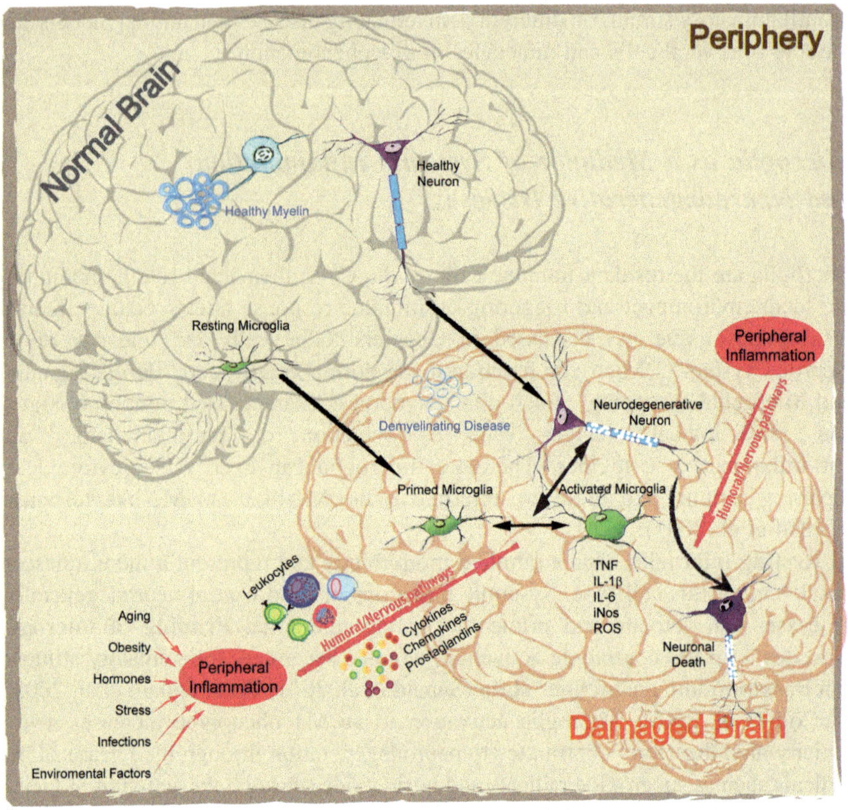


**Fig. 1** Schematic diagram showing the relationship between peripheral inflammation and demyelinating diseases. Demyelinating diseases are characterized by microglia activation; in which microglia change their morphology from resting (ramified) towards an activated round-shaped stage. The intermediate stage, "primed microglia," represents the microglial stage, which precedes a further neurotoxic microglial activation as a consequence of a secondary proinflammatory stimulus. The peripheral stimuli come from the periphery either through neural or humoral pathways and influence microglia activation. Activated microglia releases proinflammatory cytokines which can, in turn, act on myelin sheath integrity, thereby inducing demyelination, axonal loss and neurodegeneration

through the vagal afferent nerve (Perry et al. 2003; D'Mello et al. 2009; Gautron and Laye 2009; Teeling and Perry 2009; Campbell et al. 2010). The humoral pathway involves the direct action of peripheral proinflammatory cytokines (e.g., IL1β, TNFα, and IL6) and type I interferons (IFNα and IFNβ) that can initiate the synthesis of cytokines within the CNS, through blood–brain barrier (BBB) dependent or independent pathways (Perry et al. 2003; Teeling and Perry 2009).

## Demyelinating Diseases

As mentioned in Chapter "Glial cells and Integrity of the Nervous System", demyelinating diseases are a group of neurodegenerative diseases characterized by inflammatory lesions associated with demyelination, which in turn induces axonal damage, neurodegeneration, and progressive loss of function. Among them, the most important are MS, neuromyelitis optica (NMO), acute demyelinating encephalomyelitis, multifocal leukoencephalopathy, Guillain Barré syndrome, and acute disseminating encephalomyelitis. This review will mostly focus on MS and NMO, which are the most frequent in humans, and the most studied.

### *Multiple Sclerosis*

MS is a chronic inflammatory disease characterized by multifocal and repeated inflammatory events associated with demyelination–remyelination and axonal damage, which leads to poor conduction of the nervous impulse and eventual loss of sensory and motor function.

MS follows a varied clinical course, but most patients exhibit a course of repeating exacerbation and remission from the onset (relapsing/remitting MS or RRMS) eventually leading to secondary progressive multiple sclerosis (SPMS), which worsens the patients' quality of life (Playfair and Chain 1979; Neumann et al. 1998). A minority of patients exhibit primary progressive MS (PPMS), which is characterized by a constant decline from the onset with no recovery in neurological function (Playfair and Chain 1979; Loddick and Rothwell 2002).

Despite the fact that BBB breakdown is a major MS hallmark (McQuaid et al. 2009; Larochelle et al. 2011), some components of the inflammatory response contribute to the pathology even with an intact BBB (Buljevac et al. 2002; Lindquist et al. 2011). Although it has been proposed that in RRMS BBB breakdown allows the invasion of inflammatory cells, in the progressive forms inflammation remains enclosed behind an intact BBB (Playfair and Chain 1979).

#### Relapsing and Remitting MS

RRMS is the prevalent clinical type of MS and is characterized by recurrent episodes of new or worsened symptoms. Exacerbations or relapses are followed by periods of partial or complete remission, with apparent clinical stability between relapses. Relapsing episodes are unpredictable; however, peripheral inflammation may exacerbate these events (see below). Infections and other proinflammatory events have been postulated as possible triggers of the pathology and/or of relapsing episodes, and some authors have hypothesized that the autoimmune response could be a consequence of a primary central proinflammatory event (Barnett and Prineas 2004).

### Progressive Multiple Sclerosis

The progressive forms of MS lead to a continuous and irreversible evolution of the disease, inducing decline of the quality of life either from the onset (PPMS) or after a course of relapsing and remitting episodes (RRMS), named SPMS. SPMS is diagnosed as a worsening after relapsing-remitting phases, with or without acute exacerbations during the progressive stage (Wagner 1996). PPMS is a distinct, non-inflammatory, or less inflammatory pathologic form of MS. The progressive forms of MS are characterized by gray matter atrophy, which could be involved in physical and cognitive disability (Rivest et al. 2000; Pocock and Kettenmann 2007; Qian et al. 2012). Cortical lesions have peculiar inflammatory and demyelinating hallmarks, characterized by lack of BBB disruption, differential inflammatory process, and reactive microglia, suggesting different immunopathogenic mechanisms (Vitkovic et al. 2000). However, anti-inflammatory or immunomodulatory therapies have no effect on neurodegeneration and cognitive impairment in the progressive forms of MS (O'Connor et al. 2005; London et al. 2013). This could be related to the fact that in progressive MS, the inflammation creates an environment that favors retention of inflammatory cells within the lesions (Konsman et al. 1999; Godbout et al. 2005).

## *Neuromyelitis Optica*

NMO, or Devic's disease, is a demyelinating disease characterized by inflammatory demyelinating lesions mainly in the spinal cord and optic nerve, potentially leading to paralysis and blindness. It used to be considered a subtype of MS, but the pathology and clinical features make them different diseases (Mosher et al. 2001). NMO is characterized by seropositivity for immunoglobulin G (IgG) antibodies against the astrocytic water channel aquaporin-4 (AQP4), and secondary inflammation with granulocyte and macrophage infiltration, BBB disruption, and oligodendrocyte injury. Therefore, an adaptive immune response to AQP-4 underlies the chronic demyelinating in NMO.

The etiology of the disease is still unclear, but infections and BBB permeabilizing factors could be involved in triggering the overproduction of AQP4-IgG, and its access to the CNS (Schafer et al. 1999; Galiano et al. 2001). Uzawa et al. (2010) demonstrated a significant difference in the levels of some cytokines/chemokines (e.g. IL-6 for NMO) in the cerebrospinal fluid (CSF) of patients with NMO or MS, supporting the view that different immunological and pathophysiological mechanisms exist between them.

Current NMO therapies are directed toward reducing the inflammatory response and the NMO–IgG load, such as B cell depletion and plasmapheresis. However, most MS treatments, such as IFNβ, fingolimod, and natalizumab, exacerbate NMO. Therefore, it is necessary to better comprehend the diseases' underlying mechanisms and differentiate NMO from MS.

## MS and Peripheral Inflammation

MS is a neurodegenerative disease mainly characterized by inflammatory processes. Activation of systemic immunity affects primed microglia in the CNS, reactivating lesions and increasing parenchymal inflammation. Although relapsing episodes in RRMS are unpredictable, most relapses are concomitant with peripheral inflammation (Buljevac et al. 2002). RRMS patients show increased serum levels of IL1β, IL2, IL4, IL12p70, IFNγ, and TNFα during the relapse phase (Nathan 2006; Edwards et al. 2011; Trenova et al. 2011), as well as higher numbers of IL1β, IL6, and TNFα secreting cells (Ysrraelit et al. 2008), and increased levels of T helper ($T_h$)17 and Treg cells in the periphery (Edwards et al. 2011). Moreover, a change in CSF cytokine profile is observed during relapses; ranging from high levels of IL1β, TNFα, and transforming growth factor beta (TGFβ) to lower levels of IL-10 (Hauser et al. 1990; Edwards et al. 2011). However, the treatment of MS patients with TNFα inhibitors results in the exacerbation of central lesions (reviewed in Perry et al. 2003).

Differences in cytokine expression patterns are described when comparing progressive MS and RRMS (during relapses). SPMS patients present elevated levels of chemokine CC motif receptor 2 (CCR2) in T cells, increased serum/CSF levels of chemokine CC motif ligand 2 (CCL2) (Brinkmann et al. 2004), and decreased plasma/CSF values of TNFα and IL4 (Schmitz and Chew 2008). Peripheral blood mononuclear cells of both remitting RRMS and SPMS patients express low levels of IL10 mRNA, which return to basal levels during relapses in the RRMS form (Berkenbosch et al. 1987). Additionally, the progressive forms are characterized by a permanent peripheral type 1 immune activation, which could contribute to CNS damage during the progressive phase of the disease (Playfair and Chain 1979; Hampton et al. 1998). Thus, the peripheral blood of SPMS patients seems to reflect the inflammatory response accumulated in the CNS (Playfair and Chain 1979). On the other hand, RRMS is characterized by waves of T helper (Th)1 and Th17 cells, which are recruited into the brain causing the attacks (Neumann et al. 1998).

## Inflammatory Stimuli Associated with MS

Although there is a clear association between systemic inflammation and the onset or progression of different neurodegenerative pathologies, the particular nature of these inflammatory phenomena is also relevant. Numerous studies, cited below, have investigated the role of specific systemic proinflammatory stimuli including acute or chronic stimuli, physiological imbalances, or external infections and injuries. A summary of the roles of distinct proinflammatory stimuli in MS will be addressed in the following section.

## *Obesity*

During the last few years, a strong connection between metabolism, immunity, and inflammation was described. Obesity is considered an inflammatory disease, associated with metabolic and cardiovascular complications. Adipocyte tissue acts as an endocrine organ releasing adipocytokines, and is associated with increased levels of tissue and circulating inflammatory biomolecules (Oh et al. 1998). Excessive adipose tissue increases the number and activity of macrophages, mast cells, neutrophils, and lymphocytes (Ott et al. 1994; Kossmann et al. 1995). Moreover, leptin (an adipocyte-derived cytokine) has a role in regulating both innate and adaptive immunity (Bradl and Lassmann 2009), promoting the production of cytokines such as TNFβ, IL6, IL12, IL15, and granulocyte colony-stimulating factor in macrophages, and increasing their phagocytic activity, as well as inducing the chemotaxis of neutrophils (Bradl and Lassmann 2009; Lee et al. 2011; Golde et al. 2013; Procaccini et al. 2014). High levels of leptin have also been reported in both active inflammatory lesions and serum of MS patients (Batocchi et al. 2003).

Clinical data have demonstrated that obesity worsens the onset and progression of most autoimmune diseases, such as rheumatoid arthritis, systemic lupus erythematosus, inflammatory bowel disease, MS, type-1 diabetes, and psoriasis. Additionally, it impairs a positive response to the treatments usually given for these diseases (Cardona et al. 2008). Data show that 18-year old obese people are twice as likely to develop MS as their normal weight age mates (Banisadr et al. 2005). However, even if it seems quite clear that obesity and diet may influence the progression of MS, few studies have linked a caloric restriction diet to reduced MS progression (Procaccini et al. 2014).

## *Aging*

Aging processes induce a generalized proinflammatory state in the organism. This change is induced by increased immune responses in the periphery, disruption of the periphery-CNS immune communication, and an increment in "primed" microglia, which increases CNS reactivity (reviewed in Veenstra and Ransohoff 2012). Microglia in aged brains exhibit upregulated major histocompatibility complex (MHC) class II, complement receptors, toll-like receptors (TLR) 4, and cluster of differentiation (CD) 14 expression (see Chapter "Age-Dependent Changes in the Activation and Regulation of Microglia" for further reading). Therefore, peripheral innate immune stimulation induces microglial cells in aged brains to have an exaggerated inflammatory response compared with younger cohorts (Sly et al. 2001; Dilger and Johnson 2008).

PPMS and SPMS manifest around 10 years later than RRMS, therefore the timeline at which patients develop neurological deficit in PPMS and SPMS is remarkably similar, and both include aging as a major risk factor for MS progression (reviewed in Kutzelnigg et al. 2005).

## Infections

Infectious pathogens have been described as important factors involved in the development of MS. Moreover, clinical studies revealed an association between infections and relapses, which worsen neurological damage even after the infection is gone (Buljevac et al. 2002; Panitch 1994). Pathogens associated with the exacerbation include bacteria (such as *Mycoplasma pneumonia*, *Chlamydia pneumoniae*, and *Staphylococcus aureus*-produced enterotoxins), virus (Epstein-Barr virus and human herpes virus, and human endogenous retrovirus), and the protozoan (*Acanthamoeba castellanii*). Viral infections that trigger MS episodes can be reduced with IFN-γ treatment (Panitch 1994; Andersen et al. 1993).

Studies of MS patients and of animal experimental models have demonstrated the influence of these infectious agents on the development and/or exacerbation of MS (Krieger et al. 1992). However, not all infections cause progression of MS, since it has been reported that infections with some parasites, such as helminthes, can protect against the exacerbation phase of the disease (Correale and Farez 2011a, b; Krieger et al. 1992). This protection is associated with the induction of CD4+, CD25+ T cells secreting IL10, and TGFβ (Correale and Farez 2011a).

## Immune Regulation by the Neuroendocrine System

The neuroendocrine system exerts its action on the immune system through finely tune regulation. Glucocorticoids (GCs) induce the production of pro- and anti-inflammatory cytokines, specifically causing a shift from Th1 to Th2 immune response. GCs inhibit the production of Th1 related cytokines (IL1 and IL6, IL2, IL12, IFNγ) and increase the secretion of anti-inflammatory Th2 cytokines (IL4 and IL10) (Haak et al. 2009). However, GCs can increase both peripheral and central inflammatory responses to a systemic challenge if they are administered before the peripheral stimuli (Sorrells and Sapolsky 2010; Frank et al. 2010). Therefore, GCs can prime the immune response and, as a consequence, increase proinflammatory cytokine production and exacerbation of MS symptoms.

### The Hypothalamic–Pituitary–Adrenal (HPA) Axis

Clinical and experimental studies have demonstrated that abnormalities in the HPA axis, which influences the immune response, may exacerbate MS symptoms (Hofstetter et al. 2005; Seo et al. 2013). Thus high cortisol levels are often correlated with acute relapses (Hofstetter et al. 2005; Seo et al. 2013), whereas prolactin increases the peripheral production of IFNγ and IL12 by T cells (Du and Dreyfus 2002).

### The Hypothalamic–Pituitary–Gonadal (HPG) Axis

MS affects predominantly women in comparison with men, therefore, considering that gender affects the course of autoimmune diseases, the influence of sex hormones is critical (Dunn et al. 2015a, b). In particular, 17β-estradiol induces an increase of Th2 cytokines (IL10 and IL4) and a decrease of Th1 cytokines (TNFα and IFNγ) (van Riemsdijk et al. 2001; Janik et al. 1997). Estrogens, in addition to their anti-inflammatory effects, appear to be neuroprotective in CNS diseases, such as MS and Alzheimer's, disease (Nicot 2009; Gao and Tsirka 2011). Additionally, both clinical symptoms and relapse rates of MS are decreased during pregnancy, whereas the postpartum period increases the risk for exacerbation of the disease (Ling et al. 1997).

The increased secretion of estrogen, progesterone, and cortisol during pregnancy is associated with increased production of Th2 cytokines and decreased production of Th1 cytokines (Takii et al. 1992, 1994). Additionally, progesterone also inhibits NFκB and increases IL4 production, demonstrating its anti-inflammatory effect (Piccinni et al. 1995; Nishiyori et al. 1997). Male hormones, such as testosterone, also inhibit both innate and adaptive immune systems by enhancing the production of IL5 and IL10, and decreasing IFNβ secretion, thus promoting a Th2 response (Murphy and Sturm 1923).

## *Environmental Factors and Peripheral Inflammation*

Environmental factors have influence on most autoimmune diseases. Epidemiological risk factors for MS, including low vitamin D and elevated salt intake, are associated with peripheral inflammation. Recent studies have shown that components of the daily diet and gut microbiota can strongly affect the levels of effector T cells in the gut (Ransohoff et al. 2007).

On the other hand, high sodium chloride concentrations induced expression of serum glucocorticoid kinase 1 (SGK1) in T cells, which in turn stimulate the induction of Th17 cells from CD4+ T cells, promoting autoimmune diseases (Glabinski et al. 1997). However, direct correlation between salt intake and incidence of autoimmune disease is yet to be demonstrated (Tsai et al. 2002).

Vitamin D plays an important role in the regulation of the immune responses (Semple et al. 2010), modulating many inflammatory mechanisms including: (a) the regulation of inflammatory mediators, such as cytokines (IL1β, TNFα, IL6, TGF1β) and cyclooxygenases, (b) the interference with transcription factors, such as NFκB, and (c) the activation of signaling cascades, such as MAP kinases

(Xia and Hyman 2002; Bakshi et al. 2011; Semple et al. 2010; Perry and Teeling 2013). MS exacerbation correlates with low levels of Vitamin D, whereas vitamin D supplementation has a protective effect (Aubert et al. 1995; Romeo et al. 2001; Varvel et al. 2012).

## NMO and Peripheral Inflammation

There is not much evidence for peripheral inflammation affecting NMO, in contrast with MS. However, recent work shows that the peripheral immune system affects the progression of NMO. The CSF of NMO patients shows white blood cells (WBC) $\geq 50$ cells/mm$^3$ or $\geq 5$ neutrophils/mm$^3$ compared to control patients, whose counts are $<5$ WBC/mm$^3$ (Campbell et al. 2008). Additionally, removing inflammatory mediators from the blood of MNO patients alleviates the symptoms (Okada et al. 2006).

NMO can occur concomitantly with systemic autoimmune disorders such as Sjogren's syndrome and systemic lupus erythematous, which likely reflects an underlying predisposition for these patients to develop autoimmune disorders. Moreover, the presence of other systemic disease can increase the mortality rate in relapsing NMO patients (Krady et al. 2008).

Finally, the seropositivity for NMO–IgG represents a key factor for predicting future relapses; indeed, it is a prognostic marker for NMO. Additionally, humoral immune mechanisms, including the activation of B cells and the complement pathway, have been said to play a role in NMO pathogenesis (Quan et al. 2013; Kim et al. 2011).

## Experimental Models of Demyelinating Diseases

Experimental models of demyelination help in understanding the pathophysiology of such demyelinating diseases as MS (Denic et al. 2011) and NMO (Linington et al. 1992). Animal models can be divided into two groups: those which attempt to replicate the disease as accurately as possible and others that provide a reductionist approach to the diseases by studying demyelination and remyelination processes (e.g., ethidium bromide, lysolecithin, and cuprizone) (reviewed in Blakemore and Franklin 2008). For MS, the most common models have been virus-induced encephalomyelitis and various forms of Experimental Autoimmune Encephalomyelitis (EAE) (reviewed in Dai et al. 2003).

A clear distinction between NMO and MS only became possible in the past decade, and nowadays the most frequently used NMO models are NMO/EAE, NMO-IgG/complement intracerebral injection, and cytokine-injection NMO (Linington et al. 1992). Some of the main features present in these experimental models are summarized in Table 1. Consistent with the human diseases, animal

**Table 1** Summary of the main features of MS and NMO experimental models

| Experimental model | Disease features | Reference |
|---|---|---|
| Lysolecithin and Ethidium Bromide Demyelination | **MS and NMO**<br>Demyelination–Remyelination<br>NO autoimmune component<br>Microglial and Astroglial activation | Blakemore (2008) |
| Cuprizone induced Demyelination | **MS and NMO**<br>Demyelination–Remyelination<br>No autoimmune component<br>Microglial and Astroglial activation<br>Cytokine mediated inflammatory response<br>Growth factors involved in remyelination | Wilkins et al. (2001) |
| Virus- induced Encephalomyelitis | **MS**<br>BBB breakdown<br>Demyelination<br>Axon pathology<br>Cytokine upregulation<br>Central and systemic inflammatory response<br>Autoimmune component<br>Involvement of different immune cells (T cells, B cells)<br>Microglial and Astroglial activation | Grigoriadis and Hadjigeorgiou (2006) |
| EAE (active, passive, or transgenic models) | **MS**<br>Relapsing-remitting and progressive forms<br>BBB breakdown<br>Demyelination (sometimes remyelination)<br>Axon pathology<br>Cytokine upregulation<br>Central and systemic inflammatory response<br>Autoimmune component<br>Involvement of different immune cells (depending on the model): T cells, B cells, granulocytes<br>Microglial and Astroglial activation | Dai et al. (2003) |
| NMO/EAE | **NMO**<br>Autoimmune component (AQP4 IgG)<br>BBB breakdown<br>Loss of AQP4 expression<br>Presence of astrocyte destructive lesions<br>Neutrophil and T cell infiltration to the CNS<br>Activation of microglia/macrophages<br>Demyelination<br>Oligodendrocyte death | Linington et al. (1992)<br>Bradl and Lassmann (2014) |
| NMO-IgG/complement | **NMO**<br>Autoimmune component (AQP4 IgG)<br>BBB breakdown<br>Loss of AQP4 expression<br>Presence of astrocyte destructive lesions<br>Neutrophil and macrophage infiltration to the CNS<br>Demyelination<br>Neuronal death | Linington et al. (1992) |

(continued)

**Table 1** (continued)

| Experimental model | Disease features | Reference |
|---|---|---|
| Cytokine-injection NMO | **NMO**<br>Autoimmune component (AQP4 IgG)<br>BBB breakdown<br>Loss of AQP4 expression<br>Presence of astrocyte destructive lesions<br>Neutrophil infiltration to the CNS<br>Demyelination<br>Activation of microglia/macrophages | Linington et al. (1992) |

experimental models show the influence of peripheral inflammation on the progression of the disorders.

The importance of humoral components of the immune system is evident in EAE. For example, a specific cytokine profile appears during the different phases of acute the EAE model: decreased IL21 expression on the peak phase and high IL22 expression during the induction phase that decreases during recovery (Almolda et al. 2011). Additionally, systemic TNFα causes clinical signs to recrudesce and induces relapses in EAE (Crisi et al. 1995).

## MS Animal Models and Systemic Inflammation

### *Obesity*

Immunomodulatory effects of leptin, the adipocyte-derived hormone, are involved in the induction and progression of EAE (Matarese et al. 2001, 2008). In this context, the use of leptin antagonists improved the course of EAE (De Rosa et al. 2006). Moreover, the leptin-deficient (*ob/ob*) mice do not develop EAE; however, exogenous leptin treatment renders *ob/ob* mice susceptible to EAE development (Matarese et al. 2001). On the other hand, caloric restriction, (associated with low levels of leptin in plasma) can significantly increase the overall survival in several experimental animal models of autoimmune diseases (Oka et al. 2007).

### *Infections*

Peripheral infection with enterotoxin A or B exacerbates clinical signs and induces relapses in EAE (Brocke et al. 1993; Crisi et al. 1995; Schiffenbauer et al. 1993). A single dose of peripheral LPS can induce increased inflammatory, demyelinating and axonal damage in EAE lesions (Serres et al. 2009; Moreno et al. 2011) as well as CD4+ cells activation (Nogai et al. 2005). Additionally, respiratory tract pathogens (*Streptococcus pneumonia* and *Chlamydia pneumonia*) aggravate EAE symptoms (Du et al. 2002; Herrmann et al. 2006; Tauber et al. 2007).

On the other hand, some data have been published demonstrating beneficial effects of peripheral LPS. In those studies, pretreatment with LPS prior to EAE induction lead to a delay in the onset of the disease by suppressing antigen presentation and altering the expression of inflammatory mediators (Buenafe and Bourdette 2007).

The presence of blood-derived peripheral polymorphonuclear neutrophils (PMN) expressing CXC chemokine receptor type 2 (CXCR2) is requisite for oligodendrocyte death, demyelination, and BBB breakdown in both EAE and cuprizone models (Liu et al. 2010; Carlson et al. 2008). Peripheral PMN are considered the first key effector leukocytes in the pathogenesis of EAE; they produce cytokines and chemokines that in turn induce lymphocyte and monocyte activation (Carlson et al. 2008).

Moreover, the importance of PMN neutrophils for the development of a demyelinating lesion in the CNS of rats has been seen in a model of chronic neuroinflammation and demyelination in response to a sustained expression of IL1β in the CNS (Ferrari et al. 2004). Furthermore, a *relapsing-like* lesion was achieved in the same model by inducing a peripheral sustained expression of IL1β (Murta et al. 2015). Here, the involvement of CXCR2 + PMN neutrophils from the periphery was also proven central for the development of the relapse.

## Immune Regulation by the Neuroendocrine System

Estrogen inhibits clinical and histological symptoms of EAE, and pretreatment with low doses of 17beta-estradiol (E2) diminishes the symptoms of EAE by inhibiting cell migration into the CNS and promoting axon and myelin survival (Wolswijk 1998; reviewed in Murta and Ferrari 2013). Moreover, in EAE animals progesterone decreases proinflammatory cytokine secretion (IL12, IL17), increases IL10 production, and increases the CD19 + and CD8 + populations (Chang et al. 2002).

## Environmental Factors

Diet also represents an important factor in experimental animal models. In EAE mice, a high salt diet increases the number of Th17 cells, worsening the disease (Tsai et al. 2002). Conversely, vitamin D (or its metabolite 1.25-dihydroxyvitamin D3) reverses the EAE symptoms by inhibiting chemokine and inducible nitric oxide synthase (iNOS) synthesis, and CD11b + monocyte trafficking into the CNS

(Moynagh 2005). Moreover, this vitamin also suppresses EAE female selectivity (Byravan et al. 1994).

Another environmental factor associated with EAE progression is UV irradiation: several authors have shown that UV irradiation suppresses EAE by inducing immunosuppression through an alteration of dendritic and regulatory T cells, independently of vitamin D production (Hauser et al. 1984; Waxman 1998; Lappe-Siefke et al. 2003; Ng et al. 2013).

Moreover, the influence of the microbiome on different pathological conditions has been investigated. In some models of EAE, gut microflora-free animals are resistant to the induction of RR-EAE and have decreased Th17 and B cell responses (Tsunoda and Fujinami 2002; Tsunoda et al. 2003; Huitinga et al. 2000).

## Concluding Remarks

Systemic inflammatory insults are risk factors in both the etiology and progression of demyelinating diseases. The interaction between damaged brain and systemic inflammation may be responsible for the progression of neurodegenerative diseases. However, certain systemic stimuli may be beneficial for both disease progression and repair. Primed microglial cells in the diseased CNS are viewed as one of the key components in the exacerbation of central damage due to systemic inflammatory stimuli in most CNS diseases. Additionally, the peripheral immune system contributes significantly to the pathophysiology of the demyelinating diseases discussed in the present review and their animal models, and the environment appears also to be important. Better understanding of the mechanisms of CNS and immune system communication should improve therapeutics for immune mediated diseases.

**Acknowledgments** Carina C. Ferrari and Verónica Murta are members of the Research Career of the National Council of Scientific and Technological Research (CONICET), Argentina. CF is supported by CONICET (PIP 2012-2014, 11220110100560) and National Agency of Science and Technology of Argentina (ANPCyT) (PICT 2012-2014).

## References

Almolda B, Gonzalez B, Castellano B (2011) Antigen presentation in EAE: role of microglia, macrophages and dendritic cells. Front Biosci 16:1157–1171 (3781 [pii])

Andersen O, Lygner PE, Bergstrom T, Andersson M, Vahlne A (1993) Viral infections trigger multiple sclerosis relapses: a prospective seroepidemiological study. J Neurol 240(7):417–422

Arimoto T, Bing G (2003) Up-regulation of inducible nitric oxide synthase in the substantia nigra by lipopolysaccharide causes microglial activation and neurodegeneration. Neurobiol Dis 12(1):35–45

Aubert A, Vega C, Dantzer R, Goodall G (1995) Pyrogens specifically disrupt the acquisition of a task involving cognitive processing in the rat. Brain Behav Immun 9(2):129–148. doi:10.1006/brbi.1995.1013 (S0889-1591(85)71013-6 [pii])

Bakshi P, Margenthaler E, Reed J, Crawford F, Mullan M (2011) Depletion of CXCR2 inhibits gamma-secretase activity and amyloid-beta production in a murine model of Alzheimer's disease. Cytokine 53(2):163–169 (10.1016/j.cyto.2010.10.008S1043-4666(10)00703-9 [pii])

Banisadr G, Rostene W, Kitabgi P, Parsadaniantz SM (2005) Chemokines and brain functions. Curr Drug Targets Inflamm Allergy 4(3):387–399

Barnett MH, Prineas JW (2004) Relapsing and remitting multiple sclerosis: pathology of the newly forming lesion. Ann Neurol 55(4):458–468

Batocchi AP, Rotondi M, Caggiula M, Frisullo G, Odoardi F, Nociti V, Carella C, Tonali PA, Mirabella M (2003) Leptin as a marker of multiple sclerosis activity in patients treated with interferon-beta. J Neuroimmunol 139(1–2):150–154 (S0165572803001541 [pii])

Berkenbosch F, van Oers J, del Rey A, Tilders F, Besedovsky H (1987) Corticotropin-releasing factor-producing neurons in the rat activated by interleukin-1. Science 238(4826):524–526

Besedovsky HO, del Rey A (1996) Immune-neuro-endocrine interactions: facts and hypotheses. Endocr Rev 17(1):64–102

Blakemore WF (2008) Regeneration and repair in multiple sclerosis: the view of experimental pathology. J Neurol Sci 265(1–2):1–4

Blakemore WF, Franklin RJ (2008) Remyelination in experimental models of toxin-induced demyelination. Curr Top Microbiol Immunol 318:193–212

Bradl M, Lassmann H (2009) Progressive multiple sclerosis. Semin Immunopathol 31(4):455–465. doi:10.1007/s00281-009-0182-3

Bradl M, Lassmann H (2014) Experimental models of neuromyelitis optica. Brain Pathol 24 (1):74–82. doi:10.1111/bpa.12098

Brinkmann V, Reichard U, Goosmann C, Fauler B, Uhlemann Y, Weiss DS, Weinrauch Y, Zychlinsky A (2004) Neutrophil extracellular traps kill bacteria. Science 303(5663):1532–1535 (10.1126/science.1092385303/5663/1532 [pii])

Brocke S, Gaur A, Piercy C, Gautam A, Gijbels K, Fathman CG, Steinman L (1993) Induction of relapsing paralysis in experimental autoimmune encephalomyelitis by bacterial superantigen. Nature 365(6447):642–644. doi:10.1038/365642a0

Buenafe AC, Bourdette DN (2007) Lipopolysaccharide pretreatment modulates the disease course in experimental autoimmune encephalomyelitis. J Neuroimmunol 182(1–2):32–40. doi:10.1016/j.jneuroim.2006.09.004 (S0165-5728(06)00371-7 [pii])

Buljevac D, Flach HZ, Hop WC, Hijdra D, Laman JD, Savelkoul HF, van Der Meche FG, van Doorn PA, Hintzen RQ (2002) Prospective study on the relationship between infections and multiple sclerosis exacerbations. Brain 125(Pt 5):952–960

Byravan S, Foster LM, Phan T, Verity AN, Campagnoni AT (1994) Murine oligodendroglial cells express nerve growth factor. Proc Natl Acad Sci USA 91(19):8812–8816

Campbell SJ, Anthony DC, Oakley F, Carlsen H, Elsharkawy AM, Blomhoff R, Mann DA (2008) Hepatic nuclear factor kappa B regulates neutrophil recruitment to the injured brain. J Neuropathol Exp Neurol 67(3):223–230 (10.1097/NEN.0b013e318165495700005072-200803000-00005 [pii])

Campbell SJ, Meier U, Mardiguian S, Jiang Y, Littleton ET, Bristow A, Relton J, Connor TJ, Anthony DC (2010) Sickness behaviour is induced by a peripheral CXC-chemokine also expressed in multiple sclerosis and EAE. Brain Behav Immun 24(5):738–746 (S0889-1591(10)00038-3 [pii] 10.1016/j.bbi.2010.01.011)

Cardona AE, Li M, Liu L, Savarin C, Ransohoff RM (2008) Chemokines in and out of the central nervous system: much more than chemotaxis and inflammation. J Leukoc Biol 84(3):587–594 (10.1189/jlb.1107763jlb.1107763 [pii])

Carlson T, Kroenke M, Rao P, Lane TE, Segal B (2008) The Th17-ELR + CXC chemokine pathway is essential for the development of central nervous system autoimmune disease. J Exp Med 205(4):811–823 (jem.20072404 [pii] 10.1084/jem.20072404)

Combrinck MI, Perry VH, Cunningham C (2002) Peripheral infection evokes exaggerated sickness behaviour in pre-clinical murine prion disease. Neuroscience 112(1):7–11

Correale J, Farez MF (2011a) The impact of environmental infections (parasites) on MS activity. Mult Scler 17(10):1162–1169 (17/10/1162 [pii] 10.1177/1352458511418027)

Correale J, Farez MF (2011b) The impact of parasite infections on the course of multiple sclerosis. J Neuroimmunol 233(1–2):6–11 (S0165-5728(11)00005-1 [pii] 10.1016/j.jneuroim.2011.01.002)

Crisi GM, Santambrogio L, Hochwald GM, Smith SR, Carlino JA, Thorbecke GJ (1995) Staphylococcal enterotoxin B and tumor-necrosis factor-alpha-induced relapses of experimental allergic encephalomyelitis: protection by transforming growth factor-beta and interleukin-10. Eur J Immunol 25(11):3035–3040. doi:10.1002/eji.1830251108

Cunningham C, Wilcockson DC, Boche D, Perry VH (2005a) Comparison of inflammatory and acute-phase responses in the brain and peripheral organs of the ME7 model of prion disease. J Virol 79(8):5174–5184 (79/8/5174 [pii] 10.1128/JVI.79.8.5174-5184.2005)

Cunningham C, Wilcockson DC, Campion S, Lunnon K, Perry VH (2005b) Central and systemic endotoxin challenges exacerbate the local inflammatory response and increase neuronal death during chronic neurodegeneration. J Neurosci 25(40):9275–9284

Czlonkowska A, Kurkowska-Jastrzebska I, Czlonkowski A, Peter D, Stefano GB (2002) Immune processes in the pathogenesis of Parkinson's disease—a potential role for microglia and nitric oxide. Med Sci Monit 8(8):165–177

Chang A, Tourtellotte WW, Rudick R, Trapp BD (2002) Premyelinating oligodendrocytes in chronic lesions of multiple sclerosis. N Engl J Med 346(3):165–173 (10.1056/NEJMoa010994346/3/165 [pii])

D'Mello C, Le T, Swain MG (2009) Cerebral microglia recruit monocytes into the brain in response to tumor necrosis factoralpha signaling during peripheral organ inflammation. J Neurosci 29(7):2089–2102 (29/7/2089 [pii] 10.1523/JNEUROSCI.3567-08.2009)

Dai X, Lercher LD, Clinton PM, Du Y, Livingston DL, Vieira C, Yang L, Shen MM, Dreyfus CF (2003) The trophic role of oligodendrocytes in the basal forebrain. J Neurosci 23(13):5846–5853 (23/13/5846 [pii])

Dantzer R, Bluthe RM, Laye S, Bret-Dibat JL, Parnet P, Kelley KW (1998) Cytokines and sickness behavior. Ann N Y Acad Sci 840:586–590

Dantzer R, O'Connor JC, Freund GG, Johnson RW, Kelley KW (2008) From inflammation to sickness and depression: when the immune system subjugates the brain. Nat Rev Neurosci 9 (1):46–56 (nrn2297 [pii] 10.1038/nrn2297)

De Rosa V, Procaccini C, La Cava A, Chieffi P, Nicoletti GF, Fontana S, Zappacosta S, Matarese G (2006) Leptin neutralization interferes with pathogenic T cell autoreactivity in autoimmune encephalomyelitis. J Clin Invest 116(2):447–455. doi:10.1172/JCI26523

Denic A, Johnson AJ, Bieber AJ, Warrington AE, Rodriguez M, Pirko I (2011) The relevance of animal models in multiple sclerosis research. Pathophysiology 18(1):21–29 (S0928-4680(10)00022-2 [pii] 10.1016/j.pathophys.2010.04.004)

Dilger RN, Johnson RW (2008) Aging, microglial cell priming, and the discordant central inflammatory response to signals from the peripheral immune system. J Leukoc Biol 84 (4):932–939 (jlb.0208108 [pii] 10.1189/jlb.0208108)

Du C, Yao SY, Ljunggren-Rose A, Sriram S (2002) Chlamydia pneumoniae infection of the central nervous system worsens experimental allergic encephalitis. J Exp Med 196(12):1639–1644

Du Y, Dreyfus CF (2002) Oligodendrocytes as providers of growth factors. J Neurosci Res 68 (6):647–654. doi:10.1002/jnr.10245

Dunn SE, Gunde E, Lee H (2015a) Sex-based differences in multiple sclerosis (MS): part II: rising incidence of multiple sclerosis in women and the vulnerability of men to progression of this disease. Curr Top Behav Neurosci. doi:10.1007/7854_2015_370

Dunn SE, Lee H, Pavri FR, Zhang MA (2015b) Sex-based differences in multiple sclerosis (part I): biology of disease incidence. Curr Top Behav Neurosci. doi:10.1007/7854_2015_371

Edwards LJ, Sharrack B, Ismail A, Tumani H, Constantinescu CS (2011) Central inflammation versus peripheral regulation in multiple sclerosis. J Neurol 258(8):1518–1527. doi:10.1007/s00415-011-5973-5

Ferrari CC, Depino AM, Prada F, Muraro N, Campbell S, Podhajcer O, Perry VH, Anthony DC, Pitossi FJ (2004) Reversible demyelination, blood-brain barrier breakdown, and pronounced neutrophil recruitment induced by chronic il-1 expression in the brain. Am J Pathol 165(5):1827–1837

Ferrari CC, Tarelli R (2011) Parkinson's disease and systemic inflammation. Parkinsons Dis 2011:436813. doi:10.4061/2011/436813

Frank MG, Miguel ZD, Watkins LR, Maier SF (2010) Prior exposure to glucocorticoids sensitizes the neuroinflammatory and peripheral inflammatory responses to E. coli lipopolysaccharide. Brain Behav Immun 24(1):19–30 (S0889-1591(09)00386-9 [pii] 10.1016/j.bbi.2009.07.008)

Fung A, Vizcaychipi M, Lloyd D, Wan Y, Ma D (2012) Central nervous system inflammation in disease related conditions: mechanistic prospects. Brain Res 1446:144–155 (10.1016/j.brainres.2012.01.061S0006-8993(12)00164-3 [pii])

Galiano M, Liu ZQ, Kalla R, Bohatschek M, Koppius A, Gschwendtner A, Xu S, Werner A, Kloss CU, Jones LL, Bluethmann H, Raivich G (2001) Interleukin-6 (IL6) and cellular response to facial nerve injury: effects on lymphocyte recruitment, early microglial activation and axonal outgrowth in IL6-deficient mice. Eur J Neurosci 14(2):327–341 (ejn1647 [pii])

Gao Z, Tsirka SE (2011) Animal Models of MS Reveal Multiple Roles of Microglia in Disease Pathogenesis. Neurol Res Int 2011:383087. doi:10.1155/2011/383087

Gautron L, Laye S (2009) Neurobiology of inflammation-associated anorexia. Front Neurosci 3:59. doi:10.3389/neuro.23.003.2009

Glabinski AR, Tani M, Strieter RM, Tuohy VK, Ransohoff RM (1997) Synchronous synthesis of alpha- and beta-chemokines by cells of diverse lineage in the central nervous system of mice with relapses of chronic experimental autoimmune encephalomyelitis. Am J Pathol 150(2):617–630

Godbout JP, Chen J, Abraham J, Richwine AF, Berg BM, Kelley KW, Johnson RW (2005) Exaggerated neuroinflammation and sickness behavior in aged mice following activation of the peripheral innate immune system. FASEB J 19(10):1329–1331 (05-3776fje [pii] 10.1096/fj.05-3776fje)

Golde TE, Streit WJ, Chakrabarty P (2013) Alzheimer's disease risk alleles in TREM2 illuminate innate immunity in Alzheimer's disease. Alzheimers Res Ther 5(3):24 (alzrt178 [pii] 10.1186/alzrt178)

Grigoriadis N, Hadjigeorgiou GM (2006) Virus-mediated autoimmunity in Multiple Sclerosis. J Autoimmune Dis 3:1

Haak S, Croxford AL, Kreymborg K, Heppner FL, Pouly S, Becher B, Waisman A (2009) IL-17A and IL-17F do not contribute vitally to autoimmune neuro-inflammation in mice. J Clin Invest 119(1):61–69 (10.1172/JCI3599735997 [pii])

Hampton MB, Kettle AJ, Winterbourn CC (1998) Inside the neutrophil phagosome: oxidants, myeloperoxidase, and bacterial killing. Blood 92(9):3007–3017

Hauser SL, Doolittle TH, Lincoln R, Brown RH, Dinarello CA (1990) Cytokine accumulations in CSF of multiple sclerosis patients: frequent detection of interleukin-1 and tumor necrosis factor but not interleukin-6. Neurology 40(11):1735–1739

Hauser SL, Weiner HL, Che M, Shapiro ME, Gilles F, Letvin NL (1984) Prevention of experimental allergic encephalomyelitis (EAE) in the SJL/J mouse by whole body ultraviolet irradiation. J Immunol 132(3):1276–1281

Henry CJ, Huang Y, Wynne AM, Godbout JP (2009) Peripheral lipopolysaccharide (LPS) challenge promotes microglial hyperactivity in aged mice that is associated with exaggerated induction of both pro-inflammatory IL-1beta and anti-inflammatory IL-10 cytokines. Brain Behav Immun 23(3):309–317 (S0889-1591(08)00348-6 [pii] 10.1016/j.bbi.2008.09.002)

Herrmann I, Kellert M, Schmidt H, Mildner A, Hanisch UK, Bruck W, Prinz M, Nau R (2006) Streptococcus pneumoniae Infection aggravates experimental autoimmune encephalomyelitis via Toll-like receptor 2. Infect Immun 74(8):4841–4848 (74/8/4841 [pii] 10.1128/IAI.00026-06)

Hofstetter HH, Ibrahim SM, Koczan D, Kruse N, Weishaupt A, Toyka KV, Gold R (2005) Therapeutic efficacy of IL-17 neutralization in murine experimental autoimmune encephalomyelitis. Cell Immunol 237(2):123–130 (S0008-8749(05)00242-X [pii] 10.1016/j.cellimm.2005.11.002)

Huitinga I, Schmidt ED, van der Cammen MJ, Binnekade R, Tilders FJ (2000) Priming with interleukin-1beta suppresses experimental allergic encephalomyelitis in the Lewis rat. J Neuroendocrinol 12(12):1186–1193 574 [pii]

Janik JE, Curti BD, Considine RV, Rager HC, Powers GC, Alvord WG, Smith JW 2nd, Gause BL, Kopp WC (1997) Interleukin 1 alpha increases serum leptin concentrations in humans. J Clin Endocrinol Metab 82(9):3084–3086

Kim SH, Kim W, Li XF, Jung IJ, Kim HJ (2011) Repeated treatment with rituximab based on the assessment of peripheral circulating memory B cells in patients with relapsing neuromyelitis optica over 2 years. Arch Neurol 68(11):1412–1420 (10.1001/archneurol.2011.154arch neurol.2011.154 [pii])

Kim SU, de Vellis J (2005) Microglia in health and disease. J Neurosci Res 81(3):302–313

Konsman JP, Kelley K, Dantzer R (1999) Temporal and spatial relationships between lipopolysaccharide-induced expression of Fos, interleukin-1beta and inducible nitric oxide synthase in rat brain. Neuroscience 89(2):535–548 (S0306-4522(98)00368-6 [pii])

Kossmann T, Hans VH, Imhof HG, Stocker R, Grob P, Trentz O, Morganti-Kossmann C (1995) Intrathecal and serum interleukin-6 and the acute-phase response in patients with severe traumatic brain injuries. Shock 4(5):311–317

Krady JK, Lin HW, Liberto CM, Basu A, Kremlev SG, Levison SW (2008) Ciliary neurotrophic factor and interleukin-6 differentially activate microglia. J Neurosci Res 86(7):1538–1547. doi:10.1002/jnr.21620

Krieger M, Brunner T, Bischoff SC, von Tscharner V, Walz A, Moser B, Baggiolini M, Dahinden CA (1992) Activation of human basophils through the IL-8 receptor. J Immunol 149 (8):2662–2667

Kutzelnigg A, Lucchinetti CF, Stadelmann C, Bruck W, Rauschka H, Bergmann M, Schmidbauer M, Parisi JE, Lassmann H (2005) Cortical demyelination and diffuse white matter injury in multiple sclerosis. Brain 128(Pt 11):2705–2712 (awh641 [pii] 10.1093/brain/awh641)

Lappe-Siefke C, Goebbels S, Gravel M, Nicksch E, Lee J, Braun PE, Griffiths IR, Nave KA (2003) Disruption of Cnp1 uncouples oligodendroglial functions in axonal support and myelination. Nat Genet 33(3):366–374 (10.1038/ng1095ng1095 [pii])

Larochelle C, Alvarez JI, Prat A (2011) How do immune cells overcome the blood-brain barrier in multiple sclerosis? FEBS Lett 585(23):3770–3780 (S0014 5793(11)00336-X [pii] 10.1016/j.febslet.2011.04.066)

Lee M, Schwab C, McGeer PL (2011) Astrocytes are GABAergic cells that modulate microglial activity. Glia 59(1):152–165. doi:10.1002/glia.21087

Lindquist S, Hassinger S, Lindquist JA, Sailer M (2011) The balance of pro-inflammatory and trophic factors in multiple sclerosis patients: effects of acute relapse and immunomodulatory treatment. Mult Scler 17(7):851–866 (1352458511399797 [pii] 10.1177/1352458511399797)

Ling PR, Schwartz JH, Bistrian BR (1997) Mechanisms of host wasting induced by administration of cytokines in rats. Am J Physiol 272(3 Pt 1):E333–E339

Linington C, Engelhardt B, Kapocs G, Lassman H (1992) Induction of persistently demyelinated lesions in the rat following the repeated adoptive transfer of encephalitogenic T cells and demyelinating antibody. J Neuroimmunol 40(2–3):219–224

Liu L, Belkadi A, Darnall L, Hu T, Drescher C, Cotleur AC, Padovani-Claudio D, He T, Choi K, Lane TE, Miller RH, Ransohoff RM (2010) CXCR2-positive neutrophils are essential for cuprizone-induced demyelination: relevance to multiple sclerosis. Nat Neurosci 13(3):319–326 (nn.2491 [pii] 10.1038/nn.2491)

Loddick S, Rothwell N (2002) Cytokines and neurodegeneration. In: Loddick S, Rothwell N (eds) Immune an inflammatory responses in the nervous system. Oxford University Press, Oxford, pp 90–105

London A, Cohen M, Schwartz M (2013) Microglia and monocyte-derived macrophages: functionally distinct populations that act in concert in CNS plasticity and repair. Front Cell Neurosci 7:34. doi:10.3389/fncel.2013.00034

Londono D, Cadavid D (2010) Bacterial lipoproteins can disseminate from the periphery to inflame the brain. Am J Pathol 176(6):2848–2857 (ajpath.2010.091235 [pii] 10.2353/ajpath.2010.091235)

Matarese G, Di Giacomo A, Sanna V, Lord GM, Howard JK, Di Tuoro A, Bloom SR, Lechler RI, Zappacosta S, Fontana S (2001) Requirement for leptin in the induction and progression of autoimmune encephalomyelitis. J Immunol 166(10):5909–5916

Matarese G, Procaccini C, De Rosa V (2008) The intricate interface between immune and metabolic regulation: a role for leptin in the pathogenesis of multiple sclerosis? J Leukoc Biol 84(4):893–899 (jlb.0108022 [pii] 10.1189/jlb.0108022)

McColl BW, Rothwell NJ, Allan SM (2007) Systemic inflammatory stimulus potentiates the acute phase and CXC chemokine responses to experimental stroke and exacerbates brain damage via interleukin-1- and neutrophil-dependent mechanisms. J Neurosci 27(16):4403–4412

McQuaid S, Cunnea P, McMahon J, Fitzgerald U (2009) The effects of blood-brain barrier disruption on glial cell function in multiple sclerosis. Biochem Soc Trans 37(Pt 1):329–331

Minghetti L, Polazzi E, Nicolini A, Greco A, Levi G (1999) Possible role of microglial prostanoids and free radicals in neuroprotection and neurodegeneration. Adv Exp Med Biol 468:109–119

Moreno B, Jukes JP, Vergara-Irigaray N, Errea O, Villoslada P, Perry VH, Newman TA (2011) Systemic inflammation induces axon injury during brain inflammation. Ann Neurol 70(6):932–942. doi:10.1002/ana.22550

Mosher B, Dean R, Harkema J, Remick D, Palma J, Crockett E (2001) Inhibition of Kupffer cells reduced CXC chemokine production and liver injury. J Surg Res 99(2):201–210 (10.1006/jsre.2001.6217S0022-4804(01)96217-1 [pii])

Moynagh PN (2005) The interleukin-1 signalling pathway in astrocytes: a key contributor to inflammation in the brain. J Anat 207(3):265–269 (JOA445 [pii] 10.1111/j.1469-7580.2005.00445.x)

Murphy JB, Sturm E (1923) Conditions Determining the Transplantability of Tissues in the Brain. J Exp Med 38(2):183–197

Murta V, Farias MI, Pitossi FJ, Ferrari CC (2015) Chronic systemic IL-1beta exacerbates central neuroinflammation independently of the blood-brain barrier integrity. J Neuroimmunol 278:30–43 (10.1016/j.jneuroim.2014.11.023S0165-5728(14)00984-9 [pii])

Murta V, Ferrari CC (2013) Influence of Peripheral inflammation on the progression of multiple sclerosis: evidence from the clinic and experimental animal models. Mol Cell Neurosci 53:6–13 (10.1016/j.mcn.2012.06.004S1044-7431(12)00108-X [pii])

Nathan C (2006) Neutrophils and immunity: challenges and opportunities. Nat Rev Immunol 6 (3):173–182 nri1785 [pii] 10.1038/nri1785

Neumann H, Misgeld T, Matsumuro K, Wekerle H (1998) Neurotrophins inhibit major histocompatibility class II inducibility of microglia: involvement of the p75 neurotrophin receptor. Proc Natl Acad Sci USA 95(10):5779–5784

Ng RL, Scott NM, Strickland DH, Gorman S, Grimbaldeston MA, Norval M, Waithman J, Hart PH (2013) Altered immunity and dendritic cell activity in the periphery of mice after long-term engraftment with bone marrow from ultraviolet-irradiated mice. J Immunol 190 (11):5471–5484 (10.4049/jimmunol.1202786jimmunol.1202786 [pii])

Nicot A (2009) Gender and sex hormones in multiple sclerosis pathology and therapy. Front Biosci (Landmark Ed) 14:4477–4515 (3543 [pii])

Nishiyori A, Minami M, Takami S, Satoh M (1997) Type 2 interleukin-1 receptor mRNA is induced by kainic acid in the rat brain. Brain Res Mol Brain Res 50(1–2):237–245

Nogai A, Siffrin V, Bonhagen K, Pfueller CF, Hohnstein T, Volkmer-Engert R, Bruck W, Stadelmann C, Kamradt T (2005) Lipopolysaccharide injection induces relapses of experimental autoimmune encephalomyelitis in nontransgenic mice via bystander activation of autoreactive CD4 + cells. J Immunol 175(2):959–966 (175/2/959 [pii])

O'Connor JC, Satpathy A, Hartman ME, Horvath EM, Kelley KW, Dantzer R, Johnson RW, Freund GG (2005) IL-1beta-mediated innate immunity is amplified in the db/db mouse model of type 2 diabetes. J Immunol 174(8):4991–4997 (174/8/4991 [pii])

Oh JW, Van Wagoner NJ, Rose-John S, Benveniste EN (1998) Role of IL-6 and the soluble IL-6 receptor in inhibition of VCAM-1 gene expression. J Immunol 161(9):4992–4999

Oka Y, Ibuki T, Matsumura K, Namba M, Yamazaki Y, Poole S, Tanaka Y, Kobayashi S (2007) Interleukin-6 is a candidate molecule that transmits inflammatory information to the CNS. Neuroscience 145(2):530–538 (S0306-4522(06)01470-9 [pii] 10.1016/j.neuroscience.2006.10.055)

Okada S, Nakamura M, Katoh H, Miyao T, Shimazaki T, Ishii K, Yamane J, Yoshimura A, Iwamoto Y, Toyama Y, Okano H (2006) Conditional ablation of Stat3 or Socs3 discloses a dual role for reactive astrocytes after spinal cord injury. Nat Med 12(7):829–834 (nm1425 [pii] 10.1038/nm1425)

Ott L, McClain CJ, Gillespie M, Young B (1994) Cytokines and metabolic dysfunction after severe head injury. J Neurotrauma 11(5):447–472

Ottani A, Giuliani D, Mioni C, Galantucci M, Minutoli L, Bitto A, Altavilla D, Zaffe D, Botticelli AR, Squadrito F, Guarini S (2009) Vagus nerve mediates the protective effects of melanocortins against cerebral and systemic damage after ischemic stroke. J Cereb Blood Flow Metab 29(3):512–523 (10.1038/jcbfm.2008.140jcbfm2008140 [pii])

Palin K, Cunningham C, Forse P, Perry VH, Platt N (2008) Systemic inflammation switches the inflammatory cytokine profile in CNS Wallerian degeneration. Neurobiol Dis 30(1):19–29 (S0969-9961(07)00267-7 [pii] 10.1016/j.nbd.2007.11.012)

Panitch HS (1994) Influence of infection on exacerbations of multiple sclerosis. Ann Neurol 36 (Suppl):S25–S28

Perry V, Cunningham C, Boche D (2002) Atypical inflammation in the central nervous system in prion disease. Curr Opin Neurol 15:349–354

Perry VH, Newman TA, Cunningham C (2003) The impact of systemic infection on the progression of neurodegenerative disease. Nat Rev Neurosci 4(2):103–112

Perry VH, Teeling J (2013) Microglia and macrophages of the central nervous system: the contribution of microglia priming and systemic inflammation to chronic neurodegeneration. Semin Immunopathol. doi:10.1007/s00281-013-0382-8

Piccinni MP, Giudizi MG, Biagiotti R, Beloni L, Giannarini L, Sampognaro S, Parronchi P, Manetti R, Annunziato F, Livi C et al (1995) Progesterone favors the development of human T helper cells producing Th2-type cytokines and promotes both IL-4 production and membrane CD30 expression in established Th1 cell clones. J Immunol 155(1):128–133

Pitossi F, del Rey A, Kabiersch A, Besedovsky H (1997) Induction of cytokine transcripts in the central nervous system and pituitary following peripheral administration of endotoxin to mice. J Neurosci Res 48(4):287–298

Playfair JHL, Chain BM (1979) Immunology at a glance. Blackwell Science Publishing, London

Pocock JM, Kettenmann H (2007) Neurotransmitter receptors on microglia. Trends Neurosci 30 (10):527–535 (S0166-2236(07)00211-1 [pii] 10.1016/j.tins.2007.07.007)

Procaccini C, Pucino V, De Rosa V, Marone G, Matarese G (2014) Neuro-endocrine networks controlling immune system in health and disease. Front Immunol 5:143 (10.3389/fimmu.2014.00143)

Qian J, Zhu L, Li Q, Belevych N, Chen Q, Zhao F, Herness S, Quan N (2012) Interleukin-1R3 mediates interleukin-1-induced potassium current increase through fast activation of Akt kinase. Proc Natl Acad Sci USA 109(30):12189–12194 (10.1073/pnas.12052071091205207109 [pii])

Quan C, Yu H, Qiao J, Xiao B, Zhao G, Wu Z, Li Z, Lu C (2013) Impaired regulatory function and enhanced intrathecal activation of B cells in neuromyelitis optica: distinct from multiple sclerosis. Mult Scler 19(3):289–298 (10.1177/1352458512454771135245851245477l [pii])

Ransohoff RM, Liu L, Cardona AE (2007) Chemokines and chemokine receptors: multipurpose players in neuroinflammation. Int Rev Neurobiol 82:187–204 (S0074-7742(07)82010-1 [pii] 10.1016/S0074-7742(07)82010-1)

Rivest S, Lacroix S, Vallieres L, Nadeau S, Zhang J, Laflamme N (2000) How the blood talks to the brain parenchyma and the paraventricular nucleus of the hypothalamus during systemic inflammatory and infectious stimuli. Proc Soc Exp Biol Med 223(1):22–38 (pse22304 [pii])

Romeo HE, Tio DL, Rahman SU, Chiappelli F, Taylor AN (2001) The glossopharyngeal nerve as a novel pathway in immune-to-brain communication: relevance to neuroimmune surveillance of the oral cavity. J Neuroimmunol 115(1–2):91–100 (S0165572801002703 [pii])

Samad TA, Moore KA, Sapirstein A, Billet S, Allchorne A, Poole S, Bonventre JV, Woolf CJ (2001) Interleukin-1beta-mediated induction of Cox-2 in the CNS contributes to inflammatory pain hypersensitivity. Nature 410(6827):471–475 (10.1038/3506856635068566 [pii])

Schafer KH, Mestres P, Marz P, Rose-John S (1999) The IL-6/sIL-6R fusion protein hyper-IL-6 promotes neurite outgrowth and neuron survival in cultured enteric neurons. J Interferon Cytokine Res 19(5):527–532. doi:10.1089/107999099313974

Schiffenbauer J, Johnson HM, Butfiloski EJ, Wegrzyn L, Soos JM (1993) Staphylococcal enterotoxins can reactivate experimental allergic encephalomyelitis. Proc Natl Acad Sci USA 90(18):8543–8546

Schmitz T, Chew LJ (2008) Cytokines and myelination in the central nervous system. Sci World J 8:1119–1147. doi:10.1100/tsw.2008.140

Semple BD, Bye N, Ziebell JM, Morganti-Kossmann MC (2010) Deficiency of the chemokine receptor CXCR2 attenuates neutrophil infiltration and cortical damage following closed head injury. Neurobiol Dis 40(2):394–403 (10.1016/j.nbd.2010.06.015S0969-9961(10)00211-1 [pii])

Seo JH, Miyamoto N, Hayakawa K, Pham LD, Maki T, Ayata C, Kim KW, Lo EH, Arai K (2013) Oligodendrocyte precursors induce early blood-brain barrier opening after white matter injury. J Clin Invest 123(2):782–786 (10.1172/JCI6586365863 [pii])

Serres S, Anthony DC, Jiang Y, Broom KA, Campbell SJ, Tyler DJ, van Kasteren SI, Davis BG, Sibson NR (2009) Systemic inflammatory response reactivates immune-mediated lesions in rat brain. J Neurosci 29(15):4820–4828 (29/15/4820 [pii] 10.1523/JNEUROSCI.0406-09.2009)

Sly LM, Krzesicki RF, Brashler JR, Buhl AE, McKinley DD, Carter DB, Chin JE (2001) Endogenous brain cytokine mRNA and inflammatory responses to lipopolysaccharide are elevated in the Tg2576 transgenic mouse model of Alzheimer's disease. Brain Res Bull 56 (6):581–588 (S0361923001007304 [pii])

Sorrells SF, Sapolsky RM (2010) Glucocorticoids can arm macrophages for innate immune battle. Brain Behav Immun 24(1):17–18 (S0889-1591(09)00468-1 [pii] 10.1016/j.bbi.2009.10.004)

Takii T, Akahoshi T, Kato K, Hayashi H, Marunouchi T, Onozaki K (1992) Interleukin-1 up-regulates transcription of its own receptor in a human fibroblast cell line TIG-1: role of endogenous PGE2 and cAMP. Eur J Immunol 22(5):1221–1227. doi:10.1002/eji.1830220517

Takii T, Hayashi H, Marunouchi T, Onozaki K (1994) Interleukin-1 down-regulates type I interleukin 1 receptor mRNA expression in a human fibroblast cell line TIG-1 in the absence of prostaglandin E2 synthesis. Lymphokine Cytokine Res 13(3):213–219

Tauber SC, Nau R, Gerber J (2007) Systemic infections in multiple sclerosis and experimental autoimmune encephalomyelitis. Arch Physiol Biochem 113(3):124–130 (782870844 [pii] 10.1080/13813450701531227)

Teeling JL, Perry VH (2009) Systemic infection and inflammation in acute CNS injury and chronic neurodegeneration: underlying mechanisms. Neuroscience 158(3):1062–1073 (S0306-4522(08)01045-2 [pii] 10.1016/j.neuroscience.2008.07.031)

Trenova AG, Manova MG, Kostadinova II, Murdjeva MA, Hristova DR, Vasileva TV, Zahariev ZI (2011) Clinical and laboratory study of pro-inflammatory and antiinflammatory cytokines in women with multiple sclerosis. Folia Med (Plovdiv) 53(2):29–35

Tsai HH, Frost E, To V, Robinson S, Ffrench-Constant C, Geertman R, Ransohoff RM, Miller RH (2002) The chemokine receptor CXCR2 controls positioning of oligodendrocyte precursors in developing spinal cord by arresting their migration. Cell 110(3):373–383 (S009286740200 8383 [pii])

Tsunoda I, Fujinami RS (2002) Inside-Out versus Outside-In models for virus induced demyelination: axonal damage triggering demyelination. Springer Semin Immunopathol 24 (2):105–125. doi:10.1007/s00281-002-0105-z

Tsunoda I, Kuang LQ, Libbey JE, Fujinami RS (2003) Axonal injury heralds virus-induced demyelination. Am J Pathol 162(4):1259–1269 (S0002-9440(10)63922-3 [pii] 10.1016/S0002-9440(10)63922-3)

Uzawa A, Mori M, Sato Y, Hayakawa S, Masuda S, Taniguchi J, Kuwabara S (2010) Cytokine and chemokines profiles in neuromyelitis optica: significance of interleukin-6. Mult Scler 16 (12):1443–52. doi:10.1177/1352458510379247

van Riemsdijk IC, Baan CC, Loonen EH, Knoop CJ, Navarro Betonico G, Niesters HG, Zietse R, Weimar W (2001) T cells activate the tumor necrosis factor-alpha system during hemodialysis, resulting in tachyphylaxis. Kidney Int 59(3):883–892 (kid571 [pii] 10.1046/j.1523-1755.2001. 059003883.x)

Varvel NH, Grathwohl SA, Baumann F, Liebig C, Bosch A, Brawek B, Thal DR, Charo IF, Heppner FL, Aguzzi A, Garaschuk O, Ransohoff RM, Jucker M (2012) Microglial repopulation model reveals a robust homeostatic process for replacing CNS myeloid cells. Proc Natl Acad Sci USA 109(44):18150–18155 (10.1073/pnas.12101501091210150109 [pii])

Veenstra M, Ransohoff RM (2012) Chemokine receptor CXCR2: physiology regulator and neuroinflammation controller? J Neuroimmunol 246(1–2):1–9 (10.1016/j.jneuroim. 2012.02.016S0165-5728(12)00064-1 [pii])

Vitkovic L, Konsman JP, Bockaert J, Dantzer R, Homburger V, Jacque C (2000) Cytokine signals propagate through the brain. Mol Psychiatry 5(6):604–615

Wagner JA (1996) Is IL-6 both a cytokine and a neurotrophic factor? J Exp Med 183(6): 2417–2419

Waxman SG (1998) Demyelinating diseases–new pathological insights, new therapeutic targets. N Engl J Med 338(5):323–325. doi:10.1056/NEJM199801293380610

Wilkins A, Chandran S, Compston A (2001) A role for oligodendrocyte-derived IGF-1 in trophic support of cortical neurons. Glia 36(1):48–57 (10.1002/glia.1094 [pii])

Wilson EH, Weninger W, Hunter CA (2010) Trafficking of immune cells in the central nervous system. J Clin Invest 120(5):1368–1379 (10.1172/JCI4191141911 [pii])

Wolswijk G (1998) Chronic stage multiple sclerosis lesions contain a relatively quiescent population of oligodendrocyte precursor cells. J Neurosci 18(2):601–609

Xia M, Hyman BT (2002) GROalpha/KC, a chemokine receptor CXCR2 ligand, can be a potent trigger for neuronal ERK1/2 and PI-3 kinase pathways and for tau hyperphosphorylation-a role in Alzheimer's disease? J Neuroimmunol 122(1–2):55–64 (S0165572801004635 [pii])

Ysrraelit MC, Gaitan MI, Lopez AS, Correale J (2008) Impaired hypothalamic-pituitary-adrenal axis activity in patients with multiple sclerosis. Neurology 71(24):1948–1954 (10.1212/01. wnl.0000336918.32695.6b 71/24/1948 [pii])

# Regulation of Oligodendrocyte Differentiation and Myelination by Nuclear Receptors: Role in Neurodegenerative Disorders

**Adrián Sandoval-Hernández, María José Contreras, Jenny Jaramillo and Gonzalo Arboleda**

**Abstract** During development and through adulthood, differentiation of diverse cell types is controlled by specific genetic and molecular programs for which transcription factors are master regulators of gene expression. Here, we present an overview of the role of nuclear receptors and their selective pharmacological modulators in oligodendrocytes linage, their role in myelination and remyelination and their potential use as a therapeutic strategy for demyelinating diseases. We discuss several aspects of nuclear receptors including: (1) the biochemistry of nuclear receptors superfamily; (2) their role on stem cells physiology, focusing in differentiation and cell removal; (3) the role of nuclear receptor in the oligodendrocytes cell linage, from oligodendrocyte progenitors cells to mature myelinating cells; and (4) the therapeutics opportunities of nuclear receptors for specific demyelinating diseases.

**Keywords** Nuclear receptors · Oligodendrocytes · Demyelinating diseases · Alzheimer's disease

**Abbreviations**

| | |
|---|---|
| Aβ | Amyloid-β |
| AD | Alzheimer's disease |
| ALDRP | Adrenoleukodystrophy-related protein |
| AMN | Adrenomyeloneuropathy |
| aNSCs | Adult neural stem cells |
| AR | Androgen receptor |
| CAR | Constitutive androstane receptor |
| CBP | cAMP response element binding protein-binding protein |

A. Sandoval-Hernández · M.J. Contreras · J. Jaramillo · G. Arboleda
Grupo de Neurociencias y Muerte Celular, Facultad de Medicina
e Instituto de Genética, Universidad Nacional de Colombia, Bogotá, Colombia

G. Arboleda (✉)
Departamento de Patología, Facultad de Medicina, Universidad Nacional
de Colombia, Bogotá, Colombia
e-mail: gharboledab@unal.edu.co



R. von Bernhardi (ed.), *Glial Cells in Health and Disease of the CNS*,
Advances in Experimental Medicine and Biology 949,
DOI 10.1007/978-3-319-40764-7_14

| | |
|---|---|
| CNPase | 2′3′-cyclic nucleotide 3′-phosphodiesterase |
| CNTF | Ciliary neurotrophic factor |
| COUP-TF | Chicken ovalbumin upstream promoter-transcription factor |
| CNS | Central nervous system |
| DBD | DNA binding domain |
| DG | Dendate gyrus |
| EAE | Experimental autoimmune encephalomyelitis |
| ER | Estrogen receptor |
| ERR | Estrogen-related receptor |
| FXR | Farnesoid X receptor |
| GFAP | Glial fibrillary acid protein |
| GR | Glucocorticoid receptor |
| HNF4A | Hepatocyte nuclear factor 4A |
| IGF-1 | Insulin-like growth factor-1 |
| LBD | Ligand binding domain |
| LRH-1 | Human liver receptor homologue-1 |
| LXR | Liver X receptor |
| MBP | Myelin basic protein |
| MCP-1 | Monocyte chemoattractant protein 1 |
| MR | Mineralocorticoid receptor |
| MPZ | Myelin protein zero |
| MS | Multiple sclerosis |
| N-CoR | Nuclear receptor corepressor |
| NR2E1 | Orphan nuclear receptor subfamily 2 group E member 1 NPCs neural progenitor cells |
| NRs | Nuclear receptors |
| NSCs | Neural stem cells |
| Nurr1 | Nuclear receptor related 1 protein |
| OL | Oligodendrocytes |
| OPCs | Oligodendrocytes progenitor cells |
| PDGF | Platelet-derived growth factor |
| PGC-1 | Peroxisome proliferator-activated receptor-gamma coactivators 1 |
| PGJ2 | Prostaglandin J2 |
| PLP | Myelin proteolipid protein |
| PPAR | Peroxisome proliferator-activated receptor |
| PMP22 | Peripheral myelin protein 22 |
| PPREs | Peroxisome proliferator response elements |
| PR | Progesterone receptor |
| PXR | Pregnane X receptor |
| KD | Krabbe disease |
| RA | Retinoic acid |

| | |
|---|---|
| RAR | Retinoic acid receptor |
| RE | Response element |
| ROR | Retinoid-related orphan receptor |
| RXR | Retinoid X receptor |
| SF-1 | Steroidogenic factor-1 |
| SMRT | Silencing mediator of retinoid and thyroid hormone receptorsT3 triiodothyronine |
| TH | Thyroid hormone |
| TLX | A homolog of *Drosophila* tailless gene |
| TR | Thyroid hormone receptor |
| TNFα | Tumor necrosis factor alpha |
| X-ALD | X-linked adrenoleukodystrophy |
| X-CALD | X-linked cerebral adrenoleukodystrophy |
| VEGF | Vascular endothelial growth factor |
| VDR | Vitamin D receptor |

## Introduction

Transcriptional regulation determines the destiny of each stem cell during differentiation. Nuclear receptors (NRs) are transcription factors that are activated by interacting with small molecules such as steroid hormones, fatty acids and oxysterols. This superfamily of proteins regulates gene transcription through dynamic interactions with multiple protein complexes called coregulators. These coregulators are proteins with diverse enzymatic activities that facilitate or repress the gene activity by means of various mechanisms. In this way, NRs regulate the biological processes in different cell types in the central nervous system (CNS) including oligodendrocytes (OLs). In the CNS, mature OLs extend processes and ultimately form sheet-like myelin around the axons of neurons, the principal component in the white matter. Myelination involves many highly regulated steps controlling such physiological events as proliferation, migration and differentiation of oligodendrocyte progenitor cells (OPCs) (see Chapters "NG2-Glia, More Than Progenitor Cells" and "Oligodendrocytes: Functioning in a Delicate Balance Between High Metabolic Requirements and Oxidative Damage"). NRs regulate the transcription of specific genes and promoting the survival and differentiation of neuronal stem cells (NSC), the migration of OPCs and the maturation of OPCs to myelinating OLs. In this chapter, we review the current knowledge of the role of NRs in oligodendrogenesis, it also focuses on NRs as therapeutic targets for pharmacological modulation, using synthetic molecules that are candidates for treating demyelinating diseases and such other neurodegenerative disorders as Alzheimer's disease.

## Nuclear Receptors Superfamily

NRs are part of a gene superfamily of transcription factors regulated by structurally related and evolutionarily conserved ligands that act through a common mechanism (Chen 2008). These proteins are related to the regulation of many basic cellular and molecular processes, such as proliferation, differentiation and cellular homeostasis (Germain et al. 2006; Chambon 2005; Sonoda et al. 2008), and they cover a wide variety of biological functions involved in development, reproduction, physiology and pathology, therefore they are useful in therapeutics (Murphy et al. 2005). NRs modulate transcription through several mechanisms involving gene activation, repression and trans-repression. Such mechanisms can be ligand-dependent or independent, or either genomic or non-genomic (Germain et al. 2006). Additionally, such modulation depends not only on the NR per se, but on cofactors (corepressors and coactivators) that are recruited after the interaction with specific ligands and are generally tissue-specific (Chen 2008). All of this makes their signaling complex.

In humans, 48 NRs have been described, and 49 in mice. In the adult mouse brain, most NRs are expressed differentially in particular regions, having 40 and 36 NR express in the cortex and hippocampus respectively (Gofflot et al. 2007). In general, these soluble proteins are located in the nucleus or cytoplasm whose known natural agonists are steroid hormones (estrogens, progestins, androgens, glucocorticoids and mineralocortocoids), thyroid hormones, lipophilic vitamins (vitamin D and *cis*-retinoic acid) and cholesterol metabolites (bile acids, oxysterols), among others (Burris et al. 2013).

There are various ways of grouping members of the NR superfamily. Mangelsdorf et al. (1995) proposed four classes according to their ligand, their DNA binding site and their dimerization properties. Class I, or steroid receptors: some are in the cytoplasm and include estrogen receptor (ER), glucocorticoid receptor (GR), androgen receptor (AR), progesterone receptor (PR) and mineralocorticoid receptor (MR), which after binding with their respective ligand form homodimers that translocate to the nucleus to bind to a DNA response element (RE) composed of palindromic repeats. Class II, or retinoid X receptor (RXR): in contrast to the Class I receptors, these remain in the nucleus forming heterodimeric complexes with corepressor molecules, and bind to the RE formed by direct repeats. In this group are the thyroid hormone receptors (TRs), the vitamin D receptor (VDR), the retinoic acid receptors (RARs) and the peroxisome proliferator-activated receptors (PPARs). Class III: similar mechanism to Class I, but they bind to the RE formed by direct repeats as homodimers. Among them, we can find RXR ultraspiracle, chicken ovalbumin upstream promoter COUP and HNF-4A. Class IV, monomeric orphan receptors: bind to a unique sequence in the RE as monomers.

Another way to classify the NRs is according to their ligands in three categories [(Mangelsdorf 2010; Sonoda et al. 2008) and references therein]. (1) *Endocrine receptors* are those activated by hormones such as estrogen receptor, androgen receptor, progesterone receptor, glucocorticoid receptor, mineralocorticoid receptor and thyroid hormone receptor and vitamins as retinoic acid receptor and vitamin D receptor. The steroid hormone receptors function as homodimers, whereas TR, VDR, and RAR form heterodimers with RXR. (2) *Adopted orphan receptors* are NRs whose ligands were identified after being discovered. They include diet lipids and xenobiotics, and they function as heterodimers with RXR. They regulate the metabolism of lipids and glucose, thus they are potential therapeutic targets for metabolic disorders, including the farnesoid X receptor (FXR), liver X receptor (LXR), and peroxisome proliferator-activated receptor (PPAR). This group includes receptors that, despite having an identified ligand, have an unknown physiological role. They include the estrogen related receptors (ERRs), the retinoid-related orphan receptor (ROR), the constitutive androstane receptor (CAR), the steroidogenic factor-1 (SF-1), the human liver receptor homologue-1 (LRH-1), and the hepatocyte nuclear factor 4A (HNF4A). (3) Orphan receptors are those that do not yet have a known ligand and whose regulation depends mainly on co-regulators, on the receptor expression or on covalent modifications more than on the ligand itself.

NRs have additional mechanisms of action. Some NRs such as PPARs, LXRs, and FXR stimulate transcription minimally above the basal level in the absence of the ligand and in some cases may inhibit transcription in the absence of ligand by binding to a negative RE. For example, LXR knockout mice show elevated transcriptional activity (Landrier et al. 2003; Konopleva et al. 2011). Within the transcriptional regulatory mechanisms described, trans-repression is involved in the immune response (Pascual and Glass 2006). Trans-repression refers to all transcription repression mechanisms dependent on ligand-binding to a NR but do not involve direct binding to an RE. The most studied process of this type is the direct interaction of NRs with NFkB and AP-1, which results in their sequestration and decreased transcriptional activity, inhibiting a cytokine-mediated inflammatory response, involving tumor necrosis factor alpha (TNFα), interleukins and metalloproteinases (Xiao and Ghosh 2005; Karin 2006). In addition, promoters for different inflammatory genes contain REs for NFkB and AP-1, which allow them to act synergistically (Wagner and Eferl 2005; Shimizu et al. 2006; Rannou et al. 2006).

It has been observed that ligand-free PPARγ can be attached to the silencing mediator of retinoid and thyroid hormone receptors (SMRT) co-repressor, which is released upon ligand binding so it can interact with STAT3, inhibiting its transcriptional activity. STAT3 can be activated by IL6, IL10, IL11, IL21, and IL23, and it positively regulates the expression of IL17, IL23, BCL-X, BCL-2 and VEGF (vascular endothelial growth factor), which control cell proliferation, anti-apoptosis, angiogenesis and metastasis (Yu et al. 2009). These are important in carcinogenesis caused by hepatitis C or *Helicobacter pylori* infection. In a similar way, PPARβ acts with BCL-6, a transcriptional repressor factor involved in the development of B cell lymphoma. BCL-6 is bound to ligand-free PPARβ increasing the transcription of different proteins including the monocyte chemoattractant protein 1 (MCP-1) (Toney et al. 2000).

## Nuclear Receptors in Neural Stem Cells and Oligodendrogenesis

In the CNS, myelin is the main component of the white matter. As discussed in Chapters "Glial cells and Integrity of the Nervous System" and "Oligodendrocytes: Functioning in a Delicate Balance Between High Metabolic Requirements and Oxidative Damage", it is generated by highly specialized cells known as OLs, whose function is to wrap the axons of neurons and contribute to the efficiency and speed of conduction of action potentials (Sherman and Brophy 2005). In vertebrates, myelination occurs mainly during the postnatal period; it requires multiple steps and highly coordinated and regulated signals to control the proliferation, migration and differentiation of OPCs into mature or myelinating OLs. This process is called oligodendrogenesis (Emery 2010; Miller 2002).

OPCs are generated during embryonic development from multipotent stem cells of the ventral cortex, specifically the medial ganglionic eminence, the anterior entopeduncular area and the subventricular zone (SVZ) (Kessaris et al. 2006; Rakic and Zecevic 2003; Zuccaro and Arlotta 2013). During the postnatal period, most OPCs after a defined number of cellular divisions differentiate into postmitotic OLs. Approximately 5 % of OPCs maintain their potential for proliferation and migration in the adult brain, both in the white and the gray matter for the ongoing maintenance of myelin. Thus OPCs serve as a primary source of remyelination when facing demyelinating lesions, because they can migrate to the area of lesion, proliferate and differentiate into myelinating OLs (Franklin and Ffrench-Constant 2008; Kondo and Raff 2000; Nishiyama et al. 2009; Zawadzka et al. 2010). Furthermore, it has recently been proposed that OPCs could be categorized as adult stem cells (Crawford et al. 2014) because of their ability to self-renew, exhibit mitotic quiescence, and produce differentiated mature progeny (Young et al. 2013) while having multipotentiality (Kondo and Raff 2000; Zawadzka et al. 2010). Moreover, in the adult brain new OLs can emerge from adult neural stem cells (aNSCs) located in the SVZ of the lateral ventricle (Kriegstein and varez-Buylla 2009).

Recently, several cellular mechanisms that lead to oligodendrogenesis, myelination and remyelination have been described (Emery 2010; Tyler et al. 2009; Chong and Chan 2010; Fancy et al. 2009; Gibson et al. 2014). These include extrinsic factors such as growth factors (insulin-like growth factor-1, IGF-1; platelet-derived growth factor, PDGF; ciliary neurotrophic factor, CNTF), axonal surface ligands (LINGO-1, PSA-NCAM, Jagged1), secreted molecules and neuronal activity. And they include intrinsic factors such as several transcription factors (MRF, Olig1/2, Ascl1, Nkx2.2, Oct4, Sox5/6, Sox10, YY1, Tcf4, Id2/4, APC, β-catenin, among others), chromatin remodeling (HDAC) and microRNAs (differentiation: miR-219 and miR-338; OPCs: miR-17-92 via PTEN and AKT). Therefore, the search for therapies for demyelinating diseases is aimed at inducing NSCs and OPCs to generate myelinating OLs.

During development and in adults, NRs are part of the molecular network of transcription factors that control neurons, astrocytes and OL differentiation. It has

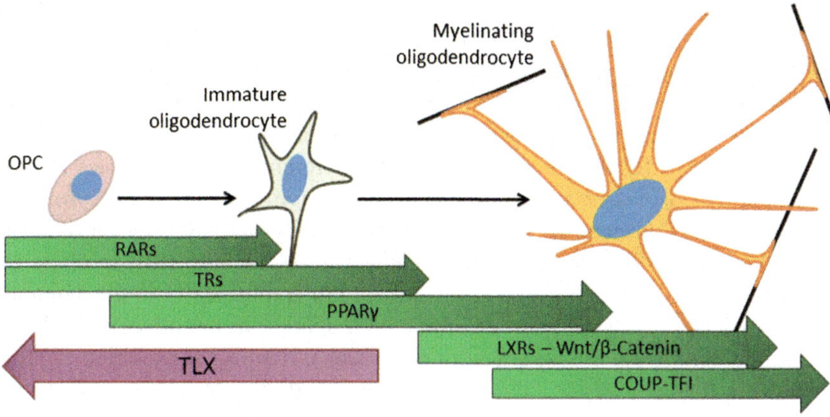


**Fig. 1** Scheme describing the role of nuclear receptors in differentiation and maturation of oligodendrocytes. *Green arrows* RARs, TRs and PPARγ are mainly involved in the maturation of OPC to immature oligodendrocytes; PPARγ, LXR, and COUP-TFI are particularly important in the full maturation of oligodendrocytes. *Pink arrow* TLX is involved in the suppression of oligodendrocyte differentiation

been shown that during embryogenesis DAX-1, ERRβ, SF-1 and LRH-1 regulate the expression of genes such as Oct3 and Oct4, which confer pluripotency characteristics of embryonic stem cells (Jeong and Mangelsdorf 2009; Tokuzawa et al. 2003). An example is the NR Errβ, which confers pluripotency characteristics independently of Oct4, Sox2, and Nanog (Ivanova et al. 2006; Luo et al. 1997), and its deficiency generates abnormalities in the blastocyst during embryogenesis (Ivanova et al. 2006). Different NR ligands have started to be explored as potential therapeutic targets in demyelinating diseases with the purpose of inducing oligodendrogenesis (Fig. 1). In this chapter some findings that involve the participation of different NRs will be described.

## *Role of RARs, RXRs and TRs*

OPC differentiation can be controlled by retinoic acid (RA) and thyroid hormone (TH) (Ahlgren et al. 1997). RA is a natural agonist for members of the RAR and RXR subfamily (Mark et al. 2006). Besides, the TH and the RA promote higher affinity interactions between members of the PPAR, RAR and RXR family (Qi et al. 1995), where RARs are essential for embryo viability (Mark et al. 2006). Goncalves et al. (2005) showed that RARα and RARγ promote the differentiation of neural progenitor cells (NPCs) to OLs and astrocytes, as RARα reduces the expression levels of SF-1, LRH-1, DAX1 and Oct4 (Gu et al. 2005; Niakan et al. 2006; Yang et al. 2007), while RARβ agonists induce neuronal differentiation (Goncalves et al. 2009). By contrast, during the early stages of embryogenesis RA was observed to inhibit OL differentiation in the spinal cord, in order to allow dispersion of OPCs in that area (Noll and Miller 1994). TH acts through binding to TR receptors (TRα and TRβ). In oligodendroglial-enriched rat cultures, triiodothyronine (T3), a type of TH, was proven to induce oligodendroglial maturation, particularly regarding the intracellular localization of mature OL markers, such as 2′3′-cyclic nucleotide 3′-phosphodiesterase (CNPase) and myelin basic protein (MBP) (Younes-Rapozo et al. 2006). This was also demonstrated in in vivo models. Hypothyroidism-induced rats have an increase in proliferation of OPCs and NSCs in the olfactory bulb and on the SVZ, and a reduction of PDGFα mRNA and MBP levels in the optic nerve, which could suggest a delay in differentiation. On the contrary, hyperthyroid rats exhibit an increase in the NG2 marker in the olfactory bulb and increased maturation of OLs (Fernandez et al. 2004). It is also suggested that TRα1 may act in a bimodal way in OL differentiation, for during early postnatal stage it promotes secretion of neurotrophic factors, which act on Purkinje neurons and astrocytes to induce differentiation of OPCs. However, in later stages, TRα1 stops OPC proliferation by cell cycle arrest (Picou et al. 2012).

Regarding RXRγ, it is proposed that its activation stimulates differentiation of OPCs to OLs and enhances remyelination in multiple sclerosis (MS) (Diab et al. 2004; Gallo and Chew 2011; Huang et al. 2011). Knockdown of RXRγ by RNA interference or RXR-specific antagonists severely inhibits OL differentiation in culture. In mice that lack RXRγ, adult OPCs efficiently repopulate lesions after demyelination, but show delayed differentiation into mature oligodendrocytes. Besides, in experimental autoimmune encephalomyelitis (EAE), agonists of RXR such as 9-cis-retinoic acid and 15-deoxy-12,14-prostaglandin J2 reduce the severity of the disease. On the other hand, Huang et al. (2011) report no differential expression of PPAR, but that RXRγ may form a heterodimer with RXRα, RXRβ, LXRα, chicken ovalbumin upstream promoter-transcription factor 1 (COUP-TFI) or Nurr1 to allow differentiation of OPCs after demyelination.

## Role of LXRs

There are two members of the LXR subfamily of NRs: LXRα and LXRβ; they share ~80 % homology (Wojcicka et al. 2007). Their natural agonists are oxidized derivatives of cholesterol called oxysterols. LXRα is expressed in liver, spleen, kidney, intestine and adipose tissue, whereas LXRβ is expressed ubiquitously at low levels (Auboeuf et al. 1997). Few synthetic agonists of LXRs are known. The most studied are T0901317 and GW3965, which bind to LXRα and LXRβ with an EC50 of approximately 20 nM. GW3965 preferentially binds to LXR while T091317 also acts as an agonist for FXR and pregnane X receptor (PXR) (Houck et al. 2004; Schultz et al. 2000; Shenoy et al. 2004).

In the CNS, LXRs have been described as inhibitors of astrogliogenesis and inducers of neurogenesis. Mice that do not express LXRs have lower midbrain development due to a decrease in the number of dopaminergic neurons.

Furthermore, in in vitro experiments it was observed that NSCs easily differentiate to dopaminergic neurons when exposed to oxysterols, (Sacchetti et al. 2009). Similarly, LXRs are important regulators of myelination, as they interact with Wnt/β-Catenin signaling, which regulates transcription of myelin protein zero (MPZ) and peripheral myelin protein 22 (PMP22) (Makoukji et al. 2011; Shackleford et al. 2013). Thus, LXRα/β$^{-/-}$ knockout mice express lower levels of MPZ and PMP22 and have fewer layers in their myelin sheaths (Makoukji et al. 2011).

Xu et al. (2014) showed that LXRβ$^{-/-}$ knockout mice have hypomyelination of the corpus callosum and the optic nerve. They also observed a deficit in the production and maturation of OLs, perhaps because LXRβ is involved in cholesterol homeostasis, an essential component of myelin, and cholesterol availability may limit the rate of maturation of OLs and myelination (Saher et al. 2005). Similarly, in primary cultures of OLs treated with LXR agonist T0901317, the ABCA1, ABCG1, ApoE, and LDLR genes, which are associated with cholesterol homeostasis, are expressed, and optimal myelination and remyelination are promoted (Nelissen et al. 2012). It is suggested that LXRβ is also essential in differentiation of radial glial cells to OPCs in the dorsal cortex, which is a newly discovered route for the production of OPCs, myelination and remyelination. Likewise, LXRβ plays an important role in brain lamination during corticogenesis (Fan et al. 2008; Xu et al. 2014).

## *Role of PPAR*

Another subfamily of NR involved in regulation of the oligodendrogenesis is that of PPARs. There are three known members: PPARα, PPARβ and PPARγ. PPARα is mainly expressed in tissues with high metabolic rates of fatty acids such as liver, muscle and heart. PPAR-γ is expressed in microglia, astrocytes, OLs and neurons. PPARs regulate gene expression by recognizing and binding to the so-called "peroxisome proliferator response elements" (PPREs) present in the promoter region of target genes. PPARs play an important role in regulating the expression of lipid metabolism genes: PPARα regulates genes involved in cholesterol metabolism, while PPARγ regulates the metabolism of fatty acids (Alaynick 2008; Huang and Schulman 2009; Makishima 2003). PPARγ agonist ligands that belong to the thiazolidinedione, or TZD (rosiglitazone and pioglitazone, among others), family are used to treat type-2 diabetes and dyslipidemias by increasing insulin sensitivity and improving glucose metabolism; PPARα agonists are another type of drug, such as fibrates (fenofibrate and gemfibrozil), which act as hypo-lipidemics and anti-atherosclerotic agents and aid in lipid metabolism (Barbier et al. 2002; Etgen and Mantlo 2003; Fajas et al. 2001; Jay and Ren 2007).

PPARs are also involved in cell differentiation in a wide variety of tissues (Barbier et al. 2002; Fajas et al. 2001), including the regulation of OL differentiation and maturation (Bernardo et al. 2009; Heneka and Landreth 2007; Roth et al. 2003;

Saluja et al. 2001). PPARβ/δ, a PPAR subtype that predominates in the CNS and is strongly expressed in immature OLs (Heneka and Landreth 2007), has been shown in vitro when activated to increase mRNA of MBP and myelin proteolipid protein (PLP) and increase survival and differentiation of OPCs, but it does not have a role in OPC proliferation (Saluja et al. 2001). PPARγ has been associated with induction of lipids synthesis, which is important in the differentiation of OPCs and myelination. PPARγ is also associated with an increase of alkyl-dihydroxyacetone phosphate synthase, a peroxisomal enzyme involved in the synthesis of myelin-rich lipid plasmalogens in B12 neural cells and isolated spinal cord OLs (Roth et al. 2003). Additionally, treatment with PPARγ agonists in rat primary cultures of OPCs promotes the differentiation of these cells and increases their antioxidant activity by increasing the levels of catalase and superoxide dismutase copper-zinc (Bernardo et al. 2009). Similarly, it was shown that pioglitazone accelerates OL differentiation and increases complex IV mitochondrial respiratory chain activity as well as the response to increased calcium and such environmental cues as ADP (De et al. 2011). PPARγ also regulates growth and differentiation of several cell types during postnatal and prenatal development. For example, Kanakasabai et al. (2012) found in vitro that mouse NSCs treated with PPARγ agonists stop proliferating and significantly increase immunoreactivity for O4 and NG2 OPC markers. Paintlia et al. (2011) in an in vitro MS model, demonstrated that agonists of PPARγ/δ protect OPCs against the cytotoxicity conferred by TNFα and IL17.

Recently, it has been confirmed that PGC-1 (peroxisome proliferator-activated receptor-gamma coactivators 1, which includes PGC-1α, PGC1-β, and PGC-1-related coactivator) interact with the NR PPARγ, PPARα, ERR, LXR and HNF-4α. Such activators are involved in mitochondrial biogenesis and in lipid and energy metabolism (Shao et al. 2010; Sugden et al. 2010; Wareski et al. 2009). PGC-1α is expressed in OLs, and its genetic deletion leads to dysfunction in the metabolism of sphingolipids and changes in the lipid composition of their cell membranes, including decreased ceramides, galactosylceramides, lactosylceramides, phosphatidylglycerol, phosphatidylethanolamine and phosphatidylserine, and accumulation of phosphatidylcholine. Although, these changes are not directly related to alterations in the structure of myelin, they do delay the maturation of OLs. Indeed, PGC-1α is able to coordinate the differentiation of OPCs, as it regulates the expression of Nkx2.2, the transcription factor related to MBP and PLP (Camacho et al. 2013). It is important to add that deletion of PGC-1α in mice causes neurodegeneration in the striatum, cortex and hippocampus, reduction of myelin-associated proteins and alteration in cholesterol homeostasis (Kiebish et al. 2012; Xiang et al. 2011).

## *Role of the COUP-TFs*

The chicken ovalbumin upstream promoter-transcription factor (COUP-TF) NR subfamily is involved in the maturation of OLs, as it directly regulates myelin-regulatory proteins. This family has two members: COUP-TFI and COUP-TFII, which share 80 % homology. It was shown in vivo that in early stages of development there is an increase in expression of these NRs in NSCs, leading to gliogenesis, while their absence keeps the glial fibrillary acid protein (GFAP) promoter epigenetically silent. Furthermore, in vitro knockdown of COUP-TF reduced neuronal phenotypes in advanced stages of neurospheres and generated mainly glial cells (Naka et al. 2008). It was also shown that animals that do not express COUP-TFI die perinatally; the mutant embryos develop alterations in cranial ganglia and nerves, with abnormal axonal projections and arborization. Therefore, COUP-TFI is required for normal fetal development (Qiu et al. 1997).

Regarding the OL lineage and myelination, COUP-TFI regulates differentiation by regulating other transcription factors, including SCIP/Oct-6/Tst-I, which is an important regulator of myelination (Yamaguchi et al. 2004). Also in the mutant COUP-TFI mice, it has been shown that there is normal proliferation and migration of OPCs. Otherwise, mature OL's markers like MBP are reduced considerably in the white matter and stratium, compared to control mice.

### *Role of TLX*

The NR homolog of the *Drosophila* tailless gene (TLX) is an inhibitor of OL differentiation and is critical for neurogenesis. TLX is exclusively expressed in the CNS (Monaghan et al. 1995). It is expressed predominantly in the neurogenic niches in the subgranular zone of the hippocampal dendate gyrus (DG) and in the SVZ of the lateral ventricles. Its maximum expression is associated with increased neurogenesis and its absence causes a thinning of the neocortex and the forebrain, as well as a reduction in the number of NSCs and progenitors (Li et al. 2008). TLX also inhibits astrogenesis and maturation during development (Land and Monaghan 2003; Li et al. 2008; Roy et al. 2004; Shi et al. 2004; Sun et al. 2007; Zhang et al. 2008).

## Nuclear Receptors as Potential Therapeutics for Neurodegeneration

### *Nuclear Receptors and Demyelinating Diseases*

As discussed in Chapters "Glial cells and Integrity of the Nervous System" and "Peripheral Inflammation and Demyelinating Diseases", demyelinating diseases include a heterogeneous group of CNS and peripheral nervous system (PNS) diseases, both from the clinical and from the pathophysiologic standpoint (Kassmann and Nave 2008; Dubois-Dalcq et al. 2008).

MS is a genetically linked autoimmune disease of the CNS that specifically affects the white matter (Hafler et al. 2005). It is characterized by a progressive deterioration of motor skills (Stathopoulou et al. 2010). The autoimmune hypothesis indicates that it is an immune reaction against myelinated axons reducing myelin and causing axonal degeneration (Dhib-Jalbut 2007).

X-linked adrenoleukodystrophy (X-ALD) is a neurodegenerative disease that affects the white matter of the brain, spinal cord, peripheral nerves, adrenal cortex and testicles (Ferrer et al. 2010). When it starts in childhood, it is known as X-linked cerebral adrenoleukodystrophy (X-CALD); when it starts later in life, it affects axon tracts and is known as adrenomyeloneuropathy (AMN) (Powers et al. 1992; Powers and Rubio 1995; Powers et al. 2000, 2001; van Geel et al. 2001). In general, X-ALD is the most common peroxisomal disorder, and it is characterized biochemically by the accumulation of very long chain fatty acids (VLCFAs) in these tissues (Singh 1997; Wanders 1999; Moser et al. 1984; Singh et al. 1984; Kemp and Wanders 2010). It is caused by mutations/deletions of the ABCD1 gene, which is responsible for transporting VLCFAs to the peroxisome for degradation (Ferrer et al. 2010).

Krabbe disease (KD) is a fatal neurological disease with autosomal recessive inheritance. The pathological features of the disease include axon loss, myelin loss, astrogliosis and the presence of PAS-positive globoid cells (Suzuki 2003). The biochemical defect responsible for the pathogenesis of KD is deficiency of galactosylcerebrosidase, a lysosomal enzyme, which leads to the progressive accumulation of psychosine in globoid cells (Suzuki 1983, 2003; Suzuki and Taniike 1995). Psychosine is highly toxic and produces pro-inflammatory agents while inhibiting plasmalogen enzymes (Haq et al. 2006; Khan et al. 2005). In addition, when it accumulates in lipid rafts it alters the architecture of the membrane and disrupts signaling pathways involved in cell survival and myelination (White et al. 2009).

PPAR has been evaluated as a potential therapeutic target in some models of neurodegenerative demyelinating diseases for its capacity to regulate lipid metabolism, oligodendrocyte proliferation and differentiation (Robinson and Grieve 2009), and for its anti-inflammatory role by inhibiting the NFκB signaling pathway (Ko et al. 2008; Remels et al. 2009).

In animal models of MS including EAE, treatment with agonists for PPARα (Gocke et al. 2009) or PPARγ (Drew et al. 2008) is effective at inhibiting the release of pro-inflammatory cytokines from microglia cells and astrocytes. Similarly, treatment with fibrates increases expression of the ABCD2 gene, which encodes for adrenoleukodystrophy-related protein (ALDRP) (Rampler et al. 2003); PPARβ agonists seem to have a protective effect against inflammatory processes that are dependent on gamma interferon (IFNy) and lipopolysaccharide (LPS) (Defaux et al. 2009). In glioma cells, treatment with PPARγ agonists reduces cell proliferation, migration and invasion. Furthermore, activation of PPARβ promotes the differentiation of rat pre-OLs (Saluja et al. 2001).

In the EAE model, PPARs ameliorates pathological manifestations through mechanisms associated with anti-inflammatory activity, reducing infiltration, migration and proliferation of Ag-specific T cells for the MBP (Lovett-Racke et al. 2004; Polak et al. 2005; Niino et al. 2001; Natarajan and Bright 2002;

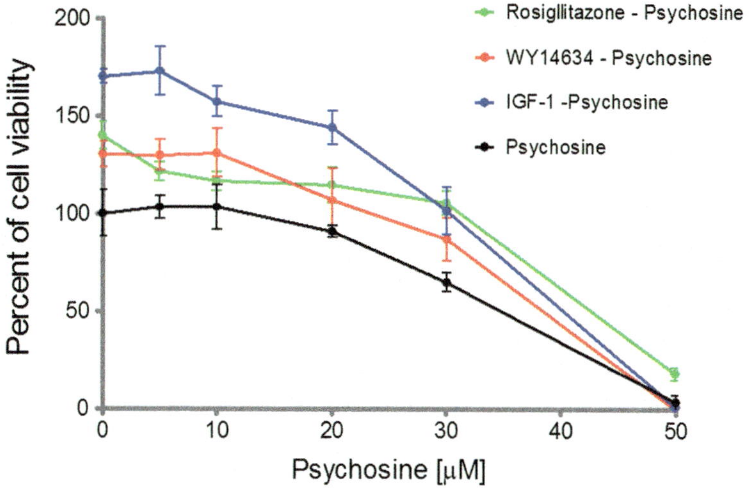


**Fig. 2** Effects of rosiglitazone and WY14643 on differentiated MO3.13 cells exposed to psycosine. IGF-1 and the PPARγ agonists, Rosiglitazone and WY14634, protect against psycosine-induced MO3.13 cell death

Diab et al. 2002; Feinstein et al. 2002; Raikwar et al. 2006; Peiris et al. 2007). In models of X-ALD oligodendrocytes, the effects of PPAR agonists have not yet been evaluated but will be important to study because of PPAR's capacity to regulate lipid homeostasis and the expression of peroxisomal proteins.

In KD, some effects of PPAR agonists are known. GalC knockdown mice, known as Twitcher, have lower levels of transcriptional activity and PPARα expression, which could explain the decrease in peroxisomal proteins (Haq et al. 2006). In our laboratory, PPARγ agonists partially protect from psychosine-induced MO3.13-cell death (Fig. 2).

## *Nuclear Receptors and Alzheimer's Disease*

Alzheimer's disease (AD) is a complex and chronic neurodegenerative disorder characterized by the presence of beta amyloid deposits, altered lipid metabolism, loss of cholinergic neurons and neuroinflammation leading to memory loss and cognitive decline (Perrin et al. 2009; Mattson 2004) (see Chapter "Age-Dependent Changes in the Activation and Regulation of Microglia"). Because NRs can control diverse physiological processes, including those associated with inflammation and lipid homeostasis, they are a plausible therapeutic target for this disease (Zolezzi et al. 2014; Burris et al. 2013; Bensinger and Tontonoz 2008).

Such PPARγ agonists as rosiglitazone that do not readily cross the blood–brain barrier improve cognition in Alzheimer's transgenic mouse models, reducing Tau phosphorylation and the amount of soluble amyloid-β (Aβ) without increasing the levels of NEP, IDE or ApoE. However, rosiglitazone slightly increases ABCA1, which is associated with a decrease in Aβ plaques (Escribano et al. 2010). Rosiglitazone is reported to reverse morphological changes in microglia, retaining their phagocytic activity needed for removing the Aβ plaques (Zhao et al. 2009). The low ability of rosiglitazone to cross the BBB suggests that it acts by increasing peripheral insulin sensitivity, which is indeed altered and a risk factor for AD (Haan 2006; Martins et al. 2006). Pioglitazone, a PPARγ agonist that readily crosses the BBB, reduced Aβ levels through a decrease in the amyloidogenic processing of APP by reducing the transcription of BACE1 (Sastre et al. 2003, 2006). It also reduced proinflammatory markers (Yan et al. 2003). Clinical studies in which PPAR agonists were used showed small cognitive improvements that depend on the ApoE genotype (Risner et al. 2006). However, these drugs have been removed from the market as they increase the risk of heart attack and bladder cancer (Chen 2008).

LXR is widely expressed in the brain. It binds oxysterols and regulates cholesterol homeostasis (Calkin and Tontonoz 2012; Schulman 2010) and inflammation (Steffensen et al. 2013). LXR agonists induce cholesterol efflux from neurons and glial cells by a positive regulation of the cholesterol transporter (ABCA1) and by induction of apolipoprotein E (ApoE) expression, which has been associated with a reduction in Aβ load (Burris et al. 2013). In transgenic animal models of AD, LXR agonist TO91317 generates cognitive improvement and histopathological changes, depending on the dose, duration of treatment and the age of the animals. These beneficial effects were associated with an increase of ABCA1 and ApoE levels and with a decrease in Aβ40 and Aβ42 amyloid load, and in some cases there was a reduction in the amyloidogenic processing favoring α-secretase cleavage (Koldamova et al. 2005; Vanmierlo et al. 2011; Riddell et al. 2007).

Treatment of 12 month old Tg2576 mice with the LXR agonist GW3965 (33 mg/kg a day for 4 months) increases levels of ApoE and ABCA1 proteins, associated with an in vitro increase in Aβ degradation and a significant decrease (approximately 65 %) in amyloid plaques (Jiang et al. 2008). Importantly, cognitive impairment improves in 20-week-old Tg2576 mice treated with GW3965 50 mg/kg a day for 6 days. From the same investigators, GW3965 modulates the immune response, and double gene deletion of LXRα and LXRβ in the humanized mouse model of AD (TgAPPswe/PS1_E9) produces a more severe amyloid pathology without interfering with the phagocytic activity of microglia. In addition, in vitro and in vivo studies show increased expression of APOE, ABCA1, ABCG1, ApoC2, and SREBP1c as well as reduced proinflammatory mediators such as TNFα and Mip1β (Zelcer et al. 2007). Additionally, the cognitive and biochemical recovery following GW3965 treatment in a mouse model of AD (APP/PS1) and its failure in knockout of ABCA1 ($ABCA1^{-/-}$), indicates that formation of lipoproteins is required for the beneficial effects of this LXR agonist, and suggesting that the anti-inflammatory component in this model may have a secondary role (Donkin et al. 2010).

With regard to myelin in AD, few studies have considered alterations in myelin as an important mechanism in the pathophysiology of this disease. The first

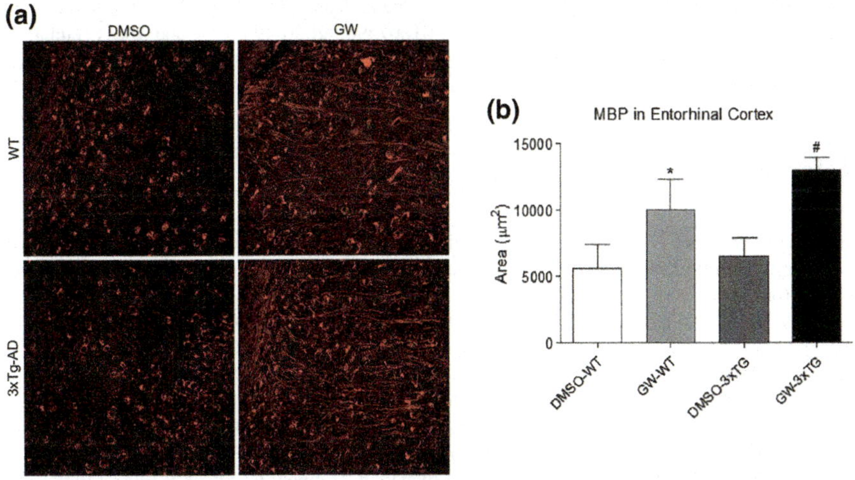


**Fig. 3** LXR agonist increases myelination in the entorhinal cortex of WT and 3xTg-AD mice. Micrograph showing myelination by using MBP staining after 12 weeks treatment with the LXR agonist GW3965 in 3xTg-AD mice. Left panel, MBP immunoreactivity increased significantly in the GW3965-treated animals as compared to untreated WT or 3xTg-AD mice. Data expressed as mean ± S.E.M. Statistical analysis performed by one-way ANOVA followed by Bonferroni post hoc test. *$P < 0.01$ compared with WT; #$P < 0.05$ compared with untreated 3Xtg-AD

detectable change in the triple transgenic mice model (3xTg) of AD is the loss of myelin markers and demyelination, starting as early as 2 months of age (Desai et al. 2010). In addition, in vitro studies show that treatment with Aβ induces OLs death by activating the neutral sphingomyelinase-ceramide pathway (Lee et al. 2004). These observations of alterations in myelin suggest the hypothesis that demyelination may underlie AD, which deserved to be further analyzed.

Recently, we have demonstrated that treatment of 3xTg-AD and wild type mice with the specific agonist of LXR (GW3965) increases OPC markers (O1 and O4) in DG of the hippocampus and also an increases myelin basic protein immunoreactivity in the entorhinal cortex (Fig. 3, unpublished results). Our results indicate that the pharmacological modulation of LXR may stimulate the remyelination process in AD.

## Conclusions and Perspectives

Nuclear receptors are emerging as a plausible therapeutic strategy in diverse disorders associated with myelin alterations, including dysmyelinating and demyelinating disorders. However, the cellular and molecular mechanisms for beneficial therapies involving these molecules need to be explored further.

**Acknowledgments** This work was supported by grants from COLCIENCIAS (110145221189), DIB-Universidad Nacional de Colombia (20101007590, 20201009689) and the Facultad de Medicina, Universidad Nacional de Colombia.

## References

Ahlgren SC, Wallace H, Bishop J, Neophytou C, Raff MC (1997) Effects of thyroid hormone on embryonic oligodendrocyte precursor cell development in vivo and in vitro. Mol Cell Neurosci 9(5–6):420–432

Alaynick WA (2008) Nuclear receptors, mitochondria and lipid metabolism. Mitochondrion 8 (4):329–337

Auboeuf D, Rieusset J, Fajas L, Vallier P, Frering V, Riou JP, Staels B, Auwerx J, Laville M, Vidal H (1997) Tissue distribution and quantification of the expression of mRNAs of peroxisome proliferator-activated receptors and liver X receptor-alpha in humans: no alteration in adipose tissue of obese and NIDDM patients. Diabetes 46(8):1319–1327

Barbier O, Torra IP, Duguay Y, Blanquart C, Fruchart JC, Glineur C, Staels B (2002) Pleiotropic actions of peroxisome proliferator-activated receptors in lipid metabolism and atherosclerosis. Arterioscler Thromb Vasc Biol 22(5):717–726

Bensinger SJ, Tontonoz P (2008) Integration of metabolism and inflammation by lipid-activated nuclear receptors. Nature 454(7203):470–477. doi:10.1038/nature07202 nature07202 [pii]

Bernardo A, Bianchi D, Magnaghi V, Minghetti L (2009) Peroxisome proliferator-activated receptor-gamma agonists promote differentiation and antioxidant defenses of oligodendrocyte progenitor cells. J Neuropathol Exp Neurol 68(7):797–808

Burris TP, Solt LA, Wang Y, Crumbley C, Banerjee S, Griffett K, Lundasen T, Hughes T, Kojetin DJ (2013) Nuclear receptors and their selective pharmacologic modulators. Pharmacol Rev 65(2):710–778. doi:10.1124/pr.112.006833 65/2/710 [pii]

Calkin AC, Tontonoz P (2012) Transcriptional integration of metabolism by the nuclear sterol-activated receptors LXR and FXR. Nat Rev Mol Cell Biol 13(4):213–224

Camacho A, Huang JK, int-Ramirez I, Yew TC, Fuller M, Lelliott CJ, Vidal-Puig A, Franklin RJ (2013) Peroxisome proliferator-activated receptor gamma-coactivator-1 alpha coordinates sphingolipid metabolism, lipid raft composition and myelin protein synthesis. Eur J Neurosci 38(5):2672–2683

Chambon P (2005) The nuclear receptor superfamily: a personal retrospect on the first two decades. Mol Endocrinol 19(6):1418–1428. doi:10.1210/me.2005-0125

Chen T (2008) Nuclear receptor drug discovery. Curr Opin Chem Biol 12(4):418–426. doi:10.1016/j.cbpa.2008.07.001

Chong SY, Chan JR (2010) Tapping into the glial reservoir: cells committed to remaining uncommitted. J Cell Biol 188(3):305–312

Crawford AH, Stockley JH, Tripathi RB, Richardson WD, Franklin RJ (2014) Oligodendrocyte progenitors: adult stem cells of the central nervous system? Exp Neurol 260:50–55

De NC, Bernardo A, De SR, Mancuso E, Magnaghi V, Visentin S, Minghetti L (2011) Peroxisome proliferator-activated receptor gamma agonists accelerate oligodendrocyte maturation and influence mitochondrial functions and oscillatory Ca(2+) waves. J Neuropathol Exp Neurol 70 (10):900–912

Defaux A, Zurich MG, Braissant O, Honegger P, Monnet-Tschudi F (2009) Effects of the PPAR-beta agonist GW501516 in an in vitro model of brain inflammation and antibody-induced demyelination. J Neuroinflammation 6:15

Desai MK, Mastrangelo MA, Ryan DA, Sudol KL, Narrow WC, Bowers WJ (2010) Early oligodendrocyte/myelin pathology in alzheimer's disease mice constitutes a novel therapeutic target. Am J Pathol 177(3):1422–1435. doi:10.2353/ajpath.2010.100087

Dhib-Jalbut S (2007) Pathogenesis of myelin/oligodendrocyte damage in multiple sclerosis. Neurology 68(22 Suppl 3):S13–S21

Diab A, Deng C, Smith JD, Hussain RZ, Phanavanh B, Lovett-Racke AE, Drew PD, Racke MK (2002) Peroxisome proliferator-activated receptor-gamma agonist 15-deoxy-Delta(12,14)-prostaglandin J(2) ameliorates experimental autoimmune encephalomyelitis. J Immunol 168 (5):2508–2515

Diab A, Hussain RZ, Lovett-Racke AE, Chavis JA, Drew PD, Racke MK (2004) Ligands for the peroxisome proliferator-activated receptor-gamma and the retinoid X receptor exert additive anti-inflammatory effects on experimental autoimmune encephalomyelitis. J Neuroimmunol 148(1–2):116–126

Donkin JJ, Stukas S, Hirsch-Reinshagen V, Namjoshi D, Wilkinson A, May S, Chan J, Fan J, Collins J, Wellington CL (2010) ATP-binding cassette transporter A1 mediates the beneficial effects of the liver X receptor agonist GW3965 on object recognition memory and amyloid burden in amyloid precursor protein/presenilin 1 mice. J Biol Chem 285(44):34144–34154

Drew PD, Xu J, Racke MK (2008) PPAR-gamma: therapeutic potential for multiple sclerosis. PPAR Res 2008:627463

Dubois-Dalcq M, Williams A, Stadelmann C, Stankoff B, Zalc B, Lubetzki C (2008) From fish to man: understanding endogenous remyelination in central nervous system demyelinating diseases. Brain 131(Pt 7):1686–1700

Emery B (2010) Regulation of oligodendrocyte differentiation and myelination. Science 330 (6005):779–782

Escribano L, Simon AM, Gimeno E, Cuadrado-Tejedor M, de Lopez MR, Garcia-Osta A, Ricobaraza A, Perez-Mediavilla A, Del RJ, Frechilla D (2010) Rosiglitazone rescues memory impairment in Alzheimer's transgenic mice: mechanisms involving a reduced amyloid and tau pathology. Neuropsychopharmacology 35(7):1593–1604

Etgen GJ, Mantlo N (2003) PPAR ligands for metabolic disorders. Curr Top Med Chem 3 (14):1649–1661

Fajas L, Debril MB, Auwerx J (2001) PPAR gamma: an essential role in metabolic control. Nutr Metab Cardiovasc Dis 11(1):64–69

Fan X, Kim HJ, Bouton D, Warner M, Gustafsson JA (2008) Expression of liver X receptor beta is essential for formation of superficial cortical layers and migration of later-born neurons. Proc Natl Acad Sci USA 105(36):13445–13450

Fancy SP, Baranzini SE, Zhao C, Yuk DI, Irvine KA, Kaing S, Sanai N, Franklin RJ, Rowitch DH (2009) Dysregulation of the Wnt pathway inhibits timely myelination and remyelination in the mammalian CNS. Genes Dev 23(13):1571–1585

Feinstein DL, Galea E, Gavrilyuk V, Brosnan CF, Whitacre CC, Dumitrescu-Ozimek L, Landreth GE, Pershadsingh HA, Weinberg G, Heneka MT (2002) Peroxisome proliferator-activated receptor-gamma agonists prevent experimental autoimmune encephalomyelitis. Ann Neurol 51(6):694–702

Fernandez M, Pirondi S, Manservigi M, Giardino L, Calza L (2004) Thyroid hormone participates in the regulation of neural stem cells and oligodendrocyte precursor cells in the central nervous system of adult rat. Eur J Neurosci 20(8):2059–2070

Ferrer I, Aubourg P, Pujol A (2010) General aspects and neuropathology of X-linked adrenoleukodystrophy. Brain Pathol 20(4):817–830

Franklin RJ, Ffrench-Constant C (2008) Remyelination in the CNS: from biology to therapy. Nat Rev Neurosci 9(11):839–855

Gallo V, Chew LJ (2011) An RXR-gamma Rx for white-matter damage. Nat Neurosci 14(1):3–5

Germain P, Staels B, Dacquet C, Spedding M, Laudet V (2006) Overview of nomenclature of nuclear receptors. Pharmacol Rev 58(4):685–704. doi:10.1124/pr.58.4.2

Gibson EM, Purger D, Mount CW, Goldstein AK, Lin GL, Wood LS, Inema I, Miller SE, Bieri G, Zuchero JB, Barres BA, Woo PJ, Vogel H, Monje M (2014) Neuronal activity promotes oligodendrogenesis and adaptive myelination in the mammalian brain. Science 344 (6183):1252304

Gocke AR, Hussain RZ, Yang Y, Peng H, Weiner J, Ben LH, Drew PD, Stuve O, Lovett-Racke AE, Racke MK (2009) Transcriptional modulation of the immune response by peroxisome proliferator-activated receptor-{alpha} agonists in autoimmune disease. J Immunol 182 (7):4479–4487

Gofflot F, Chartoire N, Vasseur L, Heikkinen S, Dembele D, Le MJ, Auwerx J (2007) Systematic gene expression mapping clusters nuclear receptors according to their function in the brain. Cell 131(2):405–418

Goncalves MB, Boyle J, Webber DJ, Hall S, Minger SL, Corcoran JP (2005) Timing of the retinoid-signalling pathway determines the expression of neuronal markers in neural progenitor cells. Dev Biol 278(1):60–70

Goncalves MB, Agudo M, Connor S, McMahon S, Minger SL, Maden M, Corcoran JP (2009) Sequential RARbeta and alpha signalling in vivo can induce adult forebrain neural progenitor cells to differentiate into neurons through Shh and FGF signalling pathways. Dev Biol 326 (2):305–313

Gu P, Goodwin B, Chung AC, Xu X, Wheeler DA, Price RR, Galardi C, Peng L, Latour AM, Koller BH, Gossen J, Kliewer SA, Cooney AJ (2005) Orphan nuclear receptor LRH-1 is required to maintain Oct4 expression at the epiblast stage of embryonic development. Mol Cell Biol 25(9):3492–3505

Haan MN (2006) Therapy Insight: type 2 diabetes mellitus and the risk of late-onset Alzheimer's disease. Nat Clin Pract Neurol 2(3):159–166

Hafler DA, Slavik JM, Anderson DE, O'Connor KC, De JP, Baecher-Allan C (2005) Multiple sclerosis. Immunol Rev 204:208–231

Haq E, Contreras MA, Giri S, Singh I, Singh AK (2006) Dysfunction of peroxisomes in twitcher mice brain: a possible mechanism of psychosine-induced disease. Biochem Biophys Res Commun 343(1):229–238

Heneka MT, Landreth GE (2007) PPARs in the brain. Biochim Biophys Acta 1771(8):1031–1045

Houck KA, Borchert KM, Hepler CD, Thomas JS, Bramlett KS, Michael LF, Burris TP (2004) T0901317 is a dual LXR/FXR agonist. Mol Genet Metab 83(1–2):184–187

Huang HJ, Schulman IG (2009) Regulation of metabolism by nuclear hormone receptors. Prog Mol Biol Transl Sci 87:1–51. doi:10.1016/S1877-1173(09)87001-4 S1877-1173(09)87001-4 [pii]

Huang JK, Jarjour AA, Oumesmar BN, Kerninon C, Williams A, Krezel W, Kagechika H, Bauer J, Zhao C, Baron-Van Evercooren A, Chambon P, Ffrench-Constant C, Franklin RJM (2011) Retinoid X receptor gamma signaling accelerates CNS remyelination. Nat Neurosci 14 (1):45–53. doi:10.1038/nn.2702

Ivanova N, Dobrin R, Lu R, Kotenko I, Levorse J, DeCoste C, Schafer X, Lun Y, Lemischka IR (2006) Dissecting self-renewal in stem cells with RNA interference. Nature 442(7102):533–538

Jay MA, Ren J (2007) Peroxisome proliferator-activated receptor (PPAR) in metabolic syndrome and type 2 diabetes mellitus. Curr Diabetes Rev 3(1):33–39

Jeong Y, Mangelsdorf DJ (2009) Nuclear receptor regulation of stemness and stem cell differentiation. Exp Mol Med 41(8):525–537. doi:10.3858/emm.2009.41.8.091

Jiang Q, Lee CY, Mandrekar S, Wilkinson B, Cramer P, Zelcer N, Mann K, Lamb B, Willson TM, Collins JL, Richardson JC, Smith JD, Comery TA, Riddell D, Holtzman DM, Tontonoz P, Landreth GE (2008) ApoE promotes the proteolytic degradation of Abeta. Neuron 58(5):681–693

Kanakasabai S, Pestereva E, Chearwae W, Gupta SK, Ansari S, Bright JJ (2012) PPARgamma agonists promote oligodendrocyte differentiation of neural stem cells by modulating stemness and differentiation genes. PLoS One 7(11):e50500

Karin M (2006) Nuclear factor-[kappa]B in cancer development and progression. Nature 441 (7092):431–436

Kassmann CM, Nave KA (2008) Oligodendroglial impact on axonal function and survival—a hypothesis. Curr Opin Neurol 21(3):235–241

Kemp S, Wanders R (2010) Biochemical aspects of X-linked adrenoleukodystrophy. Brain Pathol 20(4):831–837

Kessaris N, Fogarty M, Iannarelli P, Grist M, Wegner M, Richardson WD (2006) Competing waves of oligodendrocytes in the forebrain and postnatal elimination of an embryonic lineage. Nat Neurosci 9(2):173–179

Khan M, Haq E, Giri S, Singh I, Singh AK (2005) Peroxisomal participation in psychosine-mediated toxicity: implications for Krabbe's disease. J Neurosci Res 80(6):845–854

Kiebish MA, Young DM, Lehman JJ, Han X (2012) Chronic caloric restriction attenuates a loss of sulfatide content in PGC-1alpha$^{-/-}$ mouse cortex: a potential lipidomic role of PGC-1alpha in neurodegeneration. J Lipid Res 53(2):273–281

Ko GJ, Kang YS, Han SY, Lee MH, Song HK, Han KH, Kim HK, Han JY, Cha DR (2008) Pioglitazone attenuates diabetic nephropathy through an anti-inflammatory mechanism in type 2 diabetic rats. Nephrol Dial Transplant 23(9):2750–2760

Koldamova RP, Lefterov IM, Staufenbiel M, Wolfe D, Huang S, Glorioso JC, Walter M, Roth MG, Lazo JS (2005) The liver X receptor ligand T0901317 decreases amyloid beta production in vitro and in a mouse model of Alzheimer's disease. J Biol Chem 280(6):4079–4088

Kondo T, Raff M (2000) Oligodendrocyte precursor cells reprogrammed to become multipotential CNS stem cells. Science 289(5485):1754–1757

Konopleva M, Milella M, Ruvolo P, Watts JC, Ricciardi MR, Korchin B, McQueen T, Bornmann W, Tsao T, Bergamo P, Mak DH, Chen W, McCubrey J, Tafuri A, Andreeff M (2011) MEK inhibition enhances ABT-737-induced leukemia cell apoptosis via prevention of ERK-activated MCL-1 induction and modulation of MCL-1/BIM complex. Leukemia 26 (4):778–787

Kriegstein A, varez-Buylla A (2009) The glial nature of embryonic and adult neural stem cells. Annu Rev Neurosci 32:149–184

Land PW, Monaghan AP (2003) Expression of the transcription factor, tailless, is required for formation of superficial cortical layers. Cereb Cortex 13(9):921–931

Landrier JF, Grober J, Demydchuk J, Besnard P (2003) FXRE can function as an LXRE in the promoter of human ileal bile acid-binding protein (I-BABP) gene. FEBS Lett 553(3):299–303

Lee J-T, Xu J, Lee J-M, Ku G, Han X, Yang D-I, Chen S, Hsu CY (2004) Amyloid-β peptide induces oligodendrocyte death by activating the neutral sphingomyelinase–ceramide pathway. J Cell Biol 164(1):123–131. doi:10.1083/jcb.200307017

Li W, Sun G, Yang S, Qu Q, Nakashima K, Shi Y (2008) Nuclear receptor TLX regulates cell cycle progression in neural stem cells of the developing brain. Mol Endocrinol 22(1):56–64

Lovett-Racke AE, Hussain RZ, Northrop S, Choy J, Rocchini A, Matthes L, Chavis JA, Diab A, Drew PD, Racke MK (2004) Peroxisome proliferator-activated receptor alpha agonists as therapy for autoimmune disease. J Immunol 172(9):5790–5798

Luo J, Sladek R, Bader JA, Matthyssen A, Rossant J, Giguere V (1997) Placental abnormalities in mouse embryos lacking the orphan nuclear receptor ERR-beta. Nature 388(6644):778–782

Makishima M (2003) Lipid metabolism and nuclear receptors. Seikagaku 75(5):391–395

Makoukji J, Shackleford G, Meffre D, Grenier J, Liere P, Lobaccaro JM, Schumacher M, Massaad C (2011) Interplay between LXR and Wnt/beta-catenin signaling in the negative regulation of peripheral myelin genes by oxysterols. J Neurosci 31 (26):9620–9629. doi:10.1523/JNEUROSCI.0761-11.2011 31/26/9620 [pii]

Mangelsdorf DJ (2010) Commentary: the year in nuclear receptor control of metabolism. Mol Endocrinol 24(11):2075–2080. doi:10.1210/me.2010-0357

Mangelsdorf DJ, Thummel C, Beato M, Herrlich P, Schutz G, Umesono K, Blumberg B, Kastner P, Mark M, Chambon P, Evans RM (1995) The nuclear receptor superfamily: the second decade. Cell 83(6):835–839

Mark M, Ghyselinck NB, Chambon P (2006) Function of retinoid nuclear receptors: lessons from genetic and pharmacological dissections of the retinoic acid signaling pathway during mouse embryogenesis. Annu Rev Pharmacol Toxicol 46:451–480

Martins IJ, Hone E, Foster JK, Sunram-Lea SI, Gnjec A, Fuller SJ, Nolan D, Gandy SE, Martins RN (2006) Apolipoprotein E, cholesterol metabolism, diabetes, and the convergence of risk factors for Alzheimer's disease and cardiovascular disease. Mol Psychiatry 11(8): 721–736

Mattson MP (2004) Pathways towards and away from Alzheimer's disease. Nature 430 (7000):631–639. doi:10.1038/nature02621 nature02621 [pii]

Miller RH (2002) Regulation of oligodendrocyte development in the vertebrate CNS. Prog Neurobiol 67(6):451–467

Monaghan AP, Grau E, Bock D, Schutz G (1995) The mouse homolog of the orphan nuclear receptor tailless is expressed in the developing forebrain. Development 121(3):839–853

Moser HW, Moser AE, Singh I, O'Neill BP (1984) Adrenoleukodystrophy: survey of 303 cases: biochemistry, diagnosis, and therapy. Ann Neurol 16(6):628–641

Murphy RT, Mogensen J, McGarry K, Bahl A, Evans A, Osman E, Syrris P, Gorman G, Farrell M, Holton JL, Hanna MG, Hughes S, Elliott PM, Macrae CA, McKenna WJ (2005) Adenosine monophosphate-activated protein kinase disease mimicks hypertrophic cardiomyopathy and Wolff-Parkinson-White syndrome: natural history. J Am Coll Cardiol 45(6):922–930. doi:10.1016/j.jacc.2004.11.053 S0735-1097(04)02500-8 [pii]

Naka H, Nakamura S, Shimazaki T, Okano H (2008) Requirement for COUP-TFI and II in the temporal specification of neural stem cells in CNS development. Nat Neurosci 11(9): 1014–1023

Natarajan C, Bright JJ (2002) Peroxisome proliferator-activated receptor-gamma agonists inhibit experimental allergic encephalomyelitis by blocking IL-12 production, IL-12 signaling and Th1 differentiation. Genes Immun 3(2):59–70

Nelissen K, Mulder M, Smets I, Timmermans S, Smeets K, Ameloot M, Hendriks JJ (2012) Liver X receptors regulate cholesterol homeostasis in oligodendrocytes. J Neurosci Res 90 (1):60–71

Niakan KK, Davis EC, Clipsham RC, Jiang M, Dehart DB, Sulik KK, McCabe ER (2006) Novel role for the orphan nuclear receptor Dax1 in embryogenesis, different from steroidogenesis. Mol Genet Metab 88(3):261–271

Niino M, Iwabuchi K, Kikuchi S, Ato M, Morohashi T, Ogata A, Tashiro K, Onoe K (2001) Amelioration of experimental autoimmune encephalomyelitis in C57BL/6 mice by an agonist of peroxisome proliferator-activated receptor-gamma. J Neuroimmunol 116(1):40–48

Nishiyama A, Komitova M, Suzuki R, Zhu X (2009) Polydendrocytes (NG2 cells): multifunctional cells with lineage plasticity. Nat Rev Neurosci 10(1):9–22

Noll E, Miller RH (1994) Regulation of oligodendrocyte differentiation: a role for retinoic acid in the spinal cord. Development 120(3):649–660

Paintlia MK, Paintlia AS, Singh AK, Singh I (2011) Synergistic activity of interleukin-17 and tumor necrosis factor-alpha enhances oxidative stress-mediated oligodendrocyte apoptosis. J Neurochem 116(4):508–521

Pascual G, Glass CK (2006) Nuclear receptors versus inflammation: mechanisms of transrepression. Trends Endocrinol Metab 17(8):321–327. doi:10.1016/j.tem.2006.08.005 S1043-2760 (06)00156-1 [pii]

Peiris M, Monteith GR, Roberts-Thomson SJ, Cabot PJ (2007) A model of experimental autoimmune encephalomyelitis (EAE) in C57BL/6 mice for the characterisation of intervention therapies. J Neurosci Methods 163(2):245–254

Perrin RJ, Fagan AM, Holtzman DM (2009) Multimodal techniques for diagnosis and prognosis of Alzheimer's disease. Nature 461(7266):916–922. doi:10.1038/nature08538 nature08538 [pii]

Picou F, Fauquier T, Chatonnet F, Flamant F (2012) A bimodal influence of thyroid hormone on cerebellum oligodendrocyte differentiation. Mol Endocrinol 26(4):608–618

Polak PE, Kalinin S, Dello RC, Gavrilyuk V, Sharp A, Peters JM, Richardson J, Willson TM, Weinberg G, Feinstein DL (2005) Protective effects of a peroxisome proliferator-activated receptor-beta/delta agonist in experimental autoimmune encephalomyelitis. J Neuroimmunol 168(1–2):65–75

Powers JM, Rubio A (1995) Selected leukodystrophies. Semin Pediatr Neurol 2(3):200–210

Powers JM, Liu Y, Moser AB, Moser HW (1992) The inflammatory myelinopathy of adreno-leukodystrophy: cells, effector molecules, and pathogenetic implications. J Neuropathol ExpNeurol 51(6):630–643

Powers JM, DeCiero DP, Ito M, Moser AB, Moser HW (2000) Adrenomyeloneuropathy: a neuropathologic review featuring its noninflammatory myelopathy. J Neuropathol ExpNeurol 59(2):89–102

Powers JM, DeCiero DP, Cox C, Richfield EK, Ito M, Moser AB, Moser HW (2001) The dorsal root ganglia in adrenomyeloneuropathy: neuronal atrophy and abnormal mitochondria. J Neuropathol ExpNeurol 60(5):493–501

Qi JS, Sai-Yajnik V, Greene ME, Raaka BM, Samuels HH (1995) The ligand-binding domains of the thyroid hormone/retinoid receptor gene subfamily function in vivo to mediate heterodimerization, gene silencing, and transactivation. Mol Cell Biol 15(3):1817–1825

Qiu Y, Pereira FA, DeMayo FJ, Lydon JP, Tsai SY, Tsai MJ (1997) Null mutation of mCOUP-TFI results in defects in morphogenesis of the glossopharyngeal ganglion, axonal projection, and arborization. Genes Dev 11(15):1925–1937

Raikwar HP, Muthian G, Rajasingh J, Johnson CN, Bright JJ (2006) PPARgamma antagonists reverse the inhibition of neural antigen-specific Th1 response and experimental allergic encephalomyelitis by Ciglitazone and 15-deoxy-Delta 12,14-prostaglandin J2. J Neuroimmunol 178(1–2):76–86

Rakic S, Zecevic N (2003) Early oligodendrocyte progenitor cells in the human fetal telencephalon. Glia 41(2):117–127

Rampler H, Weinhofer I, Netik A, Forss-Petter S, Brown PJ, Oplinger JA, Bugaut M, Berger J (2003) Evaluation of the therapeutic potential of PPARalpha agonists for X-linked adrenoleukodystrophy. Mol Genet Metab 80(4):398–407

Rannou F, Francois M, Corvol MT, Berenbaum F (2006) Cartilage breakdown in rheumatoid arthritis. Joint Bone Spine 73(1):29–36. doi:10.1016/j.jbspin.2004.12.013 S1297-319X(05)00095-3 [pii]

Remels AH, Langen RC, Gosker HR, Russell AP, Spaapen F, Voncken JW, Schrauwen P, Schols AM (2009) PPARgamma inhibits NF-kappaB-dependent transcriptional activation in skeletal muscle. Am J Physiol Endocrinol Metab 297(1):E174–E183

Riddell DR, Zhou H, Comery TA, Kouranova E, Lo CF, Warwick HK, Ring RH, Kirksey Y, Aschmies S, Xu J, Kubek K, Hirst WD, Gonzales C, Chen Y, Murphy E, Leonard S, Vasylyev D, Oganesian A, Martone RL, Pangalos MN, Reinhart PH, Jacobsen JS (2007) The LXR agonist TO901317 selectively lowers hippocampal Abeta42 and improves memory in the Tg2576 mouse model of Alzheimer's disease. Mol Cell Neurosci 34(4):621–628

Risner ME, Saunders AM, Altman JF, Ormandy GC, Craft S, Foley IM, Zvartau-Hind ME, Hosford DA, Roses AD (2006) Efficacy of rosiglitazone in a genetically defined population with mild-to-moderate Alzheimer's disease. Pharmacogenomics J 6(4):246–254

Robinson E, Grieve DJ (2009) Significance of peroxisome proliferator-activated receptors in the cardiovascular system in health and disease. Pharmacol Ther 122(3):246–263

Roth AD, Leisewitz AV, Jung JE, Cassina P, Barbeito L, Inestrosa NC, Bronfman M (2003) PPAR gamma activators induce growth arrest and process extension in B12 oligodendrocyte-like cells and terminal differentiation of cultured oligodendrocytes. J Neurosci Res 72(4):425–435

Roy K, Kuznicki K, Wu Q, Sun Z, Bock D, Schutz G, Vranich N, Monaghan AP (2004) The Tlx gene regulates the timing of neurogenesis in the cortex. J Neurosci 24(38):8333–8345

Sacchetti P, Sousa KM, Hall AC, Liste I, Steffensen KR, Theofilopoulos S, Parish CL, Hazenberg C, Richter LA, Hovatta O, Gustafsson JA, Arenas E (2009) Liver X receptors and oxysterols promote ventral midbrain neurogenesis in vivo and in human embryonic stem cells. Cell Stem Cell 5(4):409–419

Saher G, Brugger B, Lappe-Siefke C, Mobius W, Tozawa R, Wehr MC, Wieland F, Ishibashi S, Nave KA (2005) High cholesterol level is essential for myelin membrane growth. Nat Neurosci 8(4):468–475

Saluja I, Granneman JG, Skoff RP (2001) PPAR delta agonists stimulate oligodendrocyte differentiation in tissue culture. Glia 33(3):191–204

Sastre M, Dewachter I, Landreth GE, Willson TM, Klockgether T, van LF, Heneka MT (2003) Nonsteroidal anti-inflammatory drugs and peroxisome proliferator-activated receptor-gamma agonists modulate immunostimulated processing of amyloid precursor protein through regulation of beta-secretase. J Neurosci 23(30):9796–9804

Sastre M, Dewachter I, Rossner S, Bogdanovic N, Rosen E, Borghgraef P, Evert BO, Dumitrescu-Ozimek L, Thal DR, Landreth G, Walter J, Klockgether T, van LF, Heneka MT (2006) Nonsteroidal anti-inflammatory drugs repress beta-secretase gene promoter activity by the activation of PPARgamma. Proc Natl Acad Sci USA 103(2):443–448

Schulman IG (2010) Nuclear receptors as drug targets for metabolic disease. Adv Drug Deliv Rev 62(13):1307–1315. doi:10.1016/j.addr.2010.07.002

Schultz JR, Tu H, Luk A, Repa JJ, Medina JC, Li L, Schwendner S, Wang S, Thoolen M, Mangelsdorf DJ, Lustig KD, Shan B (2000) Role of LXRs in control of lipogenesis. Genes Dev 14(22):2831–2838

Shackleford G, Makoukji J, Grenier J, Liere P, Meffre D, Massaad C (2013) Differential regulation of Wnt/beta-catenin signaling by Liver X Receptors in Schwann cells and oligodendrocytes. Biochem Pharmacol 86(1):106–114. doi:10.1016/j.bcp.2013.02.036 S0006-2952(13)00163-9 [pii]

Shao D, Liu Y, Liu X, Zhu L, Cui Y, Cui A, Qiao A, Kong X, Chen Q, Gupta N, Fang F, Chang Y (2010) PGC-1 beta-regulated mitochondrial biogenesis and function in myotubes is mediated by NRF-1 and ERR alpha. Mitochondrion 10(5):516–527

Shenoy SD, Spencer TA, Mercer-Haines NA, Alipour M, Gargano MD, Runge-Morris M, Kocarek TA (2004) CYP3A induction by liver x receptor ligands in primary cultured rat and mouse hepatocytes is mediated by the pregnane X receptor. Drug Metab Dispos 32(1):66–71

Sherman DL, Brophy PJ (2005) Mechanisms of axon ensheathment and myelin growth. Nat Rev Neurosci 6(9):683–690

Shi Y, Chichung LD, Taupin P, Nakashima K, Ray J, Yu RT, Gage FH, Evans RM (2004) Expression and function of orphan nuclear receptor TLX in adult neural stem cells. Nature 427 (6969):78–83

Shimizu K, Mitchell RN, Libby P (2006) Inflammation and cellular immune responses in abdominal aortic aneurysms. Arterioscler Thromb Vasc Biol 26(5):987–994. doi:10.1161/01.ATV.0000214999.12921.4f 01.ATV.0000214999.12921.4f [pii]

Singh I (1997) Biochemistry of peroxisomes in health and disease. Mol Cell Biochem 167(1–2): 1–29

Singh I, Moser AE, Moser HW, Kishimoto Y (1984) Adrenoleukodystrophy: impaired oxidation of very long chain fatty acids in white blood cells, cultured skin fibroblasts, and amniocytes. Pediatr Res 18(3):286–290

Sonoda J, Pei L, Evans RM (2008) Nuclear receptors: Decoding metabolic disease. FEBS Lett 582 (1):2–9. doi:10.1016/j.febslet.2007.11.016

Stathopoulou A, Christopoulos P, Soubasi E, Gourzis P (2010) Personality characteristics and disorders in multiple sclerosis patients: assessment and treatment. Int Rev Psychiatry 22(1): 43–54

Steffensen KR, Jakobsson T, Gustafsson JA (2013) Targeting liver X receptors in inflammation. Expert Opin Ther Targets 17(8):977–990. doi:10.1517/14728222.2013.806490

Sugden MC, Caton PW, Holness MJ (2010) PPAR control: it's SIRTainly as easy as PGC. J Endocrinol 204(2):93–104

Sun G, Yu RT, Evans RM, Shi Y (2007) Orphan nuclear receptor TLX recruits histone deacetylases to repress transcription and regulate neural stem cell proliferation. Proc Natl Acad Sci USA 104(39):15282–15287

Suzuki K (1983) The twitcher mouse. A model of human globoid cell leukodystrophy (krabbe's disease). Am J Pathol 111(3):394–397

Suzuki K (2003) Globoid cell leukodystrophy (Krabbe's disease): update. J Child Neurol 18 (9):595–603

Suzuki K, Taniike M (1995) Murine model of genetic demyelinating disease: the twitcher mouse. Microsc Res Tech 32(3):204–214

Tokuzawa Y, Kaiho E, Maruyama M, Takahashi K, Mitsui K, Maeda M, Niwa H, Yamanaka S (2003) Fbx15 is a novel target of Oct3/4 but is dispensable for embryonic stem cell self-renewal and mouse development. Mol Cell Biol 23(8):2699–2708

Toney LM, Cattoretti G, Graf JA, Merghoub T, Pandolfi PP, Dalla-Favera R, Ye BH, Dent AL (2000) BCL-6 regulates chemokine gene transcription in macrophages. Nat Immunol 1(3): 214–220. doi:10.1038/79749

Tyler WA, Gangoli N, Gokina P, Kim HA, Covey M, Levison SW, Wood TL (2009) Activation of the mammalian target of rapamycin (mTOR) is essential for oligodendrocyte differentiation. J Neurosci 29(19):6367–6378

van Geel BM, Bezman L, Loes DJ, Moser HW, Raymond GV (2001) Evolution of phenotypes in adult male patients with X-linked adrenoleukodystrophy. Ann Neurol 49(2):186–194

Vanmierlo T, Rutten K, Dederen J, Bloks VW, van Vark-van der Z, Kuipers F, Kiliaan A, Blokland A, Sijbrands EJ, Steinbusch H, Prickaerts J, Lutjohann D, Mulder M (2011) Liver X receptor activation restores memory in aged AD mice without reducing amyloid. Neurobiol Aging 32(7):1262–1272

Wagner EF, Eferl R (2005) Fos/AP-1 proteins in bone and the immune system. Immunol Rev 208:126–140. doi:10.1111/j.0105-2896.2005.00332.x IMR332 [pii]

Wanders RJ (1999) Peroxisomal disorders: clinical, biochemical, and molecular aspects. Neurochem Res 24(4):565–580

Wareski P, Vaarmann A, Choubey V, Safiulina D, Liiv J, Kuum M, Kaasik A (2009) PGC-1{alpha} and PGC-1{beta} regulate mitochondrial density in neurons. J Biol Chem 284 (32):21379–21385

White AB, Givogri MI, Lopez-Rosas A, Cao H, van BR, Thinakaran G, Bongarzone ER (2009) Psychosine accumulates in membrane microdomains in the brain of krabbe patients, disrupting the raft architecture. J Neurosci 29(19):6068–6077

Wojcicka G, Jamroz-Wisniewska A, Horoszewicz K, Beltowski J (2007) Liver X receptors (LXRs). Part I: structure, function, regulation of activity, and role in lipid metabolism. Postepy Hig Med Dosw (Online) 61:736–759

Xiang Z, Valenza M, Cui L, Leoni V, Jeong HK, Brilli E, Zhang J, Peng Q, Duan W, Reeves SA, Cattaneo E, Krainc D (2011) Peroxisome-proliferator-activated receptor gamma coactivator 1 alpha contributes to dysmyelination in experimental models of Huntington's disease. J Neurosci 31(26):9544–9553

Xiao C, Ghosh S (2005) NF-κB, an Evolutionarily Conserved Mediator of Immune and Inflammatory Responses. In: Gupta S, Paul W, Steinman R (eds) Mechanisms of lymphocyte activation and immune regulation X, vol 560. Advances in Experimental Medicine and Biology. Springer US, pp 41–45. doi:10.1007/0-387-24180-9_5

Xu P, Xu H, Tang X, Xu L, Wang Y, Guo L, Yang Z, Xing Y, Wu Y, Warner M, Gustafsson JA, Fan X (2014) Liver X receptor beta is essential for the differentiation of radial glial cells to oligodendrocytes in the dorsal cortex. Mol Psychiatry 19(8):947–957

Yamaguchi H, Zhou C, Lin SC, Durand B, Tsai SY, Tsai MJ (2004) The nuclear orphan receptor COUP-TFI is important for differentiation of oligodendrocytes. Dev Biol 266(2):238–251

Yan Q, Zhang J, Liu H, Babu-Khan S, Vassar R, Biere AL, Citron M, Landreth G (2003) Anti-inflammatory drug therapy alters beta-amyloid processing and deposition in an animal model of Alzheimer's disease. J Neurosci 23(20):7504–7509

Yang HM, Do HJ, Kim DK, Park JK, Chang WK, Chung HM, Choi SY, Kim JH (2007) Transcriptional regulation of human Oct4 by steroidogenic factor-1. J Cell Biochem 101 (5):1198–1209

Younes-Rapozo V, Berendonk J, Savignon T, Manhaes AC, Barradas PC (2006) Thyroid hormone deficiency changes the distribution of oligodendrocyte/myelin markers during oligodendroglial differentiation in vitro. Int J Dev Neurosci 24(7):445–453

Young KM, Psachoulia K, Tripathi RB, Dunn SJ, Cossell L, Attwell D, Tohyama K, Richardson WD (2013) Oligodendrocyte dynamics in the healthy adult CNS: evidence for myelin remodeling. Neuron 77(5):873–885

Yu H, Pardoll D, Jove R (2009) STATs in cancer inflammation and immunity: a leading role for STAT3. Nat Rev Cancer 9(11):798–809. doi:10.1038/nrc2734 nrc2734 [pii]
Zawadzka M, Rivers LE, Fancy SP, Zhao C, Tripathi R, Jamen F, Young K, Goncharevich A, Pohl H, Rizzi M, Rowitch DH, Kessaris N, Suter U, Richardson WD, Franklin RJ (2010) CNS-resident glial progenitor/stem cells produce Schwann cells as well as oligodendrocytes during repair of CNS demyelination. Cell Stem Cell 6(6):578–590
Zelcer N, Khanlou N, Clare R, Jiang Q, Reed-Geaghan EG, Landreth GE, Vinters HV, Tontonoz P (2007) Attenuation of neuroinflammation and Alzheimer's disease pathology by liver x receptors. Proc Natl Acad Sci USA 104(25):10601–10606
Zhang CL, Zou Y, He W, Gage FH, Evans RM (2008) A role for adult TLX-positive neural stem cells in learning and behaviour. Nature 451(7181):1004–1007
Zhao X, Grotta J, Gonzales N, Aronowski J (2009) Hematoma resolution as a therapeutic target: the role of microglia/macrophages. Stroke 40(3 Suppl):S92–S94
Zolezzi JM, Bastías-Candia S, Santos MJ, Inestrosa NC (2014) Alzheimer's disease: relevant molecular and physiopathological events affecting amyloid-β brain balance and the putative role of PPARs. Front Aging Neurosci 6. doi:10.3389/fnagi.2014.00176
Zuccaro E, Arlotta P (2013) The quest for myelin in the adult brain. Nat Cell Biol 15(6):572–575

# The Role of Galectin-3: From Oligodendroglial Differentiation and Myelination to Demyelination and Remyelination Processes in a Cuprizone-Induced Demyelination Model

**H.C. Hoyos, Mariel Marder, R. Ulrich, V. Gudi, M. Stangel, G.A. Rabinovich, L.A. Pasquini and J.M. Pasquini**

**Abstract** The aim of this work was to combine our previously published results with our new data to show how galectin-3 (Gal-3) controls myelin integrity and function, promotes oligodendroglial cell differentiation, and regulates microglial responses to limit cuprizone- (CPZ)-induced demyelination and foster remyelination. In this study, 8-week-old Gal-3-deficient (*Lgals3*$^{-/-}$) and wild type (WT) mice were fed a diet containing 0.2 % CPZ w/w for 6 weeks, after which CPZ was withdrawn in order to allow remyelination. Our results show that remyelination was less efficient in *Lgals3*$^{-/-}$ than in WT mice. Electron microscopic images from remyelinated sections in *Lgals3*$^{-/-}$ mice revealed collapsed axons with a defective myelin wrap, while remyelinated WT mice had normal axons without relevant myelin wrap disruption. MMP-3 expression increased during remyelination in WT but not in *Lgals3*$^{-/-}$ mice. The number of CD45+, TNFα+ and TREM-2b+ cells decreased only in WT mice only, with no alterations in *Lgals3*$^{-/-}$ mice during demyelination and remyelination. Therefore, Gal-3 influences remyelination by

L.A. Pasquini and J.M. Pasquini contributed equally to this work and should be considered as co-senior.

H.C. Hoyos · M. Marder · L.A. Pasquini · J.M. Pasquini
Department of Biological Chemistry, School of Pharmacy and Biochemistry, Institute of Chemistry and Biological Physical Chemistry (IQUIFIB), Buenos Aires, Argentina

H.C. Hoyos · M. Marder · L.A. Pasquini · J.M. Pasquini
School of Pharmacy and Biochemistry, University of Buenos Aires and National Research Council (CONICET), Buenos Aires, Argentina

R. Ulrich
Department of Pathology, University of Veterinary Medicine Hannover, Hannover, Germany

V. Gudi · M. Stangel
Department of Neurology, Hannover Medical School, Hannover, Germany

 
R. von Bernhardi (ed.), *Glial Cells in Health and Disease of the CNS*,
Advances in Experimental Medicine and Biology 949,
DOI 10.1007/978-3-319-40764-7_15

mechanisms involving the tuning of microglial cells, modulation of MMP activity, and changes in myelin architecture.

**Keywords** Galectin-3 · Myelination · Demyelination · Remyelination · Cuprizone · Microglia · Oligodendrocytes · MMPs

## Abbreviations

| | |
|---|---|
| MMPs | Matrix metalloproteinases |
| Gal-3 | Galectin-3 |
| *Lgals3*$^{-/-}$ | Gal-3-deficient |
| WT | Wild type |
| CPZ | Cuprizone |
| CRD | Carbohydrate-recognition domain |
| CNS | Central nervous system |
| OLG | Oligodendrocyte |
| EAE | Experimental Autoimmune Encephalomyelitis |
| CC | Corpus callosum |
| OPC | Oligodendrocyte precursor cells |
| MBP | Myelin basic protein |
| PBS | Phosphate buffered saline |
| PFA | Paraformaldehyde |
| SVZ | Subventricular zone |
| EM | Electron Microscopy |
| GFAP | Glial Fibrillary Acidic Protein |
| IOD | Integrated optical density |

V. Gudi · M. Stangel
Center for System Neurosciences, Hannover, Germany

G.A. Rabinovich
Laboratory of Immunopathology, Institute of Biology and Experimental Medicine (IBYME; CONICET), C1428 Buenos Aires, Argentina

G.A. Rabinovich
Laboratory of Functional Glycomics, Department of Biological Chemistry, Faculty of Exact and Natural Sciences, University of Buenos Aires, C1428 Buenos Aires, Argentina

L.A. Pasquini · J.M. Pasquini (✉)
Dpto. de Qca Biol, FFyB-UBA, Junín 956, C1113 Bs as, Argentina
e-mail: jpasquin@qb.ffyb.uba.ar

## Introduction

Galectins belong to a family of β-galactoside-binding lectins. Although they lack specific receptors, they can form multivalent complexes by binding to cell surface glycoconjugates containing suitable oligosaccharides and generate intracellular signals in order to regulate cell survival and differentiation (Rabinovich et al. 2007; Yang et al. 2008). In the immune system, galectin-1 and -3 (Gal-1and Gal-3, respectively) often have antagonistic roles in the modulation of adaptive immunity, with Gal-3 being predominantly proinflammatory and Gal-1 anti-inflammatory (Rabinovich and Toscano 2009). Gal-3, a chimeric protein with many actions, including modulation of innate and adaptive immunity, has unusual tandem repeats of proline and glycine-rich in short stretches fused to a carbohydrate-recognition domain (CRD) (Rabinovich and Croci 2012).

Although the function of Gal-3 in central nervous system (CNS) immunity has not been elucidated, most in vitro and in vivo studies hint at a proinflammatory role in the promotion of immune cell activation, migration and inhibition of apoptosis (Rabinovich et al. 2007), although negative regulation of lipopolysaccharide-induced inflammation has also been proposed (Li et al. 2008). Our group first studied the relevance of galectin–glycan lattices in oligodendrocyte (OLG) physiology and identified an essential role for galectin–glycan interactions in regulating OLG differentiation, leading to control of myelin integrity and function. We found that astrocytes and microglia have high expression of both Gal-1 and -3. In contrast, while immature OLGs but not differentiated OLGs highly expressed Gal-1, differentiated OLGs expressed Gal-3. Matrix metalloproteinases (MMPs) activity increased, thereby processing Gal-3 during OLG differentiation and regulating its biological activity. Recombinant Gal-3 treatment accelerated OLG differentiation in a dose- and carbohydrate-dependent manner, in accord with the "glycosylation signature" of immature versus differentiated OLGs. Furthermore, conditioned media from Gal-3-expressing, but not Gal-3-deficient (*Lgals3*$^{-/-}$) microglia, induced OLG differentiation. Supporting these findings, morphometric analyses revealed a significant reduction in the number of myelinated axons and myelin turns (lamellae), as well as in the g-ratio of *Lgals3*$^{-/-}$ mice. Moreover, myelin sheaths were more loosely wrapped around axons in *Lgals3*$^{-/-}$ mice. *Lgals3*$^{-/-}$ mice had lower anxiety levels, like those during early cuprizone (CPZ)-induced demyelination. In addition, neurospheres isolated from WT but not from *Lgals3*$^{-/-}$mice favored commitment to oligodendroglial fate. Together, these results indicate that glial-derived Gal-3, but not Gal-1, promotes oligodendroglial differentiation and thus contributes to myelin integrity and function, which has critical implications in the recovery of inflammatory demyelinating disorders (Pasquini et al. 2011).

Within the CNS, Gal-3 is upregulated by inflammatory stimuli and is harmful in prion-infected brain tissue (Mok et al. 2006, 2007; Riemer et al. 2004). On the other hand, Gal-3 mediates the activation and proliferation of microglia in a focal cerebral ischemia model in mice (Lalancette-Hébert et al. 2012). In experimental autoimmune encephalomyelitis (EAE), an animal model of CNS demyelination in which mice are immunized with myelin OLG glycoprotein, Jiang et al. (2009) reported a decreased severity in Lgals3$^{-/-}$ animals. In contrast, administration of CPZ in the diet of young adult mice causes extensive demyelination in several areas of the CNS, particularly in the corpus callosum (CC), independently of pathogenic T cells (Matsushima and Morell 2001). CPZ-induced demyelination increases the number of resident microglia and the presence of few peripheral macrophages (Masuda-Nakagawa et al. 1993; von Bernhardi and Muller 1995; McMahon et al. 2002). These microglia have been reported to phagocytize myelin, associated with the upregulation of phagocytic receptors among which TREM-2b is the most prominent (Voß et al. 2011). Remarkably, considering that myelin can inhibit the differentiation of oligodendrocyte precursor cells (OPC), this phagocytosis has been proposed to play a key role in the onset of remyelination (Kotter et al. 2006).

We have recently published a manuscript comparing CPZ-induced demyelination in 8-week-old *Lgals3*$^{-/-}$ and WT mice. Our findings showed that *Lgals3*$^{-/-}$ and WT mice are similarly susceptible to CPZ until treatment week 5, as evaluated by myelin basic protein (MBP) immunelabeling and electronic microscopy. However, OPCs generated in CPZ-treated *Lgals3*$^{-/-}$ mice showed diminished arborization, which suggests a decrease in differentiation capability. Surprisingly, while WT mice experienced spontaneous remyelination by week 5, even though the CPZ diet was maintained to week 6, *Lgals3*$^{-/-}$ mice lacked this capacity and remained demyelinated to week 6, with pronounced astroglial activation. Behavioral studies of WT and *Lgals3*$^{-/-}$ mice found lower innate anxiety after 2 weeks of CPZ treatment, but only *Lgals3*$^{-/-}$ mice had decreased locomotor activity and impaired spatial working memory. Gal-3 expression increased during CPZ-induced demyelination in microglia but not in astrocytes. Interestingly, microglial activation, ED1 expression, and phagocytic receptor TREM-2b levels increased only in CPZ-treated WT mice. In contrast, CPZ-treated *Lgals3*$^{-/-}$ mice showed an increased number of microglia with caspase-3 activation. Taken together, our results indicate that Gal-3 is expressed in microglial cells to modulate their phenotype during CPZ-induced demyelination (Hoyos et al. 2014).

Multiple sclerosis (MS) is a CNS disease leading to the demyelination of white and gray matter (Stadelmann 2011) (see Chapter "Peripheral Inflammation and Demyelinating Diseases"). Remyelination is the regenerative response to demyelination (Franklin and Kotter 2008); it is characterized by thinner axons and myelin sheaths. Although the complex biological interactions underlying remyelination are still not clear, the process is known to involve the proliferation of OPCs, their subsequent migration toward demyelinated axons and their final differentiation (Franklin and Ffrench-Constant 2008), all steps regulated by intrinsic and extrinsic factors.

MMP are a family of zinc-dependent endopeptidases which, together with endogenous tissue inhibitors of MMP, play essential roles in tissue remodeling. MMPs can degrade myelin proteins in vitro (Chandler et al. 1995, 1996; Shiryaev et al. 2009; Hansmann et al. 2012) and are suggested to be involved in the initiation of demyelination in vivo (Hansmann et al. 2012; Ulrich et al. 2006; Skuljec et al. 2011). Furthermore, both MMP and their inhibitors have been implicated in postnatal myelination, myelin maintenance, and remyelination (Skuljec et al. 2011; Ulrich et al. 2005).

In this work, to further analyze the role of Gal-3 in the control of the glial cell response to a demyelination insult, we have assessed the involvement of Gal-3 in the control of the remyelination process after CPZ-induced demyelination. To this end, we have analyzed the expression of different markers of the oligodendroglial lineage and performed morphometric analyses of electron micrographs. We have also evaluated the astroglial response, as well as changes in the microglial phenotype and the expression of MMP-3 in the CC during remyelination.

## Experimental Procedures

### Animals and Experimental Model

*Lgals3*$^{-/-}$ mice (C57BL/6 background) were generously provided by Dr. Fu-Tong Liu (University of California, Davis, USA) and generated as previously described (Hsu et al. 2000). Animals were housed in groups of 4 in a controlled environment (20–23 °C) with free access to food and water and maintained in a 12 h/12 h day/night cycle, with light on at 6 am. All animal protocols were approved by the Institutional Review Board of the University of Buenos Aires and animal experimentation was in accordance with the National Institutes of Health Guide for the Care and Use of Laboratory Animals.

Experimental demyelination was induced by feeding 8-week-old male mice with 0.2 % (w/w) CPZ (bis-cyclohexanone oxalhydrozone; Sigma Aldrich, Saint Louis, USA) mixed into standard ground rodent chow (Matsushima and Morell 2001). These mice were fed CPZ for 6 weeks and then allowed to remyelinate for 2 additional weeks. Studies, including weight measurements, were done fot 5 weeks (the point of maximum demyelination) and during the remyelination period. The demyelination–remyelination protocol is summarized in Fig. 1a.

### Slice Preparation and Brain Section Selection

Animals were anesthetized with a xylazine–ketamine mixture and intracardially perfused with 30 ml phosphate buffered saline (PBS), pH 7.4, followed by 4 % paraformaldehyde (PFA, Sigma Aldrich, Saint Louis, USA) in PBS, pH 7.4. Brains were carefully dissected out, post-fixed in the same solution overnight and later thoroughly washed in PBS and cryoprotected in 30 % sucrose in PBS. All brain slices (25-μm width) were kept at −20 °C in a PBS-glycerol solution (1:1) until used for immunofluorescence studies.

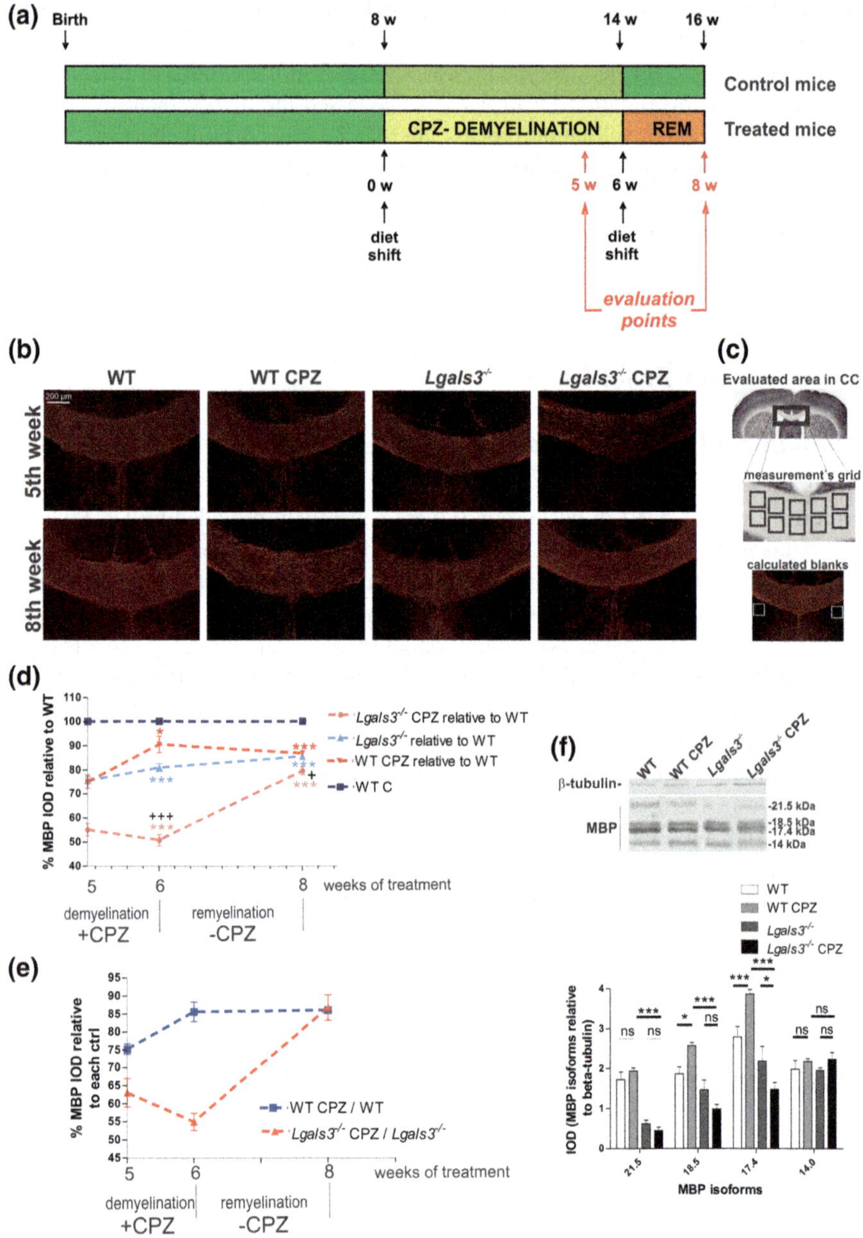


Previous to immunostaining, brain slices were selected using the Allen Mouse Brain Atlas as a reference. The corpus callosum (CC) and subventricular zone (SVZ) were studied in brain sections at the level of coronal slices 44–52 of the Atlas. After selection, brain slices were kept in PBS for immunohistochemical

◀ **Fig. 1** **a** Schematic illustration of the experimental design. Two evaluation points were determined (5th week of demyelination and 8th week—2 weeks after CPZ withdrawal from the diet). **b** Representative sections showing MBP expression in stained CC sections at the 5th week of treatment and 2 weeks after CPZ withdrawal (8th week). **c** Immunoreactive signal measured by IOD in a grid of ten 1-mm$^2$ rectangles, as detailed in the scheme (*box*) using Image Pro Plus 5.5 software. Blanks for IOD were calculated in the zones indicated by white boxes in the representative image. Images (magnification 10×) were obtained from 5 mice per treatment per time point. **d** Demyelination score at the 5th week measured by MBP IOD (WT is 100 %). **e** Remyelination rate during treatment of WT CPZ relative to WT mice and *Lgals3*$^{-/-}$ CPZ relative to *Lgals3*$^{-/-}$ mice. **f** Western blot analysis of MBP expression during remyelination. Samples were taken from CC at the 8th week. β-tubulin was used as a loading control. Values represent the mean ± SEM of five independent experiments. $*p < 0.05$, $**p < 0.01$, $***p < 0.001$ using one-way analysis of variance (ANOVA) followed by Bonferroni post hoc tests

studies. For electron microscopy (EM) and flow cytometry (isolation of microglia), the CC was dissected out from brains kept on ice using a sterile razor blade. Dissection coordinates were +1.32 mm to +0.74 mm from Bregma, according to The Mouse Brain Library. The isolated CC was immediately subjected to procedures required for each particular assay.

## *Immunohistochemistry*

Cryosections were rinsed twice with PBS (pH 7.4) and then treated with an antigen-blocking reagent composed of 5 % FCS and 0.1 % Triton X-100 in PBS. Primary and secondary antibody dilutions were prepared in 1 % FCS and 0.1 % Triton X-100 in PBS. Primary antibody incubations were done overnight at 4 °C. The primary antibodies used were rabbit anti-MBP (1/600; generously provided by Dr. A. Campagnoni, UCLA, Los Angeles, USA); rabbit anti-CAII (1/400; generously provided by Dr. W. Cammer, A. Einstein College of Medicine, New York, U.S.A.); mouse anti-CC1 (1/100, Abcam, Massachusetts, USA); goat anti-PDGFRα (1/100, Neuromics, Edina, USA); chicken anti-Glial Fibrillary Acidic Protein (GFAP) (1/100, Neuromics, Edina, USA); mouse anti-MMP-3 (1/100, Calbiochem), generously provided by Dr. Alicia Jawerbaum (CEFYBO, UBA, Buenos Aires, Argentina); and goat anti-Iba-1 (1/100, Abcam, Massachusetts, USA). Incubation with Hoechst 33342 (Sigma Aldrich, Saint Louis, USA) and various fluorescent secondary antibodies (Alexa 488, Alexa 649, Cy2 and Cy3, Jackson Immunoresearch Lab, Argentina) was done for 2 h at 37 °C with agitation. Slides were mounted and covered with Mowiol. Microphotographs were taken with an Olympus BX50 epifluorescence microscope connected to a CoolSnap digital camera.

Image Pro Plus software (version 5.5) was used for image analysis. For MBP and MMP-3, integrated optical density (IOD) was measured in CC. Figure 1c illustrates the measurement method. Ten squares (surface equal 1 mm$^2$) were displayed per image and IOD was calculated and averaged in each square for each image. For the remaining markers, the number of positive cells was counted in each case using Image J software and validated through manual count by an experimenter who was blind to the experimental design.

For CAII, CC1, PDGFRα, MMP-3, and Iba-1 images, observations were carried out using an Olympus Fluorview FV1000 MPE multiphoton microscope coupled to a Zeiss LSM 510 laser scanner. Merged versions and composite images were obtained with FV10-ASW1.7 viewer software (Olympus). Filament plots of PDGFRα and Iba-1 cells were obtained from z-stack scanning of cells at a slice distance of 0.75 μm using the IMARIS 6.3.1 program (Bitplane Sci Software) as previously described (Hoyos et al. 2014).

## *Western Blot Analysis*

MBP isoforms and GFAP were evaluated in CC cell extracts. Samples were resuspended and lysed in RIPA 1 × (NaCl 300 mM, TRIS 20 mM, pH.7.4, SDS 0.2 %) extraction buffer with a complete EDTA-free protease inhibitor cocktail (Roche). Equal amounts of protein were separated on SDS-PAGE and transferred onto PVDF membranes for Western blot analyses. Membranes were incubated with anti-MBP (1/1000) and anti-GFAP (1/1000) antibodies, followed by incubation with HRP-conjugated antibodies. Quantification was done by densitometry with the Gel Pro Analyzer 4.0 system.

## *Electron Microscopy*

Four to six animals per group were decapitated. CC were dissected out as described above, fixed and immediately prepared for EM. Ultrathin cuts were examined using a Zeiss Leo 906 E electron microscope equipped with a Zeiss Megaview III digital camera. Parameters assessed included: (a) percentage of correctly myelinated axons per field; (b) g-ratio (the ratio between the axon's diameter and the axon's diameter wrapped with myelin); and (c) number of myelin turns around an axon. Images were analyzed by experimenters who were blind to the experimental design. Eight images were obtained for each experimental condition.

## *Isolation of Microglia*

Isolation of microglia was carried out as described by Hoyos et al. (2014).

### May Grünwald Giemsa Staining

Coronal brain slices of remyelinated *Lgals3*$^{-/-}$ and WT mice were washed in distilled water to remove PBS/glycerol, preserved and mounted on glass. When slices were completely attached to glasses, they were washed twice again and incubated for 90 s with May Grünwald reagent (Biopack, Buenos Aires, Argentina). Afterwards, slices were washed four times for 2 min each, then incubated 20 min in Giemsa dye (Biopack, Buenos Aires, Argentina) and rewashed four more times for 2 min each. Both reagents were generous gifts of Dr. Miriam Lardone (Hospital de Clínicas "José de San Martín", Buenos Aires, Argentina).

### Behavioral Tests

Assessment of behavioral performance following remyelination was carried out as described by Hoyos et al. (2014).

### Statistical Analysis

Graph-Pad Prism software was used for data analysis. Results were presented as the mean ± standard error of the mean (SEM). Comparisons were performed using unpaired two-tailed Student's *t*-test or two-way analysis of variance (ANOVA), followed by Bonferroni's post hoc tests when appropriate. Values of $p < 0.05$ (*), $p < 0.01$ (**), $p < 0.001$ (***) were considered significant.

Data from behavioral assays were analyzed by two-way ANOVA, considering CPZ treatment and animal type as two main factors, and post hoc comparisons were made using Bonferroni's post-test. When a significant interaction was observed, subsequent one-way ANOVA and Newman-Keuls Multiple Comparison post hoc test were applied. A value of $p < 0.05$ was considered statistically significant.

## Results

### Remyelination Starts Earlier in WT CPZ Than in *Lgals3*$^{-/-}$ CPZ Mice but Reaches the Same Endpoint

To study the impact of Gal-3 deficiency on the remyelination process, WT and *Lgals3*$^{-/-}$ mice were submitted to demyelination through a diet containing 0.2 % (p/p) CPZ for 6 weeks and then studied after 2 weeks on a normal diet (Fig. 1a).

Animals were sacrificed at the critical point of demyelination (5th week) and after 2 weeks' remyelination (8th week).

MBP immunelabeling was carried out in WT and *Lgals3*$^{-/-}$ mice at the critical point of demyelination and at the end of remyelination (Fig. 1b). MBP IOD measurements during remyelination relative to WT showed that WT CPZ mice experienced spontaneous remyelination from the 5th week of treatment (when CPZ was still in the diet). In contrast, *Lgals3*$^{-/-}$ CPZ mice exhibited remyelination only after CPZ removal (Fig. 1c). Similar results were obtained in the analysis of each CPZ-treated group relative to its own control, i.e., WT CPZ relative to WT C mice, and *Lgals3*$^{-/-}$ CPZ relative to *Lgals3*$^{-/-}$ C mice, which indicates that data are not altered by basal hypomyelination (Fig. 1e) and that both WT CPZ and *Lgals3*$^{-/-}$ CPZ mice reach the remyelination levels of their respective C groups at the end of the process (Fig. 1e). Western blot analyses of MBP expression provided support for immunohistochemical data and highlighted that not all MBP isoforms were equally altered, the 21.5 and 17.4 kDa isoforms being most affected (Fig. 1f).

For a deeper study of the remyelination process, changes in the numbers of CAII+ and CC1+ cells were analyzed as markers of different stages in oligodendroglial differentiation. In brain coronal slices from CC (Fig. 2a, b), the number of CAII+ cells decreased in both CPZ-treated groups during demyelination, with *Lgals3*$^{-/-}$ mice starting at a lower level than WT mice. During remyelination, the number of CAII+ cells recovered only by 50 % both in WT and *Lgals3*$^{-/-}$ mice (Fig. 2b). The same tendency was observed for CC1+ cells during demyelination, although recovery during remyelination appeared to be more efficient in *Lgals3*$^{-/-}$ mice (Fig. 2c, d), with the number of CC1+ cells reaching a higher WT CPZ/WT C ratio than did *Lgals3*$^{-/-}$ CPZ/*Lgals3*$^{-/-}$ C in week 8. It is worth pointing out that although there were no significant differences between WT CPZ and *Lgals3*$^{-/-}$ CPZ in the numbers of CAII+ and CC1+ cells at the peak of demyelination (week 5), these numbers were greater in WT CPZ than in *Lgals3*$^{-/-}$ CPZ mice 2 weeks after CPZ removal from the diet (week 8).

In turn, the number of OPCs labeled for PDGFR$\alpha$ was higher in CC in *Lgals3*$^{-/-}$ CPZ compared to WT CPZ during demyelination, as well as in *Lgals3*$^{-/-}$ C compared to WT C, which suggests a higher oligodendroglial proliferative response in *Lgals3*$^{-/-}$ mice, possibly due to their basal hypomyelination and their higher degree of demyelination. When both groups were allowed to remyelinate after CPZ removal, the number of PDGFR$\alpha$+ cells decreased in both groups, but was more pronounced in WT mice, which reached levels nonsignificantly different from *Lgals3*$^{-/-}$ mice (Fig. 2e).

Interestingly, these cells were stellate, with a higher number of multipolar processes in the CC of WT mice both during demyelination and remyelination, which indicates a decreased ability of *Lgals3*$^{-/-}$ cells to differentiate. Quantitative support was obtained through the numbers and lengths of branches and terminal processes. OPCs from WT mice showed more branching, more ramification and, consequently, more ramification terminals than *Lgals3*$^{-/-}$ mice at weeks 5 and 8 (Fig. 2f).

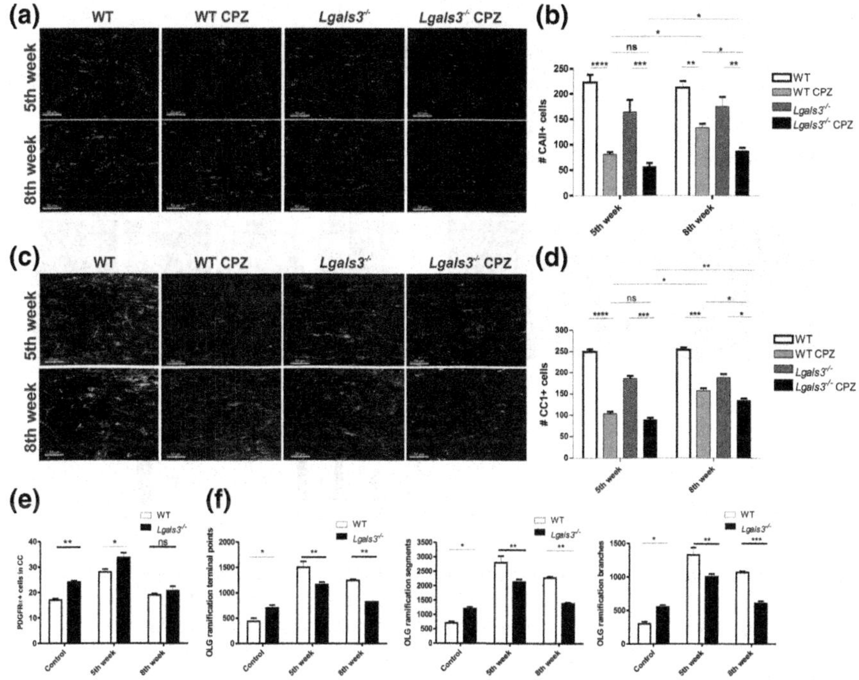


**Fig. 2** **a** Identification of CAII+ cells in CC during demyelination and remyelination. **b** Quantification shows a decrease in the number of CAII+ cells during demyelination in both experimental groups, the decrease being greater in *Lgals3*$^{-/-}$ CPZ mice. Remyelination was not complete, as values were nonsignificantly corrected in comparison to demyelination. **c**, **d** Identification and quantification of CC1+ cells following the same procedures as in (**a**, **b**), respectively. Values represent the mean ± SEM of five independent experiments. *$p < 0.05$, **$p < 0.01$, ***$p < 0.001$ and ****$p < 0.0001$ using two-way ANOVA followed by Bonferroni post hoc tests. **e** OPCs identified as PDGFrα+ cells in the CC. Quantification revealed an increase in the number of PDGFrα+ cells in CC of *Lgals3*$^{-/-}$ CPZ mice versus WT CPZ during demyelination, with an important decrease in both conditions after remyelination. **f** Filament plot was performed for each image from CC sections, comparing PDGFrα+ cell arborization between demyelination and remyelination. Quantitative assessment of the filament plot exhibited greater arborization in WT CPZ than *Lgals3*$^{-/-}$ CPZ mice, both during demyelination and remyelination. Values represent the mean ± SEM of five independent experiments. *$p < 0.05$ and **$p < 0.01$ using two-way ANOVA followed by Bonferroni post hoc tests

Ultrastructural changes evaluated by EM in CC sections (Fig. 3a) showed that myelin was more loosely wrapped around axons in *Lgals3*$^{-/-}$ than in WT mice, and that this abnormal wrapping increased in both groups during CPZ-induced demyelination. Some of these abnormalities were solved during remyelination in WT CPZ but not in *Lgals3*$^{-/-}$ CPZ mice. Morphometric analyses showed that the number of myelinated axons per field increased in both groups from demyelination to remyelination, although more significantly in WT mice. Despite this increase, WT CPZ remyelination values nearly reached those of WT ($p = 0.3710$); this did

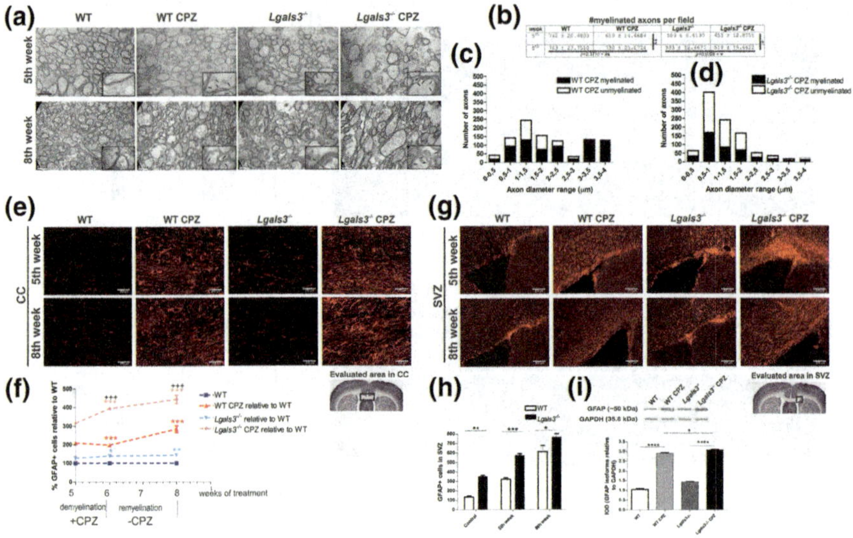

Fig. 3 **a** Representative electron micrographs at 6000× (scale bar equals 0.5 μm) from demyelinated and remyelinated axons. Each condition shows an *inset* with a highly magnified image (20000×) of a representative axon. Samples were taken in the zone highlighted in the *black box* in the scheme. As compared to that of WT mice, myelin observed in *Lgals3*$^{-/-}$ mice was loosely wrapped around the axons, an abnormality which was incremented during CPZ-induced demyelination and not solved during remyelination. *Black arrows* indicate loosely wrapped axon areas. **b** Table showing the percentage of correctly myelinated axons. Values represent the mean ± SEM *ns* non-significant, $*p < 0.05$ and $**p < 0.01$ using two-way ANOVA followed by Bonferroni post hoc tests. **c, d** Axon diameter frequency categorized by range (myelinated–demyelinated) comparing WT CPZ with WT, and *Lgals3*$^{-/-}$ CPZ with *Lgals3*$^{-/-}$ after remyelination. CPZ WT mice showed larger axons than WT, while *Lgals3*$^{-/-}$ mice exhibited similar axon size as *Lgals3*$^{-/-}$ CPZ. In both cases, there is evidence for similar patterns of axon size change between demyelination and remyelination. **e** Astrocytes immunolabeled with anti-GFAP antibody in the CC showed astrocytic activation in response to CPZ treatment in both types of animals during demyelination and its persistence in remyelination. *Bottom right* image indicates the area evaluated (*black box*, 40×). **f** Quantification of GFAP+ cells evaluated in relationship to the WT condition (set at 100 %). The increased astroglial response was particularly strong in *Lgals3*$^{-/-}$ mice in the last period of demyelination. During remyelination, the curve slopes were similar in the CPZ condition. Percentages of GFAP+ cells were relative to each WT. Both CPZ curves had a similar slope. **g** GFAP+ cells in the SVZ. Bottom right image indicates the area evaluated (*black box*, 40×). **h** Quantitative assessment of astrocytes in the SVZ. *Lgals3*$^{-/-}$ CPZ mice had more GFAP+ cells than CPZ WT mice in demyelination, but not during remyelination. **i** Western blot analysis of GFAP expression in the CC at the 8th week. The graph shows GFAP levels relative to GAPDH. Values represent the mean ± SEM of five independent experiments. *ns* non-significant, $*p < 0.05$, $**p < 0.01$ and $***p < 0.001$ using two-way ANOVA followed by Bonferroni post hoc tests

not occur for *Lgals3*$^{-/-}$ CPZ mice ($p = 0.0328$) (Fig. 3b). When data were analyzed discriminating for axon diameter and myelination status, results in *Lgals3*$^{-/-}$ mice showed an increase in the number of small axons at the expense of a decrease in the number of large ones, both among myelinated and unmyelinated axons (Fig. 3d).

## *Astrocytic Activation Persists in Remyelinated Mice Despite CPZ Removal and Was More Pronounced in Lgals3$^{-/-}$ Mice*

Astrocytes immunelabeled with anti-GFAP antibody in the CC showed there was reactive astrogliosis in response to CPZ treatment in both groups during demyelination, with more pronounced levels in *Lgals3*$^{-/-}$ mice, in agreement with previous results (Hoyos et al. 2014). This effect persisted during remyelination, and the differences between animal types became more noticeable (Fig. 3e). The quantification of GFAP+ cells, setting the WT condition at 100 %, showed an increase in astroglial response, especially in *Lgals3*$^{-/-}$ and particularly at the critical point (5 weeks) of demyelination (Fig. 3f). The analysis of GFAP+ cells in the SVZ (Fig. 3g) revealed a larger amount in *Lgals3*$^{-/-}$ CPZ than in WT CPZ mice during demyelination, a difference that remained statistically significant during remyelination (Fig. 3h). Support for these results was obtained from Western blot analyses of GFAP expression at week 8 (Fig. 3i).

## *Gal-3 Could Be Critical for MMP-3 Expression*

Immunohistochemical studies revealed a pronounced increase in the expression levels of MMP-3 in WT CPZ at 5 weeks, the critical point of demyelination, and after 2 weeks of remyelination. This increase was almost absent in *Lgals3*$^{-/-}$ mice, which indicates that Gal-3 could be critical for MMP-3 expression. MMP-3 was mainly detected in microglia, identified as Iba+ cells (Fig. 4a), and quantitative support was obtained through the determination of MMP-3 IOD in the CC in the different experimental situations (Fig. 4b).

## *Only WT Mice Exhibit Changes in the Microglial Phagocytic Phenotype During Remyelination*

The total number of microglia evaluated through Iba-1 immunelabeling decreased during remyelination in both WT and *Lgals3*$^{-/-}$ mice (Fig. 5a and b). Ramification values also decreased in both WT and *Lgals3*$^{-/-}$ mice (Fig. 5c).

However, the number of CD45+ cells during remyelination showed a significant decrease only in WT mice. Similar results were obtained for TNF and TREM-2 expression, which displayed a significant decrease after remyelination only in WT mice. In contrast, the number of CD11b$^{+}$ cells decreased during remyelination in both mouse types (Fig. 6a), in agreement with the results obtained for Iba+ cells. There were no differences across experimental groups or conditions in neutrophil migration in response to CPZ intoxication or Gal-3 depletion (Fig. 6b).

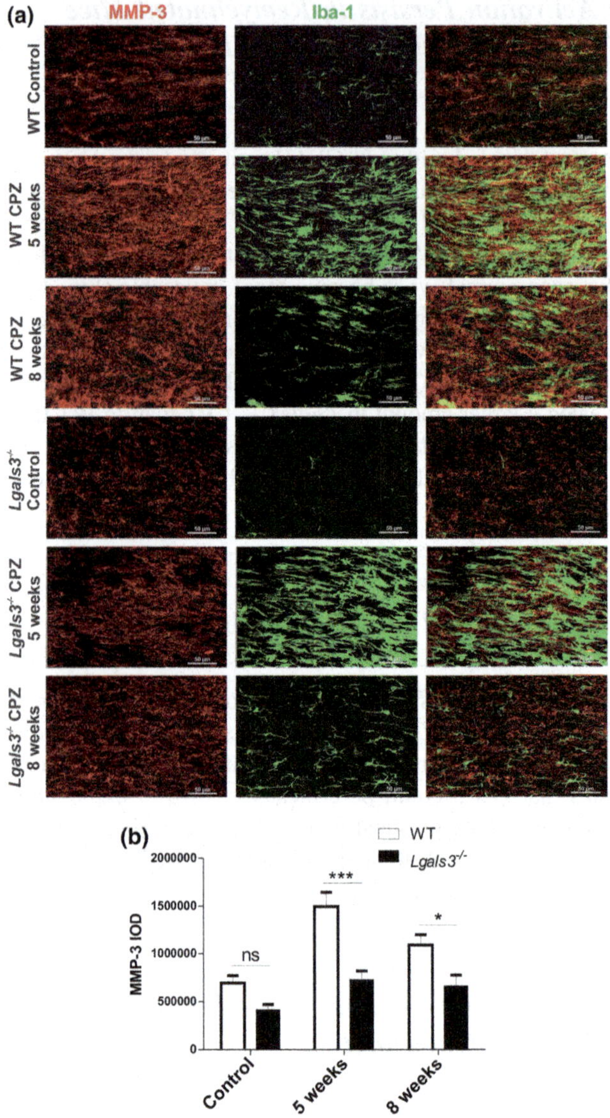


**Fig. 4** **a** Representative sections showing MMP-3 and Iba-1 in CC (40×). During demyelination, there was an increase in MMP-3 protein (*red*) in the CC of WT mice, which was even greater during remyelination. The increase in MMP-3 in *Lgals3*$^{-/-}$ mice represented 50 % of that found in WT mice. During remyelination in WT mice and demyelination in *Lgals3*$^{-/-}$ mice, MMP-3 co-localized with microglia (Iba-1+ cells in *green*). **b** MMP-3 in WT and *Lgals3*$^{-/-}$ mice during demyelination–remyelination

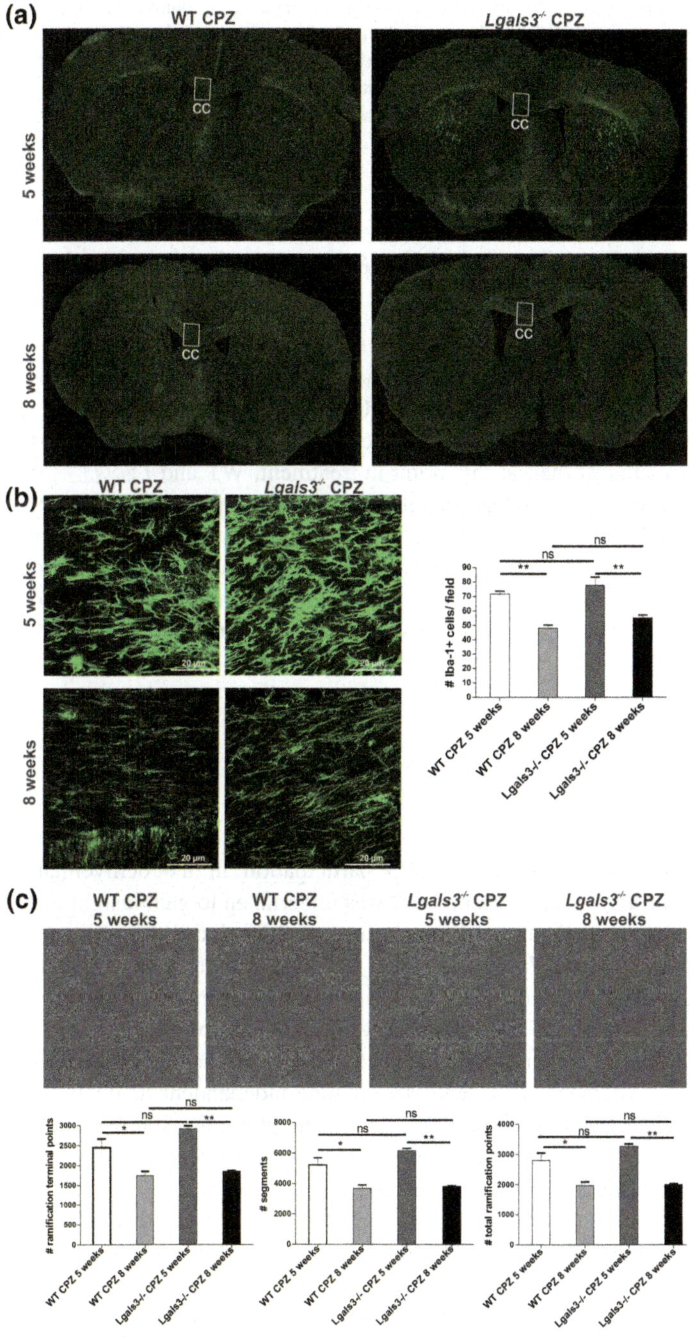
(a)
WT CPZ
Lgals3-/- CPZ
5 weeks
8 weeks
CC
(b)
WT CPZ
Lgals3-/- CPZ
5 weeks
8 weeks
20 μm
# Iba-1+ cells/ field
ns
**
WT CPZ 5 weeks
WT CPZ 8 weeks
Lgals3-/- CPZ 5 weeks
Lgals3-/- CPZ 8 weeks
(c)
WT CPZ 5 weeks
WT CPZ 8 weeks
Lgals3-/- CPZ 5 weeks
Lgals3-/- CPZ 8 weeks
# ramification terminal points
# segments
# total ramification points

◀ **Fig. 5** **a** Immunolabeled for Iba-1 in coronal brain slices during demyelination–remyelination in WT and *Lgals3*$^{-/-}$ mice. **b** Iba-1 immunohistochemistry in the CC of WT and *Lgals3*$^{-/-}$ mice during demyelination–remyelination. Quantification of Iba+ cells. **c** Filament plot for all experimental conditions. Quantification of total ramification, ramification segments and terminal points under the same conditions as mentioned above. Values represent the mean ± SEM of five independent experiments. *ns* non-significant, *$p < 0.05$, **$p < 0.01$ and ***$p < 0.001$ using two-way ANOVA followed by Bonferroni post hoc tests

## *Behavioral Performance After Remyelination Reached Pre-CPZ-Intoxication Levels in Both Mouse Strains*

After 2 weeks of recovery, both WT and *Lgals3*$^{-/-}$ mice significantly increased the number of total arm entries ($p < 0.05$), the percentage of open arm entries ($p < 0.001$) and the percentage of time spent in open arms ($p < 0.001$) (Fig. 6c). These data suggest that, at this point in treatment, WT and *Lgals3*$^{-/-}$ mice present lower levels of innate anxiety than their naïve counterparts. The WT recovery group also showed an augmented number of counts in the locomotor activity test ($p < 0.001$) (Fig. 6d), which reflects increased locomotion. On the other hand, WT and *Lgals3*$^{-/-}$ recovering mice showed no significant changes in the number of total arm entries or in the percentage of spontaneous alternations in the Y-maze test ($p < 0.05$) (Fig. 6e).

## Discussion

Following the description of Gal-3 participation in the demyelination process (Hoyos et al. 2014), the present work was undertaken to gain insight into the role of Gal-3 in the remyelination process using a CPZ-induced demyelination model in C57BL/6 control and *Lgals3*$^{-/-}$ mice. CPZ emerges as one of the most suitable models available to investigate remyelination for several reasons: first, it is simple and easy to reproduce and entails low mortality (Kipp et al. 2009); second, the absence of peripheral inflammatory cells within the demyelinated lesion, such as lymphocytes or monocytes, leads to immune-independent demyelination/remyelination (McMahon et al. 2002; Remington et al. 2007; Ransohoff and Brown 2012); third and last, nearly complete remyelination allows for the study of the mechanisms underlying successful regeneration.

When demyelination is detected in this model, increased number of microglia and astrocytes are observed within the lesion (Matsushima and Morell 2001), which indicates that glial cells play a role both in demyelination and in remyelination. Reactive glia are now recognized to mediate complex processes, including

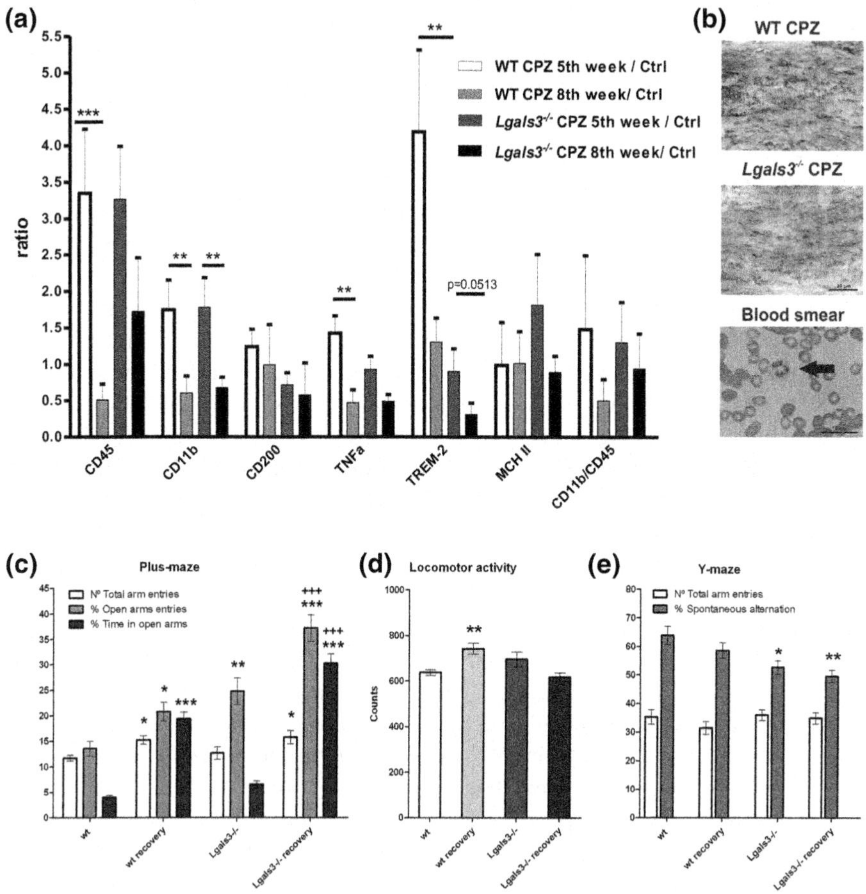


**Fig. 6** **a** Flow cytometry analysis of cell surface receptors and cytokine production by pre-sorted microglial cells (CD11b+) from CC after remyelination. CD11b+ cells were purified by magnetic micro beads conjugated to an antibody against CD11b, followed by immunolabeling with a specific antibody against cell surface receptors and cytokines, and evaluated by flow cytometry. Data were processed with Winmdi 2.8 software. Pre-immune sera were used as negative controls for polyclonal antibodies. Isotype controls were used for specific monoclonal antibodies. At least 6 mice were analyzed per experimental condition at each time point. Values represent the mean ± SEM. ns: non-significant, $*p < 0.05$, $**p < 0.01$ and $***p < 0.001$ using two-way ANOVA followed by Bonferroni post hoc tests. **b** Neutrophil evaluation was done using May Grünwald Giemsa staining in samples of CC from WT and *Lgals3*$^{-/-}$ mice treated with CPZ. Magnification 60×. **c–e** Performance of WT and *Lgals3*$^{-/-}$ mice after 2-week remyelination in the plus maze, locomotor activity and Y-maze tests. Results are expressed as the mean ± SEM of (**c**) total arm entries, percentage of open arm entries and percentage of time spent in open arms measured in the plus maze test; **d** locomotor activity counts; and **e** total arm entries and percentage of spontaneous alternation measured in the Y-maze assay. Eight to fifteen animals were analyzed per group. $*p < 0.05$, $**p < 0.01$ and $***p < 0.001$ significantly different from WT mice and $^{+++}p < 0.001$ significantly different from *Lgals3*$^{-/-}$ mice, using Newman-Keuls Multiple Comparison Test after one-way ANOVA

beneficial and deleterious effects of brain injury and in neurodegeneration. Previous studies have investigated the microglial phenotype during remyelination by performing genome-wide gene expression analyses of microglia from the CC during demyelination and remyelination in the CPZ mouse model (Olah et al. 2012; Voss et al. 2012). They provide evidence for the existence of a microglial phenotype that supports remyelination as early as the onset of demyelination and continuing throughout remyelination. Transcriptomic analyses of the remyelination-supportive microglial phenotype indicate that microglia maintain tissue homeostasis and promote regeneration.

As previously discussed in Chapters "Glial cells and Integrity of the Nervous System", "Microglia Function in the Normal Brain", and "Purine Signaling and Microglial Wrapping", microglia release either neurotoxic or pro-recovery factors, depending on whether they are differentiated toward an M1 or M2 phenotype. They are involved in many types of inflammatory processes in the brain (Hanisch and Kettenmann 2007; David and Kroner 2011) and have also been proposed to participate in the initial stages of MS. Microglia also support cell survival during tissue repair after injury to the CNS (David and Kroner 2011). Moreover, Gal-3 is required for resident microglia activation and proliferation in response to ischemic injury (Lalancette-Hébert et al. 2012). A pronounced increase in the phagocytic capacity of microglial cells during CPZ-induced demyelination is associated mainly with an upregulation of the phagocytic receptor TREM-2b in WT but not in *Lgals3*$^{-/-}$ mice (Hoyos et al. 2014). These experiments have been now replicated during remyelination, showing a significant decrease in the number of CD45+, TNFα+ and TREM-2b+ cells in WT mice but no change in *Lgals3*$^{-/-}$ mice from demyelination (5 weeks) to remyelination (7 weeks) points.

Our recently published results on demyelination (Hoyos et al. 2014) show that OPCs generated in response to CPZ-induced demyelination in *Lgals3*$^{-/-}$ mice have a decreased ability to differentiate, which could be due to the inhibitory effects of impaired phagocytosis of myelin debris in *Lgals3*$^{-/-}$ microglia (Hoyos et al. 2014). Moreover, this could also be explained by our previous findings that conditioned media from Gal-3-expressing (but not *Lgals3*$^{-/-}$) microglia promote OLG differentiation (Pasquini et al. 2011). Previous papers have demonstrated that an interruption in OPC differentiation might be the reason for remyelination failure (Franklin et al. 2008; Ulrich et al. 2008). Therefore, *Lgals3*$^{-/-}$ mice could show a delay in remyelination due to a failure in Gal-3-induced OPC differentiation. Moreover, as *Lgals3*$^{-/-}$ mouse remyelination occurs in an environment poorly conditioned by microglia, myelin generated de novo is aberrant and appears loosely wrapped around axons. This could explain the behavioral deficit observed in these animals after remyelination, which will be discussed below.

Depending on various scenarios, astrocytes can either promote neuroplasticity or secrete inhibitory matrix molecules that suppress axonal growth. Some authors have postulated that astrocytes produce chemoattractants for OPCs, allowing them to migrate from their resting positions toward the demyelination zone (Williams et al. 2007). This is probably the best explanation for the larger number of astrocytes in remyelination than in demyelination in both animal strains. However, *Lgals3*$^{-/-}$ mice reach higher levels of GFAP+ cells than WT littermates, which could reflect more severe demyelination.

While demyelination renders a significant reduction in axon caliber and a loss of small axons in *Lgals3*$^{-/-}$ mice after CPZ administration (Hoyos et al. 2014), remyelination generates a greater recovery in WT mice than in *Lgals3*$^{-/-}$ mice in small axons, which again indicates a failure in the remyelination process in the absence of Gal-3. It is presumably the more abundant presence of small-caliber axons that accounts for the incomplete remyelination observed in *Lgals-3*$^{-/-}$ mice. It has been suggested that the reduction in axon caliber after initial demyelination is largely reversible upon remyelination (Mason et al. 2001), but those axons that do not remyelinate remain small in caliber. Small-caliber axons are probably initially abundant in *Lgals-3*$^{-/-}$ mice and never become myelinated.

As already discussed in this chapter, the 21.5 kDa MBP isoform is the first to be synthesized; it promotes not only proliferation but also branching of OPCs (Smith et al. 2013). MBP isoforms including 18.2 and 14.0 kDa are involved in myelin compaction and stabilization (Chernoff 1981). During demyelination, a dramatic decrease in the 21.5 kDa MBP isoform is observed in the absence of Gal-3, which explains the deficit in normal myelin formation (Pasquini et al. 2011). During remyelination, *Lgals3*$^{-/-}$ mice do not seem to recover expression of the 21.5 and 18.2 kDa isoforms, which might explain abnormal remyelination.

In a variety of demyelinating diseases, there is a well-established role for MMPs, a family of extracellular endopeptidases for tissue remodeling and regeneration (Yong et al. 2001). MMPs cleave all components of the extracellular matrix and thereby serve important homeostatic functions. During oligodendrogenesis and remyelination, OPC can also release multiple factors to modulate neighboring cells and their environment. Gal-3 is readily cleaved by MMPs, thus altering its own carbohydrate-binding activity (Ochieng et al. 1994). We have demonstrated that Gal-3 is cleaved by MMP-2 in OPC but not in differentiated OLG (Pasquini et al. 2011). In this situation, and as described in our previous work, MMP-3 is upregulated during CPZ treatment, while Gal-3 seems to be necessary to upregulate the expression of MMP-3 and to promote microglial activation. As previously demonstrated, MMP-3 could mediate mature OLG apoptosis and microglial activation (Kim and Joh 2006). These authors have also proved that the catalytically active form of MMP-3 (actMMP-3) is released from apoptotic PC12 cells grown in serum-deprived medium. ActMMP-3 leads to the production of microglial inflammatory cytokines such as TNF through the ERK-NFκB signal transduction pathway, which in turn exacerbates neural cell degeneration.

Behavioral observations from three different approaches (plus maze, locomotor activity, and Y-maze tests) evidence changes in anxiety responses to the challenge of novelty and height, motor performance, and spatial working memory activity. We have previously demonstrated that Lgals3$^{-/-}$ mice exhibit decreased anxiety consistent with abnormalities in their myelin structure (Pasquini et al. 2011) and that CPZ treatment induces more pronounced demyelination in Lgals3-/- relative to WT mice (Hoyos et al. 2014). In the current study, after the recovery period of 2 weeks, a decrease in anxiety persisted both for WT and Lgals3$^{-/-}$ mice, as evidenced by augmented percentages of entries and time spent in open arms in the plus maze test. Locomotor activity showed an increase in the locomotor activity counts as well as in the number of total arm entries in the plus maze test. Meanwhile, spatial working memory was reestablished, as compared to their untreated counterparts. Similar behavioral performance was expected of recovering mice and their naïve counterparts. However, the recovery group showed higher open arms activity in the plus maze and augmented locomotion than untreated animals. These data suggest that some capabilities are altered even when histological analyses evidence fiber remyelination. Previous studies have already found that CPZ-fed mice display abnormal behavior during the demyelination process with partial recovery of functions during the remyelination period, with enhanced locomotor activity, improved spatial working memory, and decreased anxiety levels (Stancic et al. 2012; Franco-Pons et al. 2007; Xu et al. 2009). Our results indicate that behavioral deficits follow the course of demyelination–remyelination-induced by CPZ administration, and that some of the changes persist and seem to be irreversible even 2 weeks after CPZ withdrawal.

Remyelination is a key mechanism which restores myelin to normal conditions after demyelination and which seems to be importantly regulated by Gal-3, a molecule participating in OPC differentiation. Since Gal-3 is expressed in microglia, myelin restoration could be mediated by the Gal-3-induced M2 cell polarization, or by Gal-3 acting directly on OPC differentiation. In conclusion, our findings demonstrate that Gal-3 hierarchically governs the myelination process, at least in the CPZ model, through the modulation of microglial cells, MMP activity, and myelin architecture.

## References

Chandler S, Coates R, Gearing A, Lury J, Wells G, Bone E (1995) Matrix metalloproteinases degrade myelin basic protein. Neurosci Lett 201:223–226

Chandler S, Cossins J, Lury J, Wells G (1996) Macrophage metalloelastase degrades matrix and myelin proteins and processes a tumour necrosis factor-alpha fusion protein. Biochem Biophys Res Commun 228:421–429

Chernoff GF (1981) Shiverer: an autosomal recessive mutant mouse with myelin deficiency. J Hered 72:128

David S, Kroner A (2011) Repertoire of microglial and macrophage responses after spinal cord injury. Nat Rev Neurosci 12:388–399

Franco-Pons N, Torrente M, Colomina MT, Vilella E (2007) Behavioral deficits in the cuprizone-induced murine model of demyelination/remyelination. Toxicol Lett 169:205–213

Franklin RJ, Ffrench-Constant C (2008) Remyelination in the CNS: from biology to therapy. Nat Rev Neurosci 11:839–855

Franklin RJ, Kotter MR (2008) The biology of CNS remyelination: the key to therapeutic advances. J Neurol 255:19–25

Hanisch UK, Kettenmann H (2007) Microglia: active sensor and versatile effector cells in the normal and pathologic brain. Nat Neurosci 10:1387–1394

Hansmann F, Herder V, Kalkuhl A, Haist V, Zhang N, Schaudien D, Deschl U, Baumgärtner W, Ulrich R (2012) Matrix metalloproteinase-12 deficiency ameliorates clinical course and demyelination in Theiler's murine encephalomyelitis. Acta Neuropathol 124:127–142

Hoyos HC, Rinaldi M, Mendez-Huergo SP, Marder M, Rabinovich GA, Pasquini JM, Pasquini LA (2014) Galectin-3 controls the response of microglial cells to limit cuprizone-induced demyelination. Neurobiol Dis 62:441–455

Hsu DK, Yang RY, Pan Z, Yu L, Salomon DR, Fung-Leung WP, Liu FT (2000) Targeted disruption of the galectin-3 gene results in attenuated peritoneal inflammatory responses. Am J Pathol 156:1073–1083

Jiang HR, Al Rasebi Z, Mensah-Brown E, Shahin A, Xu D, Goodyear CS, Fukada SY, Liu FT, Liew FY, Lukic ML (2009) Galectin-3 deficiency reduces the severity of experimental autoimmune encephalomyelitis. J. Immunol 182:1167–1173

Kim YS, Joh TH (2006) Microglia, major player in the brain inflammation: their roles in the pathogenesis of Parkinson's disease. Exp Mol Med 38:333–347

Kipp M, Clarner T, Dang J, Copray S, Beyer C (2009) The cuprizone animal model: new insights into an old story. Acta Neuropathol 118:723–736

Kotter MR, Li WW, Zhao C, Franklin RJ (2006) Myelin impairs CNS remyelination by inhibiting oligodendrocyte precursor cell differentiation. J Neurosci 26:328–332

Lalancette-Hébert M, Swarup V, Beaulieu JM, Bohacek I, Abdelhamid E, Weng YC, Sato S, Kriz J (2012) Galectin-3 is required for resident microglia activation and proliferation in response to ischemic injury. J Neurosci 32:10383–10395

Li Y, Komai-Koma M, Gilchrist DS, Hsu DK, Liu FT, Springall T, Xu D (2008) Galectin-3 is a negative regulator of lipopolysaccharide-mediated inflammation. J Immunol 181:2781–2789

Mason JL, Langaman C, Morell P, Suzuki K, Matsushima GK (2001) Episodic demyelination and subsequent remyelination within the murine central nervous system: changes in axonal calibre. Neuropathol Appl Neurobiol 27:50–58

Masuda-Nakagawa LM, Muller KJ, Nicholls JG (1993) Axonal sprouting and laminin appearance after destruction of glial sheaths. Proc Natl Acad Sci USA 90(11):4966–4970

Matsushima GK, Morell P (2001) The neurotoxicant cuprizone as a model to study demyelination and remyelination in the central nervous system. Brain Pathol 11:107–116

McMahon EJ, Suzuki K, Matsushima GK (2002) Peripheral macrophage recruitment in cuprizone-induced CNS demyelination despite an intact blood–brain barrier. J Neuroimmunol 130:32–45

Mok SW, Thelen KM, Riemer C, Bamme T, Gültner S, Lütjohann D, Baier M (2006) Simvastatin prolongs survival times in prion infections of the central nervous system. Biochem Biophys Res Commun 348:697–702

Mok SW, Riemer C, Madela K, Hsu DK, Liu FT, Gültner S, Heise I, Baier M (2007) Role of galectin-3 in prion infections of the CNS. Biochem Biophys Res Commun 359:672–678

Ochieng J, Fridman R, Nangia-Makker P, Kleiner DE, Liotta LA, Stetler-Stevenson WG, Raz A (1994) Galectin-3 is a novel substrate for human matrix metalloproteinases-2 and -9. Biochemistry 33:14109–14114

Olah M, Amor S, Brouwer N, Vinet J, Eggen B, Biber K, Boddeke HW (2012) Identification of a microglia phenotype supportive of remyelination. Glia 60:306–321

Pasquini LA, Millet V, Hoyos HC, Giannoni JP, Croci DO, Marder M, Liu FT, Rabinovich GA, Pasquini JM (2011) Galectin-3 drives oligodendrocyte differentiation to control myelin integrity and function. Cell Death Differ 18:1746–1756

Rabinovich GA, Croci DO (2012) Regulatory circuits mediated by lectin–glycan interactions in autoimmunity and cancer. Immunity 36:322–335

Rabinovich GA, Toscano MA (2009) Turning 'sweet' on immunity: galectin-glycan interactions in immune tolerance and inflammation. Nat Rev Immunol 9:338–352

Rabinovich GA, Toscano MA, Jackson SS, Vasta GR (2007) Functions of cell surface galectin glycoprotein lattices. Curr Opin Struct Biol 17:513–520
Ransohoff RM, Brown MA (2012) Innate immunity in the Central Nervous System. J Clin Invest 122:1164–1171
Remington LT, Babcock AA, Zehntner SP, Owens T (2007) Microglial recruitment, activation, and proliferation in response to primary demyelination. Am J Pathol 170:1713–1724
Riemer C, Neidhold S, Burwinkel M, Schwarz A, Schultz J, Krätzschmar J, Mönning U, Baier M (2004) Gene expression profiling of scrapie-infected brain tissue. Biochem Biophys Res Commun 323:556–564
Shiryaev SA, Savinov AY, Cieplak P, Ratnikov BI, Motamedchaboki K, Smith JW, Strongin AY (2009) Matrix metalloproteinase proteolysis of the myelin basic protein isoforms is a source of immunogenic peptides in autoimmune multiple sclerosis. PLoS One 4(3):e4952
Skuljec J, Gudi V, Ulrich R, Frichert K, Yildiz O, Pul R, Voss EV, Wissel K, Baumgärtner W, Stangel M (2011) Matrix metalloproteinases and their tissue inhibitors in cuprizone-induced demyelination and remyelination of brain white and gray matter. J Neuropathol Exp Neurol 70:758–769
Smith GS, Sambroska B, Hawley SP, Klaiman JM, Gillis TE, Jones N, Boggs JM, Harauz G (2013) Nucleus-localized 21.5-kDa myelin basic protein promotes oligodendrocyte proliferation and enhances neurite outgrowth in coculture, unlike the plasma membrane-associated 18.5-kDa isoform. J Neurosci Res 91:349–362
Stadelmann C (2011) Multiple sclerosis as a neurodegenerative disease: pathology, mechanisms and therapeutic implications. Curr Opin Neurol 24:224–229
Stancic M, Slijepcevic D, Nomden A, Vos MJ, de Jonge JC, Sikkema AH, Gabius HJ, Hoekstra D, Baron W (2012) Galectin-4, a novel neuronal regulator of myelination. Glia 60:919–935
Ulrich R, Gerhauser I, Seeliger F, Baumgärtner W, Alldinger S (2005) Matrix metalloproteinases and their inhibitors in the developing mouse brain and spinal cord: A reverse transcription quantitative polymerase chain reaction study. Dev Neurosci 27:408–418
Ulrich R, Baumgärtner W, Gerhauser I, Seeliger F, Haist V, Deschl U, Alldinger S (2006) MMP-12, MMP-3, and TIMP-1 are markedly upregulated in chronic demyelinating Theiler murine encephalomyelitis. J Neuropathol Exp Neurol 65:783–793
Ulrich R, Seeliger F, Kreutzer M, Germann PG, Baumgärtner W (2008) Limited remyelination in Theiler's murine encephalomyelitis due to insufficient oligodendroglial differentiation of nerve/glial antigen 2 (NG2)-positive putative oligodendroglial progenitor cells. Neuropathol Appl Neurobiol 34:603–620
von Bernhardi R, Muller KJ (1995) Repair of the central nervous system: lessons from lesions in leeches. J Neurobiol 27(3):353–366
Voss EV, Škuljec J, Gudi V, Skripuletz T, Pul R, Trebst C, Stangel M (2012) Characterisation of microglia during de- and remyelination: can they create a repair promoting environment? Neurobiol Dis 45:519–528
Williams A, Piaton G, Lubetzki C (2007) Astrocytes-friends or foes in multiple sclerosis? Glia 55:1300–1312
Xu H, Yang HJ, Zhang Y, Clough R, Browning R, Li XM (2009) Behavioral and neurobiological changes in C57BL/6 mice exposed to cuprizone. Behav Neurosci 123:418–429
Yang RY, Rabinovich GA, Liu FT (2008) Galectins: structure, function and therapeutic potential. Expert Rev Mol Med 10:e17
Yong VW, Power C, Forsyth P, Edwards DR (2001) Metalloproteinases in biology and pathology of the nervous system. Nat Rev Neurosci 7:502–511

# Prenatal Systemic Hypoxia-Ischemia and Oligodendroglia Loss in Cerebellum

**Penha Cristina Barradas, Tiago Savignon, Alex C. Manhães, Frank Tenório, Alan P. da Costa, Marta C. Cunha-Rodrigues and Juliana Vaillant**

**Abstract** Hypoxic-ischemic (HI) injury is an important cause of death and disabilities. Despite all improvements in neonatal care, the number of children who suffer some kind of injury during birth has remained stable in the last decade. A great number of studies have shown alterations in neural cells and many animal models have been proposed in the last 5 decades. Robinson et al. (2005) proposed an HI model in which the uterine arteries are temporarily clamped on the 18th gestation day. The findings were quite similar to the ones observed in postmortem studies. The white matter is clearly damaged, and a great amount of astrogliosis takes place both in the gray and white matters. Motor changes were also found but no data regarding the cerebellum, an important structure related to motor performance, was presented. Using this model, we have shown an increased level of iNOS at P0 and microgliosis and astrogliosis at P9, and astrogliosis at P23 (up to 4 weeks from the insult). NO is important in migration, maturation, and synaptic plasticity, but in exacerbated levels it may also contribute to cellular and tissue damage. We have also evaluated oligodendroglia development in the cerebellum. At P9 in HI animals, we found a decrease in the number of PDGFRα+ cells and an apparent delay in myelination, suggesting a failure in oligodendroglial progenitors migration/maturation and/or in the myelination process. These results point to an injury in cerebellar development that might help to explain the motor problems in HI.

P.C. Barradas (✉) · T. Savignon · F. Tenório · A.P. da Costa · M.C. Cunha-Rodrigues · J. Vaillant
Depto. Farmacologia e Psicobiologia, Instituto de Biologia—UERJ,
Av. 28 de setembro, 87 fds 5° andar, Rio de Janeiro, RJ 20551-030, Brazil
e-mail: penhabarradas@gmail.com

A.C. Manhães
Depto. Ciências Fisiológicas; Instituto de Biologia Roberto Alcantara Gomes,
Universidade do Estado do Rio de Janeiro, Rio de Janeiro, RJ 20551-030, Brazil

T. Savignon
Depto. Farmacologia e Toxicologia, Instituto Nacional de Controle de Qualidade em Saúde, Fundação Oswaldo Cruz,
Rio de Janeiro, RJ 21040-900, Brazil

 
R. von Bernhardi (ed.), *Glial Cells in Health and Disease of the CNS*,
Advances in Experimental Medicine and Biology 949,
DOI 10.1007/978-3-319-40764-7_16

**Keywords** Hypoxia-ischemia · Nitric oxide synthase · PDGFα receptor · MBP · Development

**Abbreviations and Acronyms**

| | |
|---|---|
| CNS | Central nervous system |
| CP | Cerebral palsy |
| CREB | cAMP response element-binding protein |
| ED1 | Antibody that labels macrophage/microglia |
| GFAP | Glial fibrillary acid protein |
| HI | Hypoxia ischemia |
| MBP | Myelin basic protein |
| NADPH-d | Nicotinamide adenine dinucleotide phosphate reduced diaphorase |
| NADPH-d+ | Nicotinamide adenine dinucleotide phosphate reduced diaphorase positive |
| NM | Non-manipulated |
| NMDA | *N*-methyl-D-aspartate |
| NO | Nitric oxide |
| NOS | Nitric oxide synthase |
| nNOS | Neuronal nitric oxide synthase |
| iNOS | Inducible nitric oxide synthase |
| PDGFRα | Platelet derived growth factor receptor alpha |
| P0 | Postnatal day 0, here considered as the day of birth |
| P2, 7, 9, 23 | Postnatal day 2, 7, 9, and 23 |
| SHAM | Surgical control |
| SMV | Superior Medullary Vellum |
| uANOVA | Univariate analises of variance |
| WHO | World health organization |

# General Considerations

Hypoxic-ischemic (HI) brain injury is an important cause of death and disabilities around the world, both in developing and developed countries (Vannucci and Vannucci 2005; Volpe 2009). According to WHO, about one million deaths occur yearly due to birth issues (Lawn et al. 2005). Despite all efforts at neonatal care in recent decades, the number of children who suffer injury during birth has remained stable during the last decade (Nelson et al. 2003). After perinatal insults, infant brains suffer oligodendrocyte loss, hypomyelination, astrogliosis (Marín-Padilla 1997), and perturbed cortical development (Marín-Padilla 1999). The mechanisms underlying these pathological changes remain unclear.

Cerebral palsy (CP), a chronic debilitating disorder of impaired motor development, is strongly associated with perinatal brain injury (Kuban and Leviton 1994;

Volpe 2001, 2003). Various perinatal brain insults have been associated with CP, including prematurity and chorioamnionitis (Perlman et al. 1996; Verma et al. 1997; Spinillo et al. 1998; Wu and Colford 2000; Terzidou and Bennett 2001). Although full term infants can develop CP, it occurs more frequently in premature infants (Cummins et al. 1993).

Because various insults at different gestational stages induce elevated levels of cytokines and disrupt brain development, it has been proposed that aberrant cytokine expression underlies perinatal brain injury (Adlinolfi 1993). The pathogenesis of perinatal brain insults is, however, likely to involve numerous pathways associated with cytokines and oxygen-free radical species (Haynes et al. 2003; Folkerth et al. 2004), and their relative contributions have yet to be defined.

Perinatal brain injury invariably involves the gray and white matters, with the balance between them depending on the stage of cerebral developmental and vessel maturation. In order to study HI insult and its mechanisms of damage, several animal models have been proposed. Each has focused on a particular developmental stage, trying to mimic one of the many types of brain injury that occurs in humans (Fig. 1).

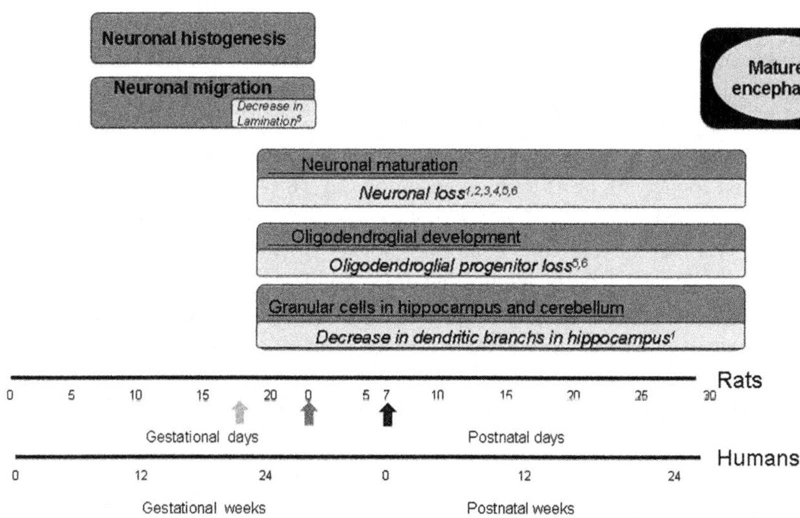


**Fig. 1** Timeline of brain development at the cellular level and the temporal relationship of rat and mouse versus human brain development. *Arrows* point to ages that HI insult in rats/mouse in prenatal life (*light gray*), at birth (*medium gray*) and postnatal life (*dark gray*) are most often performed. Light gray boxes summarize the major effects observed in rodent HI models with the numbers indicating representative references: 1—Tashima et al. (2001); 2—Grojean et al. (2003); 3—Loeliger et al. (2003); 4—Dieni and Rees (2003); 5—Robinson et al. (2005); 6—Olivier et al. (2005)

The first model was proposed by Levine (1960) using adult rats with a permanent ligation of the carotid artery. This model was particularly useful to study stroke. Rice et al. (1981) adapted this model in postnatal day (P)7 rats with the carotid ligation either permanent or temporary, creating one of the most used models in HI field. This age was chosen because it is comparable to newborn humans regarding several parameters, including cell proliferation rate, cell migration, and establishment of layering patterns in the cortex. This model has been useful for understanding several mechanisms of HI injury. However, it excludes close interaction between mother and fetus.

To include the relationship between mother and fetus, Wigglesworth proposed a model of growth restriction in 1964, in which one uterine artery was permanently ligated on embryonic day (E)17, inducing ischemia and probably hypoxia, yet the purpose of the study was exactly to show ischemia. Pups from the ligated uterine horn exhibited growth restriction at birth, in both rats and pigs (Wigglesworth 1964; Minkowski et al. 1981; Morand et al. 1982; Chanez et al. 1993; Jensen et al. 1996; Sadiq et al. 1999).

In a growth restriction model, Olivier et al. (2005) found damage to white matter like that in humans who suffer perinatal hypoxia. The growth-restricted animals did not recover weight, even in adulthood. Moreover, there were diffuse white matter lesions, increased cell death, and macrophage invasion, indicating increased inflammation. At P7, they observed a loss of pre-oligodendrocytes and deficient myelination. Those characteristics resemble what is seen in preterm infants with birth complications.

Another group in 2005 presented an HI model in which all four uterine arteries were clamped for 15, 30, or 45 min on gestation day 18 (Robinson et al. 2005). The results were similar to those of the growth restriction model. Additionally, only 45 min of HI mimicked the neuropathology of what is seen in humans (Marín-Padilla 1997, 1999): white matter astrogliosis, oligodendrocyte death, axonal injury, and altered cortical cerebral layering. Robinson et al. (2005) also described increased proinflammatory cytokines both in amniotic fluid and frontal lobe of the fetuses 4 and 24 hours after the insult. Motor performance also declined, for locomotion diminished in the open field test and steps shortened in the stride length test in adult animals.

The authors pointed out that this walking pattern is characteristic of children who develop cerebral palsy and its spastic gait. This systemic rodent prenatal HI insult accurately models human perinatal brain injury in several important ways, including functional association of altered brain development with motor delay, and consequently provides novel insights into the pathogenesis of human perinatal brain insults. As the cerebellum has the major importance in motor learning, we wish to obtain information concerning the effects of HI using a rodent systemic prenatal model. After Robinson et al. (2005) found that 45 min is the time that mimics human pathology, we ligated the four uterine arteries for this period.

## Nitric Oxide Synthase Levels and Distribution Were Impaired in a Prenatal Systemic HI Model

Enhancement of nitric oxide synthase (NOS) isoform expression has been reported in CNS areas after HI events (Kaur et al. 2006; Vexler and Yenari 2009). NO overproduction contributes to excitotoxicity, resulting in cell death and axonal damage (see Chapter "Glial Cells and Integrity of the Nervous System"). We measured the levels of neuronal (nNOS) and inducible (iNOS) isoforms at P0 (day of birth, i.e., 5 days after the HI insult). There was no difference in the level of nNOS protein in the cerebellum of HI animals compared to SHAM controls, as shown in Fig. 2a. However, the level of iNOS was significantly increased in HI animals (Fig. 2b).

Glial cells have been suggested as the major source of this NO overproduction (Kashiwagi et al. 2003). NADPH-d histochemistry labels the NOS family (all three isoforms), the enzymes responsible for NO production. The number of NADPH-d+ cells is significantly increased in cerebellar white matter of young rats (Savignon et al. 2012). At P9 there were no differences in the number of NADPH-d+ cells in the cerebellar white matter comparing non-manipulated (NM), SHAM, and HI animals. However, at P23, the number of NADPH-d+ cells decreased in NM and SHAM animals, remaining significantly higher in HI animals (as discussed in Fig. 5 of Savignon et al. 2012).

We identified NADPH-d+ cells in the white matter using specific markers for macrophage/microglia (ED1) or astrocytes (GFAP). At P9, both SHAM and HI

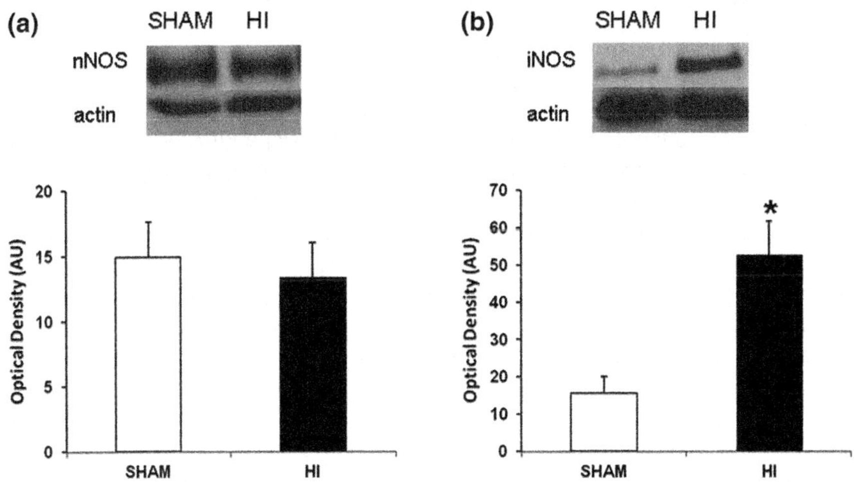


**Fig. 2** Increase in iNOS following HI injury. Both nNOS and iNOS levels are shown in the rat cerebellum at birth (P0) in SHAM and HI group. Data are represented as means ± SEM in arbitrary units (AU), resulting from 3 independent experiments. **a** nNOS–SHAM = 14.9 ± 2.7; HI = 13.4 ± 2.7, $p = 0.7036$. No significant difference was observed between groups ($p > 0.05$). **b** iNOS–SHAM = 15.6 ± 4.5; HI = 52.6 ± 9.1, $p = 0.0220$. HI group presents a significant increase in iNOS levels ($p < 0.05$)

animals presented NADPH-d+/ED1+ cells (Fig. 3a, b—arrows) and NADPH-d+/GFAP+ cells (Fig. 3c, d—arrows). In both groups, the morphology of NADPH-d+/ED1+ cells is typical of reactive microglia, i.e., small and rounded cells. At P23, both groups still presented NADPH-d+/ED1+ cells in the white matter (Fig. 3e, f—arrows), with the same morphology as in P9. However, at P23, HI animals still presented NADPH-d+/GFAP+ cells similar to reactive astrocytes (Fig. 3h—arrows), whereas SHAM animals did not present NADPH-d+/GFAP+ cells morphologically similar to reactive astrocytes, but instead showed typical GFAP+ astroglia (Fig. 3g—arrowheads), indicating that the insult has long-term effects on tissue (Savignon et al. 2012).

These results, mainly those found at P9, were not a complete surprise since the surgery procedure and anesthesia may account for an inflammation component or other damage. It is worth noting that microglia/astrocytes preferentially express the iNOS isoform when reactive, as in cases of injury and inflammation, typifying what is called microgliosis and astrogliosis (see Chapter "Glial Cells and Integrity of the Nervous System"). Thus, we have shown that the cerebellar tissue presents an environment hostile to other cells such as oligodendrocyte progenitors. It has been shown in the last two decades that NO is important in migration, maturation, and synaptic plasticity of a variety of cerebellar cells. However, it is also a contributing factor to cellular and tissue damages if that production is greatly increased, as it occurs during inflammation.

## Oligodendroglia Loss in the Cerebellum

Neurons, oligodendrocytes, and particularly their progenitors are most affected by HI (Back et al. 2002a, b). As mentioned in Chapter "Oligodendrocytes: Functioning in a Delicate Balance Between High Metabolic Requirements and Oxidative Damage", oligodendroglia progenitors do not have a mature enzymatic system to deal with the substantial free radicals delivered in HI events, particularly by microglia (Thorburne and Juurlink 1996; Le Mellédo et al. 2004). NO produced by glia expressing iNOS (You and Kaur 2000; Park et al. 2002) may also be responsible for this vulnerability. It has been demonstrated that both neurons and oligodendrocytes release considerable glutamate to the extracellular compartment, and this together with increasing NO, may cause excitotoxicity and cell death (Back et al. 2007). Activated microglia express glutamate receptors (Gottlieb and Matute 1997) and may be modulated by the excess extracellular glutamate, producing more NO.

Oligodendrocytes are derived from various subpopulations of progenitors (see Chapter "Glial Cells and Integrity of the Nervous System" for further reading on oligodendrocyte development). In the subventricular layer one arises to populate forebrain (cortex) and midbrain (thalamus and hypothalamus), while another in the ceiling of the fourth ventricle populates hindbrain (cerebellum, pons, and brain

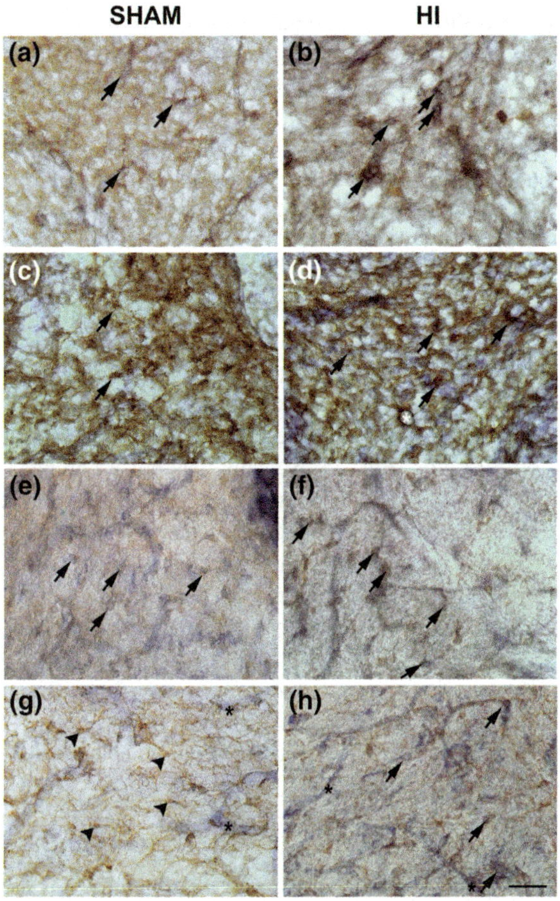


**Fig. 3** NOS activity remains associated with reactive astrocyte end feet at blood vessels in P23 HI cerebellum. Double labeling with NADPH-d histochemistry (*dark-blue*) and microglia or astroglia immunoidentification (*brown*) in the vermis region of the cerebellar white matter (0.5 mm mediolateral distance) during development. **a–d** P9; **e–h** P23. **a**, **c**, **e** and **g** (SHAM); **b**, **d**, **f** and **h** (HI); **a–b** and **e–f** double labeled with anti-ED1 antibody; **c**, **d** and **g**, **h** double labeled with anti-GFAP antibody. In both groups at P9, we can observe small, rounded NADPH-d+/ED1+ cells (**a** and **b**) or NADPH-d+/GFAP+ cells (**c** and **d**), as indicated by *arrows*. In **d**, observe a blood vessel, transversally cut, which presents NADPH-d staining, surrounded by GFAP+ astrocytic endfeet (*asterisk*). At P23, observe small rounded NADPH-d+/ED1+ cells in both groups, as indicated by arrows. HI animals display NADPH-d+/GFAP+ cells with typical reactive astrocyte morphology (*arrows*). SHAM animals do not present NADPH-d+/GFAP+ cells resembling reactive astrocytes. *Arrowheads* point to typical GFAP+ astrocytes, with no NADPH-d labeling. Notice the presence of NADPH-d+ blood vessels (*asterisks* in **g** and **h**) that are surrounded by GFAP+ astrocytic processes in HI animals (**h**) but not in SHAM animals (**g**). Calibration bar: 50 µm. Reproduced from Savignon et al. (2012)

stem). There is some disagreement regarding the timing of these events. In the forebrain it is early and in the hippocampus and cerebellum it is quite late.

Reynolds and Wilkin (1988) showed the sequential changes in oligodendroglia during development, beginning as nondifferentiated cells in the superior medullary vellum (SMV) and the base of cerebellum, which then populate the whole organ. Others described the phenotypic and antigenic changes that oligodendroglia undergo during differentiation both in vitro as in vivo (Pfeiffer et al. 1993; Baumann and Pham-Dinh 2001).

Oligodendroglial progenitors express alpha-receptor to platelet-derived growth factor (PDGFRα) (Baumann and Pham-Dinh 2001). Data from our laboratory showed that in both HI and control animals the density of PDGFRα+ progenitors at P2 is about 20 cells/100 $\mu m^2$ (unpublished data), escalating to about 50 cells per field at P9, and returning at P23 to the same levels as at P2.

At P2, there were no differences in the number of PDGFRα+ cells in cerebellar white matter in both groups (Fig. 4a, b). This was not a complete surprise, since rodent cerebellum develops rapidly postnatally. At P9, there was a significant increase in PDGFRα+ cells in both groups when compared to P2 (uANOVA; $F = 126.34$, $p < 0.001$). However, HI animals showed a significant lower number in this progenitor subpopulation compared to SHAM animals (Fig. 4c, d), indicating that a prenatal HI event somehow affected the proliferation rate and/or survival of oligodendroglial progenitors. At P23, in both groups a significant reduction in PDGFRα+ counting was observed in both groups (Fig. 4e, f). This was expected, since the rate of proliferation diminishes and the progenitors start to differentiate, downregulate PDGFRα, and form myelin. Figure 4g depicts the cell counting results for each group.

Myelin basic protein (MBP), a marker of mature oligodendrocyte and myelin (see Chapters "Glial Cells and Integrity of the Nervous System" and "Oligodendrocytes: Functioning in a Delicate Balance Between High Metabolic Requirements and Oxidative Damage"), was also impaired in the prenatal HI systemic model. At P9 in SHAM animals, MBP+ fibers were observed close to the calbindin-positive Purkinje cell layer (arrows in Fig. 5a), whereas in HI animals those MBP+ fibers were clearly located in the main white matter tracts (Fig. 5b), suggesting an apparent delay in myelination in the cerebellum. As development proceeds, oligodendrocytes/myelin were found in all extents of the granular layer in both groups. Yet, it appears that some failure occurred in the oligodendroglial progenitors migration/maturation and/or in the myelination process, since we found non-myelinated gaps in the granular layer (asterisks in Fig. 5d). This occurred only in HI animals. This pattern was maintained in HI animals until adulthood (Fig. 5f). Together, these results point to an injury in cerebellar oligodendroglia development that might help to explain the motor problems observed in HI animals.

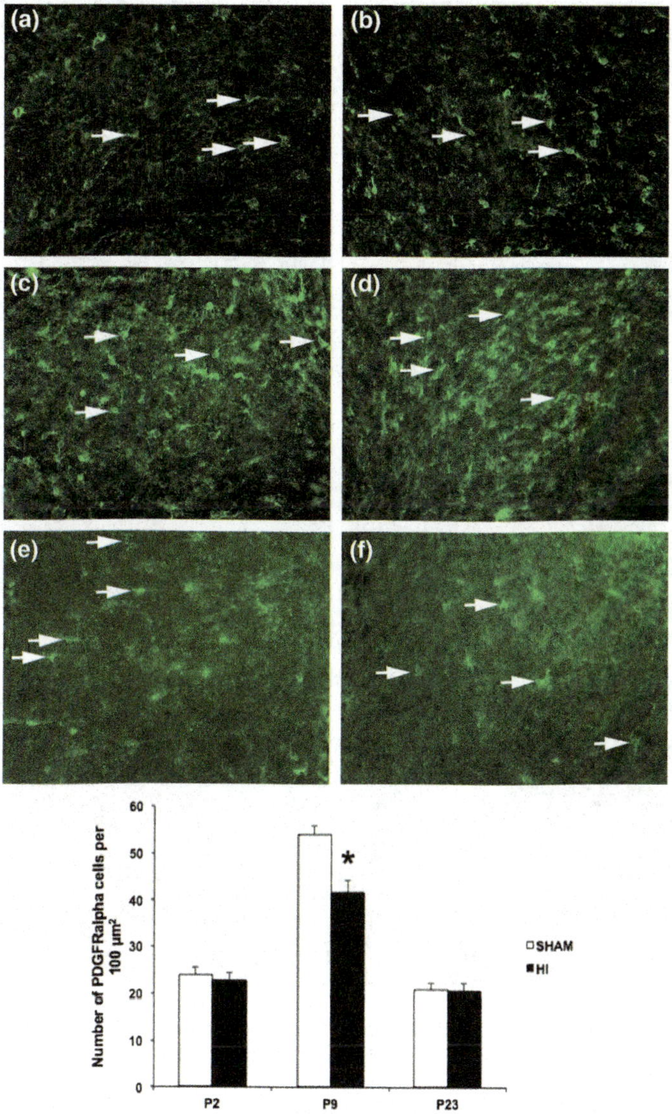


**Fig. 4** Number of PDGFRα+ cells in the cerebellar white matter of SHAM and HI animals at P2, P9 and P23. At P2, we have not observed differences in the number of PDGFRα+ cells in cerebellar white matter (**a**, **b**). At P9 (**c**, **d**) there is a significant increase in the number of PDGFRα+ cells in both groups when compared to P2 (uANOVA; $F = 126.34$, $p < 0.001$). However, HI animals have a lower number of PDGFRα+ cells than SHAM at P9 (**d**, but better shown in **g**). At P23, a significant reduction in PDGFRα+ counting was observed in both groups (**e**, **f**) when compared to P9. **g** Depicts the cell counting for each group, with density measured as number per 100 μm$^2$. Calibration bar: 50 μm

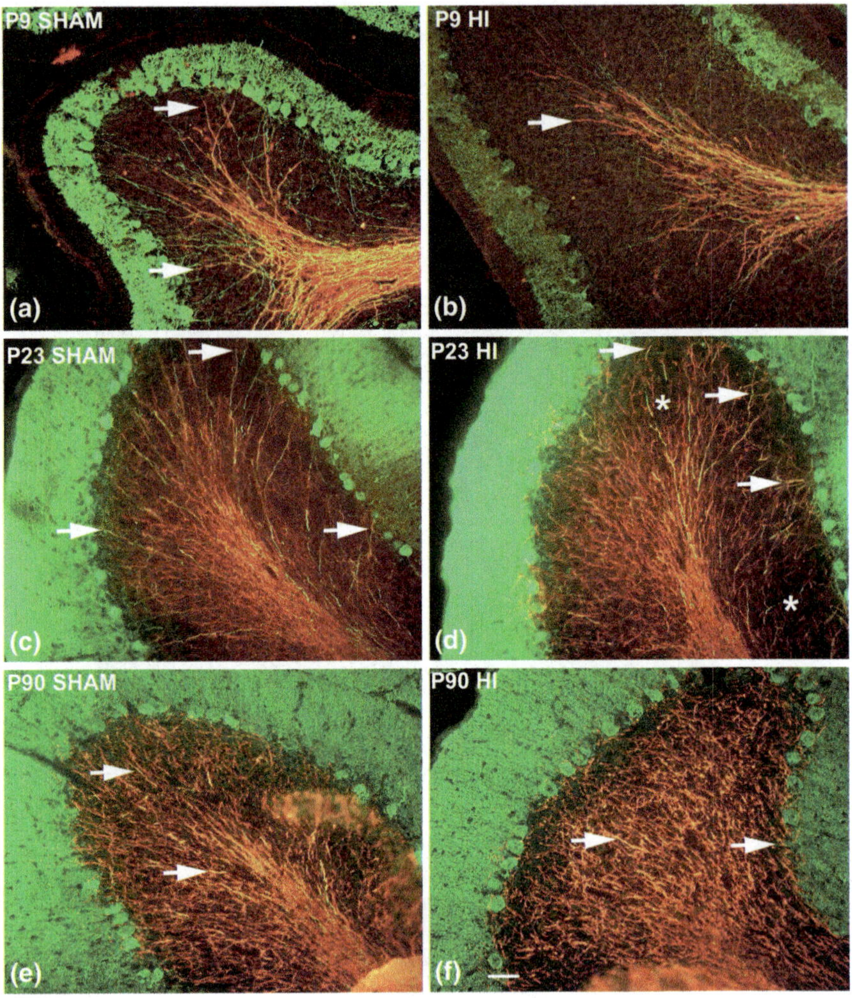


**Fig. 5** Myelination is delayed in HI animals. Myelin basic protein (MBP) is labeled in *red* and calbindin, in Purkinje cells, in *green*. At P9, there is an apparent delay in myelination in the cerebellum. In SHAM animals, some MBP+/Calbindin+ axons are close to Purkinje cell layer (*arrows* in **a**), while in HI animals those axons are not (**b**). From P23 (**c**, **d**) until adulthood (**e**, **f**), MBP+/Calbindin+ occupy all of the granular layer (*arrows*) in both SHAM and HI animals. Calibration bar: 100 μm

## Concluding Remarks

Multiple types of injury resulting from preterm birth in humans, including systemic HI, converge to hinder brain cell survival, particularly for immature oligodendrocytes and cerebral neurons (Volpe 2009). Impaired brain cell survival and differentiation continue for a prolonged period after the initial injury in animal models (Robinson et al. 2005; Mazur et al. 2010). At the time of the HI insult and in the days following, when the levels of cytokine and other inflammatory modulators are still elevated (Robinson et al. 2005), several glial and neuronal progenitor populations are entering the cerebellum parenchyma through the prospective cerebellar white matter. These progenitors, especially of oligodendrocytes, are more vulnerable to HI events because they lack the enzymatic complexes capable of dealing with the great amount of free radicals produced during HI. NO forms free radicals if produced in large amounts and is toxic to oligodendrocyte progenitors. In addition, elevated NO may trigger *N*-methyl-D-aspartate (NMDA)-mediated intracellular Ca++-influx and CREB-mediated transcription of apoptotic proteins such as Bax, Bad, and Bcl-xl, causing neuronal death (Zubrow et al. 2002a, b; Mishra et al. 2006).

Our results showed that in this systemic model of prenatal HI, oligodendroglial differentiation in the cerebellum was impaired, with a reduction in the number of PDGFR$\alpha$-cells (oligodendrocyte progenitors) and mature oligodendroglial cells, as demonstrated by reduced MBP immunostaning. This supports this model for use in devising new therapeutic strategies for HI insults.

**Acknowledgments** We would like to thank Jorge Pereira das Neves for technical assistance and Mariana Soares Magalhães for animal care.

## References

Adlinolfi M (1993) Infectious diseases in pregnancy, cytokines and neurological impairment: an hypothesis. Dev Med Child Neurol 35:549–553

Back SA, Han BH, Luo NL et al (2002a) Selective vulnerability of late oligodendrocyte progenitors to hypoxia-ischemia. J Neurosci 22:455–463

Back SA, Luo NL, Borenstein NS et al (2002b) Arrested oligodendrocyte lineage progression during human cerebral white matter development: dissociation between the timing of progenitor differentiation and myelinogenesis. J Neuropathol Exp Neurol 61:197–211

Back SA, Craig A, Kayton RJ et al (2007) Hypoxia-ischemia preferentially triggers glutamate depletion from oligodendroglia and axons in perinatal cerebral white matter. J Cereb Blood Flow Metab 27:334–347. doi:10.1038/sj.jcbfm.9600344

Baumann N, Pham-Dinh D (2001) Biology of oligodendrocyte and myelin in the mammalian central nervous system. Physiol Rev 81:871–927

Chanez C, Rabin O, Heroux M, Giguere JF (1993) Cerebral amino acid changes in an animal model of intrauterine growth retardation. Metab Brain Dis 8:61–72

Cummins SK, Nelson KB, Grether JK, Velie EM (1993) Cerebral palsy in four northern California counties, births 1983 through 1985. J Pediatr 123:230–237

Dieni S, Rees S (2003) Dendritic morphology is altered in hippocampal neurons following prenatal compromise. J Neurobiol 55:41–52. doi:10.1002/neu.10194

Folkerth RD, Haynes RL, Borenstein NS et al (2004) Developmental lag in superoxide dismutases relative to other antioxidant enzymes in premyelinated human telencephalic white matter. J Neuropathol Exp Neurol 63:990–999

Gottlieb M, Matute C (1997) Expression of ionotropic glutamate receptor subunits in glial cells of the hippocampal CA1 area following transient forebrain ischemia. J Cereb Blood Flow Metab 17:290–300. doi:10.1097/00004647-199703000-00006

Grojean S, Schroeder H, Pourié G et al (2003) Histopathological alterations and functional brain deficits after transient hypoxia in the newborn rat pup: a long term follow-up. Neurobiol Dis 14:265–278. doi:10.1016/S0969-9961(03)00082-2

Haynes V, Elfering SL, Squires RJ et al (2003) Mitochondrial nitric-oxide synthase: role in pathophysiology. IUBMB Life 55:599–603. doi:10.1080/15216540310001628681

Jensen A, Klönne HJ, Detmer A, Carter AM (1996) Catecholamine and serotonin concentrations in fetal guinea-pig brain: relation to regional cerebral blood flow and oxygen delivery in the growth-restricted fetus. Reprod Fertil Dev 8:355–364

Kashiwagi K, Iizuka Y, Mochizuki S et al (2003) Differences in nitric oxide production: a comparison of retinal ganglion cells and retinal glial cells cultured under hypoxic conditions. Brain Res Mol Brain Res 112:126–134

Kaur C, Sivakumar V, Foulds WS (2006) Early response of neurons and glial cells to hypoxia in the retina. Invest Ophthalmol Vis Sci 47:1126–1141. doi:10.1167/iovs.05-0518

Kuban KC, Leviton A (1994) Cerebral palsy. N Engl J Med 330:188–195. doi:10.1056/NEJM199401203300308

Lawn J, Shibuya K, Stein C (2005) No cry at birth: global estimates of intrapartum stillbirths and intrapartum-related neonatal deaths. Bull World Health Organ 83:409–417

Le Mellédo J-MM, Mahil N, Baker GB (2004) Nitric oxide: a key player in the relation between cardiovascular disease and major depressive disorder? J Psychiatry Neurosci 29:414–416

Levine S (1960) Anoxic-ischemic encephalopathy in rats. Am J Pathol 36:1–17

Loeliger M, Watson CS, Reynolds JD et al (2003) Extracellular glutamate levels and neuropathology in cerebral white matter following repeated umbilical cord occlusion in the near term fetal sheep. Neuroscience 116:705–714

Marín-Padilla M (1997) Developmental neuropathology and impact of perinatal brain damage. II: white matter lesions of the neocortex. J Neuropathol Exp Neurol 56:219–235

Marín-Padilla M (1999) Developmental neuropathology and impact of perinatal brain damage. III: gray matter lesions of the neocortex. J Neuropathol Exp Neurol 58:407–429

Mazur M, Miller RH, Robinson S (2010) Postnatal erythropoietin treatment mitigates neural cell loss after systemic prenatal hypoxic-ischemic injury. J Neurosurg Pediatr 6:206–221. doi:10.3171/2010.5.PEDS1032

Minkowski A, Chanez C, Priam M et al (1981) Long lasting effects of intrauterine malnutrition on neurotransmitters metabolism in the brain of developing rats. Prog Clin Biol Res 77:643–660

Mishra O, Mishra R, Ashraf Q, Delivoria-Papadopoulos M (2006) Nitric oxide-mediated mechanism of neuronal nitric oxide synthase and inducible nitric oxide synthase expression during hypoxia in the cerebral cortex of newborn piglets. Neuroscience 140:857863. doi:10.1016/j.neuroscience.2006.02.060

Morand O, Chanez C, Masson M et al (1982) Alteration in fatty acid composition of neurons, astrocytes, oligodendrocytes, myelin and synaptosomes in intrauterine malnutrition in rat. Ann Nutr Metab 26:111–120

Nelson KB, Grether JK, Dambrosia JM et al (2003) Neonatal cytokines and cerebral palsy in very preterm infants. Pediatr Res 53:600–607. doi:10.1203/01.PDR.0000056802.22454.AB

Olivier P, Baud O, Evrard P et al (2005) Prenatal ischemia and white matter damage in rats. J Neuropathol Exp Neurol 64:998–1006

Park SY, Lee H, Hur J et al (2002) Hypoxia induces nitric oxide production in mouse microglia via p38 mitogen-activated protein kinase pathway. Brain Res Mol Brain Res 107:9–16

Perlman JM, Risser R, Broyles RS (1996) Bilateral cystic periventricular leukomalacia in the premature infant: associated risk factors. Pediatrics 97:822–827

Pfeiffer SE, Warrington AE, Bansal R (1993) The oligodendrocyte and its many cellular processes. Trends Cell Biol 3:191–197

Reynolds R, Wilkin GP (1988) Development of macroglial cells in rat cerebellum. II. An in situ immunohistochemical study of oligodendroglial lineage from precursor to mature myelinating cell. Development 102:409–425

Rice JE, Vannucci RC, Brierley JB (1981) The influence of immaturity on hypoxic-ischemic brain damage in the rat. Ann Neurol 9:131–141. doi:10.1002/ana.410090206

Robinson S, Petelenz K, Li Q et al (2005) Developmental changes induced by graded prenatal systemic hypoxic-ischemic insults in rats. Neurobiol Dis 18:568–581. doi:10.1016/j.nbd.2004.10.024

Sadiq HF, Das UG, Tracy TF, Devaskar SU (1999) Intra-uterine growth restriction differentially regulates perinatal brain and skeletal muscle glucose transporters. Brain Res 823:96–103

Savignon T, Costa E, Tenorio F et al (2012) Prenatal hypoxic-ischemic insult changes the distribution and number of NADPH-diaphorase cells in the cerebellum. PLoS ONE 7:e35786. doi:10.1371/journal.pone.0035786

Spinillo A, Capuzzo E, Stronati M et al (1998) Obstetric risk factors for periventricular leukomalacia among preterm infants. Br J Obstet Gynaecol 105:865–871

Tashima L, Nakata M, Anno K et al (2001) Prenatal influence of ischemia-hypoxia-induced intrauterine growth retardation on brain development and behavioral activity in rats. Biol Neonate 80:81–87

Terzidou V, Bennett P (2001) Maternal risk factors for fetal and neonatal brain damage. Biol Neonate 79:157–162

Thorburne SK, Juurlink BH (1996) Low glutathione and high iron govern the susceptibility of oligodendroglial precursors to oxidative stress. J Neurochem 67:1014–1022

Vannucci RC, Vannucci SJ (2005) Perinatal hypoxic-ischemic brain damage: evolution of an animal model. Dev Neurosci 27:81–86. doi:10.1159/000085978

Verma U, Tejani N, Klein S et al (1997) Obstetric antecedents of intraventricular hemorrhage and periventricular leukomalacia in the low-birth-weight neonate. Am J Obstet Gynecol 176: 275–281

Vexler ZS, Yenari MA (2009) Does inflammation after stroke affect the developing brain differently than adult brain? Dev Neurosci 31:378–393. doi:10.1159/000232556

Volpe JJ (2001) Neurobiology of periventricular leukomalacia in the premature infant. Pediatr Res 50:553–562. doi:10.1203/00006450-200111000-00003

Volpe JJ (2003) Cerebral white matter injury of the premature infant-more common than you think. Pediatrics 112:176–180

Volpe J (2009) Cerebellum of the premature infant: rapidly developing, vulnerable, clinically important. J Child Neurol 24:10851104. doi:10.1177/0883073809338067

Wigglesworth (1964) Experimental growth retardation in the foetal rat. J Pathol Bacteriol 88:1–13

Wu YW, Colford JM (2000) Chorioamnionitis as a risk factor for cerebral palsy: A meta-analysis. JAMA 284:1417–1424

You Y, Kaur C (2000) Expression of induced nitric oxide synthase in amoeboid microglia in postnatal rats following an exposure to hypoxia. Neurosci Lett 279:101–104

Zubrow AB, Delivoria-Papadopoulos M, Ashraf QM et al (2002a) Nitric oxide-mediated expression of Bax protein and DNA fragmentation during hypoxia in neuronal nuclei from newborn piglets. Brain Res 954:60–67

Zubrow AB, Delivoria-Papadopoulos M, Ashraf QM et al (2002b) Nitric oxide-mediated Ca2+/calmodulin-dependent protein kinase IV activity during hypoxia in neuronal nuclei from newborn piglets. Neurosci Lett 335:5–8

# Index






R. von Bernhardi (ed.), *Glial Cells in Health and Disease of the CNS*,
Advances in Experimental Medicine and Biology 949,
DOI 10.1007/978-3-319-40764-7

CPSIA information can be obtained
at www.ICGtesting.com
Printed in the USA
LVOW02*0616240217
525239LV00002BB/26/P